ELEMENTS OF
BIOENVIRONMENTAL
ENGINEERING

ELEMENTS OF BIOENVIRONMENTAL ENGINEERING

Anthony F. Gaudy, Jr
H. Rodney Sharp Professor of Civil Engineering
University of Delaware

Elizabeth T. Gaudy
Formerly, Professor of Microbiology
Oklahoma State University

Oxford University Press

Published by Oxford University Press

Library of Congress Cataloging-in-Publication Data

Gaudy, Anthony F.
    Elements of bioenvironmental engineering.

    Includes bibliographies and index.
    1. Sanitary microbiology.   2. Water—Pollution.
3. Water—Pollution—Biological treatment.   I. Gaudy,
Elizabeth T.   II. Title.
QR48.G37   1988   576'.165   87-19896

ISBN: 978-0-910554-67-1

TO THE MEMORY OF
H. Orin Halvorson—
scientist, engineer, teacher,
and friend to both of us.

CONTENTS

PREFACE

This text is a shortened version of our text entitled *Microbiology for Environmental Scientists and Engineers*. That book, published by McGraw-Hill Book Company, is now out of print, but inquiries from various sources about its publication status indicated that we should keep the material in print. The field of bioenvironmental engineering is rather new and is growing; thus, this modified version should be increasingly useful as a text in modern courses dealing with biological aspects of pollution control. The shorter length of this version should make the book more useful as a course text than the older version, which contained more material than could be fitted into one course on the subject and was often used as a reference work rather than as a text.

The present text, *Elements of Bioenvironmental Engineering*, consists of the first eleven chapters of the original text. Chapters 12, 13 and 14, covering aspects of metabolic control mechanisms, response to environmental changes (shock load responses), and pathogenic microorganisms, have been omitted. This material, especially Chapters 12 and 13, covers more advanced and somewhat specialized topics and may be more useful as reference material. This original text is available in most libraries for those who are interested in the material in the omitted chapters. Also, a copy of this material can be obtained directly from us.

The preface to the original version, which follows, describes the philosophy and aim of the text.

Anthony F. Gaudy, Jr.
Elizabeth T. Gaudy

Prior Version PREFACE

This text is written for those whose professional interest is protection of the environment from pollution—environmental engineers and environmental scientists. Pollution control is a rapidly expanding area of interest, which is attracting persons with many different educational backgrounds, including civil engineering, chemical engineering, general biology, zoology, microbiology, chemistry, and perhaps others. All of these areas of knowledge have something to contribute to the development of a relatively new discipline that we prefer to call *bioenvironmental engineering.*

It is the present vogue to attempt to solve problems by the interdisciplinary approach. It is our belief that environmental pollution is a problem of such magnitude and urgency that it demands workers trained specifically in the areas of knowledge that are directly applicable. What is needed is a new type of multidisciplinary individual rather than a team of narrowly trained experts. The field of pollution control needs persons who consider pollution control their profession rather than individuals who consider microbiology or civil engineering, for example, their profession but are willing to apply their training to problems that are peripheral to their major interest and expertise. Everyone who wishes to participate fully in the effort to control environmental pollution should possess a common core of knowledge; that is a basic necessity for people who are to work together and communicate meaningfully with each other.

This core of knowledge contains elements of microbiology, chemistry (including biochemistry), and engineering. Microbiology is important because microbial activity is, in some cases, the result of environmental pollution and, in other circumstances, the means by which pollution is prevented. One cannot hope to control microbial activity without an understanding of microbial physiology. Chemistry is important in two ways. One cannot understand microbial physiology without some knowledge of the biochemical processes that occur in the cell. Some knowledge of chemistry is also required if one is to understand the problems involved in preventing pollution, since pollution may be defined, in most cases, as the addition of inorganic or organic chemicals in potentially harmful amounts to the environment. Finally, engineering is essential because engineers must design and plan the operation of the processes that remove pollutants from wastes that must be recycled to the environment. It is the engineer who will be held accountable for the success or failure of the pollution control strategy. Most processes (both natural and engineered) by which wastes are rendered harmless to the environment are, and probably will remain, dependent on microbial activity. Thus, the engineer designs processes that use microbes. The microbiologist contributes an understanding of the biochemistry of microbial activity, which allows these processes to be designed for optimum effectiveness. The chemist determines the nature of the pollutants and their fate. Biological scientists contribute their knowledge of the effects of environmental stress on organisms of all types, including humans. One could assemble a team composed of individuals expert in each of the areas described above. But if each of them was totally uninformed about the areas of expertise of all the others, meaningful communication would be impossible.

We envision the effective workers in this field as trained as either environmental scientists or environmental engineers. The difference in the two is one of emphasis in training and in practice, and the two must be prepared to work together to accomplish their mutual goal. The environmental scientist has a background in the biological sciences (which implies training also in chemistry and biochemistry) or primarily in chemistry with some knowledge of biology, and is also informed about the engineered processes designed to protect the environment. The environmental engineer is trained in the usual engineering curriculum with its emphasis on mathematics and applied physics, but also must have a sufficiently broad and detailed knowledge of certain aspects of biology, especially microbiology, and chemistry to facilitate intelligent design and operation of pollution control facilities and management of natural processes.

In planning this text, it was our objective to make it as useful as possible to persons with varied backgrounds. It is needed most by engineers and is addressed to their needs in particular, but we have also attempted to make it useful to science students who wish to enter this applied field. With these aims in mind, we felt that the most logical approach would be to assume that the reader needs basic information in each of the areas important to the whole. We have attempted to provide this information in the most readily understandable form and to build upon it the combination of basic and applied microbiology that is essential to the environmental scientist and engineer.

The first four chapters are designed to provide the basic information upon which the remainder of the text is built. Persons with backgrounds in biology, chemistry, or engineering will find parts of this material already familiar. Chemists and chemical engineers will find the chemistry in Chapter 3 elementary. Microbiologists and other biological scientists already will have seen the material in Chapter 4. Most civil engineers will find that parts of

Chapters 1 and 2 describe processes with which they are familiar. However, all of these individuals should find new information covering areas outside their usual training and new applications of familiar information. Engineers, other than chemical engineers, usually have little chemistry and no biology in the prescribed curriculum. Chemical engineers normally have no training in biology. Biologists learn little about the engineered processes that protect the environment. Thus, the first four chapters are written to accomplish three objectives:

1. To provide a background in unfamiliar areas for persons entering the field of pollution control.
2. To relate basic knowledge to applications of which persons in each field may be unaware.
3. To bring all readers to a point at which they possess the common background that will serve as a basis for understanding the material in Chapters 5 through 14.

While we believe that microbiology is as basic to the study of environmental engineering as are the more traditional tools of engineering, mathematics, and physics, we feel that a textbook on microbiology aimed primarily at the environmental engineer should differ in several ways from the standard microbiology text written for students of microbiology or environmental microbiology. An engineer is, by training and practice, oriented toward the application of basic knowledge and is interested most in those facts that can be useful in professional practice. We have tried to present these facts in a format that will allow the engineer to relate principles to practical applications.

First, we have attempted to limit the topics covered to those that we consider essential to the quantitative expression and mechanistic understanding of the microbial activities that occur in natural environments or in processes engineered for the purpose of exerting useful control over the natural environment. Second, our approach is more process-oriented than species-oriented. Names of genera and species are introduced where their activities are discussed, since it is as necessary to use names for microbes as for any other material in discussing its properties. While it is true that any natural population is composed of individual microbial species, it is at times advantageous to treat the entire heterogeneous population (the *biomass*) as an engineering material, with properties that are definable and controllable within certain limits. In other circumstances, it is necessary to consider the ecological relationships between individual species or metabolic groups of microorganisms. Both approaches are used, each where it is most appropriate or useful. Third, we have attempted to integrate the basic information directly with the appropriate environmental applications in each chapter rather than following the traditional format of presenting basic information followed by chapters covering various applications.

Much of the material in this text has been tried out on the many students who have passed through our courses in the past two decades. We are particularly grateful for their comments and suggestions while they were students and for their continued feedback as they progress through careers in environmental engineering or science. Their successful solving of pollution control and environmental management problems, using the practical microbiological principles we have tried to impart, was a source of encouragement that helped us to keep burning the midnight oil while writing this text during the past three years.

The help of our students goes beyond the encouragement and stimulation that good students provide for professors. It will become apparent to the reader that many of the figures used in this text are based upon experimental results which our graduate students have helped to obtain. In truth, these examples are a small sampling of the investigative efforts of student-professor research teams which have engaged in more than 300 person-years of professional investigation into the microbiological aspects of environmental engineering in the past two decades.

Specific acknowledgments are difficult to write because when we think of all those who should be acknowledged the list becomes endless. For their immediate contributions, we are grateful for the thorough review and criticism of this work by Dr. Ven Te Chow, University of Illinois, Dr. Linda Little, environmental biologist, and Dr. W. W. Umbreit, Rutgers University. Their thoughtful comments and suggestions were very valuable to us. Also special thanks are due to Drs. T. S. Manickam and Crosby Jones and to our graduate students, M. P. Reddy, Mike Hunt, and Brenda Mears for their help in assembling material used in this text. Our grateful appreciation goes to Mrs. Grayce Wynd, a friend of many years, for her expert typing of the manuscript and her helpful suggestions. Dr. D. F. Kincannon deserves special mention. His relationship with both of us, but especially with A. F. Gaudy, has progressed from that of professor and student to that of colleague and friend and we are grateful for his support and encouragement, not only during the writing of this text, but also through the many years of our association.

The development of an investigative team requires more than the vital interest and effort of professor and students. There must be help and encouragement from administrators. We are very grateful for the help given us by Dr. Jan J. Tuma, former head of the School of Civil Engineering, and Dr. Lynn L. Gee, former head of the Department of Microbiology, Oklahoma State University. We appreciate the faith they showed in us early in our professional careers and their friendship through the years.

Finally, we are grateful for the patience, encouragement and faith in us shown by our mothers and fathers throughout our lives and particularly during the last few years. Their loving interest and questions as to progress were delightfully reminiscent of their concern over our report cards many years ago.

Anthony F. Gaudy, Jr.
Elizabeth T. Gaudy

THE LIFE-SUPPORT SYSTEM

Most books on microbiology begin with a statement about the importance of microbiology and its pivotal role in the basic and applied sciences. So, too, will this text, but the point need not be belabored because, as we discuss the life-support system in this chapter, the absolute necessity for the existence (and the control) of microorganisms will become apparent.

Microbes, both the procaryotic, or nonnucleated, cell types such as bacteria and blue-green bacteria* and the eucaryotic, or nucleated, types such as protozoa, fungi, and algae, are found nearly everywhere in nature and therefore are a force in the environment. Anyone aspiring to gain an understanding of the environment with an eye toward enhancing or controlling it must study microbiology. Just as an electrical, civil, or chemical engineer comes to know that learning to use mathematics and physics as tools of the profession is necessary in solving various problems or predicting various outcomes, i.e., in designing processes, so must the environmental engineer be aware that microbiology is equally necessary. In fact, another tool—chemistry, and especially biochemistry, which is peculiar to living cells—is also required. In like manner, an aspiring environmental biologist must become aware that one needs the tools of chemistry, mathematics, and physics in order to understand and use microbiology in solving environmental problems.

Microbiology is not the only area in the biological sciences that is important in environmental technology. However, microbiology permits one

---

*The organisms previously classified as blue-green algae are now considered to be bacteria and are called *blue-green bacteria* or *cyanobacteria.*

to bring chemistry and biology together in the study of the smallest self-producing living unit, the cell. It is, therefore, a very basic and necessary tool which ranks with mathematics, physics, and chemistry.

THE ENVIRONMENT—THE LIFE-SUPPORT SYSTEM

Before discussing the role of microorganisms in the environment, it is necessary to lay down a ground rule about the meaning of the word *environment* as used in this text. It is a word that implies as many different meanings to as many different people as does *ecology*. While there is a wide array of items that could justifiably be included under the environmental umbrella, the term *environment*, stripped of all rhetoric, simply means the life-support system. All other aspects of the environment are ancillary to the life-support system, which we can define as those items that are absolutely essential to the sustenance of aerobic life: *water, air,* and *food.*

Like one type of microorganism to be studied in this text, humans can be considered aerobic organotrophs. Such microorganisms require organic food and oxygen for controlled combustion of organic foodstuff, and they must be bathed in an aqueous environment. The cells of the human body have the same requirements. This life-support system, food, air, and water, the protection and control of which are the basis for the emerging profession of environmental technology, is maintained by two major natural cycles and accompanying mineral cycles. The two major cycles are the hydrologic cycle (water source) and the carbon-oxygen cycle (food and air source). Since the organic food source contains, in addition to carbon, some nitrogen, phosphorus, sulfur, and other elements, the cyclic migrations of these elements in the biosphere (the sphere of living organisms) are also vital to the life-support system. These cycles, so important to the life of humans and microorganisms, are discussed briefly below along with some of the artificial subsystems that our species has inserted into nature.

The Water Cycle

Slightly more than 70 percent of human body weight is water, and water comprises approximately 80 percent of the weight of a microorganism. There is little wonder, then, that it is an essential ingredient in the life-support system. Water is delivered in an unending cycle of evaporation and condensation. This constant water supply has been reused for millions of years. It is well to examine this cycle, not only because it is vital to life, but also because it is responsible in large measure for moderating and maintaining the earth's climate, thus providing our living space. Water is an excellent heat reservoir, and the earth is a gigantic solar still.

Consider the simplified diagram of the hydrologic cycle shown in Fig. 1-1. Solar energy is absorbed into liquid water and causes it to vaporize. In the

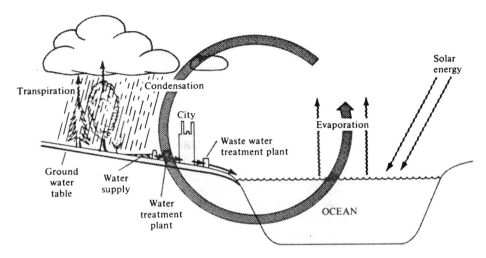

Figure 1-1 The water cycle.

diagram, the liquid bodies are designated as oceans and, while they form the major surfaces, there are, of course, others such as lakes, rivers, and ponds. Also, it must be realized that not all the water vapor is produced by evaporation from liquid surfaces. Transpiration (evaporative loss of water from macroscopic plants) accounts for some. Since the earth receives more direct sunlight in the equatorial region, this region absorbs more heat, so more water evaporates near the equator. This moisture-laden ring of heated air rises and diffuses toward the poles, and as it cools, the water vapor condenses and provides two environmental benefits. It precipitates to deliver a pure water supply, and it releases the heat absorbed during the vaporization process to warm the atmosphere. This phenomenon helps create the temperate zones in which much of the earth's human population lives.

Some of the precipitation seeps into the ground; some drains over the surface. Some of the water is used by plants and animals. Since the use of water by humans is essentially nonconsumptive, the water is generally returned to the surface water supply as waste-bearing "used" water. The waste materials, along with dissolved and suspended material eroded or leached from the earth, are carried to the oceans. There water is again purified by solar distillation. This purification and delivery process does not suffice for human needs because the water is used repeatedly in transit as surface and groundwater before it undergoes the physical purification process of evaporation. A number of biological processes intervene while the water is in transit; these can both purify and further contaminate the water. These processes involve the carbon-oxygen cycle, to be discussed subsequently. Also, we will see later that it is necessary to insert into this natural water supply–wastewater system some humanly devised technological subsystems.

These will also be described later, since they too involve microorganisms.

The tremendous heat sink function of the hydrologic cycle can be appreciated when it is realized that approximately $5 \times 10^5 \, km^3$ (120,000 mi³) of water is evaporated from the land and oceans and subsequently falls as precipitation each year. For each pound of water evaporated, approximately 600 cal/g (1000 Btu) is absorbed. This amounts to slightly less than 3×10^{20} kg·cal per year. An idea of the magnitude of this energy transfer may be obtained by comparing it with the yearly energy consumption of the United States, which in 1974 amounted to nearly 2×10^{16} kg·cal. Thus, each year the hydrologic cycle alone distributes approximately 10,000 times the energy consumed in the United States. In fact, the amount of energy stored and distributed by the hydrologic cycle amounts to nearly a quarter of the energy beamed to the earth from the sun. The tremendous amount of water evaporated and condensed in the water cycle is a minute fraction of the $1.5 \times 10^9 \, km^3$ of water estimated to be on earth. Approximately 97 percent of this water is liquid salt water in the oceans. Of the 3 percent fresh water, approximately 75 percent exists in solid form in the polar ice caps and in glaciers. At least 90 percent of the less than 1 percent remaining exists as groundwater, and the remainder as fresh surface water mainly in lakes and rivers. A minute fraction is the water vapor that stores, transports, and delivers heat and fresh water. About a quarter of the $5 \times 10^5 \, km^3$ of fresh water that is delivered falls on the land surfaces, where some of it can be used for human and plant life, as well as for the other uses to which this multipurpose resource is put.

Thus far it can be seen that the hydrologic cycle is in the main a physical phenomenon. The following information will emphasize the biological role of water, in particular the role it plays in tying together the remainder of the life-support system.

The Carbon-Oxygen Cycle

The carbon-oxygen cycle provides the major portion of the remainder of the life-support system. Figure 1-2 is a very simplified representation of this cycle. It is seen that there are two divisions or "legs"—photosynthesis and aerobic decay. Photosynthesis is shown on top because, thermodynamically speaking, it is an "uphill" process. Inorganic compounds, carbon dioxide and water, are converted to a compound consisting of carbon and water, which appropriately is called *carbohydrate*. Although the diagram shows the fixation of only one carbon atom, the organic compounds synthesized contain more than one carbon, and these are linked in carbon–carbon bonds. This type of bond is a distinguishing characteristic of organic matter. Thus, in the energy-requiring (endergonic) reaction, organic matter is synthesized. In general, syntheses require energy. In this overall reaction, water has been split and oxygen has been produced as a by-product. It is this process that permits us to live, since the reaction is responsible for the oxygen in our atmosphere. The synthesis of organic matter is, of course, the photosynthetic phenomenon

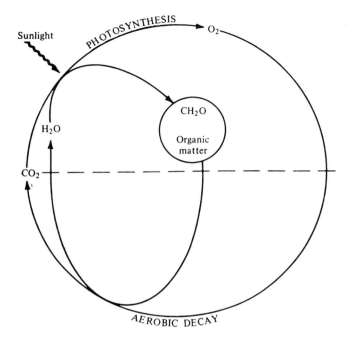

Figure 1-2 The carbon-oxygen cycle.

that creates our food supply. The source of energy that drives this biochemical process, the sun, also drives the physical process, the hydrologic cycle. How the sun's energy is captured and stored in organic compounds and is subsequently used by microorganisms is one of the subjects to be dealt with later.

The bottom leg of the cycle is labeled "aerobic decay." Here, the organic foodstuff is oxidized. The endergonic reaction (synthesis) is reversed, and in the exergonic aerobic decay reaction, energy is released. Aerobic organotrophs (such as humans) can perform this reaction, obtaining energy that can be used to synthesize organic body substances for growth and replication. Thus, solar energy flows through the living world, proceeding through photosynthesis to decay, and water, the universal solvent, is in a way the "glue" that holds it all together. The human species is part of the decay cycle and has become a potent force in it. Many of the technological processes set into the natural decay cycle (e.g., wastewater treatment plants) are engineered subsystems placed there for the purpose of providing localized control of it.

The decay reaction occurs in numerous biochemical steps and through various ecological (food) chains. Consider two general food chains: the organic matter stored in plants through photosynthesis is used as food material by various plant-eating animals (herbivores), and some of it is subject to microbial decay. Some of the herbivores die and are subject to microbial decay; some are eaten by small carnivores. Some of the carnivores die and are subject to microbial decay, and some are eaten by larger animals. This process continues. Thus, there are really two general types of food chains. In grazing food chains, plants and various animals continuously feed microorganisms, by contributing waste products while living and body material when dead. In some chains of decay, the microorganisms are the sole decaying force. Both food chains are part of the decay leg of the carbon-oxygen cycle, which leads organic matter back to carbon dioxide and water. The only input into the system is solar energy, which is continually expended in driving the cycle.

Water pollution control technologists are vitally concerned with the decay food chain, primarily because it can be overloaded easily due to localized exhaustion of the oxygen supply so that decay cannot proceed all the way to carbon dioxide and water. Various organic products are left, many of which are objectionable from the standpoint of health or aesthetics and may interfere with various other beneficial uses of water. In locations where such situations exist, due to natural (nearly always human) causes, the technological aim is either to restore and control aerobiosis or to remove the organic matter to other locations, i.e., to delocalize it. Ideally, efforts are made to try to use it for some worthwhile purpose. Although a consuming interest of technologists, this latter solution is extremely difficult to accomplish in a world striving for an affluence that has as one of its outward signs the amount one can afford to waste. The point is that until wastes become absolutely needed as raw materials, the best approach appears to be to seek ways and means to control the aerobic decay leg of the carbon-oxygen cycle via wastewater treatment processes.

Whether considering the grazing food chains or the food chains of microbial decay, the common biochemical phenomenon is the flow of energy, its use and dissipation. The energy stored in the organic material is released in oxidative reactions in which oxygen serves as the ultimate electron acceptor and is thus reduced to water. On average, no living cell completely oxidizes or respires the organic matter it assimilates. A sizable portion of the organic matter is used to form new cell substance. This synthesis requires energy, and the energy is supplied by oxidizing some of the organic food that has been consumed. As a rough rule of thumb for purposes of calculation, we may assume that about 50 percent of the organic carbon assimilated by aerobic decay microorganisms is used to build new organic body substance. In order to obtain the energy to perform this synthesis, the remaining 50 percent is oxidized to carbon dioxide and water. Thus, in the decay food chain, if a gram of organic matter were used successively as a carbon and energy source by half a dozen different species of microorganisms, nearly 99 percent of that organic matter would be totally oxidized to carbon dioxide and water and thus recycled in the carbon-oxygen cycle.

In geological terms, this continuous turnover of carbon and oxygen is rapid. It has been estimated that the carbon dioxide respired by plants and animals has an atmospheric residence time of 300 years. The oxygen that enters the atmosphere is recycled in about 2000 years. All the earth's water is split and reconstituted every 2 million years.

Carbon dioxide is converted to organic matter in two large-scale *bioreactors*, the land areas and the oceans. It has been estimated that approximately as much carbon is photosynthesized by land vegetation as is fixed by the phytoplankton in the oceans. This may amount to about 40 billion tons per year in each sphere. In the ocean, the biological photoreduction of carbon in photosynthesis is nearly balanced by its use as an organic carbon source and its subsequent decay to carbon dioxide and water, making the oceanic carbon cycle self-contained; that is, there is little or no net yearly productivity. The phytoplankton fix the carbon dioxide and enter into oceanic food chains; the major portion of this biomass consists of microorganisms. Some of these die and, for the most part, become dissolved organic matter, thus serving as a carbon source for other microorganisms. A very small fraction reaches the ocean floor as particulate organic matter and may eventually add to the supply of organic fossil deposits.

Organic matter photosynthesized on land is also recirculated to the atmosphere through the respiration process. Of the 40 billion tons of carbon fixed by land vegetation each year, approximately 25 percent is rapidly returned to the atmosphere as carbon dioxide due to the respiration of the plants that fixed the carbon. Thus, there is an annual net productivity of approximately 30 billion tons. The forests are the major photosynthesizers and account for approximately half of the net yield. To keep the land carbon cycle in balance, approximately 30 billion tons of carbon is converted to carbon dioxide through decay in the soil. Considering the entire carbon cycle, i.e., that of the oceans, wherein essentially all decay is due to microorganisms, and that on the land, where 75 percent is due to microorganisms, nearly 90 percent of the decay leg of the carbon cycle is driven by organotrophic microorganisms; most of these are aerobic. The 30 billion or so tons of organic carbon converted to carbon dioxide yearly by respiration of soil microorganisms is taken from the land "pool" of dead organic matter that is increased each year by the death of some of the living biomass.

On the land surfaces, essentially all of the carbon dioxide transferred across the land–atmosphere interface is due to biological activity (photosynthesis and respiration). However, in the oceans, carbon dioxide is transferred physically across the liquid–atmosphere interface. It has been estimated that approximately 100 billion tons of carbon dioxide is dissolved in the oceans from the atmosphere and very nearly the same amount is transferred from the oceans to the atmosphere. Oxygen is also subject to such transfer.

Thus, carbon and oxygen (as well as water) are in continuous circulation and use in the life-support system, and microorganisms have a major role in this cyclic process.

Cycling of Elements Other Than Carbon, Hydrogen, and Oxygen

Although the carbon-oxygen cycle produces the atmosphere and organic matter vital to the human species, it is well known that organic matter contains elements other than carbon, hydrogen, and oxygen. To be sure, they are the most prevalent atoms in organic material, but considerable amounts of nitrogen, sulfur, and phosphorus are found in living and dead organic matter, along with significant amounts of iron, magnesium, calcium, and potassium, as well as trace amounts of such elements as boron, copper, cobalt, manganese, zinc, and molybdenum.

Nitrogen cycle Next to carbon and oxygen, nitrogen is by weight the most abundant element in organic matter. It is contained in amino acids, which make up *protein*, one of the major classes of organic material. It is also found in a vital class of compounds called *nucleic acids*, which are responsible for the reproduction and differentiation of species, and in nucleotides, which are vital to the control of metabolism and the flow of chemical energy through living systems.

Nitrogen is abundant in the atmosphere (79 percent) in molecular form, N_2. Molecular nitrogen has the useful property of being able to be oxidized or reduced to a rather wide number of oxidation levels. Nitrogen can exist at valences of -3 to $+5$. For example, from the zero state as N_2, it can lose as many as five electrons (i.e., be oxidized to a valence of $+5$) and exist as nitrate ion, NO_3^-; or it can gain three electrons and exist in the reduced state as ammonia, NH_3, or ammonium ion, NH_4^+. This useful property permits nitrogen to be a source of inorganic chemical energy in its reduced form; ammonia nitrogen can be oxidized by certain microorganisms and the energy thus released can be trapped and used for growth. On the other hand, in oxidized form, for example, as NO_3^-, the nitrogen can serve as an electron acceptor (i.e., an oxidizing agent) for certain microorganisms and thus itself be reduced. Both of these microbial processes have a significant effect on the life-support system.

Figure 1-3 is a simplified diagram of the cyclic changes that nitrogen undergoes in the biosphere. *Nitrogen fixation* refers to the reduction of nitrogen from N_2 to the ammonia level, NH_3. This reduction takes place by three processes. First, a relatively small amount of atmospheric nitrogen is fixed by lightning and cosmic radiation. Second, an ever-increasing amount is fixed by industrial processing, which is largely responsible for the great increase in agricultural productivity in the latter half of this century. Third, nitrogen is fixed biologically by microorganisms. Microbial fixation is accomplished by two general types of organisms: those that fix nitrogen when they live in symbiotic association with higher plants, e.g., legumes such as soybeans, peas, and clover, and those that are "free living." The latter include many blue-green bacteria and several other types of bacteria.

Nitrogen at the ammonia (NH_3) or amino ($-NH_2$) level can be assimilated biologically into organic matter. The assimilation of ammonia into organic

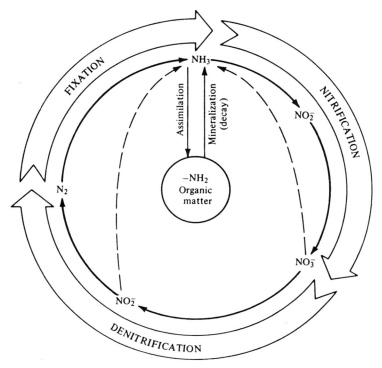

Figure 1-3 The nitrogen cycle.

matter represents neither a reduction nor an oxidation; when nitrogen is assimilated into organic matter, it retains its valence of -3. Decay of organic matter replenishes the pool of inorganic nitrogen; this process may be called *mineralization.*

Ammonia can be converted to nitrites and nitrates by aerobic autotrophic microorganisms such as the genera *Nitrosomonas* ($NH_3 \rightarrow NO_2^-$) and *Nitrobacter* ($NO_2^- \rightarrow NO_3^-$). The process has been called *nitrification.* The cycle is completed by microorganisms that can reduce nitrates and nitrites to N_2 (*denitrification*).

It is important to note the dotted arrows showing the reduction of nitrate to ammonia. Not all of the oxidized nitrogen is recycled through N_2. A large portion of the microbial nitrification that takes place in the soil provides nitrates and nitrites that are taken up by higher plants and assimilated (through the ammonia stage) into organic plant substance. Many lower plants, i.e., microorganisms, also can use nitrates and nitrites as a source of nitrogen in the synthesis of organic matter.

Of all the cycles thus far mentioned, the human species has contributed most to the nitrogen cycle through the production of commercial fertilizers in the worldwide effort to provide protein-rich food. At present, nearly the same amount of nitrogen is fixed commercially as is fixed biologically, and commercial fixation is growing as fertilizer production increases to increase food production. Some might call this unnatural fixation, but from a naturalist's point of view, it is surely within the scope of "nature," since it is brought about by a naturally occurring species. However, the human species is capable of intellectual activity, and one thought for it to ponder is whether the cycle shown in Fig. 1-3 can be maintained in balance. It is not known whether the denitrification process can keep pace with fixation in view of the massive contributions now being made by humans, who may in a few short years become the life-support system's most potent nitrogen fixers. Thus, the inventory of nitrogen in organic and mineralized form can be expected to increase in land and water masses. In regard to bodies of water, such enhancement of the aqueous environment (or fertilization of it) with one of the major constituents needed for plant life can cause excessive plant growth, choking the water body with plants, i.e., "eutrophying" it, and sharply decreasing the water supply. Eventually, water masses are converted to land masses through this process. These are extremely important considerations in efforts to gain technological control of the life-support system. Clearly, increased human fixation of nitrogen is a means of technological control of the food supply. However, this technological approach can endanger another part of the life-support system, the water supply. Solutions lie in technological-political decisions that force one to view and attack the problem from all possible directions, some of which may seem radical. For example, the needed food supply can be reduced by population control; denitrification can be enhanced by engineered subsystems placed in the decay leg of the carbon cycle, and food production could be conducted in a more contained manner so that less runoff occurs. All such approaches are of professional concern in the practice of environmental engineering science.

Sulfur cycle Sulfur, like nitrogen, forms a vital part of protein and thus of living cells. It occurs in three amino acids, and one, two, or all three are components of specific proteins. The function of sulfur in protein is often related to the spatial configuration of molecules, and this determines the reaction potential of the protein in a biological system. This important aspect of the nature of organic matter will be discussed in more detail in subsequent chapters. For now it is sufficient to emphasize that the carbon cycle, which has been cited as the source of organic matter, i.e., the food supply, does not really tell the full story. The world food supply is usually considered in relation to the amount of usable protein, and there is no known protein that does not contain nitrogen in addition to carbon, hydrogen, and oxygen, and probably few that do not contain sulfur.

Like the other four elements, sulfur cycles throughout all three parts of the biosphere, land (lithosphere), water (hydrosphere), and air (atmosphere), because it too exists in gaseous as well as soluble form. As with the elements of the carbon-oxygen and nitrogen cycles, the sulfur cycle is driven by microbial activities.

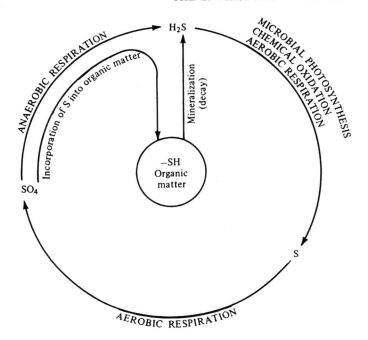

Figure 1-4 The sulfur cycle.

Figure 1-4 is a diagram of the sulfur cycle. As with carbon the incorporation of sulfur into organic matter requires an energy input; it is endergonic. Sulfur in SO_4^{2-} at a valence of $+6$ takes on eight electrons and is reduced to a valence of -2 at the sulfide level. This overall reaction is carried out by plants and microorganisms in the synthesis of sulfur-containing amino acids.

Sulfate is also reduced to hydrogen sulfide, H_2S, by sulfate-reducing bacteria that use SO_4^{2-} as a respiratory oxidant (see top curve labeled "anaerobic respiration") just as denitrifying organisms use nitrites and nitrates as an oxidant. Thus, sulfate as well as nitrates can be used as a final electron acceptor instead of oxygen. However, the sulfate reducers, unlike the denitrifiers, can use only sulfate. They are strict anaerobes, and oxygen is toxic to them. The denitrifiers use nitrites and nitrates only when there is no oxygen available. They are facultative organisms; they prefer an aerobic environment, but they can use bound oxygen in nitrites and nitrates under anaerobic conditions. Sulfate reduction and nitrate reduction are carried out by entirely different types of microorganisms. Both aerobic decay of organic matter and anaerobic respiration can contribute to the supply of H_2S.

Sulfide is oxidized to the sulfur and sulfate levels by three means—two biological and one chemical. Under anaerobic conditions, photosynthetic sulfur bacteria split H_2S in a way somewhat similar to that in which the algae and plants split H_2O (see Chaps. 7 and 8). The product of bacterial photosynthesis is elemental sulfur rather than molecular oxygen. Other bacteria in aerobic environments can use reduced sulfur as an inorganic source of chemical energy for growth, and in the process they oxidize it to sulfate. Thus, H_2S is biochemically oxidized in a manner analogous to the oxidation of ammonia, NH_3, by the nitrifying organisms. It is important to note that H_2S, unlike NH_3, can be chemically oxidized by molecular oxygen.

These oxidative processes remove H_2S and return it to the sulfate form in which it may be assimilated by microbial and plant life. It is the volatility of H_2S coupled with its relatively easy oxidation by molecular oxygen that facilitate the cycling of sulfur in the life-support system. These characteristics permit sulfur to be returned to the land through the operation of the hydrologic cycle; otherwise, one would expect that sulfur would build up continuously in the oceans. Ideally, recycling involves an interchange between all phases of the biosphere, land, oceans, and atmosphere. This occurs with carbon, hydrogen, oxygen, and nitrogen (e.g., CO_2, CH_4, O_2, N_2, NH_3, H_2O) because all can exist in gaseous forms and can be transported from the sea through the atmosphere to the land. The same possibility exists for sulfur; H_2S and SO_2 are gases. Sulfate-reducing organisms in the ocean make H_2S, which can escape to the atmosphere where it can react with oxygen and be oxidized to SO_2. Upon absorption of the SO_2 in water vapor and conversion to SO_4^{2-}, it falls to the earth in rainwater, whereupon it is available to plant life.

The nitrogen and sulfur cycles are as fundamental to the making of useful organic matter as is the carbon-oxygen cycle. All three are truly necessary to the life-support system, and they are tightly coupled to the hydrologic cycle and dependent on it. Water is not only a global air-conditioner and the medium that permits biological reactions to take place, but it also is the all-purpose solvent as well as the transport system for the elements comprising living matter.

Phosphorus and other elements Consideration of the life-support system is incomplete without the inclusion of some other elements because they, too, form a part of organic matter. Of these, phosphorus is worthy of special attention because it is required in significant amounts in the synthesis and decay of organic matter. Phosphorus in organic compounds has two major functions vital to life. It is involved in the capture and transfer of chemical energy, and it is an essential structural element in nucleic acids. The phosphorus involved in intracellular transformations of energy is highly recyclable within the cellular biomass, but some of the phosphorus incorporated in nucleic acids requires the decay of organic matter in order to be recycled. Some phosphorus in the leafy material of plants and trees is in the form of phytic acids, which decay very slowly.

The interactions of phosphorus in the biosphere are so essential to life that they are often discussed in terms of a phosphorus cycle, although phosphorus does not cycle in the sense that the aforementioned elements do. Unlike the other elements—carbon, hydrogen, oxygen, nitrogen, and sulfur—

phosphorus does not cycle with the direct aid of the hydrologic cycle. There are no volatile compounds containing phosphorus that occur commonly in nature, and the uplifting, dissolving power of the hydrologic cycle cannot assist phosphorus washed from the land into the sea to return to the land. However, living organisms in the biosphere return phosphorus from the water to the earth, and were it not for this limited recycling, the oceans would become an increasing sink for phosphorus. Phosphorus, or any other soluble but nonvolatile element of use in living matter, can be recycled due to its uptake by growing cells and its subsequent release and dissolution in water upon their death and decay, i.e., upon mineralization. It is important to realize that phosphorus is an extremely scarce element. In addition, it is not very soluble in water except under acid conditions. Since in general, the supply of phosphorus limits plant growth, phosphorus is an important component of most fertilizer mixes. Since phosphorus cannot return to the land naturally through the hydrologic cycle, it has literally been "rained" on the land artificially. Thus, the human species, in seeking to sustain its life-support system technologically, has become another "natural" link between the ocean and the land. In some instances, due to rapid runoff and overabundant applications of phosphorus, the phosphorus does not remain on the land but collects in bodies of water where its increasing abundance fosters the growth of plant life. When the body of water is used, as most are, as either a conduit or a repository for wasted organic and inorganic material of municipal or industrial origin, phosphorus may be no longer the limiting nutrient in the body of water. Other more abundant elements, e.g., nitrogen, may become the limiting nutrient and plant growth may become excessive; then the eutrophication process proceeds at an accelerated pace.

Some of the other elements needed by living organic matter may be cycled in the biosphere in the same way as is phosphorus. In some cases, a particular element may be more prevalent in the biosphere than is phosphorus, so its supply is less critical; some may not be as universally needed by living organisms as is phosphorus. Although information on other inorganic constituents is scarce, it seems reasonable that they are leached from the earth by the action of water and incorporated as needed into organic matter during the growth process. Through decay of dead organic matter, mineralization occurs and permits these elements to be cycled back to organic matter.

ASSIMILATIVE CAPACITY

In general, if decay takes place in pace with the ability of the immediate environment to cope with the organic load placed upon it, so that the products are dissipated into either the atmosphere or the hydrosphere, the decay process proceeds in accordance with the natural scheme of purification and recycling. The decay process is said not to have exceeded the assimilative capacity of the decay leg of the carbon cycle in that immediate environment or ecosystem. The term *assimilative capacity* is one that the environmental scientist ought to use with some caution, since it must be defined for each situation. The term has been used for many years by environmental engineers to refer to the amount of organic matter that can be assimilated in a body of water without causing the amount of dissolved oxygen to fall to a level that is harmful to cold-water fish (or, in general, to the higher quality game fish in the particular locale under study). However, the term is applied equally with regard to matter other than organic material, and the proper marker for measuring the stress on the assimilative capacity may not be chemical (dissolved oxygen), but rather biological, for example, a change in the distribution of species and diversity of the aquatic population. Some individuals have a tendency to discount the latter aspect because they feel that a change in species distribution is not a stress marker closely enough related to the human species. Others feel it is the most important measure of assimilative capacity because of their interest in the other species. Both attitudes need to be brought closer to a more central view, and such is the attitude of most modern professionals in the environmental control field. One does not need to think of ecological parameters of assimilative capacity because of any "in tune with nature" interest in other species, but they surely can be used as an early warning system to indicate when there has been some change from the previous conditions that might portend danger to the human species. Both approaches are best used together since they complement each other. It should be pointed out that the use of both approaches for determination and continuing assessment of the assimilative capacity of the natural ecosystem involves a massive undertaking on a national and international scale and will require a veritable army of workers exercising considerable expertise.

Generally, but not always, it is organic matter scrubbed or leached into the hydrosphere that is the material of interest in assessing the assimilative capacity of the localized natural reactor in which the decay process is proceeding. The capacity of this natural reactor is exceeded when the replenishment of the dissolved oxygen (DO) supply cannot keep pace with the biological respiratory requirement. In such cases, the use of scientific data in making technological and political decisions has generally dictated that something be done to alleviate the stress through technological control of the natural purification process. Thus, in a very significant way, human intellectual power is manifested in technological processes and devices, which are inserted into the decay leg of the carbon cycle and become a part of the natural life-support system. In the same manner, chemicals and machines are used on land to control another part of the life-support system, the food supply. The technological processes used in enhancing the natural purification process may be of a physical, chemical, biological, or combined nature.

We have married technology and politics in attempts to control pollution. The implication is that society has made a decision (passed laws) to use technology to deal with the pollution brought about largely by the ever-growing size of the society and its insatiable demand for technological produce. Perhaps more than any other technological profession, environmental science and engineering are tightly coupled to sociological needs and aspirations. Before proceeding with further aspects of science and engineering in the following chapters, it is important to give some consideration to the sociotechnological problems faced by the professional who seeks to protect and improve the environment for the benefit of society.

SOCIOTECHNOLOGICAL ASPECTS

Although some powerful, eloquent, and rather persuasive voices have attempted to arouse antagonism toward technology as a cause of environmental degradation, most people who think about it eventually come to realize that the general level of existence of the human race has been enhanced through technology, i.e., the attainment and application of scientific knowledge. To be sure, the farmer whose "bottom land" may have been permanently flooded in order to make a reservoir that provides electricity and protection against downstream floods may have valid cause to complain, but the good done for a larger number of individuals may grossly outweigh the harm done to the few. Thus, on average, that particular item of environmental technology benefited the public. On the other hand, time may show the project was ill-advised (because of ignorance and little or no technological planning); for example, the reservoir may fill with sediment.

Other examples, of course, could be cited, but one can see that even this simple example opens up a "can of worms" in relation to the environment. It surely demonstrates some of the sociotechnological facts of life. Environmental control projects affect a wide array of people, not all of whom may benefit directly; they involve human rights, individual rights, and collective rights. Thus, there are legal aspects to be considered. These projects involve considerable expenditures of funds, usually those derived from the public through taxes. The decisions regarding such technological insertions into the life-support system as dams and treatment plants are by and large made by a vote of the people, or in legislatures by pressure of the vote of the people. In a democracy, the decision adopted may be unpopular by a near majority (49.9 percent) of the individuals involved in the decision. Unfortunately, the decisions regarding these technological matters are often made by people with little or no technological knowledge.

Decisions about projects for environmental control, such as the reservoir in the example cited, are not easily reversed. Also, some time is required before the wisdom of the project, or the lack of it, can be assessed. Often, those who made the decision or benefited by the completion of the project have moved on to other fields by then. And in the example cited, they literally but unintentionally may have buried their mistake; they are seldom called to account because the passage of time has provided a new public and a diluted memory of the mistake.

Professionalism—Ethics and Accountability

Stated above are some of the reasons why it is essential for environmental professionals to be actively concerned about the social ramifications of their actions. Without their active participation in the deliberations of society, they may be good environmental technicians but never professional environmental technologists. The environmental professional must be involved at the forefront in the decision-making process rather than being simply one who effectuates a decision through technology. This concept is sometimes embraced with difficulty by people in the engineering disciplines. It was not long ago that engineers were imbued with the attitude that good ethics demanded that they remain aloof from the political sphere. Society determined what it wanted, and the technologists did their best to provide what society demanded. This attitude prevailed in other areas as well; the customer was always right. Such an attitude may be acceptable in serving the customer who really needs a 9D but insists on a 6AAA shoe. The environmental stress imparted by this strict adherence to the customer's individual right of choice is a localized one involving only that particular, unwise individual. This localization does not apply, however, to the individual decisions and attitudes that affect the environment of a wide array of people. It surely cannot apply to technological decisions about the environment. Thus, by dint of the special expertise and knowledge one may possess in environmental technology, the professional must operate on the thesis that the public or client is not always right. However, environmental professionals cannot make decisions for the public; they must function ethically as a part of it, presenting information in the form of public education so that environmental matters that require public decisions regarding technological activities can be made from a knowledgeable posture.

Under the new ethics of responsible social involvement, there will surely be times when the environmental professional's objectivity results in a position on one side of a sociotechnological argument, while the financial interest of the client or employer lies on the other side. Being able to function with technological and moral effectiveness in such dilemmas demands a new breed of professionals in any field but is especially important in that field dealing with the life-support system. This field has no parallels from which to draw examples of exemplary behavior, yet the moral behavior and actions needed here are common to many fields.

There is a common need for all to accept responsibility for their actions. This is especially true of people who possess special knowledge, since if they speak out on technical matters, they willingly or unwillingly lead or affect the thinking of the public who makes the sociotechnological decisions. Since in a democracy the responsibility for decisions is shared by the majority who makes the decision, it is easy for those who played the major role in bringing about the decision to shake off their responsibility and accountability. The ethical code demanded is not really comparable to those in medicine and law, since the theater of operation of these fields is different. In medicine, the doctor-patient relationship is a very individual and personalized one except for the area of public health, which is decidedly oriented in the direction of environmental control. In the practice of law, the lawyer-client relationship is somewhat less acutely personal than in medicine, but the question is of law and not morals or ethics. One can defend an attitude that holds that the primary responsibility of the lawyer is to put the best possible light on the client. The idea is to win the case, i.e., to defend the client's viewpoint using the laws of the land. The moralistic justice of the matter is decided by third parties (juries and/or judges) selected in a democratic manner on the basis of their specific qualifications to participate in the judgment-making process (at least this is the supposition). Lawyers are by the nature of their profession advocates of their clients. Environmental professionals are not necessarily advocates of their clients,, and although their management of the life-support system involves a life-or-death situation, the relationship is not an individual one, nor is it as immediate in time as it is in medicine. Thus, it requires the weighing of much information and much thought, because the life-support system of the individual client is shared with others, and the questions of accountability and justice fall on the shoulders of the responsible technologist rather than on third parties.

The question of accountability can be handled more easily than can that of determining what may be just. It may require extra social energies and costs, but accountability can be written into any contract. For example, groups of engineers and environmental planners who design various processes and systems may in the future also be required to operate the facility throughout its designed life, thus being accountable for a complete environmental service. This could lead to better, more complete design services and surely, in the case of the design of biological waste treatment processes, to the study of microbiology, which accounts for their functional success or failure.

The question as to what is just in regard to technological environmental control is a more difficult one. One may give a simplified answer by saying justice lies in the laws that the people pass or cause to have passed. In a sense, this is a fair way to look at it, recognizing that the system of democracy, while often declared imperfect, does accord to each individual certain rights and powers that are equal to those of every other individual. The majority rules, and in the United States the people elect legislators to make laws and executives to execute them; judges are appointed to consider their compatibility with prior laws and the Constitution. If a majority of the people disagree with what is done, the individuals are not reelected, and if necessary, the Constitution may be amended. Thus, what is just can be said to be what the majority decides is just and holds to be the law of the land. The power resides in the people, and the assumption is that therein lies knowledge and wisdom. Early fears that the latter is not necessarily true encouraged the propagation of broad-based public and private educational institutions at all levels of study. The system cannot work unless the chances for the people to make unwise decisions because of lack of knowledge are diminished through the educational process. Hence, the justice of a decision is related to knowledge of the subject, and it should be clear that the future sociological and political well-being of this and other nations is wedded to scientific and technological knowledge. Political decisions with regard to food, air, water, and energy are obviously tied to science and technology. Since the knowledgeable practitioners of this technology represent a very small minority, they must share their expertise with the public and, in doing so, must exercise moral responsibility because many citizens will have to accept or reject the conclusions and recommendations of the professionals on faith in their expertise and moral purposes. These are checked in the long run by holding the professionals strictly accountable for their advice and recommendations.

Public Law 92-500

As one example of the sociotechnological arena in which the environmental professional performs, one may examine Pub. L. 92-500, its role in the decay leg of the carbon-oxygen cycle, and the wisdom and justice of this law. The law deals with the abatement of water pollution. In general, the law is a good one because it fosters more planning and thought regarding environmental control and design of the technological means to achieve it. In the long run, it should lead to greater accountability on the part of environmental professionals. By this law, the public has recognized the essential function of technology in the life-support system, because the law mandates the use of various technological subsystems to be discussed in the next chapter. In accordance with the law, wastewater treatment is mandatory, and the stated national goal is for no polluting discharge in the next decade. By public mandate, technological control and enhancement of the environment become part and parcel of the natural environment, just as technological enhancement of the food supply has become more and more a part of "nature." Some scientists, technologists, and other individuals, however, call for an examination of the wisdom and justice of pouring billions of dollars into the construction of wastewater treatment facilities, while the money possibly could be spent for freer distribution of food, which is equally as vital as water, or education and research, which are vital to the existence of the nation.

On a technological as well as moral level, the law and its amendments demonstrate the general social tendency, after years of abuses and failure to control water pollution, to overreact to environmental problems. There does not appear to be any reasonable hope of attaining zero discharge by the 1980s. Indeed, this goal is probably as unnecessary as it is unattainable. Undeniably, it is possible to produce pure water from sewage. However, the strain on society, measured in energy, service, and capital, is so great as to place in serious jeopardy the human benefits the law was intended to safeguard. The law as now constituted is somewhat reminiscent of the couple who expended all of their resources on health insurance only to die of starvation. It seems apparent that the ultimate goals of the law will be attenuated through legislative channels or through reasonable looseness in enforcement policy (the former is, of course, to be preferred). While it is necessary to incorporate technology into the natural life-support system, it is not desirable to strangle the society whose life-support system is being "protected" by untimely or unreasonable technological restraints.

A far more reasonable approach and one consistent with the law's dependence on science and technology is to use sound technological management of the receiving streams as well as on-line and possibly in-line treatment processes, with the aim of keeping the decay leg of the carbon-oxygen cycle in reasonable balance with the photosynthetic leg and the other natural environmental cycles that have been described. This will require a high degree of wastewater treatment but hardly such an absolute as zero discharge of pollutants. Even such a chore as policing the law for nonpoint and agricultural sources seems impossible to accomplish. With people occupying more and more space and engaging in more and more activities, the difference between natural runoff and human-produced pollution becomes a nebulous distinction.

Who Pays for Environmental Control?

Previous mention was made of the billions of dollars needed for the construction of sewers and treatment plants. Vast sums of additional monies will be needed after the initial burst of construction is accomplished for corrective add-ons, operation, monitoring, and enforcement of the law. Therefore, one vital sociotechnological concern is the question of who pays for it all. Since the law applies to all, it is reasonable that the financial burden be borne by all. However, much of the population cannot adequately provide for its food supply, and it is quite incapable of complying with such a general requirement for wastewater treatment. Realization of this fact has led to social cost sharing, that is, contribution to the costs of the construction of treatment facilities through the distribution of federal tax monies. One could argue that there is equity in this because a good deal of the pollution as well as the tax monies arise from human activities that engender an affluence generally responsible for bringing about the concern over water pollution. Thus, some of the tax monies are returned to the polluters to allow them to install remedial technology. On the whole, these people have indeed given up some of the right of self-decision regarding how they expend their affluence, but on the other hand, some of their taxes are rebated for an essential purpose. And the social benefit is shared by the less affluent who also use the facility and contribute less to the general tax fund. Social scientists may wish to examine the advisability of providing the same sharing to other aspects of the life-support system, such as the food supply or important items of the life-style system, such as housing, medical care, educational opportunity, and equality of legal representation. Under Pub. L. 92-500, up to 75 percent of the initial cost of municipal wastewater treatment may be provided by federal taxes. Such a large proportion is, in general, not available in the other social services encouraged by the federal government. In planning for pollution control expenditures, it is necessary to consider the ramifications and impacts on other environmental service areas.

A large portion of the aqueous waste produced annually arises from private industrial activity. Most of the cost for pollution abatement from such sources is passed on in a form of user taxes to the consumer, for example, the higher price of the commodity whose manufacture caused the pollutional discharge, or an owner's tax in the form of less return on capital. One can reason that it is ethical to strike a balance between the two. In the long run, one may expect general social concern to come to bear through exercise of the very potent consumer power, a private affair, and through regulatory power delegated by the public to government, a political affair.

Mentioned here are but a few of the social and political concerns that must occupy the thoughts of environmental professionals. Today, there is a general realization that the problems of environmental control can be solved through activities of teams of professionals with various backgrounds of expertise. These include scientists from all four of the basic fields previously listed as vital tools, engineers of all disciplines, including those specifically prepared as environmental engineers, lawyers, economists, social scientists, naturalists, and an array of people with diverse backgrounds who classify themselves as planners and environmentalists. However, one cannot hope to solve complicated sociotechnological problems with teams composed of advocates of narrow solutions constructed around any particular discipline. Members of such teams must have a working cognizance of the special functions and capabilities of the other tools that are applied usefully in devising solutions. The technologist who is prepared to apply knowledge in mathematics, physics, chemistry, and biology to solve environmental problems is uniquely poised to lead.

However broad the array of scientific tools may be, there can be no professionalism without coupling the technological to the social considerations. While this text is devoted to technological matters with particular focus on microbiology and its application to the control of the environment, this discussion seemed vital as a reminder to the reader to step back from time to time and reassess the feasibility of current and future environmental decisions and laws with regard to their effect on society. An essential ingredient of professionalism is constructive critique and input into the making of sociotechnological policy. Nowhere is such activity a more vital part of professionalism than in matters affecting the life-support system.

PROBLEMS

1.1 Many microorganisms can use ammonia as a source of nitrogen in synthesizing organic matter. However, our species cannot do this. Go to the library (or another part of this text) for information and then discuss an important difference between microbes and animals that makes us dependent on plant life.

1.2 The three items making up the life-support system—food, air, and water—are obvious. Make a list and rank important areas of the life-style system (factors determining the quality of life) that demand and expend public funds and energies. If food, air, and water are important to an individual's life-support system, which, if any, of the items you may list as worthy of communal expense to support the life-style system would you consider to be essential to the life-support system for the political existence of your nation? Also, consider the similarities and differences between the phrases "guaranteed human rights" and "guaranteed human responsibilities."

1.3 Assume an aerobic microorganism uses 50 percent of its intake of organic food as material for making new organic matter, and that the other 50 percent of the food intake is oxidized to carbon dioxide and water. If this organism in turn is used as organic food (i.e., as a source of carbon and energy) by another species and it in turn is used by another, etc., how many turns of the food cycle are needed before nearly 99 percent of the original food (organic) material is oxidized to inorganic matter, i.e., is totally oxidized to carbon dioxide and water?

1.4 In regard to the aerobic decay leg of the carbon-oxygen cycle, the carbon dioxide produced by plants on land surfaces goes directly to the atmosphere. How does the carbon dioxide produced in lakes, rivers, and oceans get into the atmosphere?

1.5 Discuss the use of the atmosphere, the lithosphere, and the hydrosphere in the recycling of water, carbon, and oxygen. How do these three spheres making up the biosphere come into play in the cycling of nitrogen, sulfur, and phosphorus?

1.6 Why is the use of water for irrigation of crops called a *consumptive* use, whereas the use of water in homes and in industrial manufacturing constitutes in the main a *nonconsumptive* use? What is the fate of most of the water used for irrigation? What is the fate of most of the water used in homes and industry?

1.7 Everyone is aware that living things are composed largely of water and that we need it for our existence; it is a vital part of our life-support system. However, water also is used for other purposes. List some of these other uses of water and discuss them. Rank your listing and ask colleagues to do the same; then compare the uses and the rankings and defend yours to your colleagues.

1.8 Discuss how the sun drives, i.e., provides the energy for, the hydrologic cycle and the carbon-oxygen cycle. What has the sun to do with the aerobic decay leg of the carbon-oxygen cycle?

1.9 Name and discuss the two main food chains in the process of aerobic decay. How does our species relate to both? Which type of organism seems to win out in the end?

1.10 What are the two major roles of the hydrologic cycle? Explain how the hydrologic cycle tempers the earth's climate.

1.11 Considering the fact that it has taken billions of years to produce the organic matter being mined as oil, gas, and coal, and that our species probably will have used up all of this supply between the nineteenth and twenty-third centuries, what do you think are the next logical approaches in the solution of our need for this important life-style system component, i.e., our energy supply? Will such solutions require a change in the life-style to which we have become accustomed in the twentieth century?

1.12 Discuss the difference between symbiotic nitrogen fixation and free-living nitrogen fixation. Which living species fixes more nitrogen than any other? Discuss the significance of this answer in regard to the sociotechnological aspects of the environmental engineering science.

REFERENCES AND SUGGESTED READING

The biosphere. 1970. *Sci. Am.* **223**. (The entire issue is devoted to the biosphere and the life-support system.)

Chanlett, E. T. 1979. Environmental protection, 2d ed. McGraw-Hill, New York.

Davis, K. S., and J. A. Day. 1961. Water, the mirror of science. Anchor Books, Garden City, N.Y.

Heichel, G. H. 1976. Agricultural production and energy resources. *Am. Sci.* **64**:64–72.

Hospers, J. 1979. The course of democracy. *Natl. Forum* **59**:35–40. (Also see other articles in this issue.)

Hubbert, M. K. 1971. The energy resources of the earth. *Sci. Am.* **224**:60–70.

The human population. 1974. *Sci. Am.* **231**. (The entire issue is devoted to the human species and factors governing its growth and development.)

Jewel, W. J. (editor). 1975. Energy, agriculture and waste management. Ann Arbor Science Publishers, Ann Arbor.

Man's impact on the global environment. 1970. Report of the Study of Critical Environmental Problems, sponsored by the Massachusetts Institute of Technology. MIT Press, Cambridge.

McGauhey, P. H. 1968. Engineering management of water quality. McGraw-Hill, New York.

Pregel, B., H. D. Lasswell, and J. McHale (editors). 1975. Environment and society in transition: world priorities. *Ann. N. Y. Acad. Sci.* **261**.

Restoring the quality of our environment. 1965. Report of the Environmental Pollution Panel, President's Science Advisory Commitee. White House, Washington, D.C.

Revelle, R. 1976. The resources available for agriculture. *Sci. Am.* **235**:164–180.

Siever, R. 1975. The earth. *Sci. Am.* **233**:82–90.

Theobald, R. 1978. Creating a livable future. *Natl. Forum* **58**:8–12. (Also see other articles in this issue.)

U.S. Department of Agriculture. 1955. Water. The Yearbook of Agriculture. U.S. Government Printing Office, Washington, D.C.

Wagner, R. H. 1974. Environment and man, 2d ed. W. W. Norton, New York.

TWO

TECHNOLOGICAL CONTROL OF
WATER QUALITY

In this chapter, some of the technological subcycles that have become or are becoming vital components of the life-support system will be presented and some important terms defined. Later in this text, the microbiological aspects of these processes and operations will be discussed in some detail. First it is important to gain an overview of the technologically enhanced cyclic flow path of water. Then we will examine in more detail some individual processes, the operation of these subsystems, and their relationship to the microbial concepts that will be presented in later chapters.

In Fig. 1-1, which illustrates the hydrologic cycle, including the water supply and its distribution to the city, collection of the used water was shown in the path of drainage to the oceans. It is in this aqueous return flow that the major portion of the human contribution to the decay leg of the carbon-oxygen cycle occurs. Figure 2-1 shows part of that contribution to the problem (the wastewater) and to its solution (the treatment process). It will be shown that part of the problem as well as the solution involves micro-organisms. The events are cyclic; they form a vital subsystem or subcycle in the life-support system and are integrally coupled to it.

Beginning at the top and proceeding clockwise, water that has been delivered by the hydrologic cycle is extracted from some reservoir, ground or surface, natural or artificial, and is prepared for human consumption by using a series of unit operations and processes designated in the figure by a series of squares. Microorganisms play a role in preparing the water for use, but in this overall process, the primary aim is to control their amount and kind at low levels, and the operation usually terminates with a disinfection (microbial

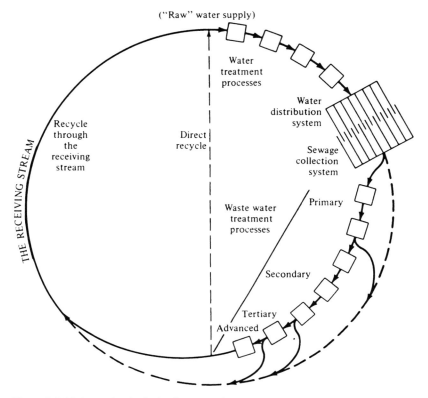

Figure 2-1 Major technological subsystems in the decay leg of the carbon-oxygen cycle.

killing) process, which is vital from a public health standpoint. Despite disinfection, microorganisms exist in the distribution system where they can react with chemicals in the water as well as those leached from the materials of which the conduits and appurtenances of the distribution system are made.

In the process of being used in various human activities, the purified water accumulates various organic and inorganic chemicals. Some of these are soluble in water, and some exist as suspended particles. Water serves as a sluicing medium, which carries wasted or unwanted material away from the human population. At this point in the path of drainage to the sea, the water has been well inoculated with a variety of microorganisms and the decay leg of the carbon-oxygen cycle is in full operation commensurate with the environment's capacity to provide the oxidant, O_2. The mineralization process usually begins in the wastewater collection system. In some cases, it proceeds at a pace so rapid that microbial activity produces highly anaerobic conditions in a part of the sewer, and both organic and inorganic products of metabolism react in an extremely adverse manner with the material of which the collection system is made. Serious air pollution problems are also caused by such metabolic activities in sewers.

Biologically enhanced corrosion of construction material of all types is an extremely important consideration in these subsystems. Major efforts are mounted either to provide an environment in the conduits and appurtenances that is favorable to controlled aerobic decay, i.e., to use the collection system as an aerobic biological treatment process, or to disrupt microbial metabolism, i.e., add microbial inhibitors to the sewers. Addition of such inhibitors, of course, needs to be done in a highly controlled manner. They must be dissipated by the time the waste reaches the treatment plant, if it is a biological plant. Depending on the nature of the disinfectant, the wastewater could be detoxified at the plant site prior to entry to the treatment process. Exercise of either approach can be called *in-line* or *on-line* treatment. Such procedures require rather large expenditures, but in locales where environmental conditions favor corrosion, the already high cost of the wastewater collection system warrants further expenditure to protect it. It is important to realize that the wastewater collection system has become a vital part of the subsystem shown in Fig. 2-1, and the subsystem is a vital part of the carbon-oxygen cycle, which in turn is vital to the life-support system. It is also well to reemphasize that the collection system is a manifestation of technological control of the environment. And while it is relatively simple to design, lay out, and construct such systems, their success or failure in accomplishing the transit function is largely related to the microbial activity within them, so that even so mundane a technological chore as the design and management of sewer systems is better accomplished with a working knowledge of microbiology.

The unit processes and operations shown downstream from the collection system are so placed to protect the natural receiving stream from excessive biological activity. The consequences of microbial activity are somewhat different in these natural conduits than in the sewer. In general, a larger supply of oxygen is available to maintain aerobic decay, since the receiving stream is more open to air circulation than is the underground collection system. However, natural aeration can accommodate only rather low concentrations of microbial carbon source and can accommodate best the carbonaceous material that is metabolized more slowly. The material collected in sewers is of higher concentration than can be accommodated by most receiving streams, and some of the organic matter contained in the wastewater is very rapidly utilized as foodstuff by microorganisms. Thus, it could cause localized conditions of anaerobiosis if it were placed directly into the natural receiving streams. Unlike the sewer, these natural streams and rivers are utilized in the biosphere for much more than carrying away the waste products of human activity. One of their most vital functions is as a water supply for the downstream population. Prevention of pollution that would hamper downstream use of the water supply requires the insertion, in the drainage path from sewer to receiving stream, of various unit processes and operations to remove as large a portion of the organic and inorganic matter as is necessary to prevent overstressing the normal function of the natural runoff channels. While there are still a number of places where the purification process may proceed satisfactorily (i.e., without dropping the DO to low levels) in the natural receiving stream without any of the treatment processes shown in the technological subsystem of Fig. 2-1, it does seem reasonable, in view of continued population growth and industrial expansion, that the line going directly from the collection system to the receiving stream be abolished.

The figure shows four return lines to the receiving stream following four levels of treatment. The first return line conducts effluent decanted from a physical process, i.e., quiescent subsidence of settleable organic matter in the wastewater. This process has been called *primary treatment*. It may be looked upon as a first line of defense against overloading the receiving stream with organic matter.

The next return line to the receiving stream emanates from sequences of processes designed to remove both suspended and soluble organic matter from the wastewater. Such processes may be physical, chemical, or biological, or combinations of these. The most successful processes, and those least understood, involve mechanisms of biological decay, i.e., microorganisms. Processes such as these advance the line of control and defense against overstress in the receiving stream, and they are called *secondary treatment processes*. These processes form the major line of defense against localized overstress of the decay leg of the carbon-oxygen cycle. In general, sociotechnological decisions have been made that wastewater should be treated at least to the secondary level prior to discharge into the receiving stream.

Often, due to localized concentration of human activities and paucity of water, normal secondary treatment may not remove sufficient organic matter from suspension and/or solution; that is, there may still be more organic matter in the effluent than can be safely treated through normal aerobic decay in the natural water course. In such cases, the effluent from secondary treatment may be channeled through additional processes for removal of greater amounts of organic matter. Secondary processes are expected to deliver organic removal efficiencies of 85 to 95 percent. It is the need for reliable delivery of such efficiencies that provides one of the incentives for gaining sufficient knowledge of microbiology to allow practical technological control of microbial metabolism.

Even when secondary processes are operating at efficiencies in the range cited above, the inflowing organic loads may be so high that the effluent still contains sufficient organic material to cause severe stress on the oxygen resource of the receiving stream. It is then that a third line of defense, which may be called *tertiary treatment*, should be inserted into the decay leg of the carbon cycle. Tertiary treatment processes may be biological, chemical, physical, or combinations of these.

In addition to removal of greater amounts of organic matter, treatment beyond the secondary level may be desirable for the removal of excessive amounts of nitrogen and phosphorus that may have been in the original wastewater stream. Their importance as plant fertilizers has previously been discussed in relation to the problem of eutrophication. Also, since nitrogen in the form of ammonium ion, NH_4^+, can be oxidized by some aerobic microorganisms, its presence may be as detrimental as organic matter in regard to utilization of the generally scanty oxygen supply in receiving streams. Thus, even if this ammonia nitrogen is not removed from solution, it may be desirable to convert it to nitrate ion, NO_3^-, i.e., to oxidize it "on shore" prior to readmitting the water to the general water resource. The conversion of NH_4^+ to NO_3^- and the removal of either nitrogen or phosphorus from solution prior to discharge constitute examples of treatment that may be called *advanced wastewater treatment*. These also may be biological, chemical, physical, or combined processes.

A word of caution is in order regarding the terms *primary, secondary, tertiary,* and *advanced* treatments. The first two have long been used in the pollution control field and are usually defined as given above. However, the terms *tertiary* and *advanced* treatment are sometimes used interchangeably. As a rule, it is preferable to provide precise statements of the process being used and the purpose for its use in the treatment scheme rather than to use the terms *advanced* and *tertiary*.

Regardless of the degree of treatment (secondary or beyond), wastewaters that contain human excrement are generally subjected to some form of disinfection prior to their reentry into the surface water resource. Chlorination has gained wide usage because of the high biochemical reactivity of chlorine with organic matter. However, this very reactivity has been the cause of recent concern. In addition to killing undersirable microorganisms, chlorinated compounds that may be deleterious to the health of downstream users may be produced when wastewaters are chlorinated. However, some form of disinfection as a safeguard against the transport of pathogenic organisms is advisable prior to release of the effluent.

After the various purification treatments, the water reenters the surface resource. As seen in Fig. 2-1, it may enter the natural drainage system (the receiving stream) where it may be even further purified. In general, one may expect that the effluent from treatments plants will benefit by residence time in the receiving streams. The receiving stream can be expected to provide a "polishing" treatment, which returns the water to a state of purity such that it can be considered as a source of raw water supply for downstream users, e.g., a community similar to the one just considered, thus completing the cycle.

The water in many natural streams that serve as water supplies consists in significant proportions of effluent from some upstream user. Thus, wastewater recycling is now practiced and will, as population and human activities increase, become increasingly necessary. It is this fact that forms the real basis of sociotechnological justice for requiring all, individually or communally, to treat wastewater prior to discharge. It also forms the logical basis for technological control and management of the natural receiving streams. In some instances, the effluent produced at the treatment plant may be treated to such a high degree that it is made suitable for more direct recycle to the water supply treatment plant. Such direct recycle is not generally practiced today. There are, however, instances when it has of necessity been done on a short-term basis, and there are presently cases in which the waste effluent, if not routinely made part of the water supply and admitted to the distribution system, is used for recreational purposes.

It should be borne in mind that, although obviously the highest use for water involves its biological role in the human life-support system, it is a multipurpose resource, and many of its uses bring it into intimate contact with the human species. In fact, most of the water supplied to individuals is used for purposes other than the vital biological life-support function. One can imagine the psychological aversion to direct water reuse for human biological needs. The extent and reliability of treatment needed to overcome such aversions are tremendous and represent a major professional responsibility of the environmental technologist. On the other hand, reasonably good treatment of wastewater coupled with complete management of the receiving bodies can help obviate the need for such direct reuse in all but exceptional cases dictated largely by geographical location and climate. The environmental professional has an equal responsibility to assure, through technological control, that the effluent discharged to the stream is of such quality that the desirable polishing treatment can take place in that important reactor and that the water resource may be usefully employed during this polishing process. This is a far better and more reasonable alternative than direct reuse. It represents the real service that society demands, and probably the only one it will be able to pay for. Thus, the technological process cycle shown in Fig. 2-1 should not revolve separately through direct recycle but should be melded with the natural cycle; the thesis here is that the circle turns more naturally than the semicircle.

It is appropriate to focus more closely on some of the technological unit operations and processes in water and wastewater treatment, since they either can affect microbial metabolism or are effected by microbial metabolism.

UNIT OPERATIONS AND PROCESSES

In a general way, the term *unit operations* is used in reference to physical processes involving little or no chemical change, e.g., sedimentation, flotation, filtration, adsorption, centrifugation, or stripping, whereas the term *unit processes* describes engineered events involving chemical changes, e.g., chemical oxidation, biological treatment processes, or chemical disinfection. Figures 2-2 and 2-3 show various operations and processes applicable to the treatment of water supplies and wastewaters, respectively. In general, they are set up to flow from inflow (left) to effluent (right), but the reader should not interpret this to indicate a fixed or typical flow scheme for water or

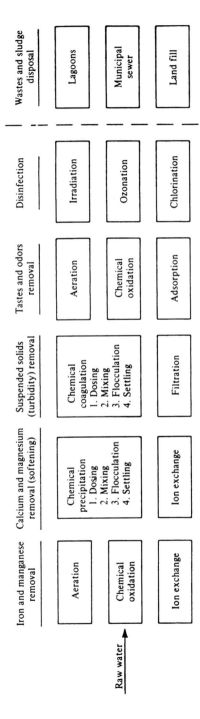

Figure 2-2 Alternative processes of water treatment and solids handling.

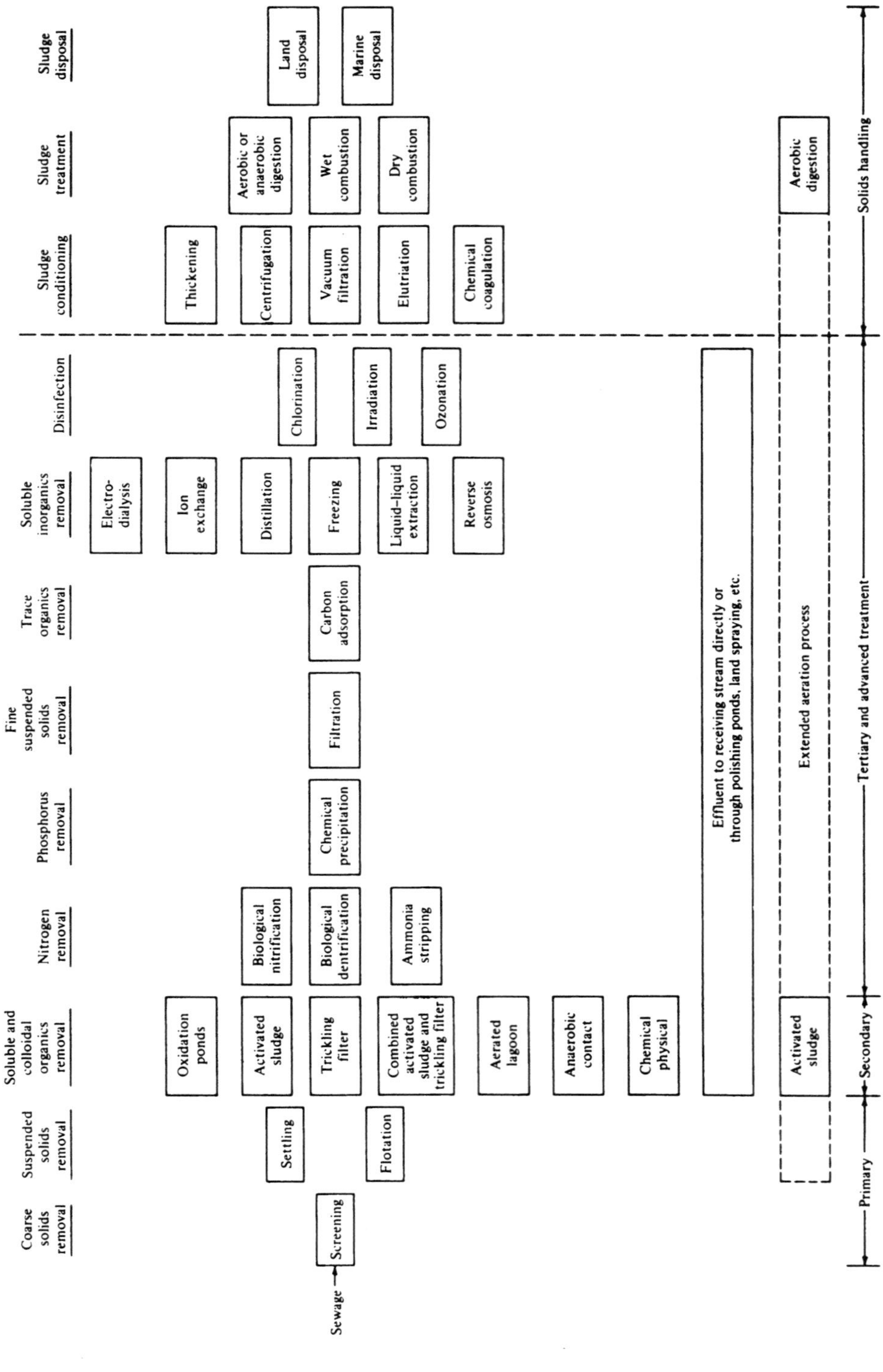

Sewage → Screening

Coarse solids removal	Suspended solids removal	Soluble and colloidal organics removal	Nitrogen removal	Phosphorus removal	Fine suspended solids removal	Trace organics removal	Soluble inorganics removal	Disinfection	Sludge conditioning	Sludge treatment	Sludge disposal
Screening	Settling	Oxidation ponds	Biological nitrification	Chemical precipitation	Filtration	Carbon adsorption	Electrodialysis	Chlorination	Thickening	Aerobic or anaerobic digestion	Land disposal
	Flotation	Activated sludge	Biological denitrification				Ion exchange	Irradiation	Centrifugation	Wet combustion	Marine disposal
		Trickling filter	Ammonia stripping				Distillation	Ozonation	Vacuum filtration	Dry combustion	
		Combined activated sludge and trickling filter					Freezing		Elutriation		
		Aerated lagoon					Liquid–liquid extraction		Chemical coagulation		
		Anaerobic contact					Reverse osmosis				
		Chemical physical									

Effluent to receiving stream directly or through polishing ponds, land spraying, etc.

Extended aeration process

Activated sludge

Aerobic digestion

Primary — Secondary — Tertiary and advanced treatment — Solids handling

30

wastewater treatment plants. In modern environmental engineering practice, there are no standardized flow schemes. Each plant's flow scheme is determined by the unique characteristics of the water in question, the particular environmental condition at the site, and the quality of the water required. Thus, the diagrams are useful in showing process functions but do not necessarily indicate a sequence of operation.

WATER SUPPLY

The raw water supply may be taken from surface or groundwater sources. The source of the water naturally can be expected to affect the quality of the water.

Removal of Inorganic Ions and Turbidity

In general, a groundwater source contains somewhat more mineral solids and less turbidity than surface waters. *Iron* and *manganese* are more commonly expected in groundwaters than in surface waters, but the bottom layers of lakes and reservoirs can contain rather high concentrations of iron and manganese, and it should be borne in mind that surface waters that receive certain industrial effluents and/or mine drainage may contain significant concentrations of these elements—more reason for not committing the error of overgeneralization and simplification regarding flow schemes. When present in groundwater, these elements exist in reduced form, Fe^{2+} and Mn^{2+}, and upon contact with air they are oxidized to Fe^{3+} and Mn^{3+}. In oxidized form, they exist as colloids and adhere to the surfaces of plumbing fixtures or other materials with which the water comes in contact. Oxidation by aeration or by the addition of stronger chemical oxidants removes these elements. Iron and manganese may also be used as energy sources by certain bacteria, and their removal helps prevent the growth of such organisms in the distribution system.

    Calcium and *magnesium* react with the fatty acids in soaps to form insoluble salts of these acids, which produce two undesirable effects: they cause greasy scums to form in the water, and they cause excessive use of cleansers. Two lines of attack have been followed to overcome these effects: (1) the replacement of soaps with other cleaning agents (detergents) that do not form insoluble compounds with calcium and magnesium, and (2) treatment of the water (softening) to remove the Ca^{2+} and Mg^{2+} (hardness). Some of the need for removing excessive concentrations of calcium and magnesium was negated after World War II by the large increase in use of synthetic deter-

---

Figure 2-3 Alternative processes of wastewater treatment and solids handling. *(Adapted from Porter, 1970.)*

gents that did not react with these elements. However, significant portions of the detergent molecules were not readily metabolized by microorganisms and they persisted in the aqueous environment. The use of detergents containing nonbiodegradable organic compounds as well as phosphorus led to severe pollution problems in some areas. Some of the sudsing properties were retained even at rather low concentrations of detergents, and the foaming of surface water was a common problem. In the case of some groundwater supplies, these foam-forming waters were delivered to the consumer's spigot. The phosphorus content enhanced the eutrophication process. A sociotechnological remedy was set in motion in which some laws were passed against the use of nonbiodegradable, phosphorus-containing detergents. In addition, a boycott by consumers of environmentally detrimental products ("consumer power") spurred industrial research on the development of biodegradable, non-phosphorus-containing detergents. Thus, concern over the use of detergents, which peaked in the 1960s, has now subsided to a large degree. Although phosphorus is a component of most detergents, the organic constituents are biodegradable. Investigation and control of the biodegradation of synthetic products in the biosphere are extremely important activities of the environmental professional.

In the process of water softening, calcium and magnesium may be precipitated by the addition of chemicals such as lime, $Ca(OH)_2$, and soda ash, Na_2CO_3, or may be removed by ion-exchange processes in which nonobjectionable cations are subsituted for calcium and magnesium. If raw water contains turbidity, chemical treatment for removal of these elements will also remove some of the turbidity, e.g., suspended clay particles. Suspended organic matter, some of which may be living (microorganisms), will also be removed by entrapment in the chemical precipitate as the floc particles subside. Chemical precipitation of calcium and magnesium is not a one-step process. The precipitating chemicals must be metered into the water at the required dosage and mixed thoroughly. This is followed by a period of formation of precipitate and floc building, then by quiescent settling. When a raw water contains turbidity, but there is no need to precipitate soluble constituents such as calcium and magnesium, the suspended particles may be removed by the addition of other chemicals such as alum or ferric chloride. These chemicals neutralize electric charges on the particles and themselves form chemical precipitates, causing the minute turbidity particles, e.g., clay and microorganisms, to agglomerate with the precipitates, forming floc particles that can be separated from the water by settling under quiescent conditions.

Usually, quiescent settling does not suffice to remove all of the floc or precipitate, and that remaining in suspension is filtered out. Particles, including microorganisms, are trapped in the interstices of the filtering medium, which often consists of specially graded sand. Microorganisms are also adsorbed to the surfaces of the filtering medium. This polishing treatment removes essentially all traces of turbidity and most of the bacteria.

Tastes and Odors

Iron and manganese, turbidity and hardness are removed to enhance the utility and desirability of the water to the consumer as well as to protect the distribution and delivery systems. The removal of tastes and odors also falls into this category in most cases, although the existence of tastes or odors in a raw water supply sometimes can be indicative of microbial contamination. Tastes and odors at rather low levels are detectable and objectionable to consumers.

Groundwater (but seldom surface water) sometimes contains dissolved gases such as hydrogen sulfide and methane. Odorous compounds in impoundment reservoirs for surface supplies arise from decaying vegetation and the activities of microorganisms and from the growth of certain blue-green bacteria. These are the most common sources, but others include compounds from certain industrial wastes (phenolic compounds are particularly objectionable) and agricultural chemicals, e.g., pesticides and defoliants. Obviously, the best means of control is to prevent the entry of odor-causing contaminants into the water supply, but this is extremely difficult in view of the wide array of possible sources and the generally low concentrations involved.

Taste and odors from the growth of blue-green bacteria in reservoirs may be controlled by adding copper sulfate or other disinfectants at various places (trouble spots) in the reservoir where the microorganisms have been observed. Potassium permanganate has also been used. It is a strong oxidant and, in addition to killing the microorganisms, it may oxidize residual odor compounds in the reservoir.

In general, procedures to remove tastes and odors at the treatment plant involve one of the processes shown in Fig. 2-2. These are aeration, which can strip out gases as well as provide a mild chemical oxidant (O_2); chemical oxidation using chlorine, chlorine dioxide, or potassium permanganate; or adsorption on activated carbon. Extreme care must be exercised in administering these treatments, especially the oxidizing compounds, which, in high enough concentrations, are toxic to humans. Proper dosages and modes of application require careful study; the improper application of chlorine can sometimes create tastes and odors in previously palatable waters, depending on the nature and concentration of chloro-organic compounds formed.

Disinfection

The final process shown, disinfection, is needed to provide a water essentially devoid of microorganisms that are potentially pathogenic to human beings. Since it is quite impractical to assay for all potentially disease-causing organisms, a class of organisms commonly found in human and animal excrement, fecal coliforms, is used for testing. If water that is devoid of this class of organisms is produced, there is a good reason to believe (but not to guarantee) that the water is not hazardous to health. It is emphasized, however, that negative results from a fecal coliform test do not indicate the absence of disease-causing animal (or plant) viruses.

Various disinfecting procedures may be used. Chemical disinfection is most widely practiced, and chlorine is the most common chemical. Chlorine has been used for many purposes besides disinfection, e.g., for taste and odor control, hydrogen sulfide removal, and iron and manganese removal. Recently, there has been much concern regarding the physiological effect on humans of certain organic by-products of chlorination, and investigative and developmental work on other disinfectants can be expected to increase in the future.

The processes shown in Fig. 2-2 do not necessarily provide a complete picture of water treatment, nor has the above discussion touched on all of the processes that may be used in preparing a raw water supply for delivery to the consumer. For example, some waters contain very high concentrations of dissolved solids other than those already mentioned; often these must be removed, and various processes can reduce the overall salts concentration.

In general, water is prepared for various consumers to very exacting standards. Some waters used in manufacturing processes and in energy-transfer operations (e.g., high-pressure boilers) are prepared to higher degrees of chemical purity than are municipal waters. However, the most important use for water is in the human life-support system. The most important environmental consideration is that the water must not be hazardous to health; this usually dictates the absence of pathogenic microorganisms. After this consideration is met, it helps if the water is appealing or appetizing. This latter consideration is balanced against allowable cost, which is determined by the segment of society being served and its willingness to bear the cost of what appeals to it.

Responsibility for the purity and potability of the treated water continues in the distribution system. The conduits can become, by design or accident, biological and chemical reactors, and this may impair their function as conduits. Thus, biological considerations in quality control can be of as much concern in the distribution system as in the operation and management of the source of supply and the treatment plant.

Disposal of Waste Products

Like most other industrial activities, the treatment of water creates waste products, which must be disposed of in a way consistent with sound technological control of pollution. Since the purpose of water treatment is the removal of suspended and dissolved solids, the waste products can be expected to contain amounts of these in proportion to the amounts present in the raw water. Depending on the processes used, chemical additives form precipitates, and the amount of solids in the waste product can be expected to exceed that in the raw water. In chemical softening using lime and soda ash, approximately two parts of sludge solids (on a dry weight basis) may be produced for each part of hardness (as $CaCO_3$) removed. Sludge consistency may vary from 5 to 15 percent solids by weight; thus, sludge represents a large volume of material. Its discharge into municipal sewers is not permitted and, like other sludges produced in the treatment of water, it may be placed in sludge lagoons, as shown in Fig. 2-2. However, the sludge does not dry rapidly, and considerable space must be given over to such sludge ponds. At larger plants, it may be economically feasible and decidedly worthwhile from the standpoint of sludge disposal to install a recalcination plant for recovery of lime, which is reused in the softening process. Recalcination requires preparatory sludge dewatering by centrifuging and/or filtering, and drying. Some of the dewatering operations are also necessary in order to use sludge for landfill.

Wastes also arise from the backwashing of filters. After a period of operation, filters become clogged with solids, and the medium is scoured and relieved of this loading by reversing the flow through the filter. The material carried away in the backwash water may in some cases be channeled to the municipal sewer; in other cases it may be put in a sedimentation basin, the supernatant may be recovered, and the sediment combined with sludges from other processes and operations for treatment as described above. Whether they are disposed of on the land or in the sewer, sludges are reintroduced into the biosphere and are fitted, in general, into the decay leg of the carbon-oxygen cycle. The organic portions are subject to decay by microorganisms, and the inorganic portions may be metabolized in part by microorganisms or may affect (for better or worse) the living environment of the microorganisms and thus the human life-support system. As would be expected, the impurities removed from raw water supplies and the waste sludges so developed have a lower potential environmental hazard than those from wastewater treatment plants.

WASTEWATER DISPOSAL

After delivery, water is used by the consumer and sent by sewer to the collection system. It is important to point out that the wastewater collection system is a much more active biological reactor than is the water distribution system. Controlled microbial metabolism in the sewer system may be advantageous as a preconditioning process before the wastewater arrives at the plant. However, rapid microbial activity that could lead to highly anaerobic conditions should be avoided, since anaerobiosis can cause very serious odors, the hazard of explosion, and the deterioration of materials of which the sewers and appurtenances are constructed. In-line treatment to either enhance or curtail microbial activity in collection systems is now receiving much more attention than in former years.

At the treatment plant, the wastewater may be subjected to a variety of operations and processes. Examples of these are shown in Fig. 2-3. Screens are usually provided to remove large objects that may be carried along in the sewer. Although there are screens of small enough mesh to remove significant amounts of particulate matter, most screens or racks are used only to protect the unit operations and the processes that follow.

Primary Treatment

Depending on the nature of the particulate matter contained in the waste, sedimentation may involve two sequential operations. In the first, dense, rapidly settling inorganic particles (e.g., sand and grit) characteristic of many municipal wastewaters may be settled out in the stilling chambers using rather low retention times of approximately 1 to 2 min. Most of the particulate organic matter remains in suspension and is removed in a second operation. Municipal wastes and some industrial wastes contain particulate organic matter that can be removed in a reasonable time (1 to 3 h) by quiescent settling in a stilling basin (primary settling tank or primary clarifier).

Physical separation by quiescent settling is a unit operation that was previously identified as primary treatment. Its major purpose is to remove settleable organic solids that were suspended or carried along in the flowing sewage but that have a settling velocity sufficiently great so that they will subside if a reasonable stilling period is provided. The settled material cannot be allowed to remain in the bottom of the basin so long that anaerobic microbial metabolism forms gases that tend to resuspend the settled material; also, odors will arise. The basins, therefore, are equipped with scraper mechanisms that gradually ease the bottom sediment to a collection hopper, and this underflow "sludge" becomes a waste product of the process. It should be remembered that a settling basin provides a quiescent chamber that permits particulate materials to separate because of differences in density or specific gravity between the particulates and the suspending fluid. A settling tank may also serve as a "flotation" tank for materials of specific gravity lower than the wastewater, e.g., grease, and surface-skimming devices are used to collect this material, which adds to the amount of waste products of the treatment process but increases the efficiency of treatment.

Flotation also may be used as a replacement for sedimentation. Particulate matter more dense than the suspending fluid can be made to float by the buoyant force imparted to the suspended solids by the attachment or entrapment of air bubbles. Air may be injected into the waste flow, followed by the application of a partial vacuum, which causes the dissolved gas to bubble off and attach to the particles, thus reducing the bulk density of the solids (*vacuum flotation*). Alternatively, in a process called *pressure flotation*, air is injected under a pressure of approximately 30 to 40 lb/in^2, and the requisite air bubbles are formed upon exposure of the wastewater to atmospheric pressure.

The efficiency that may be expected with primary treatment obviously depends on the characteristics of the wastewater. Municipal wastewaters may be expected to carry along or suspend a significant amount of particulate matter, and depending on the depth of the settling or flotation tank, retention time, and other factors, primary treatment may remove from 40 to 85 percent of these solids and perhaps 25 to 50 percent of the organic matter readily available as foodstuff for microorganisms. Some industrial wastes contain no significant amounts of settleable solids, and primary treatment is not applicable to them.

Settling basins obviously do not remove dissolved solids. In many instances, however, dissolved organic solids are the major cause of pollution since many such compounds are ready sources of carbon for microbial growth and their rapid metabolism could deplete the dissolved oxygen resource in the natural water course into which a primary effluent might be discharged. Thus, primary treatment seldom suffices from a technological point of view, and from a social point of view in the United States, primary treatment as the sole treatment of municipal wastes has been made unlawful by the passage of Pub. L. 92-500.

Secondary Treatment

Many wastes contain soluble organic matter and finely divided organic material of a size approaching colloidal. The removal of these materials from the effluent of the primary treatment process is accomplished by secondary treatment processes. There are various alternatives that may be chosen; these are listed under the column headed "soluble organics removal" in Fig. 2-3. Most of the processes shown are biological. To these may be added another type, chemical-physical treatment.

Chemical-physical treatment Chemical flocculation of organic colloids has been practiced for many years and has, for the most part, been abandoned because the degree of organic removal from many municipal and industrial wastes was not sufficiently high. The reason for rather low efficiency (50 to 75 percent) is because chemical flocculation removed little or none of the soluble organic material. More recently, chemical precipitation followed by adsorption of nonprecipitable organic compounds on activated carbon has been shown in some cases to provide removal of organic matter comparable to that in biological treatment. However, many organic compounds are not adsorbed on activated carbon; thus, its applicability is somewhat limited.

A disadvantage of chemical treatment is the production of rather large amounts of sludge consisting of chemical precipitates and flocculated organic matter. This creates a significant disposal problem not only because of the large mass of material but also because of the fact that it contains highly putrescible "raw" organic matter. The activated carbon used in the adsorption process can be regenerated by chemical treatment and by combustion of the adsorbed material, but some of the carbon is lost in the regeneration step; also, some of the adsorbing power of the carbon is lost by the regeneration procedure. Thus, it is necessary to add new carbon to the process continuously. Therefore, the two major reasons for approaching the use of chemical-physical processes for secondary treatment with caution are that they may not be able to remove significant fractions of rather easily metabolized compounds from some wastes, and that large amounts of highly putrescible sludge are developed. The other alternative, biological treatment, can partially overcome these disadvantages, but it has some others that are uniquely its own.

Biological treatment and disposal The basic advantage of biological treatment over chemical treatment lies in the fact that the aim of the process is to make the organic matter a feedstock for microorganisms, thus utilizing, but controlling the rate of, the natural decay leg of the carbon-oxygen cycle. Biological processes not only remove the organic matter that would otherwise draw down the supply of dissolved oxygen in the receiving stream, but they also stabilize a considerable portion of the organic matter by oxidizing it to carbon dioxide. Some organic compounds are not readily metabolized by microorganisms (e.g., lignin), but their removal is not the purpose of the secondary treatment plant. One might argue that it would be well to take out all traces of organic matter, whether or not it causes a biochemical oxygen demand in the receiving stream. In some cases, for example, if the water were going to be reused directly without entering the receiving stream or if organic color bodies hampered the utility of the receiving stream, removal of total organic carbon rather than only the portion that would lead to a microbial oxygen demand could be justified. However, where such treatment is needed, it is generally more successfully accomplished in sequence with secondary treatment; that is, secondary treatment is a good pretreatment for removal of nonbiodegradables or organic residuals, since it removes large amounts of organic material that could hamper or interfere with the functioning of such polishing processes as are available.

There are more variations of biological treatment processes than are shown in Fig. 2-3; only the major ones are shown. One of the most used processes involves reactors in which the growing microorganisms are held in suspension in the wastewater, i.e., fluidized reactors, Various modifications of *activated sludge* are the prime examples of this type of system. In another type, the growing cells are immobilized or fixed on a suitable medium, i.e., fixed-bed reactors. *Trickling filtration* is such a process. These processes are shown diagrammatically in Fig. 2-4, and an aerial view of an activated sludge plant is shown in Fig. 2-5.

In both processes, the organisms grown in the bioreactor are then separated from the "mixed liquor" that flows continuously from the reactor. Traditionally, separation has been brought about by sedimentation under quiescent conditions. Microorganisms approach colloidal size (e.g., *Escherichia coli* is a rod about 0.5 μm in diameter and 1 μm long) and are only slightly more dense than the liquid in which they are suspended. Thus,

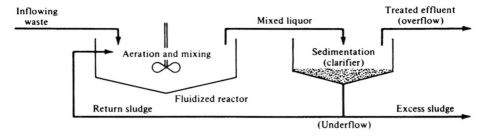

ACTIVATED SLUDGE

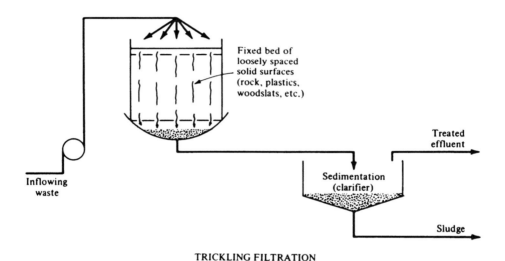

TRICKLING FILTRATION

Figure 2-4 Two major types of biological treatment processes.

quiescent sedimentation would not be expected to separate microbial cells from the mixed liquor. However, due to causes that are not well understood (one of the problems in biological treatment that the study of environmental microbiology may help to remedy), the biomass in activated sludge processes and that which is washed off trickling filters usually exists in aggregates or flocs consisting of millions of cells. These flocs are of sufficient size and density to permit them to settle in a few hours. The agglomerating tendency of the biomass in either process seems to depend on the size of the biomass in relation to the quantity of the available carbon source. Without the return of massive amounts of sludge, so that the concentration of cells in the aerator is much higher than could be produced by metabolism of the waste, the cells usually do not flocculate. On a trickling filter, cells are retained by attachment, so that the same general situation exists with regard to the ratio of foodstuff

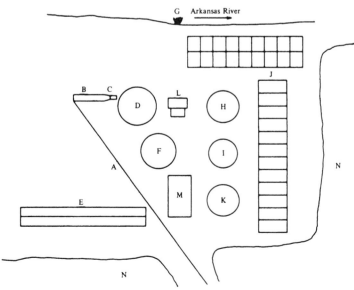

to biomass concentration, and the filter effluent contains clumps of slime organisms that are settleable in the clarifier.

The settling basins following either reactor are called *secondary settling tanks* or secondary clarifiers for obvious reasons. Autoflocculation and settling of biomass from these processes (especially activated sludge) has been considered by some engineers to be an integral part of the biological treatment process. In the extremes, flocculation and sedimentation, on one hand, have been taken for granted by designers and, on the other hand, have been considered to be such unreliable phenomena as to call for the replacement of biological treatment by other processes. However, satisfactory replacements of biological treatment have not been developed and may never be. It is emphasized that the unit process represented in the bioreactor and the unit operation accomplished in the cell separator are two distinctly different events and subject to unique considerations. Separation may be enhanced by periodic addition of flocculating chemicals. Also, separation may be accomplished by other means; for example, it may be centrifugally assisted or replaced by filtration. Quiescent sedimentation is indeed economical, but it would seem advisable to examine other ways and means of effecting separation while seeking ways to control the growth environment in the bioreactor so as to enhance autoflocculation of the biomass.

From time to time, combinations of fluidized suspended (activated sludge) and fixed-growth (trickling filter) reactors have been used. Examples of these are the use of rotating disks partially submerged in the wastewater, and processes that use some inert medium for attachment of microorganisms (e.g., sand) with the inert particles held in full or partially fluidized suspension by aeration. Also, trickling filter and activated sludge processes are sometimes used in sequence. While the treatment of organic wastes under anaerobic conditions is not unheard of, it is seldom the sole secondary treatment for reasons that will become apparent in Chap. 11.

Figure 2-5 A secondary treatment plant, utilizing the activated sludge process. (Top) Aerial photograph of the municipal wastewater treatment plant at Ponca City, Oklahoma. (*Courtesy of Blubaugh Engineering, Ponca City, Oklahoma.*) (Bottom) Schematic drawing of the treatment plant shown in the photograph. Raw sewage enters the site through the main collection sewer A, flows through grit chamber B, through a measuring flume C, to the primary settling tank (clarifier) D. The clarifier effluent then flows to activated sludge tanks E, and to the secondary clarifier F. The clarifier effluent then enters the Arkansas River at outfall G. Settled sludge from the primary clarifier is subjected to anaerobic digestion in either of the digesters H and I. The supernatant liquid from the digesters is recycled to the primary settling tank, and the digested sludge is sent to drying beds J. Part of the settled sludge from the secondary clarifier is returned to the influent end of the activated sludge tank, and excess sludge is channeled to the aerobic digester K. Aerobically digested sludge is sent to the sludge drying beds J, and supernatant liquid from the aerobic digester is recycled to the activated sludge tank. The building labeled L is the control laboratory and administration building, and M is the pump house. The pond areas bounding two sides of the plant site N are not treatment ponds but are "natural" runoff catchments resulting from borrow pits, which provided earthen fill on which the plant was constructed.

Aerobic processes differ in the ways in which aerobic conditions are maintained. In the activated sludge process, the cells are maintained in fluidized suspension, and dissolved oxygen is supplied by bubbling in either air or pure oxygen or by creating sufficient agitation to provide rapid transfer of atmospheric oxygen to the bulk medium. In a trickling filter, air is transferred across the thin layer of cells adhering to the filtration medium (rock, plastic, wood, etc.). The transfer of oxygen in rotating-disk systems is essentially the same as in a fixed-bed reactor.

Another type of aerobic treatment utilizes a different mode of aeration, i.e., *oxidation ponds*, or *stabilization basins* (see Fig. 2-3). In this type, one essentially constructs a lake into which the wastewater is channeled and in which it is retained for a considerable period of time. Retention time is usually measured in weeks, whereas in the other types of aerobic treatment it is measured in hours. An oxidation pond is an attempt to use the carbon-oxygen cycle (see Fig. 1-2) in microcosm. Microorganisms, primarily bacteria, metabolize the organic carbon, producing new cells and carbon dioxide. The carbon dioxide is used as a carbon source by algae and blue-green bacteria, which grow and produce oxygen as a by-product. The oxygen is then used by the bacteria in metabolizing the waste. The process has some utility, but its popularity is waning for reasons that are easily seen. Unless a step is added to separate the suspended biological solids (organic-utilizing microorganisms and algae) from the effluent, what has been accomplished is not movement of the organic matter through the decay path to carbon dioxide, but a swap of organic matter in the waste for organic matter in the algae that have been created in the process. This is a true carbon-oxygen cycle, but it has provided for the immediate conservation of the organic matter, not its dissipation to the larger cycle as carbon dioxide. Some of the mixed biological population may settle to the bottom of these polluted lakes where, in the absence of oxygen, the sediment undergoes anaerobic decomposition. The gases produced may, at times, cause large portions of the sediment to float to the top and escape over the effluent weir to the receiving stream or collect along the shore and create odors.

Although the terms *stabilization pond* and *oxidation pond* are often used interchangeably, the former arises, perhaps, as a recognition of the presence of anaerobic conditions. Some basins may be designed, simply by making them deep, to provide for an aerobic upper layer and an anaerobic lower layer. Soluble organic end products of anaerobic microbial metabolism may be diffused into the upper layer, where they are metabolized under aerobic conditions. It is extremely difficult for microbial-algal symbiotic aeration to provide sufficient oxygen in the upper layers under very high organic feeding (loading) conditions. The use of *aerated lagoons* has come about largely as a technological enhancement or upgrading of oxidation and/or stabilization ponds. These are large aeration basins with rather long hydraulic retention times (approximately 10 to 12 days) compared with a few hours holding time in activated sludge processes, but holding times are usually much shorter than in biologically aerated oxidation ponds. They are aerated by mechanical agitation, using floating aerators, and/or by diffusing compressed air into the bulk liquid. Like oxidation ponds and stabilization basins, aerated lagoons usually have been installed with no settling basins to separate the micro-organisms produced in them from the mixed liquid prior to reentry of the effluent into the natural drainage system. In some designs of aerated lagoons, a portion of the lagoon near the overflow weir is left unaerated (unagitated) so as to provide semiquiescent conditions that enhance settling. However, this can cause the same problems previously mentioned in regard to rising solids from the bottom of oxidation ponds. Also, without the use of extremely long retention times, little settling occurs. The biological solids concentration is rather low because cell recycling does not occur, and the cells tend to remain dispersed; i.e., they do not flocculate as in the activated sludge process. However, some wastes may contain chemicals that promote settling so that, in some cases, aerated lagoons can be upgraded by installing settling basins in series with the lagoons.

Further discussion about the microbiological aspects of the secondary processes constitutes a significant portion of subsequent chapters. If treatment stops at the secondary level, the clarifier effluent flows to the receiving stream without consideration of the other wastewater treatment processes shown in Fig. 2-3, except that in many cases the effluent is disinfected prior to discharge.

Sludge Treatment and Disposal

The secondary treatment plant produces organic sludge from the primary settling tank, if one is used to remove settleable solids, as well as biomass from the underflow of the secondary settling tank. Disposal of this waste product of the treatment process constitutes one of the most costly and difficult problems confronting environmental pollution control technologists. Some of the processes and operations used in sludge disposal are shown to the right of the dotted line in Fig. 2-3.

The general aims are: (1) reduction in volume of material, which first involves removal of some of the water that constitutes 97 to 98 percent of the sludge; (2) reduction of the volatile (organic) content of the solids, which reduces its putrescibility; and (3) ultimate disposal of the residues.

Some of the processes shown in Fig. 2-3 are used to prepare the sludge for any of the three combustion processes (reduction of organic content) shown, i.e., aerobic or anerobic digestion (biological combustion) or chemical combustion, by wet or dry incineration. The general idea is to increase the concentration of organic matter by reducing the water content prior to entry to the processes that reduce organic content.

The water content of the clarifier underflow sludge can be reduced by mechanical thickening, which involves the slow rotation of rakes through the sludge. This promotes flocculation and agglomeration of sludge particles, permitting further compaction. The released water is decanted. Thickening also may be accomplished by flotation processes and by centrifugation. Water may be removed by filtration of the sludge. The sludge may be prepared for filtration by heat treatment, which tends to coagulate the particles, or by the addition of coagulating chemicals such as ferric chloride and lime. Heat treatment can release large amounts of soluble organic matter from the sludge, and the material must be recycled to the biological treatment reactor. The amounts of chemicals required to effect satisfactory coagulation prior to filtration may be reduced by washing the sludge to remove some of the dissolved inorganic solids that might otherwise react with the added chemicals.

After the sludge has been concentrated, further treatment may be physical or biological. It may be subjected to pressure and heat treatment, thereby oxidizing a portion of the organics to carbon dioxide and reducing its putrescibility, i.e., stabilizing it. For total disposal, it may be further heat-dried until it can be burned (incinerated). Alternatively, the sludge may be subjected to biological decay under aerobic conditions, or it may be subjected to fermentation and anaerobic decay in the absence of oxygen.

Under aerobic conditions, a fairly large portion of the organic matter is oxidized biologically to carbon dioxide and water. Under anaerobic conditions, much less of the energy in the sludge is channeled into the decay leg of the carbon-oxygen cycle. Considerable portions of the organic matter can be converted to methane, CH_4; the energy is trapped in this organic gas, which can be collected above the anaerobically digesting mixed liquor in the reaction vessel. The carbon is eventually returned by chemical combustion to the decay leg of the carbon-oxygen cycle, when the gas is burned:

$$CH_4 + 2O_2 \rightarrow CO_2 + 2H_2O$$

These biological processes are of major concern in this text and will be covered in some detail in later chapters.

Residual solids from these processes may be disposed of on land or in water. Marine disposal has come under considerable criticism in recent years. It may be practiced in the future, but the determination of safe disposal sites, with the uncertainties involved, and the haul distances that will be required make it far less attractive to coastal communities than it was in the past.

There is no ideal way to dispose of the sludge produced by the waste treatment process. To be sure, it contains some valuable materials, but the cost and energy requirements for reclaiming these materials from the sludge exceed those for obtaining them from other sources, and this will probably continue to be true for many years. Currently, recycle possibilities for sludge are receiving some attention, but hygienic recycle through the decay leg of the carbon-oxygen cycle offers one of the most promising avenues of approach. Every effort should be made to extract value from the sludge, and there are now, and undoubtedly will be more, instances of successful direct recycle of components of waste sludges. By and large, these most probably will account for a small fraction of the sludge produced, and it would seem that as producers, society would be better off by facing up to the fact that the sludge is a waste product that we must dispose of, and by looking for usable treatment methods and less wasteful usage of virgin resources.

The process labeled "extended aeration" spans Fig. 2-3 from primary and secondary treatment via activated sludge to sludge disposal via aerobic digestion. In this process, usually any settleable organic solids are channeled directly to the activated sludge reactor along with colloidal and soluble organic matter. The functional biological concept for the process holds that by returning all sludge to the aeration tank and by elongating or extending the hydraulic holding time $\bar{t}$ in the aeration tank, all of the soluble organic carbon in the waste that has been converted to cells can be autodigested to carbon dioxide; i.e., it can be totally oxidized. Thus, both purification and sludge disposal are accomplished in one process. Hydraulic retention times are commonly three to five times longer than for more conventional activated sludge plants. The biochemical and ecological concepts as well as the practicability of accomplishing the aims of this process are of significant interest in environmental microbiology.

Tertiary and Advanced Processes

In Fig. 2-3, considerable space is given to listing processes that fall into the categories of tertiary or advanced treatment. These categories are listed together because some of them provide for both the aim of tertiary treatment (removal of dissolved and suspended organic matter not removed in the secondary processes) and that of advanced treatment (removal of dissolved and suspended inorganic matter).

It was seen in Fig. 2-1 that effluent from secondary treatment processes may enter the receiving stream, where small amounts of organic matter that remain in the effluent may be utilized for the growth of microorganisms and, in turn, of macroorganisms. If these materials can be metabolized in this natural tertiary treatment reactor without overstressing its many functions in the life-support system, not the least of which is its function as a life-supporting environment for other aerobic species, the receiving stream can form the natural link between wastewater discharge and raw water supply. However, there are increasing numbers of situations in which the receiving streams cannot perform this function simply because of the increasing volumes of secondary treatment plant effluent they carry. Most natural receiving streams are not designed as effluent aeration vessels; thus, dissolved oxygen becomes the limiting "nutrient." In these cases, there are two alternatives. First, ways and means may be sought to increase both the maximum efficiency and the reliability of delivery of high-efficiency secondary treatment plants. Alternatively, add-on processes may be devised (tertiary treatment facilities) through which the secondary effluent is passed for the purpose of removing the small amounts of biologically available organic matter in the secondary effluent. Artificial lakes, *tertiary ponds* or *polishing ponds*, with long liquid retention times have been used. Also, passage of the effluent through sand beds, spray irrigation with either overland (surface) flow or seepage into the soil, or combinations of these functioning as biological treatment processes have been used. Such disposal on, in, or over land prior to returning a secondary effluent to the water resource can form a very useful tertiary process depending on local soil conditions and uses. Also, since one of the existing problems with secondary treatment lies in the reliance on quiescent settling for separation of the biomass from the liquid, tertiary processes aimed solely at entrapment or flocculation of the remaining suspended biological solids have been used, e.g., sand filtration or chemical flocculation.

There is some danger that the wide use of tertiary processes for the removal of organic matter can militate against needed study of the first alternative—improvement of the secondary processes. The tertiary processes should not be looked upon simply as ways of correcting inadequacies of secondary biological processes; there is a tendency to depend on these add-ons to do the job of secondary treatment, which they cannot do since they function successfully only when organic matter is present in very low concentrations.

Often wastewater contains excessive amounts of nitrogen and phosphorus compounds. As seen previously, these are essential to the biosphere and, as such, encourage biological activity, some of which is objectionable from the standpoint of the human species. Nitrification can contribute to depletion of the dissolved oxygen in rivers; both nitrogen and phosphorus can contribute to excessive algal growth in the receiving stream. Secondary treatment processes can be operated so as to nitrify any excess ammonia, but this step may also be performed subsequent to secondary treatment. In this case, it is considered an *advanced waste treatment* process. After nitrification, the NO_3^- may be converted biologically to N_2 and stripped from the liquid to the atmosphere. Some amounts of phosphorus in excess of that normally incorporated in the biomass may be removed biologically, depending on the species present in the biomass, but excess phosphorus is more reliably removed by chemical precipitation following secondary treatment.

Advanced treatments to remove excess nitrogen and phosphorus are designed primarily to prepare wastewaters for reentry into the surface water resource. When direct reuse of the water is contemplated, further treatment is considered. Such processes are selected in accordance with the characteristics of the wastewater. Activated carbon can remove small amounts of adsorbable organic compounds. When directly recycled wastewaters constitute more than a third to half of the water supply, a buildup of inorganic salts in the water can become a serious problem. Such processes as ion exchange, reverse osmosis, freezing, and distillation have been investigated as ways to reduce the total dissolved solids concentrations. Such complete treatment is not generally practiced today and will probably never become necessary except in isolated cases, provided reliably good secondary treatment is generally practiced and the receiving streams are subjected to close surveillance and technological management.

An example of rather comprehensive treatment of water is the scheme used at the South Tahoe public utility district water reclamation plant. The effluent from the plant flows into Indian Creek Reservoir, which is a major recreational lake in this resort area. As originally installed, the treatment consisted of conventional primary sedimentation with secondary treatment provided by the activated sludge process. Advanced-tertiary treatment consisted of chemical treatment (addition of lime, flocculation, and settling) to precipitate most of the excess phosphorus and to raise the pH level sufficiently so that excess ammonium ion, NH_4^+, was converted to ammonia gas, NH_3, which was then stripped from the liquid. After adjustment (lowering) of the pH level, which assists in the recovery of lime, the water was filtered and contacted with granular activated carbon. The water was then chlorinated and finally sent to Indian Creek Reservoir, where it was used for recreation (boating, swimming, fishing). The plant has undergone some modification, particularly with respect to nitrogen removal.

It is readily apparent that many of the processes that form vital units in the technological subsystems of the carbon-oxygen cycle are biological in nature. Also, many of the chemical and physical processes are affected by biological activity. The processes have thus far been dealt with as "black boxes" or "shelf items," which the technologist can string together in various ways depending on the character of the used water and the desired character of the rejuvenated water. This text will, in significant measure, deal with fundamental biomechanics and kinetics governing these environmental control processes. It may come as a surprise to some readers that often in the past, these biological processes have been designed by individuals who have little or no knowledge of the biochemical functions taking place in the reactors. In the future, emphasis will be placed on the reliable functioning of these processes; that is, emphasis will not be on the design of a series of vessels in which "something biological" takes place, but on the control of the "something" that is taking place in the vessels. When design is properly wed to function (as it should always be) and that function involves microbial biochemistry, there is no recourse but to learn about the microorganisms. It is either that, or change jobs.

PROBLEMS

2.1 Distinguish among an aerated lagoon, an oxidation pond, an extended aeration activated sludge process, and a "regular" activated sludge process.

2.2 What types of organic matter would you expect to find in primary sludge? How does the organic composition of primary sludge compare with that of secondary sludge?

2.3 What does the extended aeration process try to accomplish? Explain the aims of the process using your knowledge of the aerobic decay leg of the carbon-oxygen cycle.

2.4 On average, the size of a bacterial cell may be $2.0 \times 0.5 \, \mu m$, and its density is so close to that of water that there is no tendency for the cell to settle; i.e., it remains suspended in the liquid medium. However, secondary settling tanks are used. What must happen if an activated sludge process is to use secondary settling tanks? How does the process come about? Do some reading and form some hypotheses before you attempt to answer this question.

2.5 Distinguish among primary, secondary, tertiary, and advanced wastewater treatment processes.

2.6 Why is the removal of suspended and soluble organic matter from wastewater so important before discharging the water to a river or a stream?

2.7 If the city of Oklahoma City dumped its wastewaters untreated into the Canadian River, there would be a serious and very apparent pollution problem. However, if the city of St. Louis put its untreated water into the Mississippi River, it would hardly change the quality of the water in the river.

(a) Discuss possible reasons why the above is true.

(b) Is it fair or unfair (a sociotechnological problem) that both cities should have to treat their wastes to the same degree?

2.8 Discuss the role of the receiving stream in the natural cycles and in the technological subcycle.

2.9 Using the references given to you, prepare a table of per capita water usage based on a total average per capita usage of 150 to 200 gal per day. Include such use categories as human consumption, i.e., water used biologically; water used for bathing, flushing, and rinsing; water used for cleaning clothes and housecleaning: and water used to irrigate lawns, gardens, etc. After completing this task, you will probably be impressed by the small volume of water used for life-support purposes (i.e., biologically required water). In view of this, discuss the reasonableness of treating all of the water supply to the same degree of purity and potability. Could we have two water supplies piped into dwellings? Is it too late for us to do this on a large scale? If so, why? Would it be a good idea if it could be done?

2.10 Discuss reasons for controlling the quality of a public water supply with respect to: iron, manganese, calcium, magnesium, tastes and odors, turbidity, color, and disinfection. Is there any relationship between these characteristics and disinfection?

2.11 Discuss ways and means to dispose of the waste products of a water supply treatment plant.

2.12 Sewers cost much more than wastewater treatment plants. What are some of the materials of which sewers are made? List all the possible things that can happen to sewers due to the fact that they carry wastewaters. List physical, biological, and chemical effects. Suggest ways to protect the sewers, i.e., the materials of construction.

2.13 What is meant by chemical-physical treatment of wastewaters? Compare the use of chemical-physical treatment with biological treatment. What are the advantages and disadvantages of each?

2.14 Differentiate between a fluidized reactor and a fixed-bed reactor. Give examples of the use of each in the purification of wastewaters and in the purification of raw water supplies.

2.15 What is an oxidation pond? What are the mechanisms that make it work? Does it accomplish its purpose as a secondary treatment of wastewaters? If not, what could be done to make it fulfill this purpose?

2.16 The conventional unit for expressing concentration used in the environmental engineering field is milligrams per liter (mg/L).

(a) If the density of a solid dissolved or suspended in water is 1.0 g/cc and the density of the medium is 1.0, i.e., it is water, express milligrams per liter in terms of parts per million parts of solvent (ppm). Express milligrams per liter in terms of pounds per million gallons of solvent. How many milligrams of solids are there in 38 L if the concentration is 25 mg/L? How many pounds of solid material are there in 1 million gallons if the concentration is 1.0 mg/L?

(*b*) If water is flowing at a rate of 1 million gallons per day, and if 1 mg/L of solids is dissolved in that water, how many pounds of solids flow by a given location in a day?

(*c*) It would seem from the answer to part (*a*) or (*b*) that if the density of the solids is the same as the density of the liquid, then 1 mg/L = 1 ppm. Suppose one knew there was 1 mg/L of a substance dissolved in water but that the density of the substance was 2 g/cc rather than 1, what would be the concentration in parts per million? Suppose the density was 0.5, what would be the concentration in parts per million? Suppose the density of the solid was 1.0, but the density of the solvent was 1.2. How many parts per million would be equivalent to a concentration of 1.0 mg/L?

REFERENCES AND SUGGESTED READING

American Society of Civil Engineers. 1959. Sewage treatment plant design (WPCF Manual of Practice No. 8). Water Pollution Control Federation, Washington, D.C.

American Society of Civil Engineers. 1969. Water treatment plant design. American Water Works Association, New York.

American Water Works Association, Inc. 1971. Water quality and treatment, 3d ed. McGraw-Hill, New York.

Azad, H. S. (editor). 1976. Industrial wastewater management handbook. McGraw-Hill, New York.

Culp, R. L., and G. L. Culp. 1971. Advanced wastewater treatment. Van Nostrand and Reinhold, New York.

Hockensmith, R. D. 1960. Water and agriculture. American Association for the Advancement of Science, Washington, D.C.

Mackenthun, K. M. 1969. The practice of water pollution biology. U.S. Government Printing Office, Washington, D.C.

McGauhey, P. H. 1968. Engineering management of water quality. McGraw-Hill, New York.

Metcalf and Eddy, Inc. 1972. Wastewater engineering. McGraw-Hill, New York.

Nemerow, N. L. 1971. Liquid waste of industry. Addison-Wesley, Reading, Mass.

Nemerow, N. L. 1978. Industrial water pollution. Addison-Wesley, Reading, Mass.

Porter, J. W. 1970. Planning of municipal wastewater renovation projects. *J. Am. Water Works Assoc.* 62:543–548.

Reid, G. K. 1961. Ecology of inland waters and estuaries. Reinhold, New York.

Schroeder, E. D. 1977. Water and wastewater treatment. McGraw-Hill, New York.

THREE

CHEMICAL COMPOSITION OF CELLS AND THE NATURE OF ORGANIC MATTER

In Chap. 1 it was evident that organic matter, its synthesis, its oxidation, and its eventual return to the biosphere as carbon dioxide and water, comprised one of the essential cyclic events permitting the human species and many others to exist. In order for this cyclic phenomenon to work, it was necessary that water both participate in the overall chemical reaction and act as a carrier of organic matter to the oceans. The overstressing of the organic carrying capacity of the earth's drainage system constituted the major reason for insertion of the technological subsystems discussed in Chap. 2. Furthermore, the most important subsystems for purifying wastewaters were themselves biological and involved the use of organic matter in wastewater streams as a source of carbon and energy for the growth of microorganisms. Thus, this organic matter was converted partially to new forms of organic matter (microorganisms) and partially to carbon dioxide. Both the natural system and the technological subsystems are concerned with this conversion of organic matter.

It cannot be said that all organic matter is made by living organisms through natural or controlled life processes; indeed, the science of organic chemistry came into being only when Friedrich Wöhler discovered in 1828 that urea could be made in the absence of living organisms. However, most organic matter is produced through life (metabolic) processes involving minute living cells. In some cases, the living unit consists of a single cell, e.g., various species of bacteria, whereas in other organisms, e.g., corn, cows, and human beings, a family of cells has evolved to perform specific functions in the life of the organism. However, whether in bacteria or corn, each of the varied living cells is characterized more by its chemical similarity than by the factors that differentiate them from one another, a manifestation of the principle of the unity of biochemistry. In this and the following chapter, the chemical (Chap. 3) and the structural-functional (Chap. 4) makeup of various microorganisms will be examined.

The study of the general chemical composition of microorganisms has immediate practical significance to the environmental control professional for two reasons. First, the biomass is a collection of single- and multiple-cell species that is to be used as an engineering material to perform the function of wastewater purification. To be sure, this biomass is not an engineering material of construction such as asphalt, concrete, or steel. Its role is even more important than the structural, insulating, or transporting function of most engineering materials. This material is responsible for the functional success or failure of the entire array of secondary treatment processes. While it may be considered a natural material, its use in the technological control of the life-support system requires that it be understood and manipulated (engineered) to bring about the desired effects. Like any other engineering material, it has characteristic chemical properties and physical behavior. Its mechanistic and kinetic behavior forms an area of inquiry that may be called the *biomechanics* of the biomass, which must form a significant portion of the knowledge of any practical professional in the environmental control field. A study of the general chemical composition of this material provides a foundation for such study.

Second, it is apparent that mainly waste organic matter has precipitated the need for technological enhancement of the decay leg of the carbon-oxygen cycle. Since both the material effecting the purification and the material being subjected to purification (mineralization) are organic in nature, the study of the composition of the biomass offers opportunity for considerable insight into the nature and character of the waste materials. The nature of the organic materials in wastes quite naturally has a bearing on the mechanism and kinetics of biomass metabolism, since these organics comprise the foodstuff (carbon sources) for orderly perpetuation of the biomass and the recycle of organic carbon to carbon dioxide.

Living cells consist largely of compounds such as carbohydrates, lipids, proteins, and nucleic acids, with molecular weights varying from less than 100 to more than 1 million. Municipal and rural effluents consisting largely of household wastes, food wastes, and human excrement contain rather large complements of these naturally occurring organism components. They also contain manufactured organic components, e.g., soaps and cleansers. However, most, but not all, of the organics of nonbiological origin consist of compounds similar to, or are components of, biologically produced compounds. Some of the compounds are not structural components of cells but are chemical products made by the cells. Many industrial wastes consist largely of one or more of these types of biological compounds; e.g., pulp and paper wastewaters contain significant amounts of carbohydrates; dairy wastes contain carbohydrates, lipids, and proteins; and meat-packing wastes contain proteins and lipids. Fermentation wastes contain some cellular components and cellular products, e.g., alcohols. Certain industrial wastes contain organics that are not biologically produced, e.g., refinery and petrochemical waste (at least these materials were not biologically produced in recent times), while others contain specific biological and nonbiological products, e.g., wastes from the production of pharmaceuticals. From the standpoint of attaining engineering control of the progression from organic to inorganic, as well as understanding the biological processes in the life-support system, some knowledge of the nature of organic matter is vital.

GENERAL NATURE OF ORGANIC MATTER

Discussion of two outstanding characteristics of organic matter will help our understanding of why it is so vital to the life-support system. First, it can be burned (i.e., it is combustible or oxidizable), and, second, its general characteristics are for the most part determined by those of its major elemental constituent, carbon.

The first characteristic has immediate practical significance and provides a means of qualitative and quantitative analysis. The fact that organic matter can be oxidized (burned) in the presence of oxygen is well known, and it is generally appreciated that in the process, the energy contained in the organic material is released as heat. It is also generally appreciated that the more reduced the organic matter is, the better fuel it makes; i.e., more heat energy is released when it is combusted to carbon dioxide and water. The heat energy is converted to other forms of energy and is used by consumers.

When the burning or oxidation takes place under physiological (biological) conditions at constant temperature and pressure, it is not the change in heat energy ΔH (enthalpy) but the available free-energy change ΔG, or the portion of it that may be stored in chemical compounds for use by the organism, that is important. The living cell is not simply a furnace. It is an entire system of combustion for energy release, capture, and use. The energy is used for many purposes, such as movement, growth, and maintenance of vitality. All of these require energy released in a form able to perform work, i.e., the free-energy change ΔG. Not all of the energy released upon oxidation of the organic matter is available to perform work. When a reaction occurs, the change in the fraction of energy available for work is dependent on the absolute temperature T at which the reaction takes place and the change in the capacity of the system for containing energy that is not available to do work ΔS (entropy). Thus:

$$\Delta G = \Delta H - T \Delta S \qquad (3\text{-}1)$$

Only a portion of the released free energy can be trapped in chemical compounds that permit the organism to do the work mentioned previously, because, as we shall see in Chap. 7, the trapping process is not 100 percent efficient and some of the free energy released is wasted as heat. In addition to oxidizing some of the carbon source to obtain energy to do work, some of the carbon in the organic foodstuff is used to form new compounds (body substance) as the organism grows. Thus, not only does the living cell oxidize organic matter to obtain energy in usable chemical form, but it is also a consumer of the energy and manufacturer or synthesizer of new organic matter from a portion of the organic carbon source with which it was fed.

MEASUREMENT OF ORGANIC MATTER

Both chemical and biochemical combustion characteristics of organic matter offer opportunities for its qualitative and quantitative measurement. Below we shall consider some chemical combustion measurements.

Equation (3-2) shows the combustion of a sample containing organic matter:

$$\left.\begin{array}{c}\text{Organic matter}\\ \text{Inorganic matter}\end{array}\right\} \xrightarrow{\text{O}_2} CO_2 + H_2O + \text{ash} + \text{heat} \qquad (-\Delta H) \qquad (3\text{-}2)$$

In this equation, the amount and type of organic matter are unknown, and it is quite impossible to operate on the equation in the usual mass balance fashion. However, it tells us what happens if we combust the organic matter to CO_2 and H_2O, i.e., if we oxidize it totally.

Volatile Solids

Since the products CO_2 and H_2O can be driven off, Eq. (3-2) shows a way to measure the amount of organic matter in a sample. The test is conducted at 600°C in a carefully prescribed manner, minimizing decomposition of inorganic salts; the loss of weight after burning may be taken as a measure of the amount of organic matter present (APHA, *Standard Methods*, 1976). This procedure provides not only a quantitative assessment of the amount of organic matter, but also the information to determine the percentage of the total dry weight of the sample that is organic. The determination of the amount of volatile solids is one of the most widely used measurements for the amount of organic matter in a sample.

Oxygen Demand

Measurement of volatile solids provides valuable information, but the weight alone tells little of the character of the organic matter; for example, compare the burning of a ton of hay with that of a ton of coal. The coal (pure carbon) would give off more heat upon combustion (greater negative ΔH), require more O_2, and produce more CO_2 during complete combustion. In short, the fuel values of the hay and coal are different. If one performed the burning in a special furnace, taking precautions to insulate it, i.e., if one used a bomb calorimeter, it would be possible to measure the heat of combustion ΔH released upon burning the sample. Usually such elaborate measurement procedures are not followed except to obtain thermochemical values for specific compounds or specific classes of organic matter; e.g., the general range of heat values for municipal refuse is often determined this way. One could also measure the amount of oxygen used up in the oxidation, and this would provide some indication of the fuel value and an indirect measure of the amount of organic matter present in the sample.

It can be argued that in nature it is not the fuel value or the O_2 used upon chemical combustion but the O_2 required for biochemical combustion that is the more critical concern. For any given population of organic food consumers, a portion of the organic matter is incorporated into new body substance; also, not all organic matter in a given sample is necessarily a usable substrate for a given population in nature. It can, however, be said that the amount of O_2 used to oxidize (burn) the organic carbon in a sample totally to CO_2 represents the upper limit of the oxygen-utilizing potential of that sample of organic carbon in nature. Thus, this oxygen demand provides extremely valuable information regarding the amount and character of a sample of waste organic matter. Special equipment must be used to measure the O_2 used in burning a sample of organic matter in a furnace. There is, however, an exceptionally good and widely used chemical oxidation procedure that can be used.

Chemical oxygen demand (COD) test. The chemical oxidizing agent need not be oxygen, and in the standard chemical oxygen demand (COD) test (APHA, 1976), the strong oxidizing agent, potassium dichromate, is used. Under acid conditions using carefully prescribed procedures, the COD test can be counted upon to oxidize nearly all organic carbon to CO_2. One can measure the amount of oxidizing agent used up in the reaction and calculate this to an equivalent amount of O_2. This method is used widely in the environmental pollution control field, and modifications of it are used in the fermentation industries as tests for the amount of total organic matter (Neish, 1952). The method lends itself to various automated and shortened analytical procedures designed to save time and increase analytical capacity. These should be used only after correlating the results to the standard COD test. The shortened reaction time can lead to some differences between the test results and those obtained using the standard COD test. It is important to note that in the COD test, all organic matter in the sample may not be oxidized. Aromatic hydrocarbons and pyridine resist oxidation. On the other hand, certain inorganic substances, e.g., ferrous iron, sulfides, sulfites, thiosulfates, and nitrites, cause an inorganic COD. The major interference in this regard is chloride ion, Cl^-, but this may be overcome by the addition of mercuric sulfate, providing Hg^{2+} ion, which combines with Cl^-.

Total oxygen demand (TOD) test While the chemical oxygen demand test provides a relatively easy and inexpensive way to measure the oxygen demand of organic matter, it is possible also to measure the amount of oxygen used during the actual combustion of a sample. The special equipment needed has been conveniently packaged in recent years. Organic matter is burned at 900°C with the aid of a catalyst, and the oxygen demand is measured by the electrolytic detection of oxygen consumption. Measurement of oxygen demand using such an apparatus has become known as the total oxygen demand (TOD) test. This method is not subject to the chloride interference; however, inorganic nitrates in a sample may contribute oxygen, thus giving falsely low results. Also, any dissolved oxygen in an aqueous sample should be purged before making the analysis. Other inorganic substances, such as hydrogen, ammonia, sulfide, and sulfite ion, exhibit an oxygen demand. Because of the high temperature, excess oxygen, and catalysts, nitrogen in nitrogen-containing organic matter is oxidized to NO; thus, the test should give higher oxygen demand values than the COD test. Some of the TOD test instruments available may not oxidize organic nitrogen, and the results in such a case would be expected to be fairly close to COD test values. Thus, while the COD test and various TOD analyzers are used to measure the same property, one must expect some differences in the results because of the different test conditions. COD values have been found to be somewhat lower than TOD values for some wastes and higher for others, while COD/TOD ratios of unity have also been observed.

Total Organic Carbon (TOC)

Referring again to Eq. (3-2), it is seen that the amount of organic matter can be determined, based upon the amount of carbon it contains, by measuring the amount of CO_2 produced upon its total oxidation. One may dry and burn a sample, then collect and measure the amount of CO_2. The CO_2 may be collected in alkali and precipitated and measured gravimetrically, or it may be measured manometrically. Fairly new devices that take advantage of the infrared-absorbing properties of CO_2 are also available. Analyses for the carbon produced upon total oxidation, TOC, are used increasingly in the environmental pollution control field, largely because of the development of the instrumentation for this analysis.

Thus, using Eq. (3-2), it is apparent that we can assess the amount of organic matter by measuring the heat liberated upon combustion ($-\Delta H$) or the loss of weight upon total oxidation of the sample (volatile solids determination). We can measure the amount of carbon in the organic matter, TOC, and use this as an assessment of the amount of organic matter, or we can measure the amount of oxygen used in the oxidation, COD and TOD. All of these methods are chemical ones and involve no biological activity whatever. They all are means of assessing the amount of organic matter in a sample, but they measure different characteristics of the organic matter.

STOICHIOMETRIC CONSIDERATIONS

All of the aforementioned characteristic properties must be determined experimentally; Eq. (3-2) cannot be balanced because the elemental composition of the organic matter is unknown. Indeed, it is first necessary for an analyst to determine what fraction of the weight of the solids in a sample is composed of organic matter (volatile solids determination). However, if one were to choose organic matter of known chemical composition, or if one determined the elemental composition and empirical formula for the organic matter in the sample, Eq. (3-2) could be balanced and one could readily see the relationships among all the types of measurement. Such calculations are helpful and, for the purposes of illustration, we shall examine the oxidation of ethane, C_2H_6, from the hydrocarbon level through alcohol (ethanol) to the acid level (acetic acid) and to the level of total oxidation (CO_2 and H_2O). Proceeding from reduced reactant to oxidized product:

$$C_2H_6 \xrightarrow[\frac{1}{2}O_2]{} C_2H_5OH \xrightarrow[O_2]{H_2O} CH_3COOH \xrightarrow[2O_2]{} 2CO_2 + 2H_2O \qquad (3\text{-}3)$$

$$\text{hydrocarbon} \quad \text{alcohol} \qquad\qquad \text{acid}$$

In the interest of brevity, a very important oxidation level between alcohol and acid has been omitted in the oxidation series—the level of aldehydes (and ketones for compounds of three or more carbons). Aldehydes and ketones occupy an important position in nature, but their inclusion here is not necessary to the example.

The terms *oxidation* and *reduction* deserve further explanation, and the simple direct statement below belies the immensity of investigational effort that has gone into this area of study by many physicists, chemists, and biochemists. The definition of oxidation and reduction given by Clark (1952) is an apt one:

> Oxidation can be defined as the addition of oxygen to a compound, the removal of hydrogen from a compound, the loss of an electron; reduction is the converse of oxidation.

All three need not occur in any specific reaction. Atoms and molecules can be oxidized without the addition of oxygen or loss of hydrogen; but in order for oxidation to occur, an electron must be lost or taken away from the compound or atom that has been oxidized. Also, it should be remembered that when an atom or molecule is oxidized, there must be a concomitant reduction in another atom or molecule. Oxidation and reduction occur concurrently; one cannot occur without the other. Examination of the oxidative series in Eq. (3-3) shows that when ethane is oxidized to ethanol, there is no loss of hydrogen, but oxygen is added to the compound. However, when ethanol is oxidized to acetic acid, there are both loss of hydrogen and gain of oxygen. In all cases, the organic compound is oxidized and the oxidizing agent, oxygen, is reduced. In these reactions, oxygen is an electron acceptor; thus, it is reduced, and the electron donor is the organic compound oxidized by the electron acceptor. It is the electron donor–acceptor relationship that characterizes oxidation-reduction reactions, and such reactions will later be seen to be of vital importance in microbial metabolism.

For the total oxidation of ethane, the balanced equation is

$$
\begin{array}{llr}
C_2H_6 + 3.5O_2 \rightarrow 2CO_2 + 3H_2O & \text{(chemical balance)} & \\
30 + 112 \quad = \quad\;\; 88 + 54 & \text{(mass balance)} &
\end{array} \tag{3-4}
$$

Both the elements and the masses are in balance, as is seen by adding the molecular weights shown in the mass balance.

If 1000 mg/L of ethane is oxidized, the concentration of O_2 required can be calculated as $(1000)(112/30) = 3733$ mg/L. This is the calculated COD (sometimes referred to as theoretical chemical oxygen demand COD_{th}) of 1000 mg/L of C_2H_6. In general, we can say that 1 mg of C_2H_6 has a theoretical COD of 3.73 mg of O_2.

Remembering that the amount of total organic carbon (TOC) was determined experimentally by measuring the CO_2 produced, the calculated TOC can be obtained by computing the carbon content of the CO_2 produced from 1000 mg/L of ethane (i.e., 800 mg/L) or, since the empirical formula for ethane is known, by computing the carbon content directly. The TOC of 1 mg of ethane is 0.8 mg of carbon. The ratio COD/TOC provides some indication of the oxidation state of the organic matter. For ethane, it is 4.63 (i.e., 3.7/0.8). The higher the ratio COD/TOC, the less oxidized is the organic matter.

The amount of energy ΔH released upon combustion can be calculated from the heats of formation of the products and the reactants from their elements. These values can be obtained from chemical data handbooks such as *Lange's Handbook of Chemistry* (Dean, 1973) or the *Handbook of Chemistry and Physics* (Weast, 1977).

$$
\Delta H^\circ = \sum \Delta H_f^\circ \text{ products} - \sum \Delta H_f^\circ \text{ reactants} \tag{3-5}
$$

The superscript $\circ$ indicates standard conditions, which are a pressure of 1 atm, a temperature of 298 K (25°C), and concentrations of 1 mol/L. For the combustion of C_2H_6 [Eq. (3-4)],

$$
\Delta H^\circ = 2(-94.05) + 3(-57.79) - (-20.24) + 0
$$

$$
= -341 \text{ kcal/mol}
$$

Calculations can also be made for the successive oxidation products of Eq. (3-4). Calculated values of COD and TOC are shown below.

Ethanol

$$
C_2H_5OH + 3O_2 \rightarrow 2CO_2 + 3H_2O \qquad \Delta H^\circ = -295 \text{ kcal} \tag{3-6}
$$

$$
\text{COD} = 2.09 \text{ mg } O_2/\text{mg } C_2H_5OH
$$

$$
\text{TOC} = 0.522 \text{ mg } C/\text{mg } C_2H_5OH
$$

Acetic acid

$$CH_3COOH + 2O_2 \rightarrow 2CO_2 + 2H_2O \qquad \Delta H° = -187.3 \text{ kcal} \qquad (3\text{-}7)$$

$$COD = 1.07 \text{ mg } O_2/\text{mg } CH_3COOH$$
$$TOC = 0.4 \text{ mg } C/\text{mg } CH_3COOH$$

It can be seen that as ethane is oxidized in successive steps from hydrocarbon to alcohol to acid, there is a decrease in energy released and a decreased COD. And in the example given, since oxygen is added, the TOC of each successive oxidized product is less than the TOC of the corresponding reactant; i.e., the ratio COD/TOC decreases.

We have referred to the calculated or "theoretical" COD, and it is amply apparent that it is the amount of oxygen required to oxidize the organic carbon and hydrogen to CO_2 and H_2O. For the compounds thus far used as examples, the calculated COD and TOD are the same, but this would not necessarily be the case for compounds containing reduced nitrogen. This difference can be seen by considering the amino acid glycine.

Under conditions of the COD test, Eq. (3-8) results; ammonia remains unoxidized. However, for conditions in the TOD analysis, Eq. (3-9) may apply; the nitrogen may be oxidized to NO.

$$NH_2CH_2COOH + 1.5O_2 \rightarrow 2CO_2 + H_2O + NH_3 \qquad (3\text{-}8)$$
$$75 \qquad\quad 48$$
$$COD_{th} \text{ glycine} = 0.64 \text{ mg/mg}$$

$$NH_2CH_2COOH + 2.75O_2 \rightarrow 2CO_2 + 2.5H_2O + NO \qquad (3\text{-}9)$$
$$75 \qquad\quad 88$$
$$TOD_{th} \text{ glycine} = 1.17 \text{ mg/mg}$$

ELEMENTAL COMPOSITION OF MICROORGANISMS

Water, as previously mentioned, comprises the major portion of the weight of a microorganism (75 to 90 percent). The cell contains the various elements discussed in Chaps. 1 and 2, which are characteristic of all organic matter in the biosphere. The elemental composition of microorganisms varies somewhat, depending on the environmental conditions and species of microorganism. The values listed in Table 3-1 for the bacterium *Escherichia coli* are representative of a widely occurring and well-studied species. Reported ash contents for microorganisms vary from 5 to 30 percent. The ash content for naturally grown heterogeneous populations of microorganisms in biological treatment plants can be much higher than 30 percent, depending on the mineral content of the medium, i.e., the carriage water for the waste organic material. The ash content of activated sludge can, at times, be more than 50 percent. There is considerable variation in the relative amounts of the elements listed after hydrogen in Table 3-1. Various species of organisms can

Table 3-1 Elemental cell composition

Element	Dry weight, %
Carbon	50
Oxygen	20
Nitrogen	14
Hydrogen	8
Phosphorus	3
Sulfur	1
Potassium	1
Sodium	1
Calcium	0.5
Magnesium	0.5
Chlorine	0.5
Iron	0.2
All others	0.3

Source: Stanier et al., 1976.

store significant amounts of phosphorus or sulfur and can concentrate some other ions, e.g., chloride. Thus, one must expect the composition of these elements to vary rather widely. However, it is seen from Table 3-1 that over 90 percent of the dry weight of the cell is composed of the elements carbon, hydrogen, oxygen, and nitrogen. Thus, if one determined the combining ratios of these elements in a sample of microorganisms, an empirical formula for cells would be approximated. One such set of values for a heterogeneous microbial population was determined by Porges et al. (1956) to be $C_5H_7NO_2$. The theoretical chemical oxygen demand of cells with this empirical formula may be calculated:

$$C_5H_7NO_2 + 5O_2 \rightarrow 5CO_2 + 2H_2O + NH_3 \qquad (3\text{-}10)$$
$$113 \qquad 160$$

Thus, 1 mg of microbial biomass = $160/113 = 1.42$ mg of COD.

Experimental values of COD for microbial populations grown under various conditions have been found to range from approximately 1.3 to 1.5 mg of COD per milligram of biomass dry weight. If the values had been based on volatile solids [as per Eq. (3-10)] rather than on the total dry weight of the biomass, these values would have been slightly higher. There is enough variation in actual COD values under various growth conditions to warrant considerable caution against the sole reliance on an empirical formula for the biomass when making energy and material balances (Gaudy et al., 1964; also see Chap. 10). A general empirical formula for heterogeneous biomass has value, and there surely can be considerable variation in biochemical composition without much change in the empirical formula. However, the relative ease with which the oxygen demand (COD) of biological sludges can be obtained experimentally indicates no need for reliance on a general empirical formula.

AN INTRODUCTION TO BIOCHEMICAL OXYGEN DEMAND (BOD)

Equations (3-2) and (3-10) show the total oxidation, i.e., the complete aerobic decay and mineralization, of organic matter. The conditions of oxidation used in these reactions are rather severe and are purposely designed so in order to make the reactions go to completion. These reactions also describe aerobic decay in nature, but the conditions are less severe; they permit life processes to proceed. Under such conditions, the oxidation takes place as part of the natural food chain, as we described in Chap. 1, and a portion of the foodstuff represented by this organic matter is used to make the vital body substances (lipids, carbohydrates, proteins, nucleic acids) of the species growing on the organic matter in the sample. Thus, total aerobic decay of a sample of organic matter under biological conditions is not a simple one-step process but occurs in multiple steps. At each step, a portion of the original organic matter is oxidized to CO_2 and H_2O, and a portion is used to produce new (microbial) organic matter. The biological incineration of a sample of organic matter can be represented as shown in Fig. 3-1, assuming that the organic material comprising the cells synthesized in each step can be metabolized by other microbial cells in the following step.

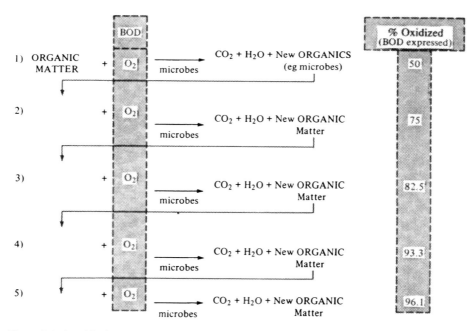

Figure 3-1 Aerobic decay—repetitive biological processing of organic matter. The oxygen utilized in this process up to any step is a measure of the biochemical oxygen demand of the original sample of organic matter. If the biological cell yield is 50 percent, then more than 95 percent of the original organic matter is totally oxidized by the end of the fifth cycle.

The fraction of organic carbon source that is channeled into new cells during growth is called the *cell yield Y* with respect to carbon source. This value varies according to environmental conditions, nature of the carbon source, species, and other factors, but for purposes of examining a quantitative example, a cell yield of 50 percent is a reasonable value. The summation of the series of gross biochemical reactions shown in Fig. 3-1 approaches accomplishment of the same overall reaction as that in the COD, TOD, and TOC tests. Measurement of the amount of O_2 used in the aerobic biological decay of organic matter is the essence of the widely used bioassay procedure called the *biochemical oxygen demand* (BOD) test.

In this bioassay procedure, a sample of waste is inoculated with a small volume of a mixed population of microorganisms capable of utilizing the organic matter as source of carbon and energy, i.e., microorganisms capable of performing step one. The initial O_2 concentration of the sample is brought to nearly the DO saturation level (approximately 9 mg/L) by aeration, and the O_2 concentration is carefully measured. In the standard test, an appropriate volume of the sample is incubated in a closed container for 5 days, and then the DO is again measured. The decrease in O_2 concentration is a measure of the BOD exerted (O_2 uptake) during that time interval.

In the series of aerobic decay steps shown in Fig. 3-1, no consideration is given to the speed or rate of reactions. Usually step one occurs rather rapidly—within the first or second day. The later steps take more time. Step two may begin before step one is completed, or it may not start until some time afterward. Also, step two may be only partially completed. Step three may not start for some time after step two has been completed. Indeed, it may not start for many days, and there is a possibility that it may never start in the incubation bottle. The reason for the uncertainty is that the steps are brought about by the microorganisms that were placed in the BOD incubation bottle. The initial seed population is selected mainly for its ability to perform step one; i.e., it has been acclimated to the use of the original organic matter as a carbon and energy source. It may or may not include organisms that can utilize the body substances of the species produced in the first step as a carbon source to accomplish the later steps. The BOD bottle is a closed microecosystem; its ultimate capabilities are determined by those of the initial inoculum. Thus, the stepwise series of aerobic decay reactions may occur quite handily in nature, e.g., in a river over an extended time, but may not occur in a BOD bottle in many days—almost never in 5 days. Thus, even if all the organic matter in a sample can be used as a source of food by microorganisms, the 5-day BOD is never expected to equal the COD or TOD, and even if the sample were incubated for a month, the terminal BOD might not equal the COD or TOD unless the microorganisms capable of forcing the food chain to approach completion (at least through step five) were present. Sometimes they are; more often they are not. So the ultimate value of the BOD recorded will depend upon the seed inoculum used. Under the best ecological conditions, it may approach the COD or TOD value. Even if it does approach total oxidation, the rate of reaction will vary with the inoculum used. One may, with considerable risk, attempt to correlate BOD values with the results of COD and TOD tests, but surely not without consideration of the differences in the determinations and the reactions on which each is based. In later chapters, the important subject of BOD (i.e., microbial respiration) will be treated in more detail at both the fundamental and applied levels.

A recent critical review of the BOD concept and the test is available (Gaudy, 1972), and the detailed procedure for performing the BOD test can be found in *Standard Methods* (APHA, 1976).

CHEMICAL NATURE OF ORGANIC MATTER
AND MICROORGANISMS

Further discussion of the nature of organic matter and the chemical composition of microorganisms requires a familiarization with the chemical properties of carbon. Many texts have been written about compounds that contain carbon and their reactions (organic chemistry) as well as the nature and reactivity of certain organic compounds in living cells (biochemistry); the reader is advised to consult the list of suggested reading at the end of this chapter for more detailed discussion than will be presented here. Regardless of the immensity of these subjects, there are certain fundamentals and key areas that we can usefully delineate to aid comprehension of the chemistry of microorganisms.

It is essential to gain an appreciation of the physical structure of the organic compounds that comprise both substrates and microorganisms. Compounds that have thus far been represented by empirical formulas (ratios of combining weights of the elements) or the slightly more descriptive molecular formulas (showing the numbers of each atom in the compound) are much more than packets of elements; they are definite space structures, and their reactions and functions depend in large measure on the shape and construction of those space structures. Like the webbed trusses of bridges, the columns and beams of skyscrapers, and power line towers, there is much more space between the elemental parts that are joined together to make the functional structure than there are solid materials. Indeed, the elemental parts of the molecules—the atoms of the various chemical elements—are themselves structures in space.

The construction of the major atoms comprising organic matter is shown in Table 3-2. Hydrogen is the simplest atom, consisting of a nucleus that contains one positively charged particle (a proton) and no uncharged particles (neutrons). The nucleus represents essentially all the mass of the atom $(1.67 \times 10^{-24} \text{ g})$ but not most of the volume or space occupied by the atom. The diameter of the nucleus is about 1/100,000 that of the whole atom. The diameter of the atom is approximately 10^{-8} cm, whereas that of the nucleus is

Table 3-2 Atomic structure of common elements in living organic matter

Element	Symbol	Electron configuration	Atomic number	Atomic weight	Valence electrons	Maximum number of covalent bonds
Hydrogen	H		1	1	1	1
Carbon	C		6	12	4	4
Nitrogen	N		7	14	5	3
Oxygen	O		8	16	6	2
Phosphorus	P		15	31	5	3
Sulfur	S		16	32	6	2

10^{-13} cm. Since the weight of the atom is so small, a relative weight scale is used. The atomic weight is approximated as the number of particles, neutrons and protons, in the nucleus. Thus, hydrogen has an atomic weight of 1. (Units of atomic weight are daltons; one dalton is the weight of the hydrogen atom.) The atomic number is the number of protons in the nucleus, and for hydrogen, both atomic weight and atomic number are 1. Considering the minute absolute weight of the hydrogen atom, 1 g atomic weight of hydrogen contains more than 10^{23} atoms of hydrogen. When expressing the atomic weight on this relative scale, it is seen that 1 g atomic weight of any element contains the same number of atoms. This number (Avagadro's) is approximately 6.02×10^{23}, and it may be used to convert relative to absolute weight scales.

Modern theory of the atom indicates that electrons, the minute negatively charged particles that move rapidly about the nucleus, exhibit wavelike properties similar to those of a beam of light. These charges exist in defined mathematical functions (orbitals), which are a measure of the probability of finding the electron in a given region in space or volume in which the electron moves. The orbital for hydrogen is represented by a sphere with its center about the nucleus. For our purposes, the orbital theory of atomic structure need not be considered further. The older concept, which considers the electron charge to exist in a planar orbit or shell about the nucleus, will suffice. It is important to remember that the electron represents an electric force in the vicinity of the nucleus, and it is the sharing and interlacing of these forces that form the bonds allowing atoms to form molecules, molecules to form macromolecules, and macromolecules to form organic structural components.

Hydrogen has one electron shell (designated as the K shell), which has a capacity to hold two electrons. Hydrogen strives to complete this shell by sharing one electron with another atom. An element is said to be more stable, i.e., less reactive, when its outer shell of electrons is completed. Thus, helium, which has two electrons, has a completed outer shell and is an inert (nonreactive) element.

Carbon has two shells for electrons; the K shell is complete, but the outer shell, designated by the letter L, has only four of the eight electrons needed to complete it. Thus, to achieve stability, it would either lose or gain four electrons; it has a valence or combining power of four. The electrons in the outer shell of an element are known as the valence electrons. It is the outer shell that is involved in chemical reactions.

Carbon has six protons and six neutrons. Thus, it has an atomic weight of 12; it is 12 times heavier than hydrogen. Nitrogen and oxygen contain five and six valence electrons, respectively, in the L shell. In phosphorus and sulfur, the L shell is complete, and each contains electrons in a third shell (M shell), which may be completed with eight electrons. Thus, phosphorus has five and sulfur has six valence electrons.

Bonding of Elements—Formation of Compounds

Atoms are joined to form molecules by electric bonds of varying strengths and types. The outer shells or valence electrons are involved in chemical bonding, and they strive to attain stability by either adding or losing electrons to complete the valence shell.

Electrovalent bonds (ionic bonding) An example of electrovalent bonding is the formation of common table salt, sodium chloride. Sodium has one electron in the M shell, and chlorine has seven. The drive toward stability indicates that sodium tends to lose its electron and drop back to the stable L level, whereas chlorine picks up this electron to complete its M shell. In the formation of the resultant compound, sodium chloride, sodium has lost its electron and has net positive charge (+1), thus becoming a positively charged ion (cation), Na^+; the chlorine atom gains an electron in the exchange, becoming a negatively charged ion (anion), Cl^-. The sodium and chloride ions are attracted to each other because of their opposite charges. The reaction can be shown as follows:

$$Na^\circ + {}_\circ\overset{\circ\,\circ}{\underset{\circ\,\circ}{Cl}}{}_\circ^\circ \longrightarrow Na^+ {}_\circ^\circ\overset{\circ\,\circ}{\underset{\circ\,\circ}{Cl}}{}_\circ^{\circ-}$$

The dots represent the electrons of the outer valence shell. The ions in the resulting *ionic compound* exert a considerable attractive force. The positively charged ions can attract several negatively charged ions, forming three-dimensional (space) aggregates of alternately charged ions as crystals. Within the crystalline structure no one pair of ions constitutes a molecule of NaCl. There is no direction to the attractive force, and when the crystalline solid is dissolved in water, the separate ions for the most part go their own ways and form electrostatic attractions with water molecules or with other cations and anions that may be present in solution. Such ionic behavior is of significance in living systems, but these bonds are not characteristic of organic (carbon) compounds. The unique space structures of complex organic compounds require more fixed and rigid bonding to provide definite shape and structural form.

Covalent bonding Many organic compounds contain some ionic bonds, but the type of bonding that predominates in organic compounds is that in which one element is bonded to another by an overlapping or interlocking of electron orbitals; that is, there is no electron exchange or capture by one of the elements but a mutual sharing of the electrons involved in the bonds so that each element completes its outer electron shell by virtue of pairing electrons. This interlocking leads to "fixed" directional bonds of varying strength and is largely responsible for the unique space structures of complex organic molecules. This type of bonding is referred to as *covalent*, and the compounds thus formed are *covalent compounds*.

Covalent bonding is also found in environmentally important inorganic

Table 3-3 Electron-dot formulas for some environmentally important nonorganic compounds

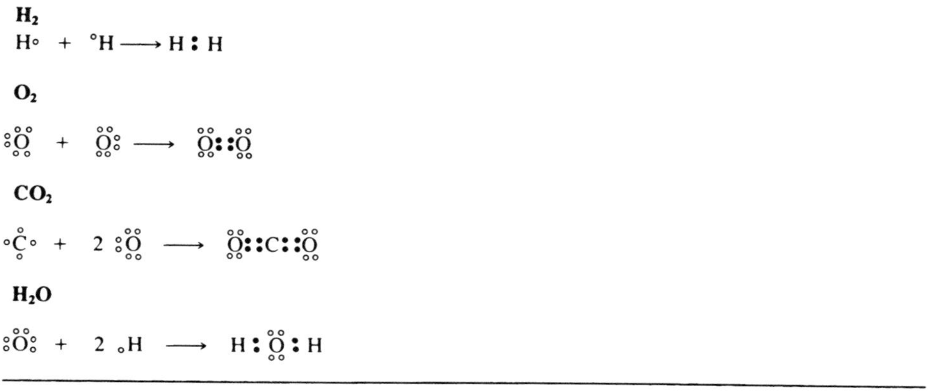

H_2

H∘ + ∘H ⟶ H **:** H

O_2

:Ö: + :Ö: ⟶ Ö::Ö

CO_2

∘C∘ + 2 :Ö: ⟶ Ö::C::Ö

H_2O

:Ö: + 2 ∘H ⟶ H **:** Ö **:** H

---

compounds, e.g., H_2, O_2, CO_2, and H_2O, shown in Table 3-3. In each case, the outer shell is stabilized by sharing, not exchanging, the electrons (K shell for hydrogen, L shell for carbon and oxygen). The filled circles indicate the shared electrons.

Polarity In molecules consisting of two atoms of the same element, such as H_2 and O_2, the electrons are shared equally by each atom. However, some elements that form covalent bonds have a greater attraction for electrons than do others; for example, oxygen and nitrogen have a stronger attractive force for electrons than do carbon and hydrogen, and they exhibit a slight negative charge. In a water molecule, oxygen has a stronger attraction for the electrons shared with the hydrogen atoms; therefore it exhibits a slight negative charge, whereas the hydrogen exhibits a slight positive charge. This phenomenon is known as *polarity*, and water is a polar compound. In addition to the O^-—H^+ bond, another pair exhibiting polarity is that between N^- and H^+; thus, covalent bonds may exhibit charge characteristics (but not structural characteristics) like ionic compounds.

Coordinate covalent bonding Covalent bonds may be found in which one element donates both electrons in the binding pair. This type of bond is found more often in inorganic than in organic compounds, e.g., in H_2SO_4 and H_3PO_4. Organophosphorus compounds are particularly important in biological systems, and phosphoric acid illustrates this type of bonding:

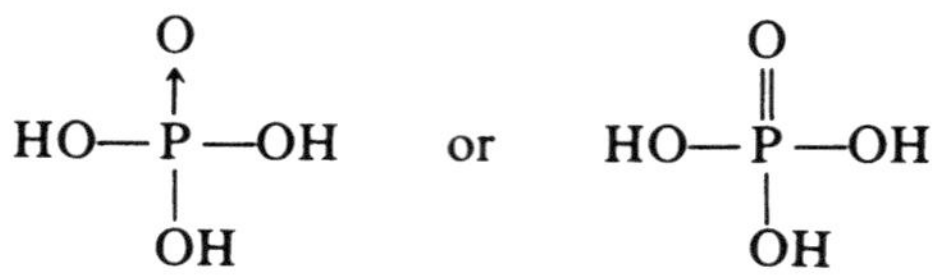

$$
\begin{array}{ccc}
\text{O} & & \text{O} \\
\uparrow & & \parallel \\
\text{HO—P—OH} & \text{or} & \text{HO—P—OH} \\
\mid & & \mid \\
\text{OH} & & \text{OH}
\end{array}
$$

Phosphorus has five valence electrons, two of which form a covalent bond with oxygen; that is, phosphorus donates two electrons and oxygen none to the bond. This type of bond is sometimes designated with an arrow (left) or, more often, with simply a double bond (right). This type of covalent bonding is sometimes called a *semipolar bond* because oxygen, which still has all of its six electrons and in addition a share in the pair from phosphorus, now has a negative charge, whereas phosphorus has donated electrons to the bond without gaining a share of the oxygen electrons and is left with a positive charge.

Carbon–carbon bonding Carbon is the major element in organic matter and, although it forms covalent bonds with many elements, e.g., hydrogen, oxygen, nitrogen, sulfur, phosphorus, and the halogens (fluorine, chlorine, bromine, iodine), the key to the diversity and complexity of organic compounds is the ability of carbon to form covalent bonds with other carbon atoms. The carbon–carbon bond provides the mechanism that permits carbon to form the skeletal framework for all the organic compounds. The spatial model of the carbon atom (Fig. 3-2) was suggested 100 years ago, and modern orbital concepts have substantiated it. The atom is represented by a sphere at the center of a tetrahedron, and the four valence bonds of equal length extend to the corners of the tetrahedron (see Fig. 3-2A). In diagram B, the tetrahedron has been removed and the bonds have been satisfied with hydrogen, resulting in a spatial representation of methane. This ball-and-stick model is very useful in studying the structures of organic compounds. It clearly shows the directional aspect of covalent bonding that leads to the definite or fixed-space structure of the molecule. More representative modeling could be provided by making the carbon sphere larger than those representing hydrogen and by showing electron clouds representing the orbitals rather than indicating the bonds by tie sticks. In Fig. 3-2C, the molecule has been rotated out of the plane of the paper so that the top hydrogen atom has come toward the reader, and the elements are projected to a plane, yielding the normal or standard representation of the structural formula for methane. The projected structural formula is similar to the electron-dot formulas used in Table 3-3, except that the dash has replaced the two dots representing the shared electron pair. Often the dashes are omitted, e.g., in CH_4.

Because of the tetrahedron-like space structure of the valence bonds in the carbon atom, it is apparent that chains of carbon atoms that are usually written –C–C–C– in structural formulas are by no means as linear as the representation implies. In Fig. 3-3 are shown various spatial forms of carbon skeletons. This stereochemical property of organic molecules is important in determining the reactivity and biological function of the compounds.

When we consider the variety of shapes and lengths of compounds that can be formed by carbon–carbon bonding and the variety of elements besides hydrogen and carbon with which carbon can form bonds, there is little wonder that there are many more organic than inorganic compounds. For the smaller

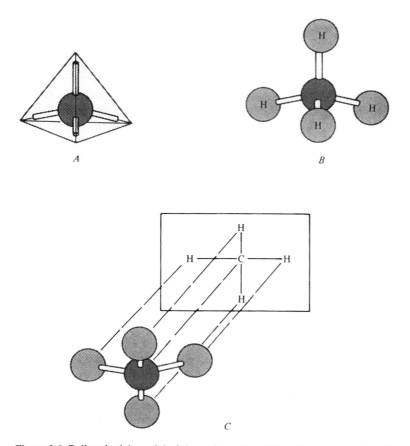

Figure 3-2 Ball-and-stick model of the carbon atom: (*A*) carbon atom enclosed in a tetrahedron to show its four valence bonds; (*B*) methane; (*C*) structural formula for methane based on projection lines from its ball-and-stick model. (*Adapted from Liener, 1966.*)

organic compounds, the types of reactions in which they participate are determined by attachment to the carbon skeleton of certain functional groups or the presence of certain characteristic bondings. For example, alcohols, aldehydes, ketones, and acids possess certain key chemical groups. Esters, ethers, and peptides represent important bondings involving functional groups of small molecules and are involved in joining them into large "poly" or "macro" molecules. Some of the most important functional groups and bondings are given in Tables 3-4 and 3-5.

The functional groups are responsible for the chemical behavior typical of classes of organic compounds such as alcohols, aldehydes, and acids; these are groups common to those particular types of compounds and form the basis of a means of classification of the organic compounds. The functional groups of molecules are locations of either high or low electron density. Since

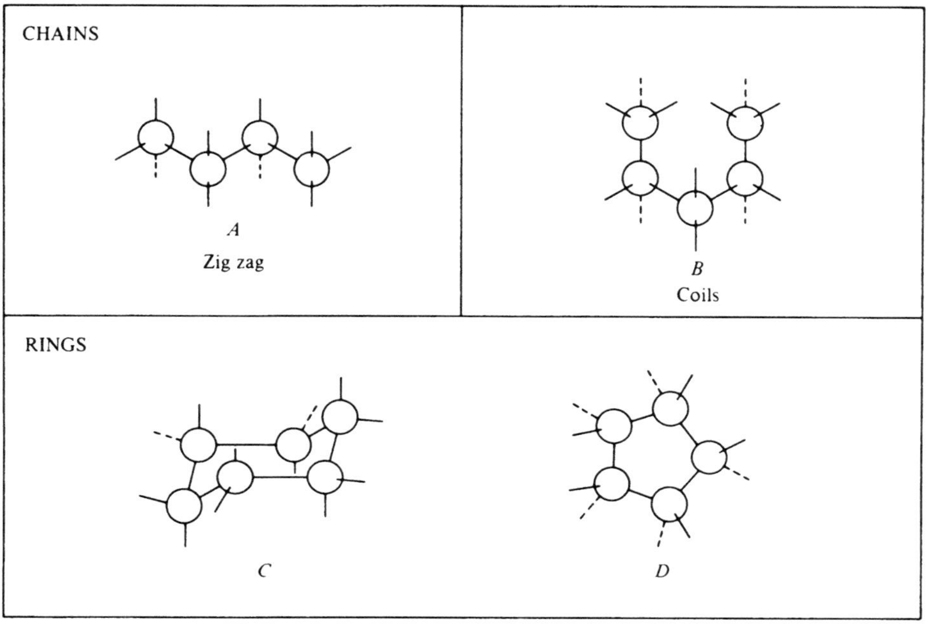

Figure 3-3 Spatial structures formed by $(C-C)_n$ compounds.

bonds are made by sharing electrons that are not already being shared, it may be deduced that functional groups mark locations in molecules that may become involved in chemical reactions with other compounds. For example, nitrogen has five valence electrons, but in the amino group $-NH_2$, only three of the five are used in bond formation:

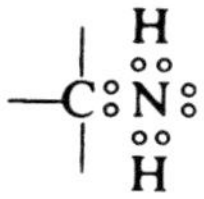

$$-\overset{\displaystyle |}{\underset{\displaystyle |}{C}} : \overset{\circ\circ}{\underset{\circ\circ}{N}} :$$

Thus, it is rich in electrons and a prime location for reaction.

The reactivity of the functional groups is emphasized by considering some of the more complex molecular bondings shown in Table 3-5. Esters are formed by a reaction between the carboxyl (acid) group of one compound and the hydroxyl group of another. Such bonding is of paramount importance in biological systems. The acid does not have to be an organic compound (in the table, the R and R' indicate that there are other parts of the molecule that are not being shown). In the formation of thiol esters, a sulfhydryl (rather than a hydroxyl) group is involved. The most distinguishing feature of protein is not that it contains significant amounts of nitrogen but that it consists of amino acids linked together by peptide bonds. In this bond, the carboxyl group of one acid reacts with the amino group of another acid. Glycosidic bonds might

Table 3-4   Important functional groups in organic compounds

Functional group		Type of molecule	Example	
Name	Formula		Name	Formula
Methyl	$\overset{\displaystyle H}{\underset{\displaystyle H}{-C-H}}$	Hydrocarbon and others	Methane	$H-\overset{H}{\underset{H}{C}}-H$
Hydroxyl	—OH	Alcohols	Propanol	$H-\overset{H}{\underset{H}{C}}-\overset{H}{\underset{H}{C}}-\overset{H}{\underset{H}{C}}-OH$
Carbonyl	$-\overset{\displaystyle O}{\overset{\displaystyle \|}{C}}-$	Aldehydes and ketones	Glyceraldehyde	$H-\overset{H}{\underset{H}{C}}-\overset{H}{\underset{H}{C}}-\overset{O}{C}-H$
			Dihydroxy Acetone	$H-\overset{H}{\underset{H}{C}}-\overset{O}{C}-\overset{H}{\underset{H}{C}}-H$
Carboxyl	$-\overset{\displaystyle O}{\overset{\displaystyle \|}{C}}-OH$	Acids	Acetic acid	$H-\overset{H}{\underset{H}{C}}-\overset{O}{C}-OH$
Amino	$-N\overset{\displaystyle H}{\underset{\displaystyle H}{}}$	Amino acids and amines	Glycine	$H-\overset{N\!H_2}{\underset{H}{C}}-\overset{O}{C}-OH$
Sulfhydryl	—S—H	Certain amino acids, mercaptans	Cysteine	$H-S-\overset{H}{\underset{H}{C}}-\overset{N H_2}{\underset{H}{C}}-\overset{O}{C}-OH$

be considered a special form of ether bond (i.e., C–O–C). They involve the carbonyl group of one carbohydrate and the hydroxyl group of another.

Table 3-5 Characteristic bonding of important types of biochemical compounds

Name of bonding	Structure	Source of R groups	Type of molecules
Ester	$$\begin{array}{c} O \\ \| \\ R-C-O-R' \end{array}$$	R, acid R', alcohol	Lipids (fats, waxes)
Phosphate ester	$$\begin{array}{c} O \\ \| \\ HO-P-O-R \\ \| \\ OH \end{array}$$	Phosphoric acid R, hydroxyl carbon compound	Nucleic acids Metabolic intermediates (energy capture)
Thiol ester	$$\begin{array}{c} O \\ \| \\ R-C-S-R' \end{array}$$	R, acid R', sulfhydryl carbon compound	Metabolic intermediates (energy capture and utilization)
Glycosidic bond	R—O—R'	R, carbohydrate R', carbohydrate	Polysaccharide
Peptide	$$\begin{array}{c} O \\ \| \\ R-C-N-R' \\ \| \\ H \end{array}$$	R, amino acid (acid end) R', amino acid (amino end)	Protein
Acid anhydride	$$\begin{array}{c} O \quad\ O \\ \| \quad\ \| \\ R-C-O-P-OH \\ \| \\ OH \end{array}$$	Phosphoric acid R, acid	Metabolic intermediates (energy capture and utilization)

Note: R indicates a part of the molecule (residue) which is not shown. R and R' are different residues. For example, an ester might be formed from acetic acid and propanol by a reaction in which water is removed; see Table 3-4 for structures which would be represented by R and R' in this ester.

Hydrogen bonding The bonds of carbon with various other elements provide a wide variety of strong bonds for building the space structures of the small molecules, and the functional groups permit a wide variety of reactions, including special types of bonds joining small molecules to form large molecules. However, the variety and great specificity of reaction of macromolecules in the biological world are provided by distinct spatial configurations made possible by weaker bonds than those already discussed. These weak bonds are made possible in large measure by the phenomenon of polarity, which was previously defined as the unequal sharing of electrons in covalent bonds, leading to slightly positive and negative centers in the molecule. Polarity fosters the creation of electrostatic forces of attraction by which polar molecules arrange themselves in loose lattice structures.

Water molecules are a good example. They do not exist in a haphazard array but are aggregated together as shown in Fig. 3-4. A coordinate bond is formed between oxygen and hydrogen in which the electron-rich oxygen donates both electrons. This type of bonding between positively charged hydrogen and a negatively charged donor atom such as oxygen or nitrogen is called a *hydrogen bond*. It is a relatively weak bond, requiring only one-twentieth the energy needed to make or break a carbon–carbon or carbon–oxygen bond. The fact that water molecules are bonded together accounts for the high boiling point (low volatility) of water relative to that of other materials of similar molecular weight. Since hydrogen bridging can occur not only between water molecules but also between water and other polar compounds, we have another clue as to why water functions as the *universal solvent* for other polar compounds, since a polar group of the compound can form a hydrogen bridge with water. This aspect is important in regard to the accessibility of organic nutrients to microorganisms in the aqueous environment but, more important, it is hydrogen bonding and other weak attractive forces mentioned later that account for the intricate and specific spatial configurations of biochemical macromolecules. Envision a long chain of hundreds of small molecules joined, for example, by strong ester or peptide linkages. Some of the small molecules so joined may contain R–OH and R–NH$_2$ groups, i.e., those exhibiting polarity. How does that chain exist in the microscopic aqueous universe in which it is suspended? Is it a jumbled pile, a straight chain, a spiral, or some other configuration? It may be any of these,

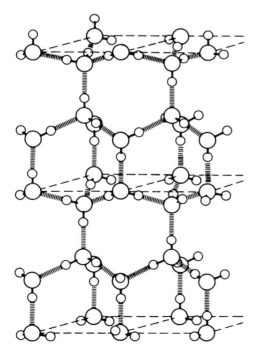

Figure 3-4 Schematic diagram of a lattice formed by water molecules. The energy gained by forming specific hydrogen bonds (|||||) between water molecules favors the arrangement of the molecules in adjacent tetrahedrons. Oxygen atoms are indicated by large circles and hydrogen atoms by small circles. Although the rigidity of the arrangement depends on the temperature of the molecules, this structure is, nevertheless, predominant in water as well as in ice. *(From Watson, 1977.)*

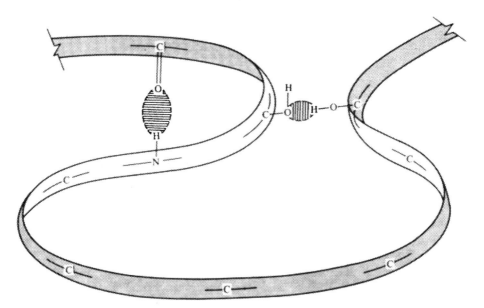

Figure 3-5 Representation of a long-chain organic molecule held in a particular space structural form by hydrogen bonding.

but whatever it is, it has a characteristic shape and spatial structure without which it cannot perform its function in the living cell. Various macromolecules that formerly functioned as parts of living cells but which are no longer functional because of "getting out of shape" play a role in the microbial environment; they are generally fated to become food for macromolecular aggregates (organisms) that are still functioning.

The space structure of the macromolecules is provided mainly by hydrogen bonding. The folded structure diagramed in Fig. 3-5 represents a chain of small molecules bonded together; the hydrogen bridges hold the chain in a specific three-dimensional shape. If these bonds were disrupted, this specific space structure would no longer exist, and the ability of the molecule to engage in its usual reactions would be destroyed.

The Classes of Biological Compounds

The major classes of biological compounds, i.e., those found in living cells, are lipids, carbohydrates, proteins, nucleic acids, and their derivatives. To the environmental scientist, many other organic compounds are also important. For example, hydrocarbons are not usually found in living organisms but they are made by them, and they can serve microorganisms as a carbon source from which the biological compounds referred to above are synthesized. However, the major portion of organic matter in the environment does quite naturally consist of these compounds, and a more detailed survey of them not only helps define the composition of living cells but also tells us much about the nature of organic substrates.

We shall discuss the compounds in the order listed above. Lipids are a good starting point because they involve hydrocarbon-like properties, and such a starting point is common in texts on organic chemistry, which should be among those consulted in studying cell composition and the nature of organic substrates. As we describe each major class of compounds, we shall also discuss its functional groups and some of the reactions in which it may be involved.

Lipids Lipids are organic compounds that are soluble in organic solvents such as ether, benzene, acetone, chloroform, hexane, carbon tetrachloride, and petroleum ether, and are only sparingly soluble in water. This practical definition provided by biological chemists is paralleled by the definition of grease provided by sanitary chemists, except that in the latter case the organic solvent of choice is freon. This solvent has been selected after much experimentation as the one best suited for use with water, wastewaters, and sludges common to the pollution control area. As with the methods used for measurement of total organic matter and its combustibility, there will be differences in values for lipid analyses because of differences in reagents and analytical conditions.

The lipid content of bacteria varies from 10 to 15 percent of the dry weight; it is slightly greater in fungi. Since lipids constitute a significant portion of the human diet, they may be expected to form a significant portion of the organic matter in municipal wastes.

Although lipids include a highly heterogeneous mix of organic compounds, it can be seen in Table 3-6 that, for the first two groupings at least, there is a common chemical characteristic: simple and compound lipids contain fatty acids and alcohols and, for the most part, these are bonded with ester linkages.

Table 3-6 Classes of lipids

---

1. Simple lipids.
   a. Fats and oils: esters of various fatty acids and the alcohol glycerol.
   b. Waxes: esters of various fatty acids and various long-chain monohydroxyl alcohols.
2. Compound lipids.
   a. Phospholipids: esters of various fatty acids and glycerol; they also contain phosphorus and nitrogen compounds.
   b. Glycolipids: esters of various fatty acids and alcohols; they also contain a carbohydrate.
3. Nonsaponifiable lipids: compounds that are soluble in organic solvents but do not contain fatty acids (do not yield soaps when subjected to hydrolysis with strong alkali such as NaOH or KOH, hence, nonsaponifiable). Sterols, fat-soluble vitamins, and plant pigments are included in this subclass.

Simple lipids The fats and oils are the most common and the commercially important lipids. They are formed as shown below:

Alcohol + 3 fatty acids ⇌ ester + $3H_2O$
(glycerol) (triglyceride)

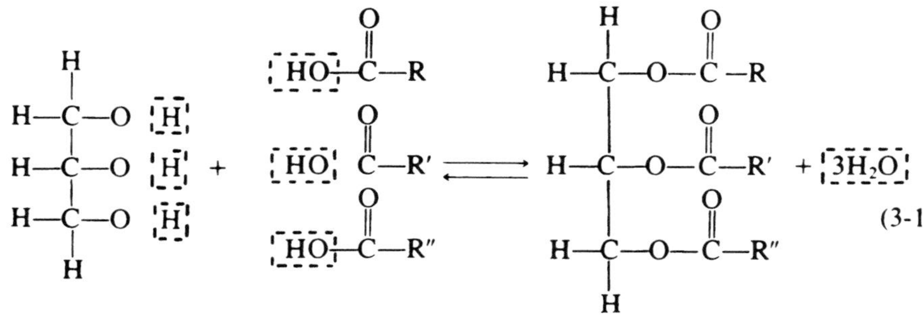

$$(3\text{-}11)$$

Glycerol is a fairly small molecule and contributes little to the average weight of fat and oil molecules. Generally, the fatty acids are even-numbered, straight-chained molecules containing 12 or more carbons and one carboxyl group. It might be expected then that the physical and chemical behavior of the lipids is determined primarily by the fatty acids. Because of this and the fact that they are metabolically important compounds in the aqueous environment, it is well to examine them in more detail.

The fatty acids exist in nature as products of the microbial metabolism of various compounds or as partial products of the breakdown of decaying animal and vegetable fats. It is seen above that when an ester linkage is made, a hydrogen atom from the alcohol and a hydroxyl group from the carboxyl group of the acid are removed as water. When the reverse reaction occurs, i.e., when the fat is broken down into its constituent parts or hydrolyzed, water is added.

Glycerol is easily oxidized to the carbonyl level (ketone or aldehyde) and is readily incorporated into metabolic schemes, which will be discussed later. The fatty acids are also metabolized, but their orderly utilization is somewhat more involved, and the process is complicated by the fact that all but the very short-chain fatty acids are insoluble in water. Only the first four shown in Table 3-7 are miscible in water, and solubility thereafter decreases very sharply with chain length. β-Hydroxybutyric acid is soluble, but a macromolecular polymer consisting of chains of this compound, called poly-β-hydroxybutyric acid (PHB), is insoluble. Some microorganisms store PHB as a source of reserve carbon as animals store fats.

The low molecular weight acids, particularly acetic, propionic, and butyric, are collectively known as *volatile acids* because they can be vaporized (distilled) at atmospheric pressure. Volatile acid determinations are important in environmental and pollution control work because the acids are prominent microbial products in the decay of some macromolecules, being produced

Table 3-7 Some important fatty acids

Name	Formula
Formic	$H-\overset{\overset{\displaystyle O}{\|\|}}{C}-OH$
Acetic	$CH_3-\overset{\overset{\displaystyle O}{\|\|}}{C}-OH$
Propionic	$CH_3-CH_2-\overset{\overset{\displaystyle O}{\|\|}}{C}-OH$
Butyric	$CH_3-(CH_2)_2-\overset{\overset{\displaystyle O}{\|\|}}{C}-OH$
β-Hydroxybutyric	$CH_3-\underset{\underset{\displaystyle OH}{\|}}{CH}-CH_2-\overset{\overset{\displaystyle O}{\|\|}}{C}-OH$
Valeric	$CH_3-(CH_2)_3-\overset{\overset{\displaystyle O}{\|\|}}{C}-OH$
Caproic	$CH_3-(CH_2)_4-\overset{\overset{\displaystyle O}{\|\|}}{C}-OH$
Caprylic	$CH_3-(CH_2)_6-\overset{\overset{\displaystyle O}{\|\|}}{C}-OH$
Capric	$CH_3-(CH_2)_8-\overset{\overset{\displaystyle O}{\|\|}}{C}-OH$
Myristic	$CH_3-(CH_2)_{12}-\overset{\overset{\displaystyle O}{\|\|}}{C}-OH$
Palmitic	$CH_3-(CH_2)_{14}-\overset{\overset{\displaystyle O}{\|\|}}{C}-OH$
Palmitoleic	$CH_3-(CH_2)_5-CH=CH-(CH_2)_7-\overset{\overset{\displaystyle O}{\|\|}}{C}-OH$
Stearic	$CH_3-(CH_2)_{16}-\overset{\overset{\displaystyle O}{\|\|}}{C}-OH$
Oleic	$CH_3-(CH_2)_7-CH=CH-(CH_2)_7-\overset{\overset{\displaystyle O}{\|\|}}{C}-OH$
Linoleic	$CH_3-(CH_2)_4-CH=CH-CH_2-CH=CH-(CH_2)_7-\overset{\overset{\displaystyle O}{\|\|}}{C}-OH$
Linolenic	$CH_3-CH_2-CH=CH-CH_2-CH=CH-CH_2-CH=CH-(CH_2)_7-\overset{\overset{\displaystyle O}{\|\|}}{C}-OH$

Numbering system: (a) for carbons, from right to left, carboxyl group = 1; (b) for substituent groups, from right to left beginning with first carbon after carboxyl: α, β, etc., e.g., β-hydroxybutyric acid.

both aerobically and anaerobically but mostly as a result of anaerobic decay. They are also sources of food for aerobic microorganisms. The higher molecular weight fatty acids occur in microorganisms and in higher plants and animals largely as constituents of simple and compound lipids.

It is seen that some fatty acids contain one, two, or three double bonds between carbon atoms. Like substituent functional groups, double bonds are locations of chemical reactivity in molecules, and they also permit a form of isomerism known as *geometrical isomerism*. The term *isomerism* applies to compounds that have the same chemical formula, and in some cases the same structural formula, but different space structures. For example, consider the formula for oleic acid. Two configurations (*cis* and *trans*) around the double bond are possible (see Fig. 3-6). Oleic acid and elaidic acid have the same structural formula but decidedly different physical and chemical properties. Various types of isomerism are important in metabolic reactions, since only specified spatial structures react in living systems. The great majority of unsaturated fatty acids in nature are of the *cis* form.

Properties of the Double Bond Two major types of reactions occur at the sites of unsaturation (double bonds). Halogenation is the addition of a halogen, e.g., chlorine; and hydrogenation involves the addition of hydrogen. Other addition reactions, such as the addition of water, also are common. Oxidation occurs readily at the double bond. The oxidation may or may not lead to cleavage of the bond.

Halogenation

$$-\overset{\displaystyle |}{\underset{\displaystyle H}{C}}=\overset{\displaystyle |}{\underset{\displaystyle H}{C}}- \ + \ Cl_2 \longrightarrow -\overset{\displaystyle \overset{Cl}{|}}{\underset{\displaystyle \underset{H}{|}}{C}}-\overset{\displaystyle \overset{Cl}{|}}{\underset{\displaystyle \underset{H}{|}}{C}}- \qquad (3\text{-}12)$$

Hydrogenation

$$-\overset{\displaystyle |}{\underset{\displaystyle H}{C}}=\overset{\displaystyle |}{\underset{\displaystyle H}{C}}- \ + \ H_2 \longrightarrow -\overset{\displaystyle \overset{H}{|}}{\underset{\displaystyle \underset{H}{|}}{C}}-\overset{\displaystyle \overset{H}{|}}{\underset{\displaystyle \underset{H}{|}}{C}}- \qquad (3\text{-}13)$$

The halogenation reaction contributes to the so-called *chlorine demand* when chlorine is used as a microbial disinfectant, i.e., the tying up of disinfectant that otherwise would be available to react with biochemically functional molecules in the cell. It also can produce chlorinated hydrocarbon compounds that may be hazardous to humans. Hydrogenation increases the melting point and decreases the solubility of fatty acids and thus of the fats of which they are a part.

Oxidation Fatty acids and unsaturated hydrocarbons (alkenes) generally can undergo varying degrees of oxidation. For example, consider the partial

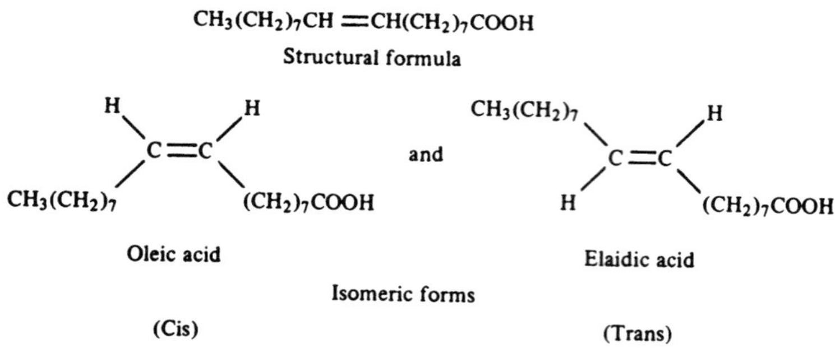
CH3(CH2)7CH = CH(CH2)7COOH
Structural formula
H
H
C = C
CH3(CH2)7
(CH2)7COOH
and
CH3(CH2)7
H
C = C
H
(CH2)7COOH
Oleic acid
Elaidic acid
Isomeric forms
(Cis)
(Trans)

Figure 3-6 Geometrical isomers with same structural formula made possible by the double bond. Oleic acid and most other naturally occurring unsaturated fatty acids have the *cis* configuration.

oxidation of ethylene to ethylene glycol (both are important petrochemicals):

$$\underset{\text{ethylene}}{CH_2=CH_2} \xrightarrow[H_2O]{KMnO_4} \underset{\text{ethylene glycol}}{\underset{OH\quad\; OH}{CH_2-CH_2}} \tag{3-14}$$

Further oxidation of double bonds can cause cleavage of the carbon–carbon bond between carbons bearing the hydroxyl groups. Upon such cleavage, the carbons can react further with oxygen, leading to production of acids and/or carbon dioxide. The reactive shorter acids with four to six carbons exhibit offensive odors. A rancid fat or oil develops a disagreeable odor, which is due to the oxidative cleavage of higher molecular weight unsaturated fatty acids to form lower molecular weight ("stench") acids. On the other hand, some fats such as butter already contain a variety of short-chain fatty acids, and these can be released upon simple hydrolysis, i.e., oxidation need not occur. This is called *hydrolytic rancidity* as compared with the *oxidative rancidity* described above.

Properties of the Carboxyl Group One of the major reactions of the carboxyl (acid) group is the reaction with alcohols to form esters (esterification) already described. Also, the acid group may participate in hydrogen bonding. This group exhibits weak acid properties in aqueous solution, since, unlike the strong inorganic acids, they are only slightly ionized:

$$R-\overset{O}{\overset{\|}{C}}-OH \rightleftharpoons R-\overset{O}{\overset{\|}{C}}-O^- + H^+ \tag{3-15}$$

The degree of dissociation as measured by the dissociation constant (ionization constant) K_a is defined in accordance with the law of mass action; at equilibrium:

$$K_a = \frac{[H^+][R-\overset{\overset{\displaystyle O}{\|}}{C}-O^-]}{[R-\overset{\overset{\displaystyle O}{\|}}{C}-OH]} \tag{3-16}$$

or more generally

$$K_a = \frac{[H^+][A^-]}{[HA]} \tag{3-17}$$

where [HA] is the activity (molar concentration for practical purposes) of the undissociated acid, and [A$^-$] is the concentration of the anion or conjugate base. Since the fatty acids are weakly ionized, the value of the denominator [HA] will be large and the K_a values small. For example, the K_a of acetic acid is 1.8×10^{-5}. Knowing the value of K_a, one can calculate the concentration of hydrogen ion and conjugate base for any given initial concentration of acid. For example, if one prepared a 0.05-molar solution of acetic acid, the concentrations of the three species would be in proportions according to the K_a value. Letting the hydrogen ion concentration be X, the value of the numerator in Eq. (3-16) is X^2, and the concentration of undissociated acid is the solution molarity, 0.05, minus that which has dissociated, i.e., $0.05 - X$:

$$1.8 \times 10^{-5} = \frac{X^2}{0.05 - X}$$

Solving for X using the quadratic form,

$$X = [H^+] = 9.4 \times 10^{-4}M$$

Hydrogen ion concentration is often expressed more conveniently in terms of pH rather than molarity. pH is the negative logarithm of the hydrogen ion concentration:

$$pH = \log\frac{1}{[H^+]} = -\log[H^+] \tag{3-18}$$

Using Eq. (3-17) and the H$^+$ molarity just calculated, we can find the pH of a $0.05M$ solution of acetic acid:

$$pH = -\log(9.4 \times 10^{-4})$$
$$= -\log 9.4 + (-\log 10^{-4})$$
$$= -0.97 + 4 = 3.03$$

The slight degree of ionization can be expressed as a percentage:

$$\left(\frac{9.4 \times 10^{-4}}{0.05}\right)100 = 1.9\%$$

Thus, less than two molecules per 100 are ionized in a 0.05-molar solution of acetic acid, whereas a strong inorganic acid such as HCl at the same molar concentration is totally ionized.

Resistance to Changes in pH-*Buffering Action* Equation (3-17) can be used to provide a practical explanation of the buffering characteristics of some solutions, i.e., their ability to resist changes in pH. The control of pH in biological systems is of extreme importance and, fortunately, many polluted waters contain chemical constituents that form natural buffering systems. We can solve Eq. (3-17) for the hydrogen ion concentration:

$$[H^+] = \frac{K_a[HA]}{[A^-]} \qquad (3\text{-}19)$$

In order to express $[H^+]$ in terms of pH, we take the negative logarithm of both sides of the equation:

$$-\log[H^+] = -\log K_a - \log \frac{[HA]}{[A^-]}$$

$$pH = pK_a + \log \frac{[A^-]}{[HA]} \qquad (3\text{-}20)$$

Note that the negative logarithm of K_a is now designated pK_a (a designation parallel to that for $-\log[H^+]$), and the ratio of undissociated acid $[HA]$ to ionized conjugate base $[A^-]$ has been reversed in order to remove the minus sign. Equation (3-20) is known as the *Henderson-Hasselbach equation*, and it is useful in work with microorganisms when it is desired to maintain the pH within certain limits. For a given pH and ionization constant (pK_a) for the acid involved, one can calculate the ratio of anion to undissociated acid, $[A^-]/[HA]$, required to yield the desired pH. One must obtain a source of anion in order to prepare a solution containing the needed ratio. Such a source is the salt of the acid. For example, in the case of acetic acid, sodium acetate is such a salt. It is easily formed by adding NaOH to the acid:

$$\begin{matrix} O & & O \\ \parallel & & \parallel \\ CH_3COH + NaOH \rightarrow CH_3CONa + HOH \end{matrix} \qquad (3\text{-}21)$$

Sodium acetate ionizes:

$$\begin{matrix} O & & O \\ \parallel & & \parallel \\ CH_3{-}CONa \rightarrow CH_3CO^- + Na^+ \end{matrix} \qquad (3\text{-}22)$$

thus supplying the anion, A^-, needed to provide the required pH. This system of anion or salt, A^-, and undissociated acid, HA, resists changes in pH because of the H^+-capturing tendency of the anion when the concentration of hydrogen ion is increased and the H^+-donating tendency of the acid when the concentration of hydrogen ion is effectively decreased. For example, using acetic acid, reaction (3-23) is the action of the anion when acid is added:

$$CH_3\overset{\displaystyle O}{\overset{\|}{C}}O^- \underset{\displaystyle H^+}{\nearrow} CH_3\overset{\displaystyle O}{\overset{\|}{C}}OH \qquad (3\text{-}23)$$

Reaction (3-24) is the action of the acid when hydroxyl ion is added, effectively lowering the value of $[H^+]$ since $H^+ + OH^- \rightleftharpoons H_2O$:

$$CH_3\overset{\displaystyle O}{\overset{\|}{C}}OH \underset{\displaystyle OH^-}{\nearrow} CH_3\overset{\displaystyle O}{\overset{\|}{C}}O^- + H_2O \qquad (3\text{-}24)$$

The presence of both the undissociated acid and its salt has prevented the pH from changing; i.e., they have provided a buffer system against any change in pH by preventing a change in the concentration of hydrogen ions. Obviously, if one kept adding H^+ (or OH^-), the buffer capacity would become exhausted because all of the anion, or the acid, could be used up. Or, if the ratio of $[A^-]/[HA]$ was rather one-sided, the system would be poised to resist a change in only one direction. For example, if the ratio of $[A^-]/[HA]$ were $1/10$, the system's ability to buffer against added H^+, i.e., resist a drop in pH, would be near exhaustion. However, the acid has the capacity to form large amounts of hydrogen ion to resist the addition of OH^- (increase in pH). Often one wishes to provide approximately equal protection from changes on both sides of the desired pH. Thus, nearly equal concentrations of A^- and HA are sometimes used. Examination of Eq. (3-20) shows that if $[A^-]$ and $[HA]$ are equal, then $pH = pK_a$. Thus, one should choose an acid with a pK_a value that is fairly close to the pH desired. The ionization constant of acetic acid of 1.8×10^{-5}, or pK_a of 4.7, is usually too low, so acetic acid should not be used in controlling pH during microbial growth. Also, many organisms can utilize acetic acid as a carbon and energy source, thus decreasing the capacity of the buffer system. For most applications, phosphoric acid (pK_a of $H_2PO_4^- \rightleftharpoons HPO_4^{2-}$ is 6.8) or carbonic acid (pK_a of $H_2CO_3 \rightleftharpoons HCO_3^-$ is 6.1) is suitable. The phosphate buffer system is widely used in microbial growth studies. It is emphasized that this added buffer system is used to gain control of the pH in laboratory experiments, but it would be too costly and would cause eutrophication of the receiving stream if such buffers were used in secondary treatment plants. Therefore, pH control in the field is provided by continuous measurement of pH and automatic addition of acid or alkali to maintain it within the desired limits.

Soaps Fatty acids containing more than four or five carbon atoms are insoluble, but the sodium and potassium salts of long-chain fatty acids are soluble. Any salt of a fatty acid may be called a soap, but not all are soluble. For example, sodium and potassium palmitate are soluble and are major constituents of commercial soaps. However, calcium and magnesium salts of fatty acids are insoluble and their formation in "hard" waters, i.e., waters containing high concentrations of these divalent cations, causes a greasy scum in the water and excessive use of soap.

Whereas fatty acids are made by the acid hydrolysis of neutral fats, i.e., the reverse of reaction (3-11), soaps are made by alkaline hydrolysis:

$$
\begin{array}{l}
\text{H} \quad\quad \overset{\text{O}}{\overset{\|}{}} \\
\text{HC—O—C—R} \\
\;|\quad\quad\; \overset{\text{O}}{\overset{\|}{}} \\
\text{HC—O—C—R} + \text{KOH} \rightarrow 3\,\text{R—}\overset{\overset{\text{O}}{\|}}{\text{C}}\text{—OK} + \text{HC—OH} \\
\;|\quad\quad\; \overset{\text{O}}{\overset{\|}{}} \\
\text{HC—O—C—R} \\
\text{H}
\end{array}
\qquad (3\text{-}25)
$$

with products:
$$
\begin{array}{l}
\text{H} \\
\text{HC—OH} \\
\;| \\
\text{HC—OH} \\
\;| \\
\text{HC—OH} \\
\text{H}
\end{array}
$$

Alkaline hydrolysis of lipid material is called *saponification*. If, upon being subjected to alkaline hydrolysis, water-soluble products are not produced, the sample of lipid material is categorized as belonging to the *nonsaponifiable fraction*. Thus, alkaline hydrolysis serves to distinguish lipids of types 1 and 2 from the type 3 lipids of Table 3-6.

Much has been written about the cleansing action of soaps, since it has important ramifications in environmental pollution as well as bacterial metabolism. The salts of long-chain or high molecular weight fatty acids are unique in that the physical-chemical properties of acids and hydrocarbons are so different, and both sets of properties are combined in these molecules. The hydrocarbon end is insoluble in water and soluble in lipid materials, such as fats, oils, and grease, whereas the carboxyl end is soluble in water and insoluble in fats. The soap cleans essentially by dissolving (more correctly, emulsifying) greasy materials, which are responsible for preventing water from rinsing dirt away from the surfaces to be cleaned. If the surface to be cleaned (e.g., the fabric of clothing) is agitated, the grease or oil surfaces are mechanically broken up into small droplets, which would rapidly coalesce were it not for the fact that the hydrocarbon ends of soap molecules dissolve into the droplet. The ionized carboxyl end of the soap, with its negative charge, repels other droplets of grease, thus preventing the droplets from coalescing, and dissolving or emulsifying them in the water, which can then be flushed away (see Fig. 3-7).

The power of soaps to dissolve or emulsify lipid materials has important ramifications in microbial growth. The membranes of all microorganisms (and the walls of some) contain lipid materials (see Chap. 4), and the dissolution by soaps and other surface-active agents of this essential structure may play a role in their disinfecting power. Also, the protruding charge of the carboxyl group can change the electric charge on the cell surface, and this could affect the agglutinating or flocculating characteristics of the cells. Microbial cells generally carry a negative charge, and protruding carboxyl groups can make the charge on the cell surface more negative, thus hampering flocculation.

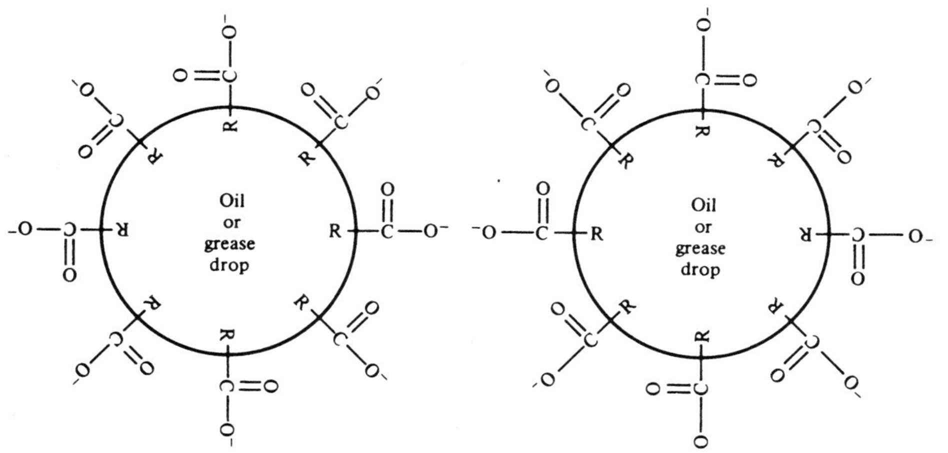

Figure 3-7 Oil droplets in aqueous suspension are prevented from coalescing; i.e., they remain emulsified due to the outward orientation of the negatively charged hydrophilic carboxyl ends of the soap molecules.

Anything that tends to stabilize microbial cells in suspension is at odds with the process in the secondary settling tank in biological treatment plants.

Waxes Waxes are found in animal and plant materials. In nature, one of their major roles is the formation of a protective coating on animal surfaces such as feathers, hair, and wool, and in plants they are found on stems, leaves, and fruits. They are much more resistant to microbial decay than are other lipids, hence their role as protective coatings. One of their major commercial uses is as protective coatings of a sort, e.g., floor and auto waxes. Carnauba wax from palm leaves is an ester of the 28-carbon montanyl alcohol and the 26-carbon cerotic acid and is an important ingredient in commercial waxes. Waxes are not found as constituents of microorganisms, but they do serve as a carbon source and are subject to microbial decay.

Compound lipids The compound lipids are of two general types: phospholipids and glycolipids. The phospholipids are found in all microorganisms, plants, and animals, but the glycolipids are found primarily in animals and occur in only a few types of microorganisms.

Phospholipids (Phosphatides) The phospholipids commonly found in microorganisms are derivatives of glycerol phosphate and can be represented by the general structure shown in Fig. 3-8A. They contain many different fatty acids, represented by R_1 and R_2, most frequently with 16 or 18 carbons, saturated or unsaturated. In bacteria, the substituent attached to phosphate by an ester bond is most frequently ethanolamine. Choline, serine, and inositol are attached to phosphate in other types of phospholipids. These substituents give the groups of phospholipids their names; e.g., a phospholipid containing serine is a phosphatidylserine.

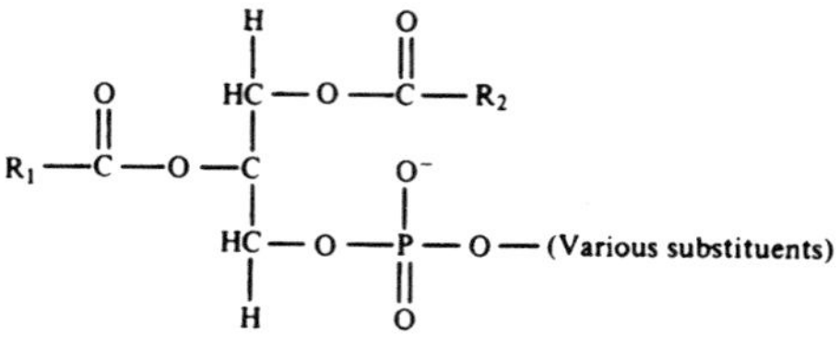

Phospholipids (phosphatides)

A

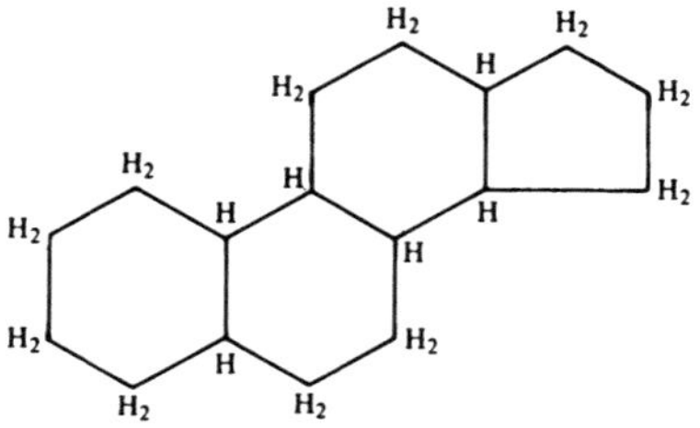

Cyclopentanoperhydrophenanthrene, the basic
4-membered ring structure of steroids

B

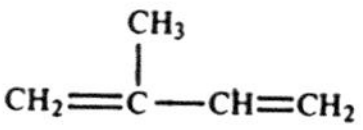

$$CH_2{=}\overset{\displaystyle CH_3}{\underset{\displaystyle}{C}}{-}CH{=}CH_2$$

Isoprene

C

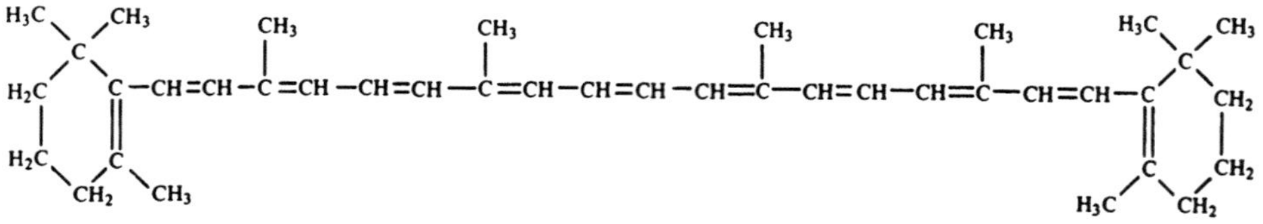

β-carotene

D

Figure 3-8 Essential structural features of phospholipids, steroids, and carotenes.

Glycolipids These are a class of lipids that contain sugars and fatty acids. One type of glycolipid contains no glycerol but may contain a long-chain alcohol. The sphingolipids contain the 18-carbon amino alcohol sphingosine. These compounds are found in brain tissue (hence, they are often called *cerebrosides*) and in nerve tissue. They can serve as a microbial carbon source in the decay leg of the carbon–oxygen cycle but are not constituents of microorganisms.

Glycolipids of a different type are found in the membranes of algae and cyanobacteria. These are glycerol-containing lipids with two fatty acids and one or two molecules of galactose.

Nonsaponifiable Lipids This class includes a wide variety of compounds that constitute a relatively small percentage (approximately 2 per cent) of all lipids found in the biosphere. They include the steroids, plant pigments, and fat-soluble vitamins.

Steroids are built upon a four-membered ring system called cyclopentanoperhydrophenanthrene (see Fig. 3-8*B*). Many different types of steroid compounds are found in nature. They occur in animals and large plants and in various types of microorganisms, yeasts, and fungi, but not in bacteria. Steroids include such compounds as cholesterol, the bile acids (e.g., cholic acid), sex hormones (e.g., testosterone, progesterone, estradiol), and adrenocortical hormones (e.g., cortisone).

The *carotenoids* are a type of lipid derived from β-carotene, which is shown in Fig. 3-8*D*. They are found in many bacteria, usually associated with the cell membrane (see Chap. 4), and they are responsible for the orange color of carrots. Carotenoids are related to the compound isoprene (See Fig. 3-8*C*), which is seen in Fig. 3-8*D* to be a repeating unit in β-carotene. Dozens of variations of carotenoids similar to carotene are found in the biosphere.

Carbohydrates It will be recalled that in the photosynthetic leg of the carbon-oxygen cycle, carbon dioxide and water are converted first into carbohydrate. Thus, the pivotal role of carbohydrates as the building material for the carbon structure of all other organic matter in the biosphere is apparent. We shall see later that carbohydrate also plays a central role in the metabolism of the organotrophic organisms as well as the photosynthetic organisms. While the carbohydrate content of microorganisms may vary from 15 to 35 percent of the dry weight, the carbohydrate content of macroscopic plants may vary between 60 and 90 percent of their dry weight. Carbohydrates comprise well over 50 percent of the average human diet, and they are important natural and synthetic materials in construction and in various manufacturing industries.

Carbohydrates contain large numbers of hydroxyl groups. Also, they are centrally situated in regard to oxidation level; i.e., they may be reduced to alcohols (called *sugar alcohols*) or oxidized to acids (called *sugar acids*). They are either aldehydes (in which case they are called *aldoses*) or ketones (*ketoses*). Thus, a precise definition of carbohydrates is that they are *poly-*

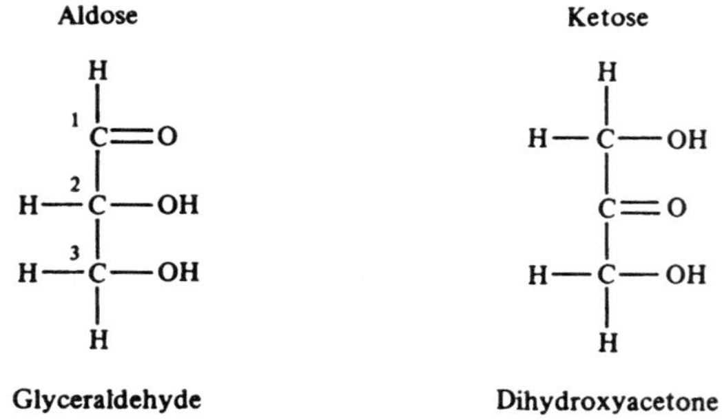

Aldose

Ketose

Glyceraldehyde   Dihydroxyacetone   **Figure 3-9** Trioses (Fischer projections).

hydroxyl aldehydes or ketones. Since the smallest ketone must contain at least three carbon atoms, the smallest monosaccharides are trioses (see Fig. 3-9).

By adding carbon atoms to each of the trioses, a series of sugars is expanded from trioses (C_3) to tetroses (C_4), pentoses (C_5), hexoses (C_6), and heptoses (C_7). There are only a few naturally occurring C_7 sugars, and these are ketoses. Sedoheptulose is the metabolically significant C_7 sugar. These small carbohydrate molecules are called *monosaccharides*, and they comprise the monomers from which complex carbohydrates are made.

The monosaccharides are joined together by glycosidic bonds (see Table 3-5). The general class name to describe molecules consisting of two or more monosaccharide units is *oligosaccharides*. They may be disaccharides, trisaccharides, tetrasaccharides, etc. The disaccharides are by far the most prevalent of these intermediate-length carbohydrates. Most of the carbohydrates in nature and in microorganisms consist of a very large number of monosaccharides joined by glycosidic bonds, and these are referred to as *polysaccharides*.

Monosaccharides Like the lipids, the carbohydrates are three-dimensional structures that may be represented by elements and functional groups connected by "girders and beams," the covalent bonds. Thus, glyceraldehyde can be represented by the three-dimensional structure shown in Fig. 3-10.

The stick formula for glyceraldehyde given in Fig. 3-9 is a horizontal (two-dimensional) projection of the space model shown to the left in Fig. 3-10, i.e., D(+)-glyceraldehyde. If one were viewing not the model at the left but its reflection in a mirror, the image would be the model shown on the right side. The only difference is the reversal of H and OH groups. Both forms of these sugars exist in nature, and this small difference can affect the chemical reactivity of the compound. The small structural difference also affects the way in which a beam of polarized light is bent when passed through a solution of the compound. Compounds with this property are called *optically active*; if the polarized light is bent to the right, the compound is called *dextrorotatory* (indicated by +), and if to the left, *levorotatory* (indicated by −). If one had a solution containing equal numbers of molecules of both forms, the effects

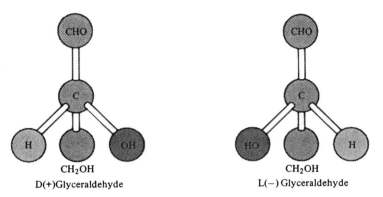

CH₂OH
D(+)Glyceraldehyde

CH₂OH
L(−)Glyceraldehyde

Figure 3-10 Model of glyceraldehyde. The L form is a mirror image of the D form.

would cancel out; there would be no rotation of light, and the solution would be called a *racemic mixture.* The two forms (mirror images) are called *enantiomorphs.* They rotate light to the same degree but in opposite directions.

All compounds that have one or more carbons to which four different substituents are attached, thus permitting more than one geometrical shape, can rotate a beam of polarized light. Such a carbon atom is called an *asymmetric carbon.*

Examination of the formula for glyceraldehyde shows that it contains one asymmetric carbon. The corresponding ketotriose, dihydroxyacetone, has no such asymmetric carbon.

It is seen in Fig. 3-10 that the letters D and L are used in addition to (+) and (−) in labeling enantiomorphs. Glyceraldehyde begins a series of sugars of increasing carbon-chain length. Each member of the series has its enantiomorphic counterpart (mirror image). Arbitrarily, all sugars with the OH written to the right side on the asymmetric carbon farthest removed from either the aldehyde or the ketone group are designated D. If the OH is written on the left, it is an L sugar. For all of the sugars, this asymmetric carbon is the next to last (the penultimate) carbon in the horizontal projection.

After deciding on the D and L designation, it was found that some sugars labeled D were levorotatory and vice versa, so it has become necessary to use both D and L and (+) and (−) to designate series derivation (D or L) and light rotation, dextrorotatory, +, or levorotatory, −.

In the biosphere, the D series of simple sugars (monosaccharides) is important. Figure 3-11 shows the development of this series. The highlighted sugars are those of greatest metabolic importance. In the series derived from D(+)-glyceraldehyde, there are two tetroses, four pentoses, and eight hexoses. Also, a count of asymmetric carbons reveals two in tetroses, three in pentoses, and four in hexoses. Note that the OH group on the penultimate carbon is to the right in each sugar.

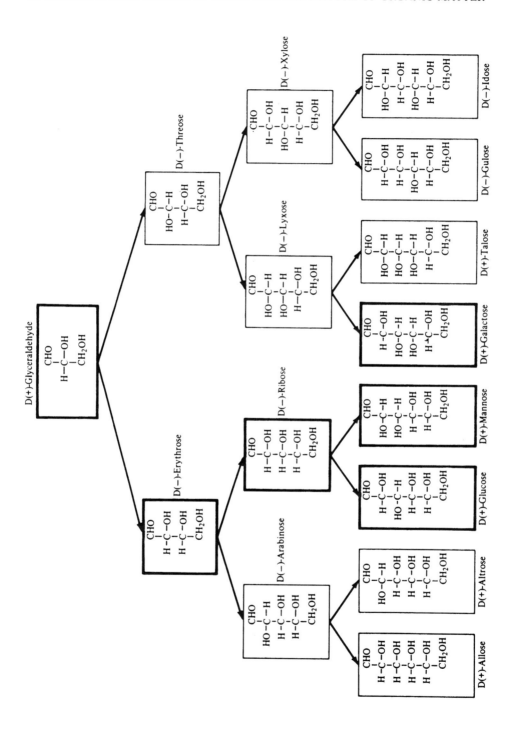

A series of ketoses also exists; there are some commonly occurring, metabolically important ketomonosaccharides that will be discussed later.

Cyclic (Ring) Structures of Monosaccharides Carbohydrates exhibit properties of aldehydes and ketones as well as alcohols, and the carbonyl group can react with water and with alcohols. The products are known as *acetals* and have the general formula shown to the right in Eq. (3-26):

$$\underset{\text{aldehyde}}{\text{R}-\overset{\displaystyle \text{H}}{\text{C}}=\text{O}} + \underset{\text{alcohol}}{2\text{HOR}'} \overset{\text{H}^+}{\rightleftharpoons} \underset{\text{acetal}}{\text{R}-\overset{\displaystyle \text{H}}{\underset{\displaystyle \text{OR}'}{\text{C}}}-\text{OR}'} + \text{H}_2\text{O} \qquad (3\text{-}26)$$

The above reaction occurs in the presence of an acid catalyst and excess alcohol. In the absence of excess alcohol and a catalyst, an unstable intermediate is produced, as shown in Eq. (3-27):

$$\underset{\text{aldehyde}}{\text{R}-\overset{\displaystyle \text{H}}{\text{C}}=\text{O}} + \underset{\text{alcohol}}{\text{HOR}'} \rightleftharpoons \underset{\text{hemiacetal}}{\text{R}-\overset{\displaystyle \text{H}}{\underset{\displaystyle \text{OR}'}{\text{C}}}-\text{OH}} \qquad (3\text{-}27)$$

Since the monosaccharides contain both carbonyl and alcohol groups, there is a possibility for the formation of an intramolecular hemiacetal, a ring structure. The modes of designation of ring structures are illustrated using glucose in Fig. 3-12. The space model (*B*) shows the proximity of the oxygen of the carbonyl carbon, C_1, and the OH group on C_5. Hence, the formation of the six-membered ring structure involving C_1 and C_5. This ring structure is typical of the other aldohexoses as well. Sugars that form such six-membered rings are called *pyranoses* because of their relation to the compound pyran.

The sugar fructose, which is a ketohexose, can exist in a five- as well as a six-membered ring form. The five-membered ring, which occurs more commonly, is shown in Fig. 3-13. Sugars that form such five-membered rings are called *furanoses* because the ring form resembles the compound furan. Aldopentoses also form five-membered (furanose) structures; the aldopentose ribose is shown in Fig. 3-14.

It can be seen from the space model representation in Fig. 3-12*B* that the two-dimensional structural formula of Fig. 3-12*A* showing a straight chain is useful but does not really describe the molecule. It is particularly ineffective when we consider the ring structure. The ring structure shown in the lower right portion of the figure represents a bird's-eye view of a hexagonal transparent tabletop or slab. Carbons 1 through 5 and the oxygen atom in the ring lie in the plane of the slab. The hydroxyl group on carbon-2 lies below and the hydrogen above the plane of the slab. Similarly, the hydroxyl is above

Figure 3-11 The D-aldoses, triose through hexose. Aldoses in heavy boxes are those most commonly found in nature and the most metabolically significant.

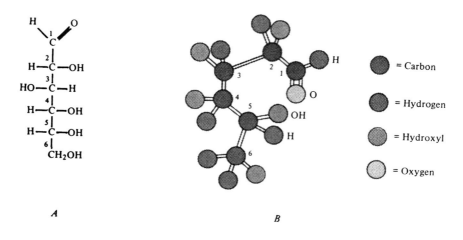

A

B

= Carbon

= Hydrogen

= Hydroxyl

= Oxygen

C

D

Figure 3-12 Structural representations of D(+)-glucose. Open-chain (Fischer) structural formula (A), open-chain space model (B), cyclic form (C), ring form (D).

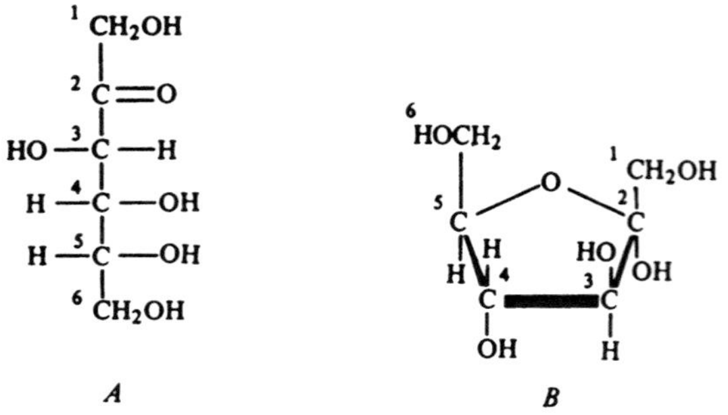

A

B

Figure 3-13 Open-chain (A) and ring form (B) D-fructose.

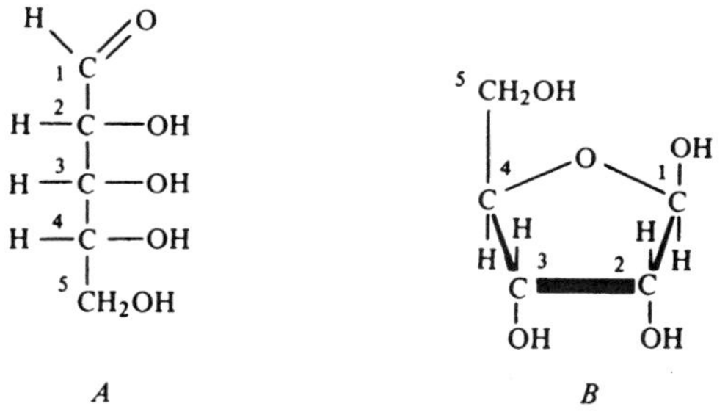

H —C —OH (structure A)

$$^1 C \overset{H}{\underset{}{\diagup}} \overset{O}{\diagdown}$$

H —^{2}C —OH
H —^{3}C —OH
H —^{4}C —OH
^{5}CH$_2$OH

A

^{5} CH$_2$OH

B

Figure 3-14 Open-chain (*A*) and ring form (*B*) D-ribose.

and the hydrogen below carbon-3. This, too, is not as true a representation of the structure of the molecule as a space model because, for one thing, carbons 1 through 5 do not really lie in the same plane. However, it is more representative than the horizontal projection or structural formula in Fig. 3-12*C* to its left. In order to abbreviate such formulas, neither the carbons nor the hydrogens in the ring are usually written. Thus, the ring form may be written as shown in Fig. 3-15. This is in accord with the convention suggested by Haworth.

Mutarotation Examination of the structural ring formula of glucose reveals that the formation of the ring has created an additional asymmetric carbon. Carbon-1 is now asymmetric; therefore, the ring form should exist in two stereoisomers. When D(+)-glucose is dissolved in water, it exists mainly as a mixture of the two isomeric ring structures with a very small fraction in the open-chain (aldehyde) form as shown in Fig. 3-16. One of the ring structures, α, has an optical rotation of +112°, and the other, β, an optical rotation of +19°. After a time, an equilibrium mixture is attained with an optical rotation of +53°. At equilibrium, the mixture contains traces of the aldehyde form but consists essentially of the two ring forms, 63 percent β and 37 percent α. In writing these structures, the OH on C$_1$ is written on the downward side to designate the α form and on the upward side to designate the β form. It can be seen that the structure of a simple compound like glucose is a specific yet variable group of three-dimensional entities in space. In a relatively weak solution of 100 mg/L of glucose in water, there would be more than 3×10^{20} molecules [i.e., $(6.02 \times 10^{23})/1800$] milling about, proportioned between the α, β, and open-chain space structures described above. For these molecules to be used as a source of nutrition by microorganisms in

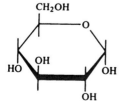

Figure 3-15 Haworth projection of D-glucose. The C and H atoms in the ring are omitted.

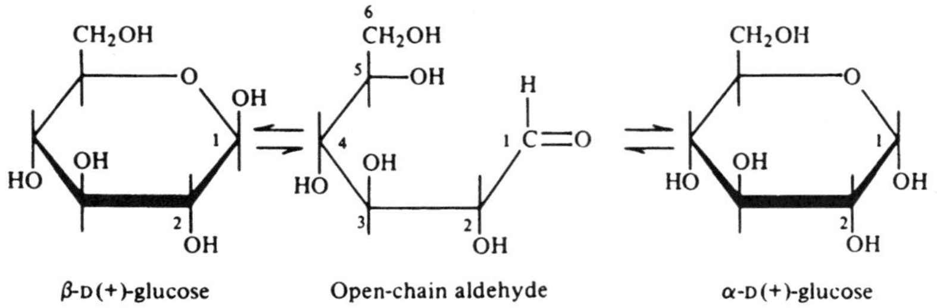

β-D(+)-glucose   Open-chain aldehyde   α-D(+)-glucose

Figure 3-16 Mutarotation of glucose.

that space, there must be contact and specific binding (reaction) with specific molecules of the microorganisms. These reactions occur only if the molecules are in the correct spatial configuration and the correct juxtaposition to permit the reactions to occur. In some cases, only the β form will react. In such cases, all of the glucose is used only because the removal of some of the β form forces the reaction to be pulled away from the equilibrium condition described above toward the β form.

Sugar Derivatives While monosaccharides can be changed to sugar alcohols and acids by reactions of the aldehyde and ketone groups, other forms of derived carbohydrates can be produced by reactions of the hydroxylated carbons. Examples of some important sugar derivatives found in nature are shown in Fig. 3-17.

Oxidation may take place at the aldehyde or keto group, as in the case of gluconic acid as shown, or at other positions. Reduction of the aldehyde group of glucose yields sorbitol. Substitution of an amino group for a hydroxyl leads to the production of an amino sugar, e.g., glucosamine. Substitution of H for the OH on the C_2 of ribose yields deoxyribose, which is one of the carbohydrates essential to all living organisms because it is a component of the genetic material.

Oligosaccharides—disaccharides By far the most important oligosaccharides in nature are the disaccharides, and these are the only ones we shall consider. The identity of a disaccharide derives from the specific monosaccharides of which it is composed and the nature of the bond between them. All the important ones considered here have the molecular formula $C_{12}H_{22}O_{11}$; i.e., they are the sum of two hexoses minus H_2O. They are structural isomers and all are composed of hexoses. When a disaccharide is hydrolyzed, it yields two molecules of hexose.

It will be recalled that the hemiacetal ring form of monosaccharides is an unstable intermediate in the formation of acetals [see Eqs. (3-26) and (3-27), and Fig. 3-16]. If one reacts a hexose hemiacetal, e.g., glucose, with an alcohol such as methanol, an acetal is formed at carbon-1, the potential aldehyde position, as shown in Fig. 3-18; such an acetal is called a *glucoside*. The C–O–C bond that joins the two parts of the acetal is called a *glycosidic linkage*. In a disaccharide, an alcoholic group of another monosaccharide is

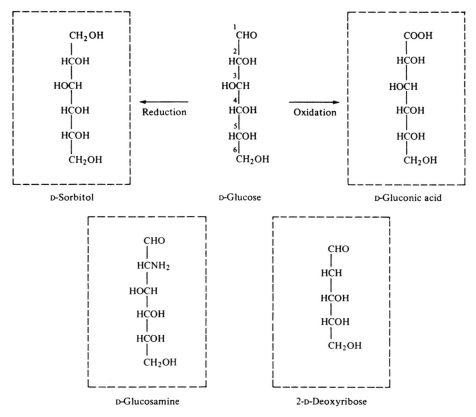

Figure 3-17 Some important sugar derivatives.

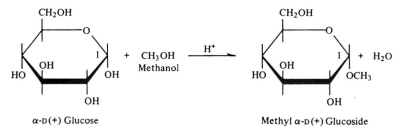

Figure 3-18 Formation of methylglycoside.

joined in glycosidic linkage rather than that of methanol. Thus, all glycosidic linkages involve the potential aldehyde or ketone carbon of one monosaccharide and a hydroxyl carbon of another monosaccharide.

Important Disaccharides. Maltose is an important disaccharide mainly because it is a breakdown product of a much longer polysaccharide, starch,

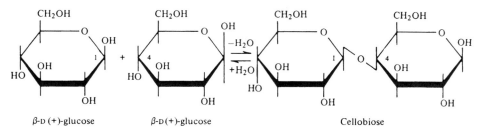

α, D(+) Glucose   α, D-(+) Glucose   Maltose

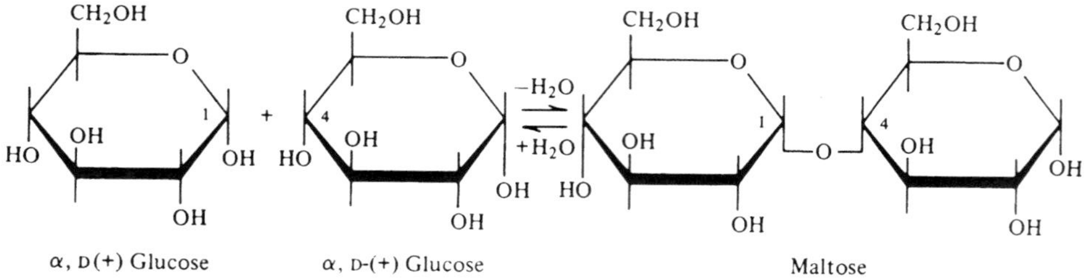

Figure 3-19 Formation of maltose, $\alpha,1 \rightarrow 4$ glucoside.

which is composed of glucose molecules. Maltose consists of two molecules of α-D(+)-glucose joined in a glycosidic bond, as shown in Fig. 3-19.

If we call carbon-1 the head of the molecule, we note that both of the α-D(+)-glucose molecules are facing to the right in Fig. 3-19; that is, they are lined up head to tail. When these are joined as shown, the C_1 of one glucose molecule is joined to the C_4 of the other, and water has been released. Note that the OH on the C_1 involved in the glycosidic bond is in the α (downward) position. The bond is named from left to right, so it is an $\alpha,1 \rightarrow 4$ glycosidic bond uniting the two molecules of glucose. This bond is characteristic of maltose. Note that the maltose formed is α-maltose, because the OH on the C_1 that is not involved in the bond is in the α position. Maltose, like glucose and other carbohydrates, is subject to mutarotation; that is, the potential aldehyde at the head of the molecule can exist in the open chain and the chain can be closed to the β or the α position.

Cellobiose is an important disaccharide mainly because it is a breakdown product of a much longer polysaccharide, cellulose, which is composed of glucose molecules. It is similar to maltose, with the exception that the glycosidic bond between the two glucose units is a $\beta,1 \rightarrow 4$ bond (see Fig. 3-20).

Lactose is an important disaccharide found naturally in mammalian milk (approximately 5 percent). It is therefore quite commonly found in dairy wastes and in household wastes as well. Its nutritive value is especially useful in the manufacture of infant foods, and it is a useful substrate in the

β-D (+)-glucose   β-D(+)-glucose   Cellobiose

Figure 3-20 Formation of cellobiose, $\beta,1 \rightarrow 4$ glucoside.

production of pharmaceutical products. It is composed of a molecule of β-D(+)-galactose and one of α-D(+)-glucose joined in a $\beta,1 \rightarrow 4$ glycosidic linkage, as shown in Fig. 3-21. Like maltose, α-lactose exists in equilibrium with its β form.

Sucrose is familiar to all as table sugar. Its occurrence in municipal wastes is therefore to be expected. Wastewaters from sugar refinery operations consist almost entirely of sucrose. The juices of sugarcane and sugar beets contain 15 to 20 percent sucrose. Wastes from fruit and vegetable canning operations also contain sucrose. It is composed of α-D-glucose and β-D-fructose united in an $\alpha,1 \rightarrow 2$ linkage (see Fig. 3-22).

Sucrose differs from the disaccharides just described in that the monosaccharides are aligned head to head. As a result, the potential aldehyde group of the glucose and the keto group of the fructose portion are tied up in the bond. Therefore, sucrose does not exhibit mutarotation and there are no α and β forms.

Since there can be no open chain for exposing the aldehyde or ketone group, sucrose does not give the typical reactions of aldehydes and ketones that are often used in analyses of sugars. The open-chain forms are easily oxidized by cupric ion under mildly alkaline conditions; i.e., they reduce the cupric to cuprous copper. Thus, they are called *reducing sugars*, and both qualitative and quantitative methods of analysis utilize this property. Sucrose is not a reducing sugar.

Polysaccharides Thus far we have been discussing fairly small molecules with molecular weights of a few hundred. Polysaccharides are extremely large

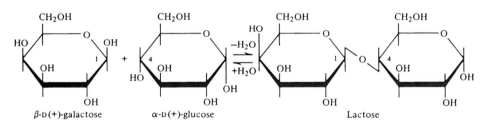

Figure 3-21 Formation of lactose, $\beta,1 \rightarrow 4$ galactoside.

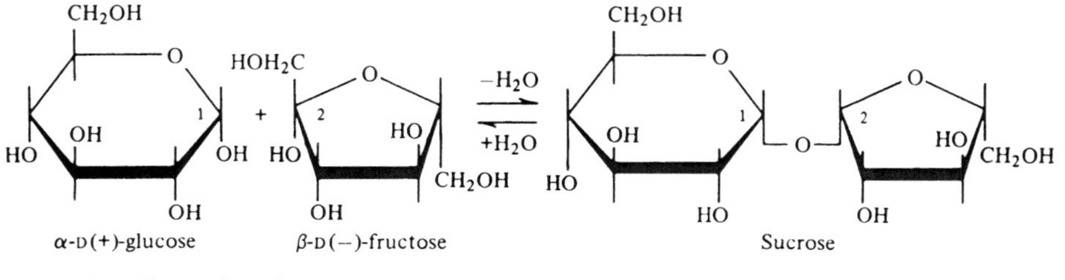

Figure 3-22 Formation of sucrose, $\alpha,1 \rightarrow 2$ glucoside.

molecules—macromolecules. Most of the carbohydrates in the biosphere exist as such macromolecules, with molecular weights ranging from 25,000 to 15 million. They consist for the most part of simple sugars and derived sugars linked together by glycosidic bonds. They are insoluble and form colloidal suspensions in water. They do not taste sweet, and they exhibit little or no reducing power. There are polysaccharides consisting of pentoses (pentosans) and hexoses (hexosans), and others contain several different monosaccharides as well as sugar derivatives. There is a wide variety of polysaccharides, but the most abundant are those consisting of glucose units. The most important of these are starches, glycogen, and cellulose.

Starch is made and stored in large amounts by plants, which use it as a reserve food store. Starches, in turn, are a major food material for plant-eating animals and human beings. Most of the glucose units are linked head to tail in $\alpha,1 \to 4$ glycosidic bonds; some are linked $\alpha,1 \to 6$, thus providing branching points in the molecule. Figure 3-23A shows the branching structure; each glucose unit is represented by a circle. The details of the $\alpha,1 \to 4$ and $\alpha,1 \to 6$ glycosidic bonding of glucose units are shown in Fig. 3-23B. Ap-

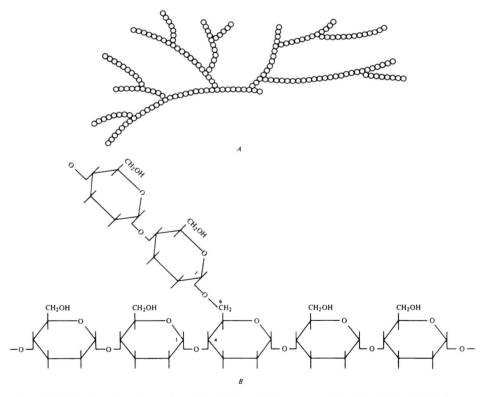

Figure 3-23 Starch molecule, amylopectin: (A) branched structure, (B) $\alpha,1 \to 4$ glucosidic linkages and $\alpha,1 \to 6$ branching points.

proximately 10 to 20 percent of naturally occurring starch exists as a linear, but coiled, chain of $\alpha, 1 \rightarrow 4$ linked glucose units. This type of starch is called *amylose*. Most of the starch in nature is of the branched type shown in Fig. 3-23 and is called *amylopectin*. Amylose molecules may vary in length from 200 to 3000 glucose units, whereas amylopectin molecules are larger, with 300 to 6000 glucose units; thus, the molecular weight of amylopectin may approach 1 million.

Glycogen is the name given to a class of polysaccharides synthesized by animals and some microorganisms as a source of stored energy and carbon. It is similar to the amylopectins but is more branched and has a higher molecular weight, in the range of 1 to 4 million.

Cellulose is the most abundant carbohydrate in the biosphere; in fact, it is the most abundant source of organic carbon. It is a fibrous carbohydrate found mainly in plant cell walls. Wood is approximately 50 percent cellulose, and cotton is essentially pure cellulose. Cellulose is not easily subject to decay, hence its value as a material of construction. As with other polysaccharides, it exists in various chain lengths and molecular weights (2000 to 3000 glucose residues). It is a nonbranched polymer of glucose units linked in $\beta, 1 \rightarrow 4$ bonding; thus, it may be considered as made of units of cellobiose.

The major source of cellulose in municipal wastewaters is paper. Paper is pulped wood, which is subsequently re-formed into sheets and forms of varying thickness and strength. Pulping may be accomplished by purely mechanical means (grinding of logs), or the wood may be cooked with various chemicals (chemical pulping) that loosen the cellulose fibers from the lignin that binds them together in the wood. This is a harsher treatment than mechanical pulping and frees the cellulose from the remainder of the wood. The very finest papers consist almost wholly of cellulose, so approximately half of the tree is solubilized and dissolved in the cooking liquor. If it were not for the fact that the water is evaporated and organics burned off in recovery systems designed to reuse or recover some of the cooking chemicals, a pulp plant effluent would contain a pound of waste organic matter for each pound of dry paper produced. Relatively low concentrations of these organics do escape into the effluent from pulp-washing operations. The organic matter consists largely of lignin, which is not readily biodegradable and therefore does not stress the natural pace of aerobic decay, but some sugars and low molecular weight acids are also washed from the pulp, and these must be subjected to secondary treatment prior to discharge of the effluent to the natural water course. Thus, it can be seen that in the manufacture of this cellulosic product, it is not so much loss of product in the effluent that causes a potential environmental pollution problem but discarding of an unwanted portion of the raw material. This is true of some other industries as well, e.g., slaughterhouse and meat-packing wastes. However, in others, the waste organic materials subject to aerobic decay may be unusable by-products of the manufacturing process, not unwanted portions of the original raw material, e.g., wastes from the fermentation industries.

These aspects are mentioned here to emphasize that the nature of the organic matter that finds its way into the decay leg of the carbon-oxygen cycle is dependent largely on the nature of the original raw material and the process by which the waste is produced. Thus, waste materials from one industry will differ greatly from those of another. Fewer differences in the composition of various municipal wastes are to be expected since the major "processes" producing these wastes are the normal living habits of people. Although there are vast differences and inequalities among us, the processes by which we produce aqueous wastes are pretty much the same. To understand why the composition of municipal waste is what it is and how it can vary, we study biochemistry and sociology, because they relate to the processes by which the waste is produced. On the other hand, to gain knowledge of the reasons for the composition and variable nature of industrial wastes, it is essential to study the manufacturing processes by which they are produced. This is an essential point of beginning for any environmental professional who hopes to solve industrial waste pollution problems. Useful information on the process industries and their wastes can be found in the appropriate references given at the end of this chapter (Nemerow, 1971, 1978; Shreve, 1967).

A few species of microorganisms can produce cellulose, but they account for a very small fraction of the cellulose in the biosphere. Also, a few species can metabolize cellulose anaerobically, and fewer yet can metabolize it aerobically. However, the fact that cellulose is not readily attacked by a wide variety of microorganisms is little consolation to the sports enthusiast who discovers that his or her tent has rotted after storage in the basement or, more important, to service organizations that discover that the cotton fabric of clothing and other supplies has rotted in storage. Such biological corrosion occurred in alarming proportions during World War II in the Pacific theater of operations and led to extensive research as to its cause and prevention. Various strains of fungi produce large amounts of enzyme(s) called *cellulases* that hydrolyze cellulose to glucose, which is then metabolized by the fungi and many other microorganisms as well.

Polysaccharides other than those described above are widely distributed in plants and microorganisms. The pentosan xylan consists of units of xylose. It occurs in wood along with cellulose. Araban is a pentosan consisting of arabinose units. Levan (levulan) is a polymer of fructose units and is produced by some bacteria. Dextran is a branched bacterial polysaccharide consisting of glucose units like amylopectin or glycogen but linked $\alpha, 1 \rightarrow 6$ rather than $1 \rightarrow 4$ with occasional branching points linked $1 \rightarrow 4$ and $1 \rightarrow 3$. Other polysaccharides, called *mucopolysaccharides*, contain amino sugars.

Analytical determination of carbohydrates Various analyses are available for the quantitative estimation of various types of carbohydrates, e.g., hexose, pentose, aldose, and ketose. Also, there are enzymic analyses for specific compounds, e.g., glucose, galactose. All such analyses can be useful for specific investigations, but the environmental scientist is usually interested in assessing the total amount of carbohydrate. For example, one would like a test comparable to the extraction procedure for measuring the total amount of lipids, i.e., fats, oil, and grease, previously described in *Standard Methods* (APHA, 1976). *Standard Methods* does not yet include a way to measure amounts of carbohydrate; however, such a test has been successfully adapted by investigators in the environmental area. The method involves the use of the compound anthrone. Under strongly acid conditions, all carbohydrates in a sample can be hydrolyzed to monosaccharides. Pentoses are then converted to furfural and hexoses to 5-hydroxymethylfurfural. These products form a blue complex in the presence of anthrone. The intensity of the color is proportional to the concentration of carbohydrate present in the sample (Neish 1952; Ramanathan et al., 1969).

Proteins Proteins constitute the most complex types of organic matter found in the biosphere, and they are an essential part of all living matter. The protein content of microorganisms can vary from species to species and with environmental conditions, but in general the protein content of procaryotic cells ranges between 40 and 60 percent of the dry weight.

Some proteins function as structural components, e.g., in cell walls and membranes in microorganisms, in the internal and external membranes of animal cells, and as major components of skin and hair. Other functions are motility, e.g., flagella and cilia in microorganisms, and contraction as in muscles. Antibodies and hormones are proteinaceous; protein is the major constituent of animal and plant viruses; blood contains a large proportion of protein. Also, the biological catalysts that control the rates of biochemical reactions in all living cells, the enzymes, are proteins. Since protein is such a major component of animals and is the most prized dietary constituent, it is also a major component of municipal wastewaters and certain industrial wastes, e.g., slaughterhouse and meat-packing wastes and dairy wastes.

From the above comments it can be appreciated that proteins represent a wide array of molecules of different types, and classification is rather difficult. In general, proteins can be divided into two broad categories: simple proteins and conjugated proteins.

Simple proteins All simple proteins are polymers of α-amino acids joined together by peptide bonds (see Table 3-5). The length of the polymers varies, so that molecular weights of proteins can vary from a few thousand to several million. The chemical characteristics common to all simple proteins are peptide bonds and nitrogen content. All proteins contain approximately 16 percent nitrogen. Simple proteins have traditionally been subclassified according to their solubility in water and various solvents; thus, the *albumins* (found in egg whites and blood), which are soluble in water and in dilute salt solution, are distinguished from *globulins*, which are insoluble in water but soluble in dilute salt solutions. Categorizing proteins on the basis of solubility in water and various solvents dates back to early studies on proteinaceous materials, and these older classifications are used less and less in protein nomenclature.

Conjugated proteins Conjugated proteins consist of simple proteins chemically bound to various nonprotein constitutents. These are subclassified in accordance with the nature of the nonprotein component.

Lipoproteins are proteins bound to lipids. The lipids are generally cholesterol esters and phospholipids.

Glycoproteins consist of carbohydrates or derived carbohydrates bound to protein.

Phosphoproteins contain phosphoric acid. Casein, a constituent of milk, is an example of this type of conjugated protein.

Chromoproteins contain a pigmented compound; hemoglobin, found in blood, consists of a simple protein portion (globin) and heme, an iron-containing pigment.

Structure and properties of proteins Even though proteins exhibit tremendous variety, they are all built up of the relatively few amino acids shown in Table 3-8. A general formula for an amino acid is shown below:

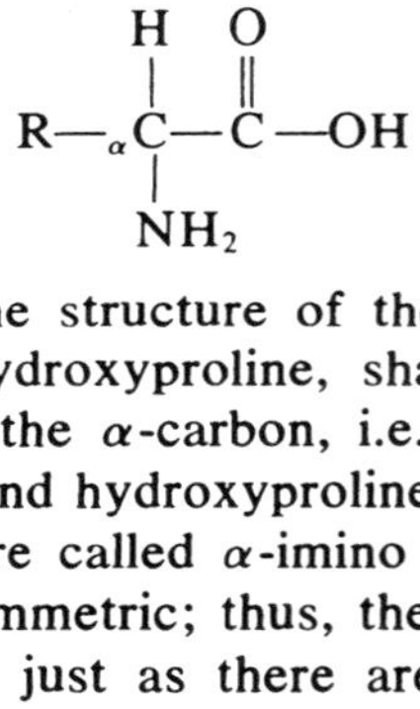

$$R\overset{H}{\underset{NH_2}{-_\alpha C}}\overset{O}{-}\overset{\parallel}{C}-OH$$

The amino acids differ in the structure of the side chain R but, with the exception of proline and hydroxyproline, share a common feature: they contain an amino group on the α-carbon, i.e., the carbon adjacent to the carboxyl carbon. In proline and hydroxyproline, the α-nitrogen is in the ring (see Table 3-8), and these are called α-imino acids. With the exception of glycine, the α-carbon is asymmetric; thus, there are optical isomers, enantiomorphs, for amino acids, just as there are for other compounds, e.g., carbohydrates. Consider the two forms shown below and compare them with D- and L-glyceraldehyde.

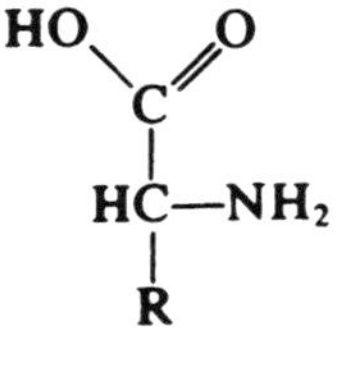

D-amino acid 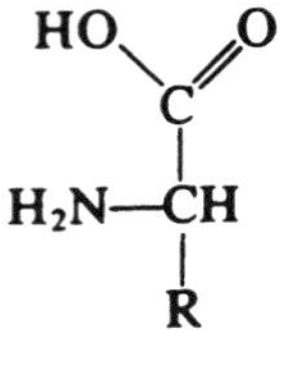   L-amino acid

Unlike the sugars, the naturally occurring amino acids in proteins belong to the L series. However, many D-amino acids occur in nature—in microbial cell walls, in some antibiotic compounds produced by microorganisms, and in the capsular materials of some cells.

Properties and Reactions of Amino Acids Amino acids are soluble in water and insoluble in nonpolar organic solvents. We shall consider two reactions and the ionizing characteristics of aqueous solutions of amino acids.

The most important reactions are those between the carboxyl group of one amino acid and the α-amino group of another to form the peptide bonds characteristic of proteins:

$$
\begin{array}{c}
\underset{H}{HN}-CH-\overset{O}{\overset{\|}{C}}-OH + H-N-CH-\overset{O}{\overset{\|}{C}}-OH
\end{array}
$$

$$
H_2O \quad \quad H_2O \tag{3-28}
$$

$$
\underset{H}{HN}-CH-\overset{O}{\overset{\|}{C}}-N-CH-\overset{O}{\overset{\|}{C}}-OH
$$

peptide bond

Protein consists of many amino acids bound together by peptide bonds. The peptide formed in reaction (3-28) could react with other amino acids, building up a polypeptide; proteins are such polypeptides. The reverse of the reaction is recognized as hydrolysis of the peptide. Amino acids can also undergo reactions typical of organic acids. Thus, they can form esters with alcohols. The amino acids are not very volatile, but their esters are to varying degrees, offering one means of separating them for analysis.

The ionizing characteristics of amino acids also offer a means of separating them. The following discussion of this characteristic will help to explain the behavior of proteins and will aid in the understanding of similar behavior by nonorganic compounds of environmental interest.

As with other organic acids, the carboxyl group of amino acids ionizes in aqueous solution:

$$
R-\overset{O}{\overset{\|}{C}}-OH \rightleftharpoons R-\overset{O}{\overset{\|}{C}}-O^- + H^+ \tag{3-29}
$$

The basic amino group also ionizes in solution:

$$
R-NH_2 + H^+ \rightleftharpoons R-NH_3^+ \tag{3-30}
$$

Table 3-8 The amino acids

Classification	Name	Structure	Abbreviation	Remarks
I. Aliphatic Hydrocarbon in side chain	Glycine	$\boxed{H}-CH-COOH$ with NH_2	Gly	Only amino acid without an asymmetric carbon
	Alanine	$\boxed{CH_3}-CH-COOH$ with NH_2	Ala	
	Valine	$\boxed{\begin{array}{c} CH_3 \\ CH \\ CH_3 \end{array}}-CH-COOH$ with NH_2	Val	Dietary essential in humans
	Leucine	$\boxed{\begin{array}{c} CH_3 \\ CH-CH_2 \\ CH_3 \end{array}}-CH-COOH$ with NH_2	Leu	Dietary essential
	Isoleucine	$\boxed{CH_3-CH_2-CH \atop CH_3}-CH-COOH$ with NH_2	Ileu	Dietary essential
Hydroxy group in side chain	Serine	$\boxed{HO-CH_2}-CH-COOH$ with NH_2	Ser	OH group available for ester formation

Category	Name	Structure	Abbr.	Notes
	Theonine	$CH_3-CH-CH-COOH$ with OH and NH_2	Thr	Dietary essential
Carboxyl group in side chain	Aspartic acid	$HOOC-CH_2-CH-COOH$ with NH_2	Asp	Acid reaction
	Glutamic acid	$HOOC-CH_2-CH_2-CH-COOH$ with NH_2	Glu	Acid reaction
Amide group in side chain	Asparagine	$H_2N-C(=O)-CH_2-CH-COOH$ with NH_2	AspNH$_2$	
	Glutamine	$H_2N-C(=O)-CH_2-CH_2-CH-COOH$ with NH_2	GluNH$_2$	
Basic group in side chain	Lysine	$H_2N-(CH_2)_4-CH-COOH$ with NH_2	Lys	Dietary essential, basic reaction
	Arginine	$H_2N-C(=NH)-NH-(CH_2)_3-CH-COOH$ with NH_2	Arg	Dietary essential, basic reaction, contains a guanido group

Table 3-8 (*Continued*)

Classification	Name	Structure	Abbreviation	Remarks
Sulfur atom in side chain	Cysteine	$HS-CH_2-CH-COOH$; NH_2	CySH	SH group easily oxidized
	Cystine	$S-CH_2-CH-COOH$; NH_2 $S-CH_2-CH-COOH$; NH_2	CyS–SCy	Disulfide linkage important in determining protein structure
	Methionine	$CH_3-S-(CH_2)_2-CH-COOH$; NH_2	Met	Dietary essential
II. Aromatic Benzene ring in side chain	Phenylalanine	$C_6H_5-CH_2-CH-COOH$; NH_2	Phe	Dietary essential, phenolic group
	Tyrosine	$HO-C_6H_4-CH_2-CH-COOH$; NH_2	Tyr	Phenolic group

III. Heterocyclic
Heterocyclic ring in side
chain

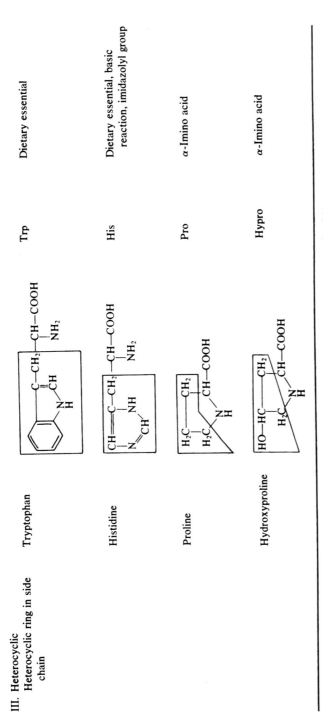

Tryptophan	Trp	Dietary essential	
Histidine	His	Dietary essential, basic reaction, imidazolyl group	
Proline	Pro	α-Imino acid	
Hydroxyproline	Hypro	α-Imino acid	

Source: Adapted from "Organic and Biological Chemistry", I. E. Liener, Ronald Press, New York, 1966.

At certain pH values, both the acidic and basic groups are ionized and the molecule exists as a dipolar ion (zwitterion form, or hybrid ion):

$$R-\underset{\underset{NH_3^+}{|}}{\overset{\overset{H}{|}}{C}}-\overset{\overset{O}{\|}}{C}-O^-$$

If such an ion were placed in an electric field, i.e., were subjected to electrophoresis, there would be a pH value at which it would migrate to neither the cathode (−) nor the anode (+). Its net charge would be 0, and this pH value is called the *isoelectric point*. Its negative and positive charges are equal, and at this point of electric neutrality, there is no tendency to move toward either pole. The net charge of the amino acid shifts with the pH. Under acid conditions, it exists in cationic (+) form, and at alkaline pH values in anionic (−) form:

$$R-\underset{\underset{NH_3^+}{|}}{\overset{\overset{H}{|}}{C}}-\overset{\overset{O}{\|}}{C}-OH \underset{H^+}{\overset{OH^-}{\rightleftharpoons}} R-\underset{\underset{NH_3^+}{|}}{\overset{\overset{H}{|}}{C}}-\overset{\overset{O}{\|}}{C}-O^- \underset{H^+}{\overset{OH^-}{\rightleftharpoons}} R-\underset{\underset{NH_2}{|}}{\overset{\overset{H}{|}}{C}}-\overset{\overset{O}{\|}}{C}-O^- \quad (3\text{-}31)$$

low pH   isoelectric   high pH
              point
              (zwitterion)

Each amino acid has its unique isoelectric point. For basic amino acids, e.g., lysine and arginine, this pH value is rather high—in the range of 8 to 11— and for the acidic ones, e.g., aspartic and glutamic, it is rather low—pH 2 to 3. For those with only one carboxyl and one amino group, the isoelectric point is near neutral pH.

Examination of Eq. (3-31) shows an important fact about amino acids; they are compounds that buffer against changes in pH. They act as both donors and acceptors of hydrogen ions, as both acids and salts. Each ionizing group has its characteristic pK value. Glycine, with one amino and one carboxyl group, exhibits two pK values, whereas aspartic acid exhibits three, one for each carboxyl group and one for its amino group.

It is well known that in the biological purification of some industrial wastewaters there is need to control the pH (see the discussion of the effect of pH on microbial growth in Chap. 5). However, in the biological treatment of municipal wastes there is seldom need to control the pH to remain within the range 6 to 8, because municipal sewage contains inorganic and organic buffering systems. The protein content of sewage is usually such that one can expect significant buffering action from its presence. The peptide bonds of proteins tie up the α-amino and carboxyl groups, but there are both amino groups and acid groups as well as other ionizable groups in the side chains of many acids. Ionizable groups other than –NH₂ and –COOH are shown in Fig.

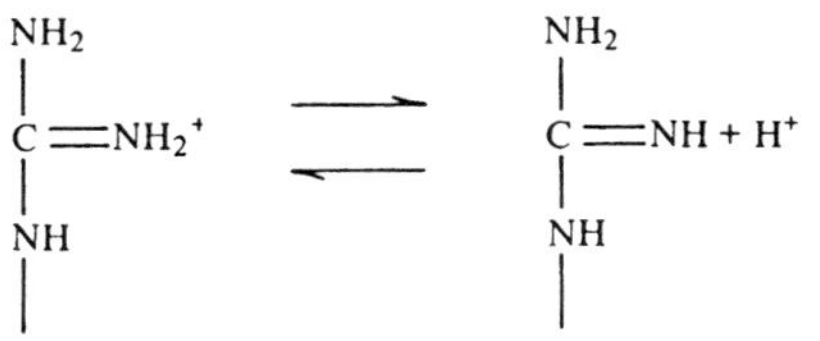

$$NH_2 \qquad NH_2$$
$$| \qquad\qquad |$$
$$C=NH_2^+ \longrightarrow C=NH + H^+$$
$$| \qquad\qquad |$$
$$NH \qquad NH$$
$$| \qquad\qquad |$$

Guanido group, arginine, pK = 12.5

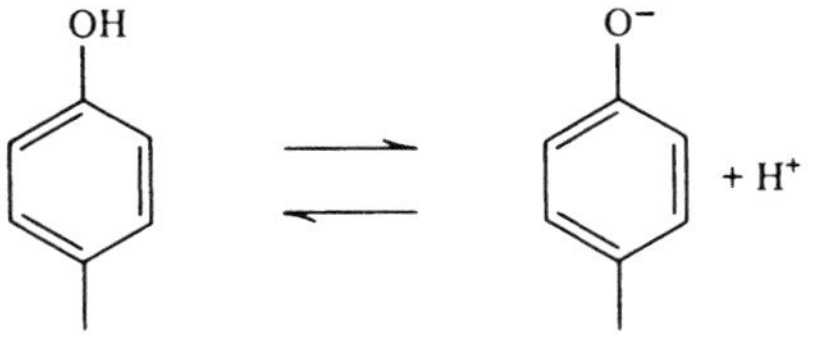

Phenolic group, tyrosine, pK = 10

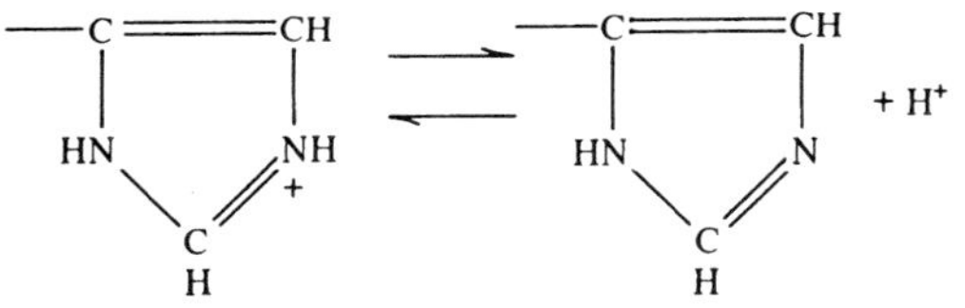

Imidazolyl group, histidine, pK = 6.0

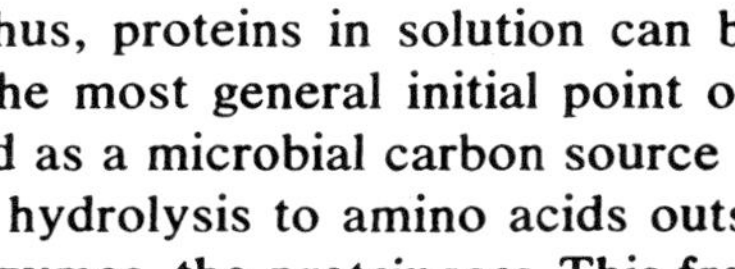

$$SH \qquad S^-$$
$$| \qquad |$$
$$CH_2 \longrightarrow CH_2 + H^+$$
$$| \qquad |$$

Sulfide group, cysteine, pK = 8.3

Figure 3-24 Ionizable groups in amino acid side chains other than amino and carboxyl groups.

3-24. Thus, proteins in solution can buffer against changes in pH. Furthermore, the most general initial point of attack when proteinaceous materials are used as a microbial carbon source in the process of biological purification is their hydrolysis to amino acids outside of the microbial cells by extracellular enzymes, the *proteinases*. This frees the –NH$_2$ and –COOH groups of the peptide bond. So even as the proteins are consumed, they can contribute to the ability of the system to resist changes in pH. When the majority of ionizable groups in the protein or its constituent amino acids in wastewaters exhibit pK values near neutrality, the system may exhibit sufficient buffering action to maintain a neutral pH without the addition of acid or alkali for pH control. It should also be apparent that sludges from secondary treatment processes, which consist essentially of microbial cells at approximately 50 percent protein content, can exhibit considerable buffering capacity when subjected to aerobic and anaerobic biological treatment, i.e., sludge digestion.

Proteins, amino acids, and other dipolar ions are known as *amphoteric* substances. When a solution containing a specific protein is brought to the isoelectric pH value, the net charge on the molecule is 0, and most proteins are least soluble and tend to precipitate at the isoelectric pH value. Students of environmental science will recognize the similarity of behavior between proteins and inorganic amphoteric substances. For example, $Al(OH)_3$, the precipitate formed during chemical coagulation of turbidity in water by addition of hydrated aluminum sulfate [filter alum, $Al_2(SO_4)_3 \cdot 14H_2O$], exhibits a rather restricted pH range, 5 to 7, for good clarification because of the fact that it is an amphoteric hydroxide. It ionizes, going back into solution at pH values below and above this range. Note the similarity between reactions (3-32) and (3-31):

$$Al^{3+} + 3OH^- \underset{H^+}{\overset{OH^-}{\rightleftharpoons}} \underset{(ppt)}{Al(OH)_3} \underset{H^+}{\overset{OH^-}{\rightleftharpoons}} H^+ + AlO_2^- + H_2O \qquad (3\text{-}32)$$

$$\text{low pH} \qquad\qquad \text{pH 5–7} \qquad\qquad \text{high pH}$$

Primary Structure of Proteins In describing proteins, we have said that they are polymers of amino acids joined by peptide bonds. The properties and chemical functioning of each of the many different proteins found in the biosphere are dependent on the spatial structure of the molecule. Of primary consideration in this regard is the sequence of specific amino acids in the polymer. The great diversity that is possible can be appreciated by assuming that if upon hydrolysis the particular protein in question were found to contain one each of 20 different amino acids, the possible number of sequencing combinations would be 20! or 2×10^{18}. When one considers that the proteins found in nature consist of many more than 20 amino acid residues, and that each of the amino acids may be repeated many times in the same molecule, the great variety of possible primary structures is evident.

Regardless of the variety of amino acid sequences, all protein molecules have a common feature; one end has an amino acid in which the α-amino group is free of peptide bonding (the N terminal), and the other has a free carboxyl end (the C terminal). Figure 3-25 shows N- and C-terminal amino acids of a polypeptide chain. Only five amino acids are shown, but one could insert 50 or 500 and the backbone of a protein would be the same, that is, a strand of peptide linkages from which the various R groups of the amino acids protrude into space. This backbone is purposely not drawn as a straight line, because the molecule can fold and twist, bringing into play forces leading to secondary, tertiary, and quaternary spatial structures.

Examples of primary structures of specific proteins are shown in Fig. 3-26. The upper portion shows the amino acid sequence for the enzyme lysozyme. This protein is the catalyst that breaks down bacterial cell walls in certain species; thus, it can be expected to be found in extended aeration activated sludge plants and in aerobic and anaerobic digesters. It consists of a single-stranded polypeptide of 129 amino acid residues. Note the N and C terminals and that the amino acid residues are numbered beginning with the N terminal. Some proteins consist of two polypeptide strands, as seen in the

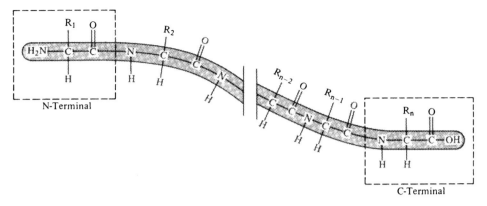

Figure 3-25 Primary structure of proteins. A backbone of peptide linkages forms the polypeptide strand.

lower portion of Fig. 3-26. Represented here is the structure of the bovine hormone insulin, which regulates the metabolism of carbohydrates in cows. In human insulin there is only a slight change in primary structure; amino acids 8, 9, and 10 of the A chain are thr–ser–ileu. The two peptide strands are bonded together by disulfide bonds formed between cysteine molecules. Also in lysozyme, the single strand is folded in a specific way due to disulfide bonds. These are features of tertiary structures to be discussed later. The primary structure of proteins refers only to the sequence of amino acids in the peptide-linked chain.

Secondary Structure of Proteins As indicated above, there is more to protein structure than the sequence of amino acids. Each polypeptide backbone has a specific structural conformation in space; that is, the chain is not free to bend randomly but has a specific dimensional shape. The peptide-bonded backbone may be coiled in springlike fashion or folded and bent in various ways. Whatever the shape of the backbone, the forces that give it its shape are the relatively weak hydrogen bonds referred to earlier in this chapter. The backbone can assume various specific shapes. Some polypeptide chains are held in a specific conformation in space by hydrogen bonding with another polypeptide chain. Such interchain hydrogen bonding is shown in Fig. 3-27. This figure is an isometric representation showing only two polypeptides of a pleated sheet. The peptide bonds are in the plane of the sheet, and the R groups protrude upward and downward from the folded planes. The sheet may consist of many strands of peptide-bonded amino acids. Both N terminals may be at the same end (parallel pleated sheet), or the N and C terminals may alternate.

Many protein backbones are shaped in the form of an α helix, like a screw or a spiral staircase or a coiled spring. This shape is maintained by intrachain hydrogen bonds. Figure 3-28 shows a right-handed α helix; imagine the helical peptide backbone twisting around an imaginary pole through the

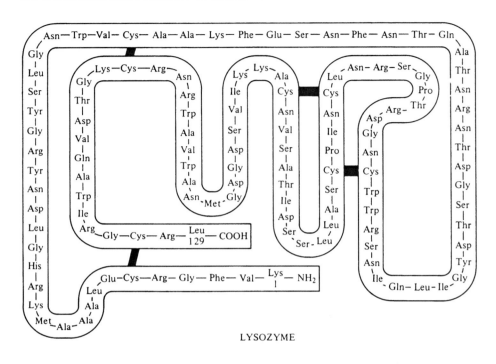

LYSOZYME

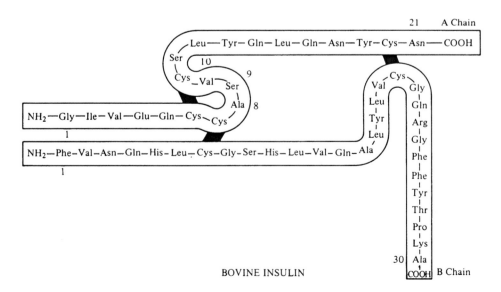

BOVINE INSULIN

Figure 3-26 Lysozyme and bovine insulin.

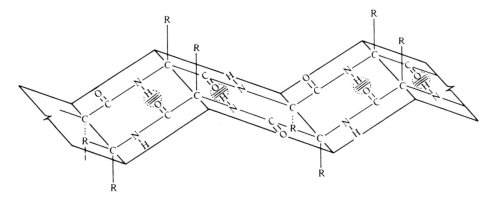

Figure 3-27 Folded sheet structure of proteins. Polypeptide strands are linked by hydrogen bonds.

hollow core of the helix. The hydrogen bonds (broken lines) are formed between the amino group of one acid and the carbonyl group of the third amino acid above it.

Tertiary Structure of Proteins Thus far we have discussed structures contributed by the sequence and number of amino acids (primary) and that contributed by hydrogen bonding (secondary). The side chains protruding from the peptide helix or pleated sheet also contribute to the special space structure of each protein strand. Were it not for reactions of the side chains, the space structure of the individual proteins could not be as specific as it has been found to be, and the unique functions of certain proteins, e.g., catalysis of specific biochemical reactions, could not occur. In Fig. 3-26 it was seen that the single strand of lysozyme is held in a unique folded configuration by disulfide bonds that result from the reaction of the –SH groups of the cysteine side chain, as shown in Fig. 3-29*A*. Strands can also be given specific conformations by hydrogen bonding involving the side chains of the amino acids. Such a bond between aspartic and glutamic acids is shown in Fig. 3-29*B*. Various other combinations are possible, e.g., tyrosine and histidine, serine and aspartic acid.

Salt linkages can occur in which the amino group of one of the basic amino acids reacts with a carbonyl in a side chain of one of the acidic amino acids. A salt linkage between lysine and aspartic acid is shown in Fig. 3-29*D*. The side chains of some amino acids are hydrocarbon in nature and tend to be repelled by water (hydrophobic). Isoleucine, phenylalanine, and valine are examples. These have a tendency to become associated, and the attraction is called a *hydrophobic bond*. Such a bond between two leucine side chains is shown in Fig. 3-29*C*.

Quaternary Structure of Proteins Some proteins consist of aggregates of subunits, i.e., multiple polypeptides held together in specific conformations by noncovalent bonds or surface interactions. Each member of the aggregate

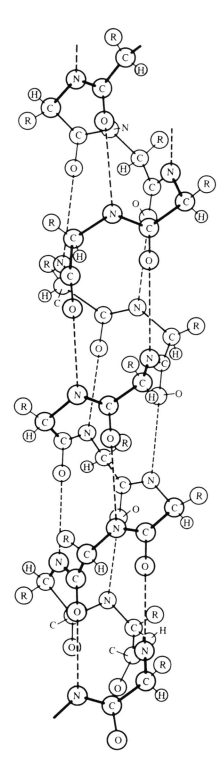

Figure 3-28 Alpha helix of a protein molecule is a coiled chain of amino acid units. The backbones of the units form a repeating sequence of atoms of carbon (C), oxygen (O), hydrogen (H), and nitrogen (N). The R stands for the side chain that distinguishes one amino acid from another. The configuration of the helix is maintained by hydrogen bonds (broken lines). The hydrogen atom that participates in each of these bonds is not shown. *(From Kendrew, 1961.)*

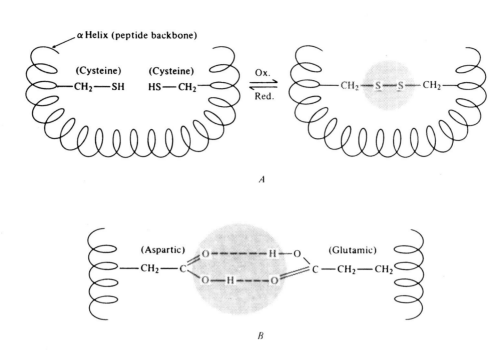

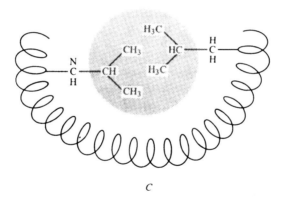

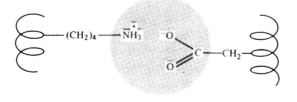

Figure 3-29 Types of bonding forces that contribute tertiary structure to proteins. (*A*) Formation of a disulfide bond; (*B*) hydrogen bonding of R groups; (*C*) hydrophobic bond; (*D*) salt linkage.

possesses its unique primary (sequence of amino acids), secondary (hydrogen-bonded α helix or folded sheet), and tertiary (fixed three-dimensional shape) structure, but in addition, each fits together in a unique spatial arrangement, thus providing a fourth level of structure that is important to its function. For example, hemoglobin consists of four peptide chains, each of which contains a heme prosthetic group. These chains have complementary tertiary structures and fit closely together. The enzyme lactic acid dehydrogenase, which catalyzes the oxidation of pyruvic acid to lactic acid, also consists of four subunits.

Denaturation of Proteins The denaturation of proteins is best understood by first remembering that proteins possess their specific spatial shapes (quaternary, tertiary, secondary) because these particular configurations are needed for the molecule to perform its unique function in the living world. Thus, when denaturation is defined as the altering of the three-dimensional shape by breaking down the quaternary, tertiary, or secondary structure of the protein, it is also fair to say that it represents changing the molecule from its natural or native state, hence *denaturation*. Denaturation, then, represents a breaking of those forces and bonds that provide the secondary and tertiary structures, i.e., hydrogen bonds, disulfide bonds, salt linkages, and others. Agents that do this are important to the biochemist because they provide ways to study the nature of the protein. They are important to the environmental scientist because such agents can kill living systems. Some common ways in which protein can be denatured are discussed below.

The ionization of amino acids, as previously seen, is dependent on pH; thus it is expected that *changes in pH* would disrupt salt linkages. For example, consider the salt linkage shown in Fig. 3-29. Under highly acid conditions, the carboxyl group of aspartic acid would accept a hydrogen ion and the attractive force would no longer exist. Similarly, under highly alkaline conditions, lysine would lose a hydrogen ion and the attractive force would no longer exist.

Heat and other forms of energy such as ultraviolet radiation are sources of kinetic energy that bring about violent agitation of atoms and rupture weak linkages such as salt linkages and hydrogen bonds. The usual result is coagulation and precipitation of the protein, e.g., the solidifying of egg white on heating. Since the breaking of hydrogen bonds opens up or uncoils the polypeptide strands of the protein molecules, they are usually more easily attacked and used by organisms as foodstuff in the denatured state than in the native state. This is also a nutritive basis for cooking meat for human consumption.

Heavy metals, such as Ag^+, Pb^{2+}, and Hg^{2+}, become tightly bound to acid groups in the side chains of the amino acids in the polypeptide chain such as glutamic and aspartic acids. Salt linkages are thus broken and the proteins are precipitated. Addition of *organic solvents*, which are more capable of forming hydrogen bonds with protein molecules (intermolecular bonds) than are the proteins themselves (intramolecular bonds), cause disruption of the intramolecular hydrogen bonds. Ethyl acohol is particularly effective in denaturing proteins and is used as a skin disinfectant. *Detergents* can also be bound to proteins, occupying sites needed for the intramolecular bonds that maintain spatial structures and functions. They can cause centers of hydrophobic bonding not characteristic of native protein.

In some cases, denaturation may be reversed by the removal of conditions that caused it (reversible denaturation); in others, the severity of denaturing conditions are such that there is a permanent derangement of the secondary, tertiary, or quaternary structure (irreversible denaturation). In any case, the primary structure of the protein remains intact; hydrolysis is required to destroy primary structure.

Measurement of Protein Estimation of the amount of protein in a sample can be made by measuring the nitrogen content of the organic matter in the sample, assuming it is all due to protein, and assuming that protein contains 16 percent nitrogen (i.e., organic nitrogen $\times 6.25 =$ protein content). Organic nitrogen is determined by using the Kjeldahl procedure outlined in *Standard Methods* (APHA, 1976). An even more reliably unique feature of protein is the peptide bond. Quantitative assessment of the peptide bond content of a sample can be made using the chelating properties of copper in alkaline solution. A complex is formed between an atom of copper and four nitrogen atoms of peptide bonds. The intensity of the purple complex is proportional to the number of peptide bonds (Neish, 1952; Ramanathan et al., 1969). The test is named the *biuret method* because the characteristic purple color is given by the compound biuret,

$$NH_2-\overset{\overset{\displaystyle O}{\|}}{C}-NH-\overset{\overset{\displaystyle O}{\|}}{C}-NH_2$$

Enzymes Enzymes are protein molecules so vital to the life processes that they deserve separate discussion. The study of enzymes (enzymology) comprises an entire discipline in biochemistry. They are important to the environmental scientist because they are affected, as already seen, by environmental changes and because they control the rates of biochemical reactions, e.g., aerobic or anaerobic decay in nature and in treatment plants. The brief coverage here will characterize the mechanism and kinetics of action of enzymes, thus helping the understanding of growth and growth rates of microorganisms. However, the reader is advised to consult some of the texts dealing with enzymes that are listed at the end of this chapter. The major intent here is to gain familiarity with the general nature of enzymes, the types of enzymes, mechanism of action, factors affecting kinetics, and inhibitors of enzyme activity.

General nature Enzymes possess some of the same general properties as the inorganic catalysts (cobalt, platinum, etc.); i.e., they accelerate the rate of reaction and are entirely recoverable and renewable after the reaction is completed. Since they are reusable, they are effective in very small concentrations. However, unlike inorganic catalysts, these organic catalysts are often many times larger than the molecules, or substrates, on which they act. They are all either simple or conjugated proteins, and each exhibits a high degree of specificity in regard to the substrate on which it acts. Thus, the enzymes provide a means by which living systems can exert precise control of biochemical reactions.

Classification Enzymes may be classified into six general groups according to the types of reactions they catalyze.

Group classification	Type of reaction
1. Oxidoreductases	Oxidation-reduction reactions
2. Transferases	Transfer of a chemical group from one molecule to another
3. Hydrolases	Hydrolysis of various molecules, e.g., lipids, carbohydrates, proteins
4. Lyases	Nonhydrolytic addition or removal of substituent groups
5. Isomerases	Internal rearrangement of substituent groups, i.e., making of isomers
6. Ligases or synthetases	Joining of two molecules coupled with the cleavage of a pyrophosphate bond

Note that the names of all six groups end with -*ase*. Although there is no reference in the above classification to substrate, many enzymes are named by adding the suffix -*ase* to the name of the substrate on which the enzyme acts. Within each group there are both correct systematic names and trivial names for subgroups. For example, within group 1 there are, among others, dehydrogenases, which use a molecule other than oxygen as the hydrogen acceptor, and oxidases, which use oxygen as the electron acceptor. Complete familiarization with enzyme nomenclature is beyond the scope of this chapter; while it is an important subject, an understanding of the mechanism of action is more important.

Enzyme action Whenever a chemical reaction occurs, an energy change accompanies it, as was explained earlier [see Eqs. (3-1) and (3-2)]. For a reaction to proceed to the right, the amount of energy contained in the products must be less than that in the reactants; i.e., thermodynamically, all reactions must follow a downhill energy flow with negative ΔH or ΔG—they are exergonic. For an uphill flow of energy (endergonic reaction) to occur, energy must be put into the system. In biological systems many such reactions take place in synthesizing new biomass components. To make them proceed, biological systems couple these energy-consuming reactions to various exergonic reactions, as will be discussed in later chapters.

Even exergonic reactions do not proceed without overcoming an energy barrier or without an initial excitation of the reactants. For example, reaction (3-2) is exergonic, but merely bringing organic matter together with oxygen does not cause combustion. In order to start such a reaction, it is necessary to bring the organic matter to "kindling" temperature. This is done by adding energy in the form of heat. The energy that has to be added before the reaction can proceed spontaneously is called the *threshold energy* or the *activation energy E** for the reaction. It represents the excess energy that molecules must have upon collision in order to react with each other. Activation energy is therefore an energy barrier or an activated energy state through which reactants must pass along the path to the final energy state of the products of the reaction. Figure 3-30 is a representation of the path through the activation energy level that is taken as the initial reactant(s) proceed to the lower energy level of the products. Activation energy can be measured by performing a reaction at various temperatures and then plotting the log of the reaction velocity versus the reciprocal of the absolute temperature in accord with the Arrhenius equation:

$$\log K = \log A - \frac{E^*}{2.3RT} \tag{3-33}$$

In Eq. (3-33), K is the specific rate constant for the reaction, A is a constant in units of moles per liter, E^* is the activation energy, R is the gas constant (1.987 cal per degree per mole), and T is the temperature in degrees Kelvin. When one compares experimentally determined activation energies

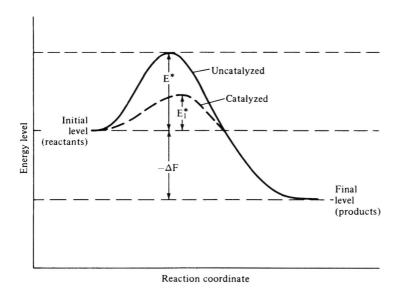

Figure 3-30 Reaction profile showing a plot of the energy of the reagents as a reaction proceeds. The activation energy E^* is higher for the uncatalyzed reaction than for the enzyme-catalyzed reaction E_1^*.

for a reaction with and without catalysis by an enzyme, it is invariably found that the activation energy for the enzyme-catalyzed reaction is lower. For example, E^* for the acid hydrolysis of sucrose is 20.6 kcal/mol, whereas the same reaction catalyzed by yeast invertase has an activation energy of 11.5 kcal/mol. For many other enzymes, the difference is even greater.

The exact mechanism by which enzymes lower activation energy and bring about the degree of excitation needed to disrupt substrate bonds is not fully known, but a number of facts do seem evident. Enzymes possess highly specific space structures and at specific sites on or within the space web the substrate molecule becomes bound to the enzyme. The chemical and spatial nature of enzyme and substrate must match in order for this enzyme-substrate complex, ES, to be formed. In addition, there is in the vicinity of this site of action (termed the active site of the enzyme) a location of high electrical charge such as a metallic ion or charged chemical grouping, a place where ample excitation is generated between the enzyme and the substrate atoms to bring about an activated state and, consequently, to cause a reaction to occur.

The conditions for catalysis can be illustrated using the enzyme lysozyme and its action on bacterial cell walls. A great deal is known about this enzyme and its mode of action which serves to demonstrate that these molecules are indeed intricate space structures. Figure 3-31 is a three-dimensional rendering of the lysozyme molecule which was diagramed in Fig. 3-26. Figure 3-31 is a half-tone reduction of a large, full-color illustration which shows even more clearly the three-dimensional structure of the molecule (see Phillips). The figure is shown here because it provides the correct representation of a complicated molecule as a space structure and emphasizes the "structural engineering" aspect of a macromolecular organic compound. By convention, in this method of representing molecular models, the atoms lie at the intersections and elbows of the structure. The electron shells of the atoms are not shown since, if they were shown as spheres or electron clouds surrounding the nuclei, they would meld together, giving the molecule its true solid, elliptical or egg-shaped appearance. It is recalled that the electron shell consists mostly of space, not of solid matter; therefore, the space frame representation of Fig. 3-31 gives a rather accurate representation of the molecule. The straight line connectors represent bonds between the atoms. The longer bonds (for example, see the three parallel and nearly vertical lines in the lower left portion of the molecule) are hydrogen bonds between portions of the peptide chain, which are important in maintaining the tertiary structure, or spatial configuration, of the molecule.

Looking down into the molecule, the portion of the peptide backbone closest to the reader's eye, i.e., the topmost part of the molecule, is the darkest portion of the chain on the left side of the dashed line. Proceeding to the right, the lines in the drawing become lighter because they are at a lower elevation. To the right of the dashed line, portions of the peptide chain become darker again. Thus, there is a valley, or cleft, in the molecule. The

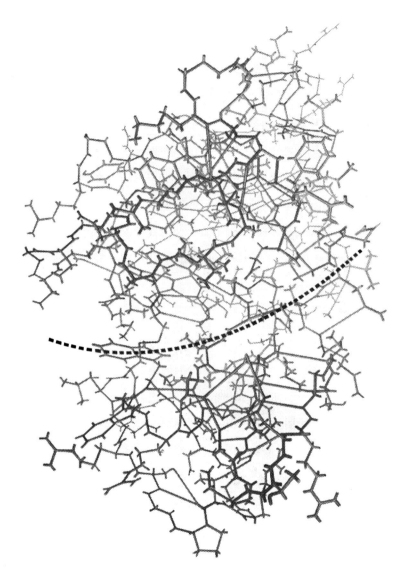

Figure 3-31 A space model of the lysozyme molecule. The enzyme is an elliptical protein with dimensions of $45 \times 30 \times 30$ Å. The peptide chain is composed of 129 amino acids (see Fig. 3-26). In the model shown, atoms are located at intersections of lines representing the bonds between them. The three-dimensional depth of the model is represented by differences in shading, with the darker portions of the peptide chain being those closest to the reader's eye. The dashed line indicates the valley, or cleft, in the enzyme molecule into which fit six sugar residues of the polysaccharide substrate. *(Illustration copyright by Irving Geis. From Cantor and Schimmel: Biophysical Chemistry, W. H. Freeman and Company, 1979.)*

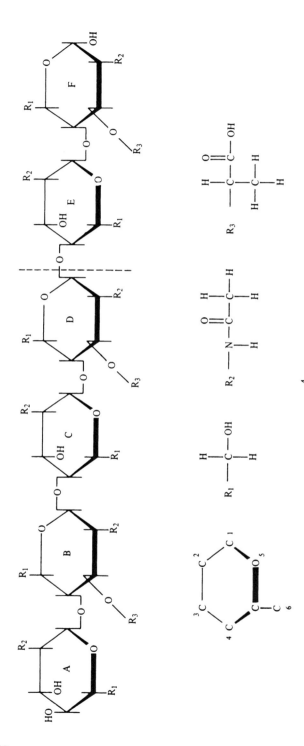

Figure 3-32 The substrates of lysozyme. (*A*) A conventional representation of a hexasaccharide fragment of the polysaccharide found in bacterial cell walls. The hexose residues are derivatives of glucose and are joined by $\beta,1\rightarrow4$ glycosidic bonds. Rings A, C, and E are N-acetylglucosamine (NAG) residues and rings B, D, and F are N-acetylmuramic (NAM) acid residues. The six rings represent the portion of the longer polysaccharide chain which fits into the cleft of the enzyme. The dotted line indicates the site of bond cleavage. (*From Phillips, 1966.*) (*B*) A space model of the polysaccharide chitin, which occurs in the exoskeleton of insects and crustacea. The polymer is composed of N-acetylglucosamine units in $\beta,1\rightarrow4$ linkage. The six rings are shown in the configuration which they assume upon binding to the enzyme; i.e., ring D is distorted and the H-bond between rings D and E (see dashed lines representing intramolecular H-bonds between other rings) is absent. The breakage of the glycosidic bond is indicated by the dotted line. (*Illustration copyright by Irving Geis. From Cantor and Schimmel: Biophysical Chemistry, W. H. Freeman and Company, 1979.*)

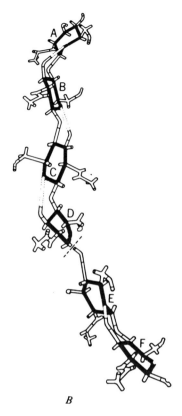

B **Figure 3-32b** (*Continued*).

dashed line shows the axis of the valley into which the portion of the substrate on which the enzyme acts can be fitted.

The enzyme causes lysis of bacterial cell walls, which are complex molecules with a backbone consisting of alternating units of N-acetyl-glucosamine (NAG) and N-acetylmuramic acid (NAM), both derivatives of glucose. Further discussion of cell wall structure can be found in Chap. 4 (also see Figs. 4-7 and 4-8). Chitin, a polymer of NAG and containing no NAM, is also a substrate for lysozyme. Figure 3-32*A* shows a short fragment of the polysaccharide backbone of the rigid layer of bacterial cell walls. Six glucose rings can fit into the valley or cleft in the lysozyme molecule, and the dashed line shows the bond which is hydrolyzed. Figure 3-32*B* shows a model space structure of a chitin-like polymer consisting only of NAG. Again, the dashed line between rings D and E indicates the site of cleavage. The substrate molecule of Fig. 3-32*B* is shown fitted into the cleft of the enzyme in Fig. 3-33. It is held there by hydrogen bonds which are shown tying it down or docking it, as it were, to the enzyme. For example, there is one hydrogen bond each for rings A and B, four associated with ring C, and one tying down ring D. Intramolecular hydrogen bonds between the NAG residues are also

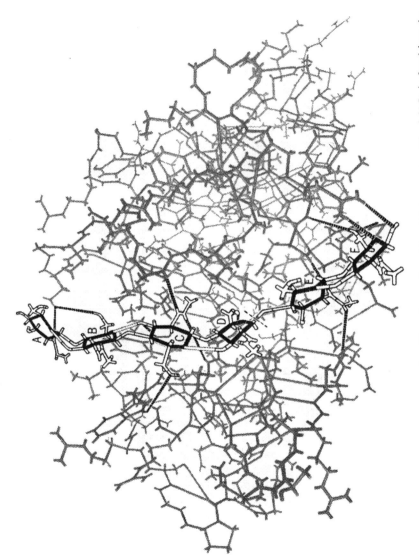

Figure 3-33 The enzyme-substrate complex. The substrate hexa-N-acetylglucosamine (Fig. 3-32*B*) is fitted into the cleft of the lysozyme molecule and is anchored in place by H-bonds between the substrate and the enzyme. *(Illustration copyright by Irving Geis. From Cantor and Schimmel: Biophysical Chemistry, W. H. Freeman and Company, 1979.)*

shown. The combined substrate and enzyme may be referred to as the enzyme-substrate complex.

The bonding between enzyme and substrate molecules brings about enough excitation and strain to enhance reaction. Detailed discussion of just how the cleavage comes about is somewhat beyond the scope of this text, and it must be emphasized that the exact mechanism is still a subject of research. There is evidence that binding of the substrate units A, B, and C to lysozyme by the hydrogen bonds shown in Fig. 3-33 causes a change in the shape of ring D, which is accompanied by a movement of certain parts of the enzyme molecule, tending to close the cleft in the enzyme. This shifting of atomic location in the enzyme and in the substrate, due to straining of the bonds, puts the reactive site into the position in space required for the reaction. The mechanism which has been postulated involves the participation of amino acid residues 52 (aspartic acid) and 35 (glutamic acid). These residues lie close to the bond which is broken. A hydrogen ion from the OH group of glutamic acid may attach itself to the oxygen atom in the glycosidic bond joining rings D and E, thus breaking the bond. This would leave carbon atom 1 of ring D with a positive charge (a carbonium ion). This charge is momentarily stabilized by the negative charge on the side chain of residue 52. A water molecule could then donate an OH ion to combine with the carbonium ion and an H ion to replace the one lost by residue 35. Thus, the hydrolysis is completed by the addition of water to the product molecules. The enzyme and substrate then separate and the enzyme is free to complex with another segment of the polysaccharide chain.

The details of the mechanism described above may not be entirely correct. They were deduced as reasonable possibilities, based upon demonstrated facts in regard to enzyme structure (primary, secondary, and tertiary), and upon knowledge of the bond which is broken and of the site of attachment of substrate to enzyme (Phillips). While not all of the details are known, this example illustrates the salient features of mechanism of enzyme action. There is an attachment at specific sites made possible by specific tertiary conformation of the enzyme. There is some conformational change in the enzyme or substrate, or both, which permits reaction between substrate and one or more sites on the enzyme that are situated in such a way as to promote transfer of the atoms needed to complete a reaction. Completion of the reaction leads to the breaking of the binding site bonds and the enzyme is freed for recycling in the system.

Kinetics of enzyme-catalyzed reactions While discussing these important catalysts and their general mechanistic behavior, it is important to relate the mechanism to one of its important ramifications, kinetics, i.e., control of the velocity or rate of biochemical reactions. This also will aid in understanding some of the kinetic aspects of microbial growth.

Let it be assumed that an enzyme has been purified and a small amount placed in a reaction vessel with a known concentration of the substrate on which it acts, e.g. lysozyme hydrolyzing a bacterial cell wall or sucrase hydrolyzing sucrose. Further, assume we can analyze the substrate S and the product P, e.g., sucrose as substrate and either glucose or fructose as product. Water is also a reactant, but its concentration is so high that no measurable change will occur. If we measured the course of disappearance of the reactant and appearance of one of the products, under ideal conditions, we would generate a data curve something like that shown in Fig. 3-34A. In the early portion of the experiment, the data taken to measure the disappearance of substrate fall along an essentially straight or nearly straight line; i.e., the decrease in sucrose concentration appears to be linear; its rate of disappearance, $-\Delta S/\Delta t$, is constant. The term *zero order* describes this phase. The same kinetic mode applies to the appearance of product, $\Delta P/\Delta t$. However, after some time, some lower concentration of S is reached, after which substrate is removed at an ever-decreasing rate and product appears more slowly. If it can be shown that, in this phase of both curves, a constant fraction of the substrate undergoes reaction in each succeeding equal interval of time, the reaction is said to be in a first-order, decreasing-rate kinetic phase.

In general, it can be expected that in an enzyme study such as the one just described, the rate of reaction proceeds at a decreasing rate from the start, but for a short period the change in slope of the curve is so small that the measured data for all practical purposes fall along a straight line. If we measured the slope of tangents to the S curve at various points along the time scale, i.e., measured the velocity v after the linear portion, and determined what concentration of S existed at each corresponding value of v, we would have the information to plot the curve shown in Fig. 3-34C. In this portion of the figure, the concentration S corresponding to the slopes in the upper portion, is plotted against these values of v. The data to make a plot of v versus S usually are not obtained by measuring slopes all along one such curve but by setting up a number of reaction vessels, each containing equal amounts of enzyme but varying concentrations of substrate. The reaction is allowed to run for only a short time, during which the velocity appears to stay constant; i.e., one measures the initial velocity as shown in Fig. 3-34B. The initial velocity is usually found to increase for higher values of initial substrate concentration S_0; S_{0_1} is the lowest substrate concentration in Fig. 3-34B. These velocity values are plotted against the corresponding S_0 values in Fig. 3-34C. This method overcomes some of the pitfalls inherent in letting an isolated enzyme reaction run too long. For example, with some enzymes a buildup of reaction products can inhibit the enzyme, giving a falsely low velocity.

In Fig. 3-34C, at high S, the velocity becomes constant or is approaching a constant upper or maximum value V. At lower concentrations of S, the velocity is ever decreasing. Such a plot of data suggests that the substrate concentration can control the rate of enzyme kinetics; as substrate concentration changes, so does the value of v. Furthermore, the plot shown in Fig.

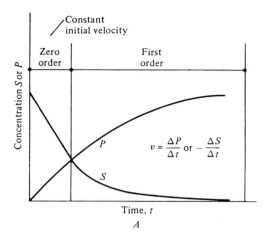

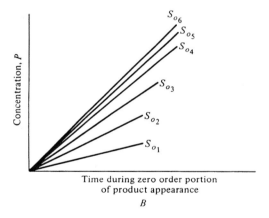

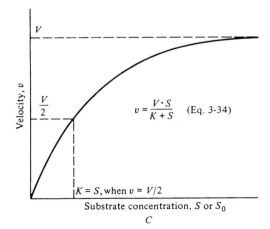

Figure 3-34 Hypothetical plots of enzyme reaction data. (*A*) Time course of an enzyme-catalyzed reaction. (*B*) Plot of initial (zero-order) portion of reactions for increasing substrate concentrations S_0, increasing from S_{0_1} to S_{0_6}. (*C*) Plot of initial velocity v at various initial substrate concentrations S_0 for data fitting a rectangular hyperbola, e.g., the Michaelis-Menten equation.

3-34C often can be shown to fit the equation of a rectangular hyperbola:

$$v = \frac{V(S)}{K + S} \tag{3-34}$$

K is a constant that controls the shape of the curve; low values give a very sharply breaking curve and higher values give a slowly breaking curve as the maximum velocity V is approached. The value of the shape factor K is numerically equal to the substrate concentration at which v is half the maximum velocity V. An experimental curve for enzymatic hydrolysis of sucrose of the type shown in Fig. 3-34C was observed at the turn of the century. The dependence of v on S at low concentrations and its independence at high concentrations were reasoned by Henri in 1902 to be due to the formation of an intermediate compound between enzyme and substrate prior to the making of the products of the reaction. In discussing mechanisms of action, we made such a statement, but it was based on nearly 75 years of subsequent research, most of which tended to prove that Henri was right.

Ten years after Henri's work, Michaelis and Menten provided a mathematical analysis explaining the shape of the curve. Let it be assumed that the enzyme-catalyzed reaction proceeds in accordance with the following reaction sequence:

$$E + S \underset{k_2}{\overset{k_1}{\rightleftharpoons}} ES \underset{k_4}{\overset{k_3}{\rightleftharpoons}} E + P \tag{3-35}$$

E is recyclable

In reaction (3-35), E is the amount of enzyme in the system that is free to react, and ES is the amount of enzyme tied up in complexes with substrate, both activated and nonactivated. If activated, the reaction proceeds to the right at rate k_3 to make product and free enzyme for recycle; if not activated, ES breaks down at rate k_2. The total amount of enzyme E_T in the system is $E + ES$. With reference to Eq. (3-35), we now assume that very shortly after the reaction begins, the levels of E and ES vary only slightly; i.e., they attain a steady state concentration and the rate of formation of complex ES is equal to its rate of breakdown. In accordance with this assumption, we may now equate the rate of formation of ES to the rate of its breakdown.

Net rate of formation of ES = net rate of breakdown of ES

$$k_1(E)(S) - k_2(ES) = k_3(ES) - k_4(E)(P)$$

Factoring and rearranging:

$$\frac{k_4 P}{k_2 + k_3} + \frac{k_1 S}{k_2 + k_3} = \frac{ES}{E}$$

The above expression may be simplified by neglecting the back reaction from P to ES, since initially P is very small. Also, the remaining velocity constants can be lumped into one; i.e., let

$$\frac{k_2 + k_3}{k_1} = K_m$$

then

$$\frac{E}{ES} = \frac{K_m}{S}$$

Unfortunately, the amounts of free enzyme E and enzyme–substrate complex ES are difficult to measure and, in any event, we would like to put the equation in terms of reaction velocity v and maximum velocity V, which are measurable quantities. We would also like to determine whether the above expression, derived by assuming a gross mechanism of enzyme action, does indeed represent a curve that best fits one derived from experimental reaction results.

It can be reasoned that the maximum velocity occurs when there is no free enzyme, i.e., when it is all tied up in ES. The maximum velocity would then equal:

$$V = k_3 E_T$$

The above is true only at high substrate concentrations; at all other concentrations, the measurable velocity v is:

$$v = k_3 ES$$

Now we can relate the measurable velocities v and V to the ratio E/ES. Substituting $(E_T - ES)$ for E:

$$\frac{E_T - ES}{ES} = \frac{K_m}{S}$$

Rearranging,

$$\frac{E_T}{ES} = \frac{K_m}{S} + 1$$

Substituting for E_T and ES the terms involving V and v,

$$\frac{V/k_3}{v/k_3} = \frac{K_m}{S} + 1$$

Canceling like terms and rearranging,

$$v = \frac{V(S)}{K_m + S} \tag{3-36}$$

Equations (3-34) and (3-36) have the same form, but we now have assigned a physical meaning to the shape constant; that is,

$$\frac{k_2 + k_3}{k_1} = K_m$$

in accordance with the hypothesized sequence of reactions shown in Eq. (3-35) and the assumptions we made in deriving Eq. (3-36).

There are several points we should make about this equation. First, it is not necessary to follow precisely this line of reasoning in order to derive Eq. (3-36). As with K in Eq. (3-34), the constant K_m is numerically equal to the substrate concentration at which the velocity v is half the maximum velocity V. The Michaelis-Menten constant K_m is not necessarily indicative of the affinity of the enzyme for its substrate, because it includes also the speed with which the enzyme–substrate complex breaks down to enzyme plus product. The affinity of E for S deals only with the relative values of k_1 and k_2.

It can be seen that if one knows the K_m for the enzyme reaction, it is possible to predict the velocity relative to V for any value of S, and if one knows V as well as K_m, the actual velocity of the reaction can be predicted for any value of S. It must be remembered that since the velocity of an enzyme reaction also is dependent on the concentration of enzyme present, the maximum velocity V is expressed per unit amount of enzyme.

Also, it is emphasized that V and K_m are affected by the environmental conditions, e.g., temperature and pH, at which the reaction takes place. Furthermore, they are not affected to the same degree by a particular condition. The values of V and K_m can be estimated from a plot of v versus S (or S_0) as in Fig. 3-34C, but it is best to apply some analytical geometry to Eq. (3-36) and put it into one of several straight-line forms. Various forms of the equation are shown in Fig. 3-35. We have found the form given in Fig. 3-35C to be a good one to use in determining the numerical values of the two constants of a rectangular hyperbola, but it is always best to try several forms. After obtaining the constants from a linear plot of the data, one should use them in Eq. (3-36) to calculate values of v at various values of S. If the calculated curve plots a good fit to the data, one has evidence that (1) the constants were determined correctly, and (2) the data do follow the hyperbolic curve.

The rectangular hyperbola can be used to describe a number of phenomena that take place due to quite different mechanisms of action. For example, although we have just derived the Michaelis-Menten equation on the basis of an hypothesized mechanism, there are other mechanisms that give rise to the same kinetic equation. For example, the Langmuir adsorption isotherm:

$$S_{\text{adsorbed}} = \frac{KS_0}{K' + S_0} \tag{3-37}$$

which describes the adsorption of a gas or solute onto an adsorptive surface, has the same form as Eqs. (3-34) and (3-36). For a given amount of surface, the concentration of solute adsorbed is proportional to the concentration of solute S (or gas) in solution. If the rate at which adsorption takes place is proportional to the space on the surface that is unoccupied by solute (which is related to the amount of S adsorbed), then the rate law for the process has the same form as Eqs. (3-34) and (3-36). Thus, the same type of equation can

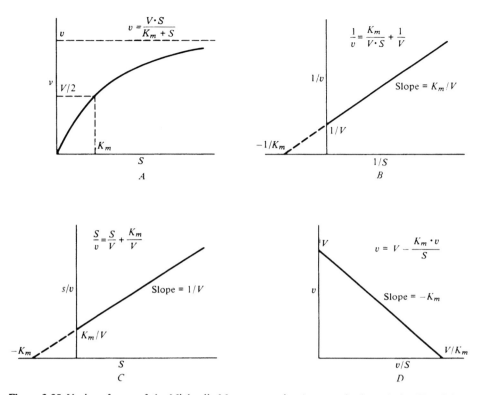

Figure 3-35 Various forms of the Michaelis-Menten equation (rectangular hyperbola) [Eq. (3-36)].

describe the rates of phenomena with widely differing mechanisms of action. This same form of equation can apply in describing the growth of microorganisms; these aspects will be discussed in Chap. 6. It is important to emphasize here that sameness in the law of kinetic expression should in no way imply sameness in the mechanism causing the observed kinetic.

Cofactors and prosthetic groups Some enzymes require, in addition to the "simple" protein part (the apoenzyme), another nonprotein part, or prosthetic group, in order to perform their function as catalysts; these fall into the category of conjugated proteins. These prosthetic groups are often called *cofactors*. Most cofactors function as temporary or intermediate carriers of various chemical groupings in group transfer reactions. The most prominent cofactors are a class of compounds called *nucleotides*, which will be described in some detail later in this chapter. One of the most important of these is adenosine triphosphate (ATP), which will be shown later to be used primarily to drive endergonic reactions. Other important cofactors include nicotinamide adenine dinucleotide, coenzyme A, lipoic acid, glutathione, thiamine pyrophosphate, pyridoxal phosphate, tetrahydrofolic acid, flavin, and heme, biotin, and metal ions, as well as a few others. In general, these are necessary in the respiratory and synthetic activities of microorganisms.

Environmental factors affecting enzyme function and kinetics It is fairly common knowledge that microorganisms as well as other types of living matter are affected by environmental factors, such as pH, temperature, and the presence of toxic or inhibitory substances. Much attention will be given to these factors, particularly pH and temperature, in Chap. 5. One important reason that these factors exert such profound effects on the rate of growth and selection of species in an ecosystem is their effect on the enzymes that control the metabolism of the organisms. Some of the reasons for these effects should be apparent in view of the information already presented.

pH *and Temperature* We have seen that enzyme function depends on the maintenance of a special space configuration and access to special chemical groups that bind the enzyme and substrate together as well as special groups at which the reaction takes place. Anything that causes a change in ionization of these sites or conformation of the space structure is bound to have an effect on enzyme function. The enzyme protein contains many ionizable groups, and this ionization, as we have seen, is dependent on the pH of the solution. Usually there is a narrow range of pH values at which activity is maximum, i.e., the optimum pH. Below or above this range, activity is severely diminished. The curve of velocity versus pH observed for most enzymes is the bell-shaped type shown in Fig. 3-36.

Changes in pH value affect the protein structure of the enzyme irreversibly, i.e., denature it. In addition, changes in the ionization of the substrate may be brought about by changes in pH values. Less severe changes in pH levels may cause changes in the mode of binding of enzyme and substrate, not sufficient to prevent the reaction entirely but severe enough to slow the velocity.

The velocity of a chemical reaction increases with increased temperature in general accord with the Arrhenius equation [Eq. (3-33)]. However, increases in temperature also denature proteins; thus, there is a condition in which the increased rate is counterbalanced by the decrease in activity due to

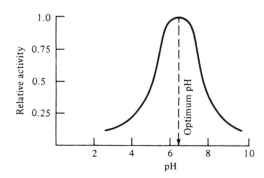

Figure 3-36 Example of a bell-shaped curve of enzyme activity versus pH exhibited for many enzyme-catalyzed reactions.

denaturation. Usually, if one runs an enzyme-catalyzed reaction at varying temperatures, the initial velocity increases with temperature but, as time progresses, the velocity falls away more rapidly at higher temperatures. This type of result is shown in Fig. 3-37A. At 80°C the initial slope is high, but activity ceases after a very short time, and the velocity measured between t_0 and t_1 is rather low. If one were to assess the velocity by the amount of substrate used or the product produced at time t_1, the optimum temperature would be 70°C. However, if one performed the assay at t_2, the apparent optimum temperature would be significantly lower (55°C), as shown in Fig. 3-37B.

Most enzymes are inactivated very rapidly at temperatures ranging from 70 to 100°C; however, a few enzymes can withstand temperatures of 100°C. Although the rates of most enzyme-catalyzed reactions are greatly slowed at low temperatures and are nearly zero at 0°C, this depressed temperature does not denature the enzyme. In fact, enzymes are quite commonly frozen for storage. The rates of reactions for most enzymes increase two to three times for each 10° rise in temperature, up to temperatures at which the rate falls off due to denaturation.

Effect of Inhibitors Since all biochemical reactions are controlled by enzymes, anything that interferes with or inhibits the functioning of enzymes can exert a tremendous effect on the biological activity in treatment plants and streams. The presence of enzyme inhibitors in biological treatment processes is of obvious interest because inhibitors interfere with the needed technological control of metabolic function. Their effect on the biota in a

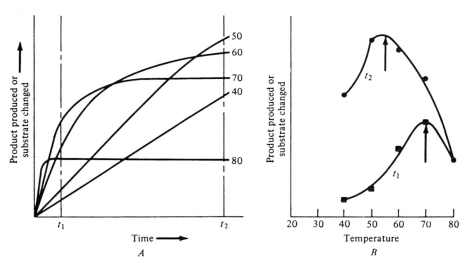

Figure 3-37 (A) Effect of temperature on enzyme reactions and (B) differences in apparent optimum temperature depending on time period for measuring rate.

receiving stream can be disastrous. Many of the toxic components of industrial and commercial wastes are in fact enzyme inhibitors. Many of the pesticides used in agricultural activities function due to their inhibitory action on specific enzymes. Other important applications of enzyme inhibitors are their use in the pharmacological and medical fields. Also, they have a military application and consequently an important environmental significance.

Any change in the chemical or physical environment that leads to an alteration in the spatial conformation of the enzyme (or its substrate) can be considered as an inhibitor. For example, changes in pH values and temperature may be inhibitory. However, in a more definitive way, inhibitors can be considered as those chemical substances that lead to the prevention or loss of enzyme function.

Some substances affect a wide variety of enzymes, and these can be called *nonspecific inhibitors*. For example, cyanide can be classified in this category because, while it is particularly effective in inhibiting cytochrome oxidase, which controls the transfer of electrons to oxygen in aerobic respiration (50 percent inhibition at $CN^- = 10^{-8}$ molar), it is also an effective inhibitor of more than 40 other enzymes. Other chemicals such as azide and sulfide, which also function as respiratory poisons, effectively inhibit a rather wide array of enzymes. Their effectiveness is due to the fact that these compounds can form stable complexes with metals such as iron and copper, and in many enzymes these metal atoms play an essential mechanistic role in catalysis. Also, among other modes of action, cyanide may act as a reducing agent, thus breaking disulfide links in the enzyme.

Chlorine and ozone, which are powerful oxidizing agents, probably owe a large part of their effectiveness as microbial disinfectants to their ability to inactivate many different enzymes; i.e., they are probably nonspecific enzyme inhibitors. The action of nonspecific inhibitors is usually not reversible. Various other inhibitors are more specific in the reaction they inhibit, and many owe their inhibitory capacity to their similarity in structure to the normal substrate for the enzyme.

Most enzymes exhibit a high degree of specificity toward the substrate on which they act, but specificity is not perfect and compounds with a structural configuration very similar to that of the substrate can compete for the binding site, thus blocking it for the substrate. This type of *competitive inhibition* depends in the main on the relative affinities of the substrate and the inhibitor for the enzyme as well as the relative concentrations of substrate and inhibitor. Since the inhibitor effect can be overcome by providing the substrate with a competitive advantage, e.g., by increasing the substrate concentration, a competitive inhibition shows no effect on the maximum velocity V. Such a competitive inhibitor does have the effect of increasing K_m; i.e., it causes a greater dependence of v on S and flattens the curve of v versus S. If one plots the reciprocals $1/v$ versus $1/S$, the slope changes (increases), but the intercept on the $1/v$ axis remains the same, regardless of inhibitor concentration.

Another type of inhibition that is readily demonstrated by a change in the plot of $1/v$ versus $1/S$ is called *noncompetitive inhibition*. This type is manifested by a change in the intercept on the $1/v$ axis, i.e., a change in V. The value of V decreases as inhibitor concentration is increased, but there is no change in the slope K_m.

As the name implies, *noncompetitive inhibitors* do not compete with the substrate; thus, this type of inhibition cannot be reversed by increasing the substrate concentration. One may surmise that the bindings of inhibitor and substrate are not necessarily at the same location. Thus, noncompetitive inhibition may be a manifestation of an influence exerted over a significant distance in the molecule.

In noncompetitive inhibition, the maximum velocity V is decreased in proportion to the inhibitor concentration because the inhibitor effectively reduces the enzyme concentration available for reaction. Also, the inhibitor-enzyme–substrate complex is converted to products at a slower rate than is the enzyme–substrate complex; hence, the K_m is increased for increased inhibitor concentrations. Heavy metal ions often cause noncompetitive inhibition.

Other types of inhibition are also known. Some enzyme reactions can be inhibited by high concentrations of substrate and some by a buildup of products of the reaction. Feedback inhibition, in which the product of a metabolic pathway inhibits the initial enzyme of the pathway, is a special type of inhibition and will be discussed in Chap. 12.

Nucleotides and nucleic acids Nucleotides and nucleic acids are important constituents of microorganisms and all living cells. Nucleic acids, which are polynucleotides, may constitute 5 to 30 percent of the dry weight of microorganisms, depending on the species and the environmental conditions. Animal and vegetable cells contain approximately as much nucleic acid as do microbial cells. Thus, municipal wastes can also be expected to contain them. The nitrogen content of nucleic acid is about the same as it is for protein (15 to 16 percent) but, unlike protein, nucleic acids contain a significant amount of phosphorus (9 to 10 percent). Much of this section will be deveoted to the description of the spatial structure of nucleic acids. The general mechanism of their function in cell replication and protein synthesis will be discussed in Chap. 12. In addition to the nucleic acids, certain mono- and dinucleotides perform special vital functions in metabolism; they were mentioned previously as coenzymes. The spatial structures of these compounds will also be described in this section.

Nucleotides All nucleotides found in nucleic acids consist of three smaller molecules: (1) a purine or pyrimidine, (2) ribose or deoxyribose, and (3) phosphoric acid.

The common monomeric constituents of nucleic acids are shown in Fig. 3-38. The pyrimidine bases are derivatives of the compound pyrimidine (see

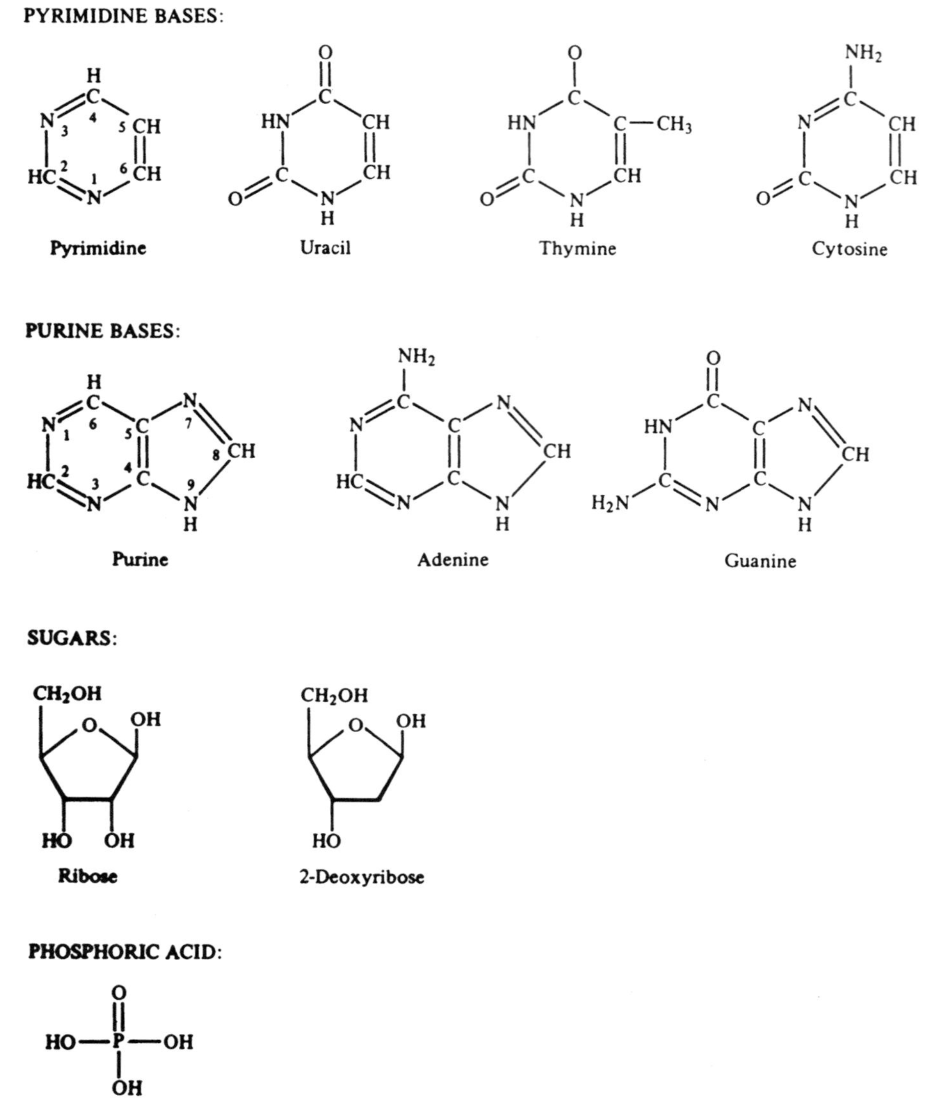

PYRIMIDINE BASES:

| Pyrimidine | Uracil | Thymine | Cytosine |

PURINE BASES:

| Purine | Adenine | Guanine |

SUGARS:

Ribose   2-Deoxyribose

PHOSPHORIC ACID:

Figure 3-38 Components of nucleic acids.

the numbering system in Fig. 3-38), and the purine bases are derivatives of the compound purine. The purines adenine and guanine and the pyrimidines thymine, cytosine, and uracil are the bases found in almost all nucleic acids; a few other bases occur in certain types of nucleic acids.

In nucleotides, the purine or pyrimidine bases are bonded to the sugars. If either ribose or deoxyribose is bonded to a pyrimidine, the bond is formed between C_1 of the sugar and N_1 of the pyrimidine. If bonded to a purine, the link is C_1 to N_9. Purines and pyrimidines so bonded to either of these sugars are called *nucleosides*. The nucleosides are named for the base; thus, adenine bonded to ribose is called adenosine, etc. (see Fig. 3-39).

To form nucleotides, phosphoric acid is linked to the sugar moiety of a nucleoside in an ester bond, usually at the C_5 position. Thus, the names adenosine 5'-phosphate, adenine nucleotide, adenylic acid, and adenosine monophosphate (AMP) could be used to designate such a nucleotide (see Fig. 3-39).

Nucleotides as enzyme cofactors Certain nucleotides were listed previously among the enzyme cofactors. Important cofactors containing nucleotide moieties are shown in Fig. 3-40, and their functions are described briefly below. They will be mentioned many times in this text.

The key to the importance of adenosine triphosphate (ATP) lies in the two phosphate bonds designated by the squiggle in Fig. 3-40. The squiggle is used to indicate that when these bonds break by hydrolysis or transfer of one or both phosphate moieties, a considerable amount of energy is released. ATP is called an *energy-rich* compound, and the bonds are often referred to as *high-energy bonds*. The latter term should not be construed as meaning that a high amount of energy is required to break these bonds. Quite to the contrary, these bonds are rather labile and tend to break easily; when either of them does, it releases approximately 7000 cal/mol of ATP ($\Delta G° = -7000$ cal). Organisms couple the energy released upon hydrolysis of ATP with endergonic reactions required in the synthesis of cellular macromolecules, e.g., the making of proteins, lipids, polysaccharides, and nucleic acids. ATP is not the only high-energy compound used for this purpose, but it is by far the major one. The general mode of action of ATP in providing energy for synthesis is demonstrated in Fig. 3-41. In this example, compound A–B is to be synthesized from compounds A and B. This synthesis requires 5000 cal/mol. The needed energy input is supplied by the hydrolysis of ATP to ADP (adenosine

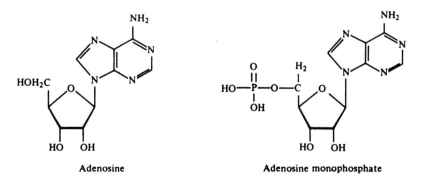

Adenosine   Adenosine monophosphate

Figure 3-39 A nucleoside (left) and a nucleotide (right).

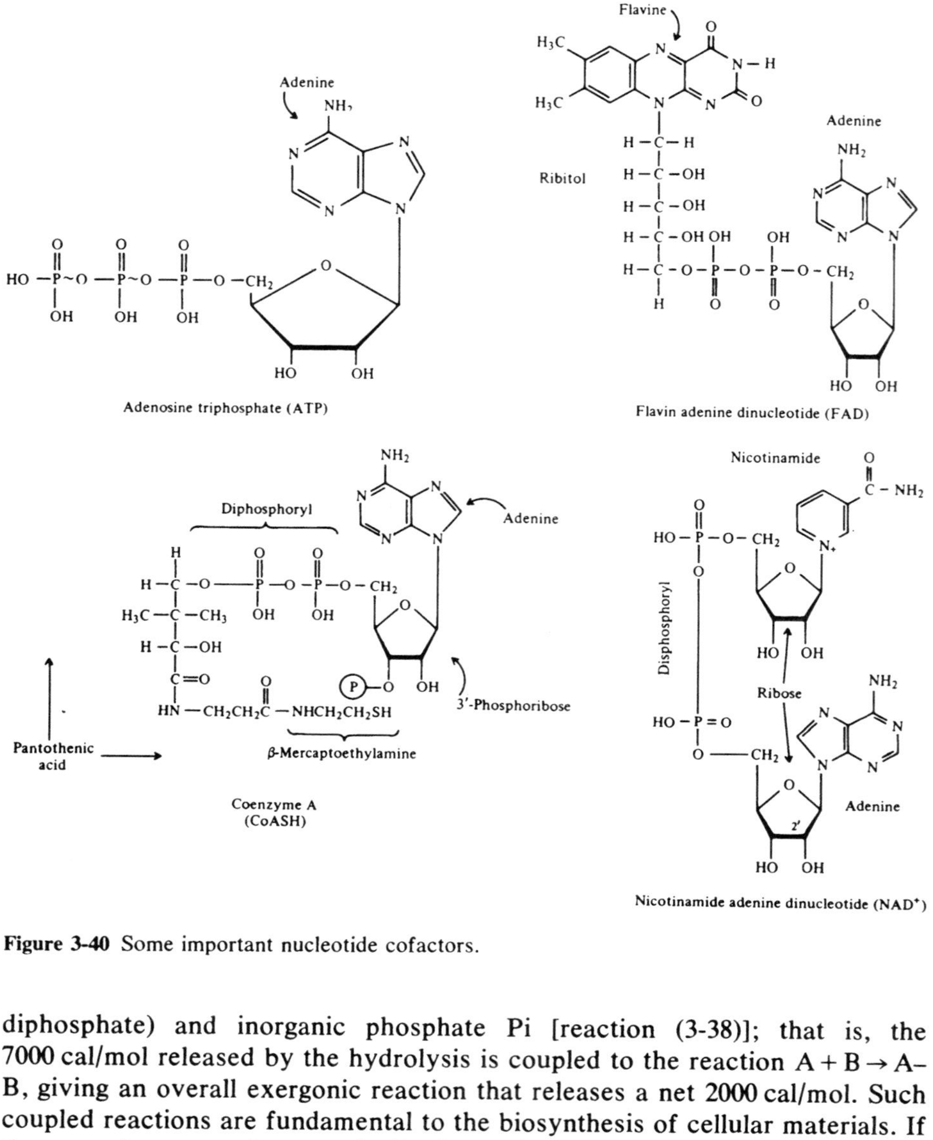

Figure 3-40 Some important nucleotide cofactors.

diphosphate) and inorganic phosphate Pi [reaction (3-38)]; that is, the 7000 cal/mol released by the hydrolysis is coupled to the reaction $A + B \rightarrow A-B$, giving an overall exergonic reaction that releases a net 2000 cal/mol. Such coupled reactions are fundamental to the biosynthesis of cellular materials. If the organisms are those typically found in aerobic secondary wastewater treatment processes, compounds A and B are molecules derived from the organic matter in the waste. The organisms first must have synthesized ATP by oxidizing other molecules in the waste. The basic reason for aerobic respiration, i.e., oxygen utilization or exertion of biological oxygen demand of the organic matter, is to obtain ATP for driving reactions that require energy.

As will be shown later, the electrons removed during oxidation of the organic compound are passed along by various electron transport molecules;

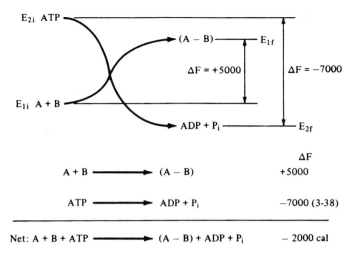

		ΔF
A + B $\longrightarrow$	(A – B)	+5000
ATP $\longrightarrow$	ADP + P_i	–7000 (3-38)

Net: A + B + ATP $\longrightarrow$ (A – B) + ADP + P_i – 2000 cal

Figure 3-41 Driving an endergonic reaction with an exergonic reaction. The coupling of a synthesis with hydrolysis of ATP.

eventually they are passed to oxygen, making water. Both flavin adenine dinucleotide (FAD) and nicotinamide adenine dinucleotide (NAD) are among the electron-transporting coenzymes. The passing of an electron from one compound to another represents an oxidation of the donor compound. If (1) the oxidation releases sufficient energy, and (2) the organism possesses a mechanism to couple this energy release to run reaction (3-38) or its equivalent reactions in reverse, ATP can be synthesized, and the energy released from the oxidation is trapped in the ATP for later use by the organism. In many organisms, the oxidation of the electron carriers NAD and FAD is coupled to the synthesis of ATP and represents the primary mode of ATP production. ATP and other high-energy compounds can be made in other ways as well, as will be discussed later.

Coenzyme A (CoASH) is used in many acyl (acid) group

$$(R—\overset{\overset{\displaystyle O}{\displaystyle \|}}{C}—)$$

transfer reactions. The group to be transferred is generally connected to CoASH by the formation of a thiol ester

$$(R—\overset{\overset{\displaystyle O}{\displaystyle \|}}{C}\sim SCoA).$$

The ester synthesis is coupled to the hydrolysis of ATP. The energy released is contained in the acyl~SCoA and is used to transfer the acyl group to another compound. In this transfer, the CoASH is released and recycled. An example of such a reaction sequence is given below. This is an important sequence in many organisms for the oxidative metabolism of acetic acid (found in various wastewaters, e.g., municipal sewage, pulp and paper wastes). The tricarboxylic acid citrate is an intermediate compound in an important cyclic oxidation scheme, the tricarboxylic acid (TCA) cycle, a metabolic pathway common to many microorganisms. Note that three enzymes and two coenzymes are involved in this reaction sequence. Other nucleotide coenzymes include guanosine triphosphate (GTP) and inosine triphosphate (ITP), which transfer energy in certain reactions in the same way as does ATP; uridine diphosphate (UDP), which is important in some carbohydrate reactions; and cytidine diphosphate (CDP), which is involved in the biosynthesis of phospholipids.

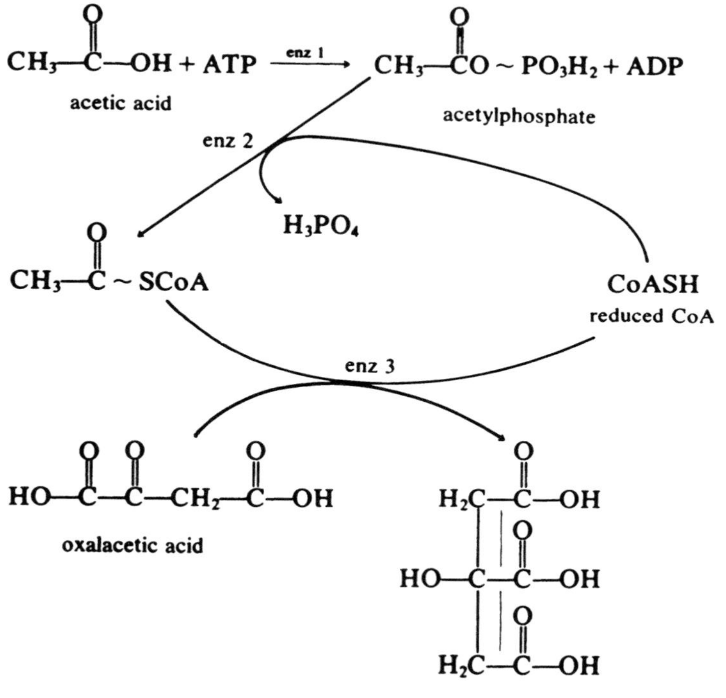

$$CH_3\!-\!\overset{O}{\underset{}{C}}\!-\!OH + ATP \xrightarrow{\text{enz 1}} CH_3\!-\!\overset{O}{\underset{}{C}}O \sim PO_3H_2 + ADP \qquad (3\text{-}39)$$

acetic acid   acetylphosphate

enz 2

$$CH_3\!-\!\overset{O}{\underset{}{C}} \sim SCoA \qquad\qquad H_3PO_4 \qquad\qquad CoASH \qquad (3\text{-}40)$$

reduced CoA

enz 3

$$HO\!-\!\overset{O}{\underset{}{C}}\!-\!\overset{O}{\underset{}{C}}\!-\!CH_2\!-\!\overset{O}{\underset{}{C}}\!-\!OH \qquad H_2C\!-\!\overset{O}{\underset{}{C}}\!-\!OH \qquad (3\text{-}41)$$

oxalacetic acid

$$HO\!-\!\overset{}{\underset{}{C}}\!-\!\overset{O}{\underset{}{C}}\!-\!OH$$

$$H_2C\!-\!\overset{O}{\underset{}{C}}\!-\!OH$$

citric acid

Nucleic acids Most of the nucleotides in organisms are present as polynucleotides in nucleic acids. The constituents of ribonucleic acid (RNA) and deoxyribonucleic acid (DNA) are given in Table 3-9. Other than the occurrence of D-2-deoxyribose in DNA and D-ribose in RNA, the only difference is that RNA contains uracil, whereas DNA contains thymine. Each contains four different nucleotides that are defined by the four bases they contain. The nucleotides are linked together by ester bonds between the phosphate moiety

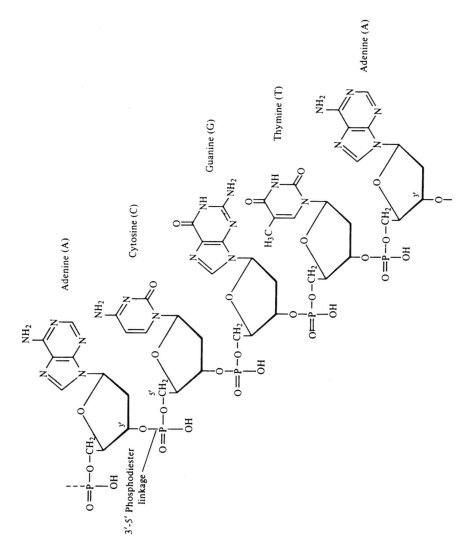

Figure 3-42 A portion of a polynucleotide chain, e.g., one strand of a DNA molecule. Note the sugar–phosphate linkages that form the backbone of the molecule.

139

Table 3-9 Composition of RNA and DNA

	RNA	DNA
Purines	Adenine, guanine	Adenine, guanine
Pyrimidines	Cytosine, uracil	Cytosine, thymine*
Sugars	D-Ribose	D-2-Deoxyribose
Acid	H_3PO_4	H_3PO_4

*Some viral DNAs have cytosine derivatives rather than thymine.

on C_5 of the sugar of one nucleotide and the C_3 hydroxyl of the sugar of another nucleotide. A segment of the polynucleotide chain of a DNA molecule is shown in Fig. 3-42. The backbone of the polynucleotide consists of a repeating chain, sugar–phosphate; the phosphate forms a 3',3'-phosphodiester linkage. Along this backbone are arranged the purine and pyrimidine bases of each nucleotide.

Deoxyribonucleic Acid (DNA) Deoxyribonucleic acid is the molecule in a cell that is unique to it and its kind (species). This molecule contains all the genetic information that is passed on to the progeny cells and all the information and directions (blueprints) for synthesizing the new cell. For the life of a species to continue, this molecule must be capable of self-replication. Its content in a given bacterial cell at any time generally varies from one to two molecules, depending on whether the cell has just divided or is about to divide. Some cells undergoing very rapid growth may contain four molecules of DNA, since the rate of DNA replication may be more rapid than the rate of cell division. In general, in a growing biomass consisting of bacterial cells in various stages of replication, the DNA content is rather constant and accounts for perhaps 2 to 4 percent of the dry weight of the culture.

Except in some viruses, the DNA molecule occurs as a double strand of polynucleotides in a spiral or helical arrangement. The backbone of each

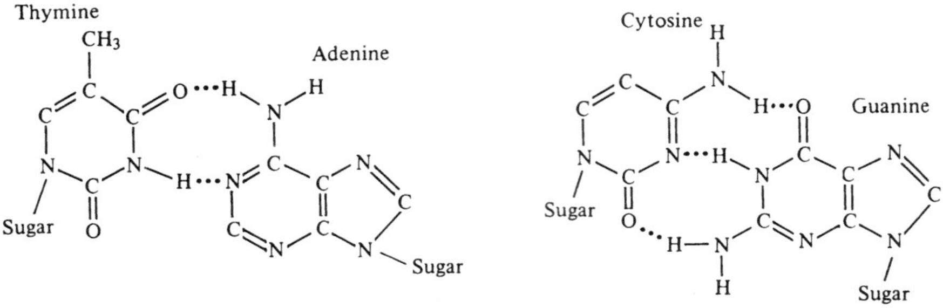

Figure 3-43 Hydrogen bonds pairing thymine and adenine, cytosine and guanine in a DNA molecule.

strand is the sugar–phosphate chain seen in Fig. 3-42. The bases are turned toward the inside of the helix. The polynucleotide strands are held together by hydrogen bonds that form between pairs of the inwardly turned purine and pyrimidine bases. The base pairing is such that cytosine in one strand is hydrogen-bonded to guanine in the other; adenine and thymine are also paired. The hydrogen bonds are shown in Fig. 3-43. The strands are oriented in opposite directions (antiparallel), as shown in Fig. 3-44. For clarity of presentation, the strands are not shown in twisted form. It is noted that in the left-hand strand the phosphorus at the top is on the C_5, whereas for the

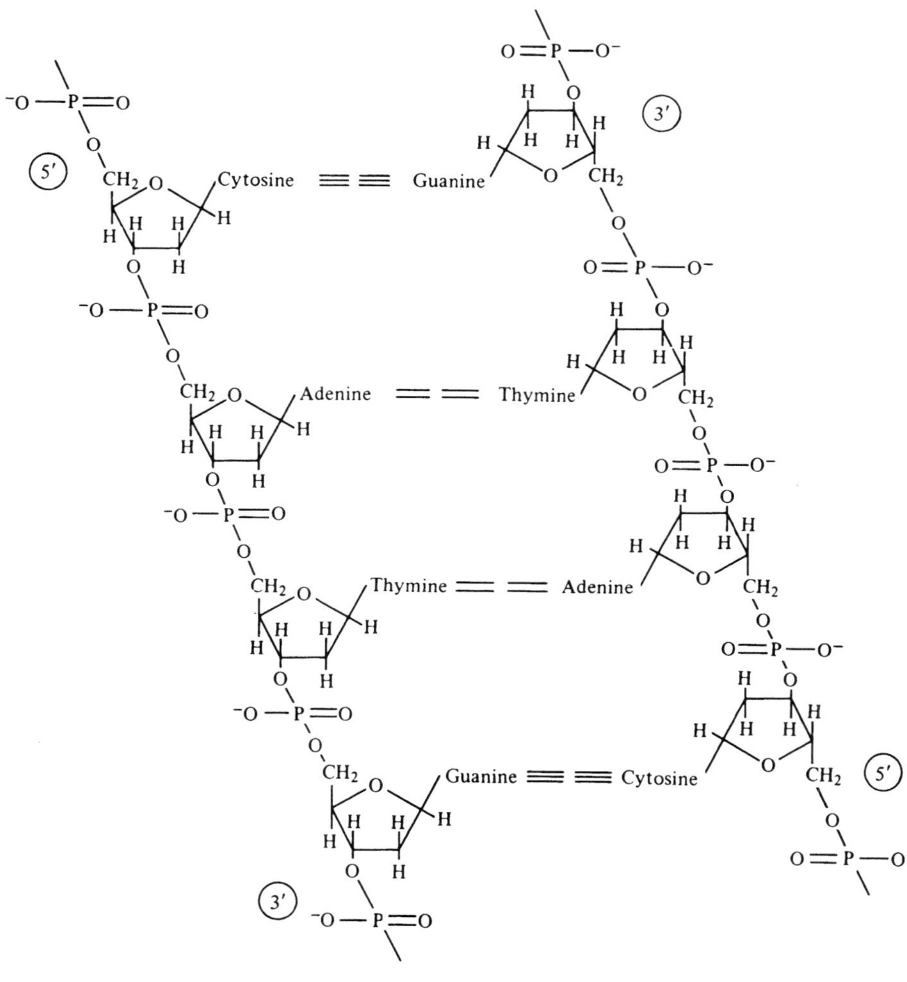

Figure 3-44 Segment of a DNA molecule showing hydrogen bonding of C≡G and A=T and antiparallel alignment of the polynucleotide strands.

right-hand strand it is on C_3 of the deoxyribose moiety; i.e., the strands are running in opposite directions. Figure 3-45 shows the double-stranded helix. The ribbons are formed by the sugar–phosphate backbone; the ties between the ribbons represent the hydrogen-bonded base pairs. These are shown in more detail for a portion of the segment of DNA shown (see the four base pairings). The uniqueness of each species of living matter is determined by the sequence of nucleotides in the polynucleotide chain.

Ribonucleic Acid (RNA) Ribonucleic acid is a single-stranded poly-nucleotide. RNA is needed by the cell to transcribe the genetic information

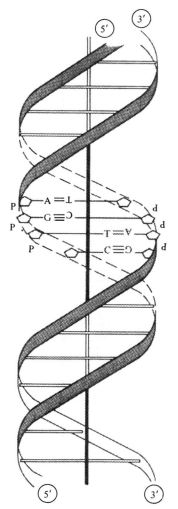

Figure 3-45 Segment of a DNA molecule showing the antiparallel helical arrangement of polynucleotide strands wound around an imaginary axis. The strands are held together by base pairing, A=T and G≡C.

contained (coded) in DNA so that the cell can make the enzymes needed for reactions yielding chemical energy as well as for using the chemical energy to make all of the organic constituents of the organism. The RNA so copied now carries the message contained in the DNA and is appropriately called *messenger* RNA (mRNA). The structural configuration of mRNA has not been as well defined as that of DNA, but there does not appear to be a great amount of hydrogen bonding within the molecule. If one makes an analogy considering the DNA to be the master set of plans for the cell, the mRNA can be considered a set of shop drawings. The mRNA becomes bound to ribonucleoprotein particles called ribosomes. These are the sites of protein synthesis for all of the enzymes the cell requires. Most of the RNA in cells is rRNA, *ribosomal* RNA (80 to 90 percent). In bacteria, ribosomes are found scattered in the cell cytoplasm and attached to the cell membrane (see Chap. 4 for discussion of cell structure). The mRNA serves as a template on the ribosomes where specific amino acids are joined together in the sequence dictated by the template to make specific proteins. Continuing with the analogy, the ribosomes may be looked upon as a workbench for making the proteins. Still another kind of RNA is necessary to this process. In addition to mRNA and rRNA, there is a lower molecular weight polyribonucleotide that transports or transfers the needed amino acids from the pool of free amino acids in the cell to the ribosomes. This type of RNA is called *transfer* RNA (tRNA). There are at least as many types of tRNA as there are amino acids to be transferred. A considerable amount of information is available in regard to its structure (see Chap. 12). The replication of DNA and the synthesis of enzyme protein will be discussed in Chap. 12.

Another type of RNA which is vitally important is that found in some viruses that contain no DNA and use RNA as their genetic material. In a few cases, this RNA is double-stranded.

PROBLEMS

3.1 (*a*) Calculate the pH of a 0.05-molar solution of acetic acid.

(*b*) What is the pH of a 1.0-molar solution of hydrochloric acid?

3.2 What is meant by the term *rancidity*?

3.3 What is meant by the term *denaturation*? What are the conditions under which it may occur?

3.4 (*a*) What is an amphoteric compound?

(*b*) Discuss the significance of aluminum hydroxide and proteins as examples of amphoteric compounds.

3.5 (*a*) Write the Fischer and Haworth projections for fructose and for glucose.

(*b*) Using toothpicks or balsa wood, construct the space model for the ring forms of these compounds.

3.6 Give examples of some general types of polysaccharide found in wastewaters and in nature. Discuss their role and their significance to a water pollution control specialist.

3.7 The following data were collected in a study of a particular enzyme acting on its substrate:

Substrate concentration, mol	Velocity (rate of disappearance of substrate, mol/min)
0.14	0.15
0.22	0.17
0.30	0.25
0.58	0.33
0.80	0.40

Determine the maximum velocity V_{max} and the Michaelis-Menten constant K_m for the enzyme.

3.8 Calculate the free energy of oxidation of methane.

3.9 (a) Distinguish among the following types of bonds: ionic, covalent, and coordinate.

(b) Discuss these types of bonding in relationship to hydrogen bonding. In what types of materials do hydrogen bonds play an important role?

3.10 (a) Calculate the theoretical COD, TOC, and TOD of the following compounds: glucose, arginine, cysteine, palmitic acid.

(b) Why do the COD and TOD differ for some of the compounds?

3.11 (a) Under anaerobic conditions, 1 mol of glucose can be fermented to 2 mol of lactic acid. Calculate the COD, TOC, and TOD values of the products and compare them with those of the original glucose for a glucose concentration of 1000 mg/L.

(b) Under aerobic conditions, 1 mol of glucose can be metabolized to 6 mol of carbon dioxide and 6 mol of water. Calculate the COD, TOC, and TOD values of these products and compare them with those of the original glucose for a glucose concentration of 1000 mg/L.

(c) Under anaerobic conditions, some microorganisms can ferment 1 mol of glucose to 2 mol of ethanol and 2 mol of carbon dioxide. Calculate the COD, TOC, and TOD of the products and compare them with those of glucose for a glucose concentration of 1000 mg/L.

3.12 One of the empirical formulas for biomass is $C_5H_7NO_2$.

(a) Calculate the theoretical COD of 1 mg of cells.

(b) Calculate the TOC and TOD of 1 mg of cells.

3.13 Based on your knowledge to date of the concept of biochemical oxygen demand, discuss whether a theoretical BOD could be calculated for glucose. If your answer is yes, discuss the implications and consequences of the assumptions made in such a calculation. If yes is not an entirely satisfactory answer, state why, and suggest possible ways in which you might approximate a theoretical BOD for the compound.

3.14 (a) What is an unsaturated compound, and what reactions generally occur at the locations of such unsaturation?

(b) Discuss one important reason why unsaturated compounds are important in the pollution control field.

3.15 Determine how many grams of KH_2PO_4 and K_2HPO_4 are required to prepare a 1-molar buffer solution at pH 7.5 ($pK_a = 6.8$).

REFERENCES AND SUGGESTED READING

American Public Health Association. 1976. Standard methods for the examination of water and wastewater, 14th ed. APHA, Washington, D.C.

Baum, S. J. 1978. Introduction to organic and biological chemistry, 2d ed. Macmillan, New York.

Brock, T. D. 1979. Biology of microorganisms, 3d ed. Prentice-Hall, Englewood Cliffs, N.J.

Clark, W. M. 1952. Topics in physical chemistry. Williams & Wilkins, Baltimore.

Dean, J. D. (editor). 1973. Lange's handbook of chemistry. McGraw-Hill, New York.

Dickerson, R. E., and I. Geis. 1969. The structure and action of proteins. Harper & Row, New York.

Fruton, J. S., and S. Simmonds. 1958. General biochemistry, 2d ed. John Wiley, New York.

Gaudy, A. F. Jr. 1972. Biochemical oxygen demand. *In* Water pollution microbiology, R. Mitchell (editor). John Wiley, New York.

Gaudy, A. F. Jr., M. N. Bhatla, and E. T. Gaudy. 1964. Use of chemical oxygen demand values of bacterial cells in wastewater purification. *Appl. Microbiol.* **12**:254–260.

Geissman, T. A. 1977. Principles of organic chemistry, 4th ed. W. H. Freeman, San Francisco.

Harper, H. A. 1975. Review of physiological chemistry. Lange Medical Publications, Los Altos, Calif.

Kendrew, J. C. 1961. The three-dimensional structure of a protein molecule. *Sci. Am.* **205**:96–110.

Lehninger, A. L. 1975. Biochemistry, 2d ed. Worth, New York.

Liener, I. E. 1966. Organic and biological chemistry. Ronald Press, New York.

Mahler, H. R., and E. H. Cordes. 1966. Biological chemistry. Harper & Row, New York.

Neilands, J. B., and P. K. Stumpf. 1958. Outlines of enzyme chemistry, 2d ed. John Wiley, New York.

Neish, A. C. 1952. Analytical methods for bacterial fermentations, Publication NRC 2952. National Research Council of Canada, Saskatoon.

Nemerow, N. L. 1971. Liquid waste of industry. Addison-Wesley, Reading, Mass.

Nemerow, N. L. 1978. Industrial water pollution. Addison-Wesley, Reading, Mass.

Phillips, D. C. 1966. Three-dimensional structure of an enzyme molecule. *Sci. Am.* **215**:78–90.

Porges, N., L. Jasewicz, and S. R. Hoover. 1956. Principles of biological oxidation. *In* Biological treatment of sewage and industrial wastes, B. J. McCabe and W. W. Eckenfelder, editors, pp. 25–48. Reinhold, New York.

Ramanathan, M., A. F. Gaudy, Jr., and E. E. Cook. 1969. Selected analytical methods for research in water pollution control, Manual M-2. Center for Water Research in Engineering, Bioenvironmental Engineering, Oklahoma State University, Stillwater.

Sawyer, C. N., and P. L. McCarty. 1967. Chemistry for sanitary engineers. McGraw-Hill, New York.

Shreve, R. N. 1967. Chemical process industries. McGraw-Hill, New York.

Stanier, R. Y., E. A. Adelberg, and J. Ingraham. 1976. The microbial world, 4th ed. Prentice-Hall, Englewood Cliffs, N.J.

Stryer, L. 1975. Biochemistry. W. H. Freeman, San Francisco.

Watson, J. D. 1977. Molecular biology of the gene. W. A. Benjamin, Menlo Park, Calif.

Weast, R. C. (editor-in-chief). 1977. Handbook of chemistry and physics. CRC Press, Cleveland.

White, E. H. 1970. Chemical background for the biological sciences, 2d ed. Prentice-Hall, Englewood Cliffs, N.J.

FOUR

THE MICROORGANISM

Microorganisms, as the term implies, may be defined as those organisms that are too small to be visible without the aid of a microscope. They are also sometimes defined as organisms that exist as individual cells. Neither definition is totally satisfactory, although either is applicable to the majority of the entities that microbiologists recognize as falling within their area of study. In general, *protozoa, algae, fungi, bacteria,* and *viruses* are included in this broad grouping. However, the viruses, the smallest of all microorganisms, are not cells and differ in so many respects from other organisms that it is doubtful whether the word *organism* should be used in describing them. Some fungi and algae may exist as aggregates of many thousands of cells, forming plainly visible structures such as mushrooms (fungi) or seaweeds (algae). One reliable criterion by which such cellular aggregates may be distinguished from the higher plants and animals is the absence of differentiation into tissues that have distinct functions and that are composed of cells with recognizable structural modifications related to their function. Thus, in spite of the great diversity of cell types and modes of life found among microorganisms, it is possible to recognize the forms of life that belong to this category.

All living things are composed of cells, and all living organisms, whether unicellular or multicellular, perform the same indispensable functions. A single cell (or the individual cells of an undifferentiated aggregate) must of necessity perform all these functions, whereas in higher organisms certain functions may be carried out primarily by specialized cells. The following functions are indispensable to all organisms:

1. Protection from the environment, i.e., the establishment of a boundary that separates the organism from the outside world
2. Capture of nutrients
3. Production of energy in a biologically usable form
4. Metabolic conversion of foodstuff to cellular material
5. Excretion of waste products
6. Preservation and replication of genetic information

It is largely the variation in the mechanisms by which these (and other dispensable functions) are carried out that gives rise to the structural differences that are the primary basis for the classification of organisms into the traditional groupings of taxonomy.

Since many microorganisms could not be classified unequivocally as belonging to either of the two kingdoms, plant and animal, to which all other living forms belonged, Haeckel proposed in 1866 that a third kingdom be recognized—the protists. This kingdom would include protozoa, algae, fungi, and bacteria (viruses were unknown in 1866) and would be distinguished from the plant and animal kingdoms on the basis of a lack of differentiation of cells and tissues. With the advent of the electron microscope, it became obvious that there was a very fundamental difference in internal cellular structure between the bacteria and the blue-green algae on the one hand, and all other living organisms, whether plants, animals, or protists, on the other. All cells, except bacteria and blue-green algae, have complex, well-differentiated internal structures, or organelles, bounded by membranes similar to those that surround the cells. Perhaps the most important of these internal structures is the nucleus, a membrane-enclosed organelle that contains, in addition to other components, the genetic material of the cell, the DNA, organized into readily recognizable structures called *chromosomes*. In bacteria and blue-green algae, the DNA is not organized into structures recognizable as chromosomes and is not separated from the cytoplasm by a nuclear membrane. Therefore, the most basic division of living organisms is between the *eucaryotes*, either unicellular or multicellular organisms that have a true nucleus, and the *procaryotes*, which have no nucleus. The latter category includes only bacteria and blue-green algae, and the latter are now generally referred to as blue-green bacteria or cyanobacteria; thus, procaryotes and bacteria are synonymous terms. Protozoa, algae, and fungi are grouped as protists, and viruses, which are noncellular, are included in none of the above groupings. Microorganisms, then, include the *eucaryotic protists*, the *procaryotes*, and the *viruses*.

We shall consider, in this chapter, the general structural features that characterize each of these types of microorganisms and allow them to be distinguished from one another. Specialized structures of many types are found among the eucaryotic microorganisms, and these are generally important in the identification of genera and species. However, such structures are beyond the scope of this text, and we shall describe only the structures common to all eucaryotic cells or to all the organisms of major groups. Since structure is important in relation to the activities of microorganisms primarily from the viewpoint of the function performed by each structural component, the relationship between structure and function will be emphasized.

STRUCTURE AND FUNCTION IN EUCARYOTES

The typical eucaryotic cell has a complex internal structure not found in the procaryotic cell (Fig. 4-1). The characteristic organelle of the eucaryotic cell is the *nucleus*, which contains the DNA of the cell associated with basic proteins to form the chromosomes. A nucleolus, containing RNA, is often visible as a dense body within the nucleus. Two membranes perforated by pores separate the nucleus from the cytoplasm. The outer nuclear membrane is probably continuous with an extensive, complex internal membrane system, the *endoplasmic reticulum*, which forms channels throughout the cytoplasm. Parts of the endoplasmic reticulum (the rough ER) are covered with *ribosomes*, small bodies containing protein and RNA, which are the site of protein synthesis.

The apparatus responsible for trapping chemical energy in an aerobic cell is contained in the *mitochondrion*, an organelle bounded by a double membrane and containing an internal membrane system (*cristae*), within which are the components of the electron transport system and a number of enzymes associated with respiratory metabolism. Mitrochondria are absent in cells that grow anaerobically. A somewhat similar organelle, the *chloroplast*, is found in photosynthetic eucaryotes, i.e., algae. This membrane-bounded organelle also contains an internal, lamellar membrane system, which houses the pigments involved in photosynthesis as well as the photosynthetic electron transport

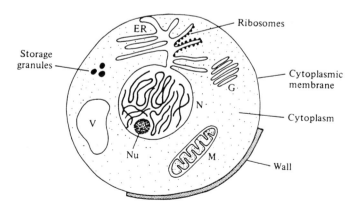

Figure 4-1 A simple diagrammatic representation of a eucaryotic cell showing commonly occurring organelles. N, nucleus; Nu, nucleolus; M, mitochondrion; V, vacuole; G, Golgi body; ER, endoplasmic reticulum.

system. Both mitochondria and chloroplasts contain small circular DNA molecules similar to those found in procaryotes, ribosomes, and other components of protein synthesis, which suggests that they may have evolved from symbiotic procaryotic cells. Both organelles, at least in some eucaryotes, have been shown to be self-replicating.

Many eucaryotic cells contain Golgi bodies, stacked membranes that function in secretion. The *Golgi apparatus* may package and transport material synthesized by the cell from one area of the cell to another or to the cell's exterior. Membrane-bounded *vacuoles* of a variety of types are found in eucaryotes. Vacuoles may contain food, water, storage products, or wastes.

All of these organelles are embedded in the *cytoplasm*, a clear semiliquid material containing in aqueous solution or suspension all the molecules synthesized and used by the cell: the soluble enzymes, vitamins, amino acids, small carbohydrates, lipids, nucleotides, coenzymes, metabolic products, and synthetic intermediates. Surrounding the cytoplasm in all cells is the *cytoplasmic membrane* or plasma membrane. This membrane, like those surrounding most cellular organelles, appears in the electron microscope as a three-layered structure called the *unit membrane* (Fig. 4-2). The two outer layers are electron-dense, whereas the inner layer is transparent to the electron beam. The total structure has a width of 7.5 to 10 nm, and the three layers have approximately equal thicknesses. The principal components of the membrane are phospholipid and protein, which constitute approximately 25 and 50 percent, respectively, of the dry weight of the membrane. It is believed that the phospholipids that have polar (hydrophilic) and nonpolar (hydrophobic) ends are oriented with the polar regions at the outer surfaces of the membrane, forming a lipid bilayer in which various proteins are embedded, whereas the hydrophobic, or nonpolar, regions of the phospholipid extend inward. The major function of the cytoplasmic membrane in eucaryotes is the regulation of the passage of molecules into and out of the cell. The membrane forms the effective boundary between the organism and its environment. Although there may be additional layers outside the membrane, a cell wall or pellicle and often a slime layer, the membrane forms the semipermeable barrier to the passage of ions and molecules. While water and some small molecules may diffuse through the membrane, most nutrients are transported into the cell by specific proteins located in or on the membrane. Such specific transport systems increase the rate of passage into the cell manyfold as compared with the rate of simple diffusion, and they also allow the cell to take up nutrients against a concentration gradient. The membrane is able to exclude materials in the environment; i.e., the cell may be impermeable to certain ions or molecules because these cannot penetrate the membrane.

In algae and fungi, a *true cell wall*, which gives strength, shape, and rigidity to the cell, surrounds the cytoplasmic membrane. Both algal and fungal cell walls contain polysaccharides of various types as the major component. The ground structure of algal walls is generally cellulose microfibrils arranged in layers, and with this may be associated proteins and polymers of mannose

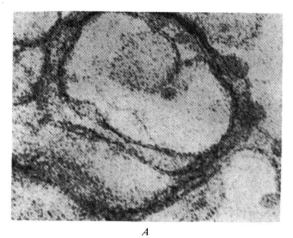

A

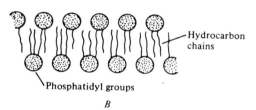

Hydrocarbon chains

Phosphatidyl groups

B

Figure 4-2 The unit membrane. (A) Electron micrograph of a preparation of purified membranes from the Gram-positive bacterium *Micrococcus lysodeikticus*, ×500,000. (*Courtesy of Edward A. Grula, Oklahoma State University.*) (B) A diagrammatic representation of the membrane as a phospholipid bilayer, which appears in the electron micrograph as two electron dense (dark) lines separated by a clear space. The electron-dense layers are the phosphate-containing polar groups oriented toward the inner and outer surfaces of the membrane. The space between is occupied by the fatty acid side chains of the lipids, which are hydrophobic.

(mannans), xylose (xylans), or uronic acid (pectin). The walls of some fungi also contain cellulose, although chitin, a polymer of *N*-acetylglucosamine, is the more common fibrous layer in fungi. As in the algal wall, other polysaccharides and proteins are associated with the cellulose or chitin. Protozoa, like the cells of higher animals, have no true cell wall. Varying degrees of rigidity are imparted to different species of protozoa by pellicles of varying thickness. The pellicle may be composed of protein in association with lipids or polysaccharides and may have a very complex structural organization in certain groups of protozoa. Both algae and protozoa include species that form shells of inorganic material such as silica or calcium carbonate. Diatomaceous earth, which has many industrial uses, is composed of the deposited shells of diatoms, algae that form intricate siliceous shells to serve as the cell wall. The most vital function of the cell wall is protection of the membrane from rupture due to osmotic pressure. The aqueous habitat of microorganisms almost always has much lower concentrations of ions and molecules than does the interior of the cell, and the difference in concentration across the membrane creates an osmotic pressure within the cell that would cause it to burst without the support of the rigid wall. Microorganisms with no supporting wall can survive in surroundings that contain high concentrations of solutes. In some protozoa that have no rigid external layer, e.g., some of the amoeboid types, the buildup of internal pressure is prevented by the contractile vacuole, which continually gathers water entering the cell and periodically expels it through the membrane.

The *flagellum* is the organ of locomotion in eucaryotic microorganisms, except in one group of protozoa that possess *cilia*. Flagella are longer than cilia and occur singly or in small numbers on each cell, whereas a single ciliated cell, e.g., *Paramecium*, may have more than 10,000 cilia. Both cilia and flagella are made up of bundles of microtubules (hollow cylinders composed of protein) surrounded by an outer membrane that is continuous with the cytoplasmic membrane. Both are attached to basal bodies within the cell.

Some eucaryotic cells form an extracellular *capsule* or *slime layer* composed of polysaccharide. The capsule of *Cryptococcus neoformans*, a pathogenic yeast, is typical of a heteropolysaccharide capsule that contains more than one type of sugar. The capsule of *C. neoformans* is composed of two hexoses, D-galactose and D-mannose; a pentose, D-xylose; and a hexuronic acid, D-glucuronic acid. The distinction between capsules and slime layers cannot be drawn clearly, but slime layers are larger and more diffuse, whereas capsules generally are more compact and conform to the shape of the cell. Capsules and slime layers are thought to provide protection against desiccation since they have a very high water content. They may also protect against attack by viruses. Pathogenic microorganisms are protected against the body's defense mechanisms, antibodies and phagocytic cells, by their capsules.

EUCARYOTIC MICROORGANISMS

The eucaryotes occur in such a variety of forms that we shall not attempt to describe them. We shall discuss briefly some of the more common forms of each major group of eucaryotes. The reader should consult one of the specialized texts listed at the end of this chapter for information on individual genera and species.

The Fungi

The fungi are, in general, the most structurally uniform group of eucaryotes. The predominant form of growth is filamentous. Individual filaments are called *hyphae*, which are very long, branching tubular structures that elongate at the tip. A hypha may have no crosswalls or it may be divided at irregular intervals by crosswalls that have pores through which cytoplasm and even nuclei may move. The cytoplasm is thus continuous throughout the mass of hyphae, which is called the *mycelium*. Nuclei and other organelles are found throughout the mycelium. Specialized hyphae differentiate into spore-bearing

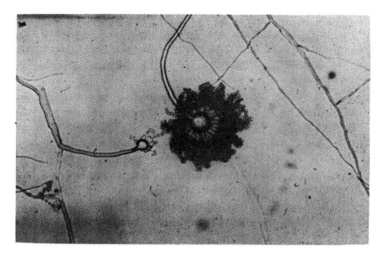

Figure 4-3 Spore-bearing aerial hypha of the fungus *Aspergillus. (Phase contrast photomicrograph, courtesy of Lynn. L. Gee, Oklahoma State University.)*

filaments (Fig. 4.3). These aerial hyphae give the fungi the familiar fuzzy appearance typical of the molds found on bread, fruit, and other surfaces. In certain fungi, large masses of mycelium aggregate to form a plantlike structure, the mushroom or toadstool, which arises from an extensive network of underground hyphae.

The *yeasts* are nonfilamentous fungi that reproduce by budding (Fig. 4-4). A small protuberance forms on the surface of the cell and enlarges until it is

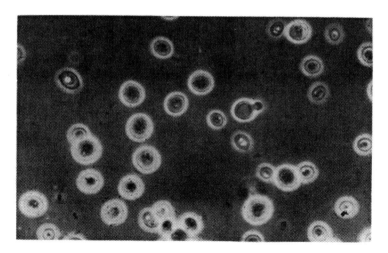

Figure 4-4 Yeast cells. Buds of different sizes are attached to the mother cells. *(Phase contrast photomicrograph, courtesy of Lynn L. Gee, Oklahoma State University.)*

approximately the same size as the mother cell. The nucleus divides as the bud enlarges and one nucleus passes into the bud. A crosswall forms between the two cells and they are physically separated.

Some fungi are *dimorphic*; i.e., they can exist either in the filamentous form or in a yeastlike or nonfilamentous form. *Geotrichum candidum* (Fig. 4.5)

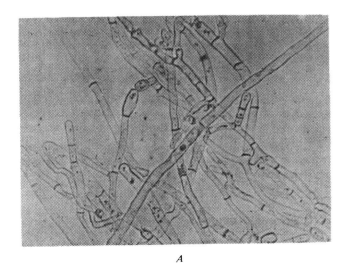

A

B

Figure 4-5 *Geotrichum candidum*, a fungus commonly found in activated sludge. (*A*) The filamentous form with branching hyphae. (*B*) Arthrospores from a pure culture of *G. candidum* added to an activated sludge. The arthrospores are readily distinguishable as the large round to rectangular bodies. *(Courtesy of Micheal Hunt, Oklahoma State University.)*

is one of the most predominant fungi in trickling filter slimes, where its growth habit is filamentous. It is also one of the filamentous forms most frequently encountered in activated sludge, and it has been implicated in sludge bulking, i.e., failure of the sludge to settle in the clarifier. *G. candidum* can grow in the form of *arthrospores*, individual, nonfilamentous cells, and this growth form is encouraged by agitation. The presence of this form of the fungus in relatively large numbers in activated sludge could account for the rapid appearance of filaments when conditions favor growth in the filamentous form.

The Algae

Many thousands of species of algae have been described. In form, these range from unicellular microscopic cells to the large aggregates of filamentous cells that resemble plants, the seaweed or kelp usually found in coastal waters. Algae are primarily aquatic organisms but are able to survive periods of desiccation, so they are found in soil, on the bark of trees, and on rocks in very moist climates or near the sea where spray wets the surface. A symbiotic association of an alga and a fungus is often found on rocks and trees. These associations, called *lichens*, are able to grow in environments that are unfavorable to the growth of either organism alone.

The unicellular algae are often motile by means of flagella or, in some forms with hard external shells, by extrusion of a portion of the cell through a groove in the shell, allowing a sort of amoeboid movement. Other unicellular species are immotile. Cells of some species aggregate into colonies that may be immotile or motile by a concerted movement of the flagella. The cells of a colony are often embedded in a mucilagenous capsule. Filamentous algae may be branched or unbranched and may or may not be divided into distinct cells by crosswalls. The life cycle of algae that are not flagellated frequently involves the formation of motile, flagellated *zoospores*, several of which may be formed and liberated from a single cell.

The algae are generally divided into three large groups based on the color imparted to the cells by the chlorophyll and other pigments involved in photosynthesis. The *green* algae are primarily fresh water inhabitants. These include unicellular, filamentous, and colonial forms. The *brown* algae are most frequently marine organisms and include several groups that are extremely abundant in the surface layers of the ocean. One such group, the *dinoflagellates*, are unicellular, flagellated organisms covered with an armor of interlocking plates. These organisms comprise the "red tides" sometimes seen in large areas of the sea. When these massive growths occur, shellfish, which consume the algae, cannot be eaten because of toxins produced by some species of dinoflagellates. These toxins affect humans but not shellfish. The *diatoms* mentioned previously are also members of this group. They produce beautifully patterned shells that fit together like the halves of a petri dish. Diatoms are said to be the most widespread form of algae, occurring in both fresh and salt water, on many surfaces and in soil. The *red* algae include a few unicellular forms but are mostly filamentous. Seaweeds are included in this group.

The Protozoa

Most protozoa are single-cell organisms (Fig. 4-6), although these individual cells often have a very complex and highly organized structure. Many also undergo complex life cycles. These life cycles, particularly in the parasitic protozoa, may involve several changes in form and structure. Life cycles and means of locomotion (flagella, cilia, and pseudopodia) form the basis for classifying the major groups of protozoa.

The flagellated protozoa include a number of organisms that are clearly protozoa and others, like *Euglena*, that share characteristics with both protozoa and algae and could be classified as either. Many of these organisms can combine the chemoheterotrophic mode of life with the photoautotrophic; i.e., they possess chloroplasts, like algae, and can utilize photosynthesis while simultaneously behaving as osmotrophic or phagotrophic protozoa, utilizing soluble or particulate organic food material. Some microbiologists (including ourselves) prefer to classify these organisms with the algae, but they represent a transitional group that does not fit neatly into either category. Other organisms classified as protozoa have obvious structural resemblances to algae and are considered to be descendants of algae that have lost their chloroplasts.

In addition to the flagellates, two other groups have been classified on the basis of means of locomotion. These are the amoeboid protozoa, which move by means of pseudopodia, and the ciliated protozoa. The amoeboid protozoa possess a flexible membrane that allows the organisms to "flow" along a

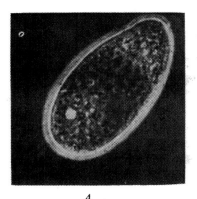
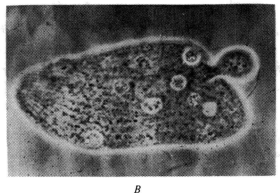

A   *B*

Figure 4-6 Phase contrast photomicrograph of two types of protozoa (unidentified) commonly found in activated sludge. *(Courtesy of Lynn L. Gee, Oklahoma State University.)*

surface by the movement of cytoplasm into an extension of the membrane. These organisms may be naked, as in *Amoeba*, or may have shells, or tests, usually of chitin, through which pseudopodia can protrude to allow movement. Since some protozoa may utilize both pseudopodia and flagella or may have both amoeboid and flagellated stages during their life cycle, the amoeboid and flagellated protozoa are now grouped together as the *Sarcomastigophora*. This group also includes some ciliated organisms that are more similar in organization and life cycle to the amoeboid and flagellated forms than to the major group of ciliates, the *Ciliophora*. The latter group includes organisms with nuclei of two types, a macronucleus and a micronucleus. Many of these organisms are very complex in structure. Sessile forms attach to surfaces by a stalk and utilize cilia for only food capture rather than for locomotion.

The third major grouping is the *Sporozoa*. These are all parasitic organisms characterized by the formation of spores as one stage in the life cycle. *Plasmodium*, perhaps the best known representative of this group, causes malaria. Other members of the group cause a variety of diseases of humans and animals.

The protozoa are important both as disease-causing organisms and as a vital link in the food chain from bacteria to higher organisms. Diseases of humans and animals are caused by amoeboid and flagellated protozoa as well as by the Sporozoa. The phagotrophic protozoa feed on bacteria and in turn serve as food for larger organisms. The larger ciliates consume not only bacteria but also eucaryotes, fungi, algae and other protozoa, and even small multicellular organisms such as rotifers. The ability of some protozoa to form cysts and of the shelled amoebae to retreat into their shells allows them to resist desiccation and thus to inhabit environments subject to periods of little moisture or drying. The free-living protozoa are abundant in soil and in fresh and salt waters. Protozoa are the most complex single-cell eucaryotic microorganisms in both structure and life cycle, and the degree of specialization within these single cells probably represents the ultimate degree of differentiation possible for a unicellular microorganism.

STRUCTURE AND FUNCTION IN PROCARYOTES

Many of the structural features of eucaryotic cells, particularly the larger ones, are readily discernible when cells are examined in a wet mount under a light microscope. However, when one views procaryotic cells under the same conditions, it is unusual to see any distinguishable cellular structures. In very large cells, such as some *Bacillus* species, one may observe storage granules, and spores are readily visible when these occur. However, other structural features can be observed only in the electron microscope or sometimes by staining. This is not surprising, since there is a tenfold difference in the average sizes of procaryotic and eucaryotic microorganisms, and many eucaryotic cells are much more than ten times larger than the average procaryote.

Since the fundamental difference between the procaryotes and the eucaryotes is the membrane-bounded organelles that perform various functions in the eucaryotic cell, it is apparent that these functions either must be absent in procaryotes or must be carried out by less complex structures. An example of a difference in functions due to lack of a specialized structure is the utilization of particulate food. Protozoa with flexible outer surfaces can surround and enclose food particles, forming a food vacuole in which digestion takes place. Many protozoa possess specialized structures for capture and ingestion of food, mouths, and sometimes even tentacles. Bacteria, having no such specialized structures and being unable to phagocytize particulate matter because the membrane is surrounded by a rigid wall, are limited to the use of soluble small molecules. Other functions for which eucaryotes possess specialized organelles are performed simply in bacteria without benefit of a specialized structure. Many fungi produce elaborate spore-bearing structures. Bacteria that produce spores do so by forming a spore within the cell; the remainder of the cell then disintegrates. The bacterial cytoplasmic membrane performs the functions assigned to mitochondria and chloroplasts in the eucaryotic cell. We shall see other examples as we describe the structure of the procaryotic cell.

The *essential* components of the typical procaryotic cell make up a very brief list: a cell wall, a cytoplasmic membrane, a single molecule of DNA, ribosomes, and the cytoplasm. Other components are dispensable, even to cells in which they are found, and are never found in some groups of bacteria. Even the wall is absent in one group, the mycoplasmas.

All procaryotic cells are divided into two major groups, Gram-positive and Gram-negative, on the basis of the Gram stain reaction that correlates with the structure of the cell wall. The distinguishing characteristic of procaryotic cell walls is the presence of a peptidoglycan (mucopeptide) layer. This material forms the rigid layer of all procaryotic cell walls, including those of the blue-green bacteria, but has not been found in any eucaryotic cell. (A few unusual bacteria lack this layer.) Two acetylated amino sugars form the basic glycan, in which the monomers N-acetylglucosamine and N-acetylmuramic acid alternate (see Fig. 4-7). A short peptide chain is attached to the carboxyl group of N-acetylmuramic acid. This tetrapeptide contains two unusual amino acids, D-alanine and D-glutamic acid; it will be recalled that L isomers of these amino acids occur normally. In many bacteria, another unusual amino acid found only in procaryotes occurs in the tetrapeptide. This is diaminopimelic acid (Fig. 4-7). This may be replaced by L-lysine in other bacteria. The presence of a dicarboxylic amino acid and a diamino acid in the tetrapeptide is essential, since these allow crosslinking of adjacent peptidoglycan chains. This crosslinking may be direct through peptide bond formation between the free amino and carboxyl groups of diaminopimelic acid or lysine and the terminal D-alanine, or additional short peptide chains

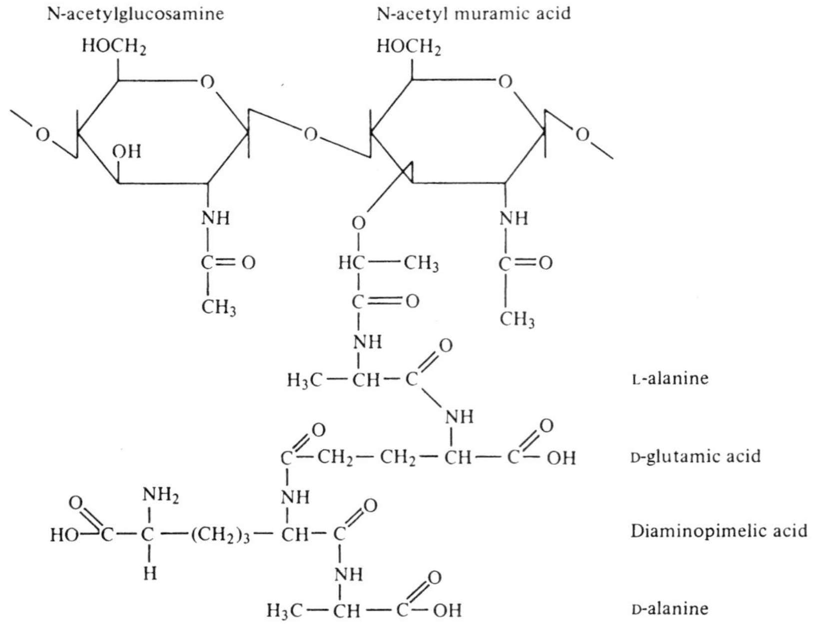

N-acetylglucosamine   N-acetyl muramic acid

HOCH₂   HOCH₂

(chemical structure diagram)

L-alanine

D-glutamic acid

Diaminopimelic acid

D-alanine

Figure 4-7 The repeating unit of the cell wall mucopeptide of *Escherichia coli.*

may be used as crosslinks, attaching to the tetrapeptides at the same points. In *Escherichia coli* the tetrapeptides are joined directly, whereas in *Staphylococcus aureus* the interbridge is a pentapeptide of glycine units. Since all parts of the wall are connected by covalent bonds, the entire structure may be considered to be one gigantic molecule whose size and shape are those of the cell and give the cell its shape and rigidity (Fig. 4-8).

We have said that almost all procaryotes may be classified as either Gram-positive or Gram-negative (some do not stain well), and this is an important consideration in bacterial taxonomy. In the Gram stain, cells are stained with crystal violet and then flooded with iodine solution. The iodine

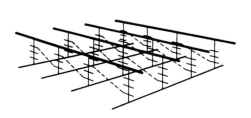

Figure 4-8 A suggested three-dimensional model of the mucopeptide layer of a Gram-positive coccus, *Staphylococcus aureus.* The solid lines represent the polysaccharide chain made of alternating molecules of *N*-acetylglucosamine and *N*-acetylmuramic acid. The vertical lines represent the peptide side chains (see Fig. 4-7). In *S. aureus*, L-lysine replaces diaminopimelic acid. The dotted lines represent the pentaglycine interbridge. *(From Rogers, 1965.)*

and crystal violet form a water-insoluble complex that is extracted readily from Gram-negative cells by an organic solvent such as alcohol but is retained by Gram-positive cells. After treatment with alcohol, Gram-positive cells are blue because they retain the blue stain, whereas Gram-negative cells have no color and may be stained with a simple stain, an aqueous solution of safranin. Gram-negative cells are then pink to red, while Gram-positive cells are blue to purple.

Much research has been devoted to an investigation of the exact mechanism of the Gram reaction, but it remains a purely empirical method for differentiating the two major groups of procaryotes. Although both the Gram-positive and Gram-negative bacteria have walls that contain peptidoglycan, there is much more of this material in the wall of Gram-positive bacteria than in that of Gram-negative bacteria. Peptidoglycan accounts for 50 to 90 percent of the dry weight of the walls of Gram-positive bacteria but amounts to 10 percent or less of the weight of the Gram-negative bacterial wall. The walls of Gram-positive bacteria also contain some polysaccharide and up to 50 percent teichoic acids, polymers containing ribitol or glycerol units joined by phosphodiester bonds. Other constituents of teichoic acids in different species may include glucose, galactose, N-acetylglucosamine or N-acetylgalactosamine, and D-alanine. The variety of amino acids found in the wall of Gram-positive bacteria is quite limited, but the Gram-negative cell wall contains protein and therefore a full complement of the amino acids. In addition to protein, the Gram-negative cell wall contains a lipopolysaccharide layer and a lipoprotein layer. The walls of the two major groups of bacteria thus are easily distinguished by chemical analysis and, in most cases, the results of chemical analyses correlate with the results of the Gram stain. Caution must be taken in interpreting the Gram stain reaction, however, since some Gram-positive bacteria become Gram-negative with age. Some species of *Bacillus*, for example, must be stained when the culture is only a few hours old to demonstrate a positive stain. In examining a Gram stain, if one finds both Gram-positive and Gram-negative cells, if these cells are of the same size and morphology, and if one has good reason to believe the culture is pure, it may be concluded that the organism is Gram-positive even though there may be a preponderance of Gram-negative cells. This is because Gram-positive cells may lose their ability to retain the crystal violet–iodine complex at a relatively early age, but aging does not convert Gram-negative to Gram-positive cells.

The correlation between Gram staining and the chemical composition of the wall suggests that the thick peptidoglycan layer of the Gram-positive cell wall is responsible for retention of the crystal violet–iodine complex. If the wall is removed by treatment with lysozyme, an enzyme that specifically hydrolyzes the backbone of the peptidoglycan (see Chap. 3), the dye complex is easily extracted with alcohol; i.e., neither the membrane nor the cytoplasm retains the dye. Whatever the mechanism, the Gram stain has been a very useful tool in the classification of bacteria, and it continues to be an important criterion for identification of bacterial genera.

Three groups of bacteria form no peptidoglycan. We have previously mentioned that the mycoplasmas have no cell wall. These are very small, irregularly shaped cells that differ from other bacteria in a number of properties. They are bounded by a single, three-layered membrane, which is given stability in some species by the presence of sterols, an unusual membrane component in bacteria. The organisms are generally parasitic or pathogenic in animals and plants, and this type of environment (within a host) contributes to their ability to survive without a rigid wall. The genus *Halobacterium* can survive only in highly concentrated salt solutions containing approximately 15 percent or more sodium chloride. These organisms have a wall of lipoprotein that is stabilized by Na^+, Cl^-, and Mg^{2+} ions and disintegrates at salt concentrations of less than about 12 percent. The methane-forming bacteria have recently been reported to have no peptidoglycan in their walls.

The cytoplasmic membrane of the procaryotic cell is basically identical in structure with that of the eucaryotic cell, i.e., a phospholipid bilayer in and on which are located various proteins. Phospholipids account for up to 40 percent of the dry weight of the membrane and form its primary structure, as we have seen previously for eucaryotic membranes. The remainder of the weight of the membrane is composed of proteins. The variety of proteins associated with the membrane is undoubtedly greater in procaryotes than in eucaryotes because, as was pointed out above, the membrane of procaryotic cells performs some functions that are assigned to special organelles in eucaryotic cells. Both eucaryotic and procaryotic membranes control the permeability of the cell, and both contain specific proteins required for the transport of ions and molecules through the membrane. In addition to these proteins, the procaryotic membrane contains enzymes that catalyze some of the reactions in the synthetic pathways for components of the membrane and the cell wall. The bacterial membrane is also the major site of energy metabolism in the cell. The electron transport system and associated enzymes are located in a separate organelle, the mitochondrion, in eucaryotic cells, but in bacteria the electron transport system and at least some dehydrogenases are located in the cytoplasmic membrane. In photosynthetic bacteria, the membrane is the site of photosynthetic energy metabolism, performing the function of the chloroplast in photosynthetic eucaryotes. In some highly aerobic bacteria, particularly autotrophs which get energy from the oxidation of inorganic compounds, and in photosynthetic bacteria, the cytoplasmic membrane is folded extensively into the cytoplasm, forming multiple layers that presumably provide much greater areas of membrane surface for the reactions of photosynthesis and oxidative phosphorylation. These internal membrane systems are apparently continuous with the cytoplasmic membrane and therefore are not considered to be separate organelles, although the distinction seems somewhat unimportant from the viewpoint of function at least. In the blue-green bacteria, the folded membranes (*thylakoids*) that function in photosynthesis may not be continuous with the cytoplasmic membrane. One additional function that has been proposed for the bacterial membrane substitutes, in a way, for the function of the nucleus in eucaryotes. It is believed that the DNA of the bacterial cell is attached to the bacterial membrane at specific sites, probably the locations of enzyme complexes involved in the replication and functioning of the DNA. The membrane would thus serve to segregate copies of the DNA into each daughter cell at the time of division.

The *ribosomes* of the procaryotic cell serve the same function as those of the eucaryotic cell; they are the site of protein synthesis. The basic architecture of both types of ribosomes is the same, but those of the eucaryotic cell are larger. In the procaryotic cell, the ribosome is made up of two subunits, a smaller 30 S subunit and a larger 50 S subunit.* Together they compose the 70 S ribosome, which functions in the synthesis of proteins. The corresponding subunits in the eucaryotic cell are 40 S and 60 S, and the functional ribosome is an 80 S particle. The 30 S subunit contains a single molecule of RNA, 16 S, and about 20 different protein molecules. The 50 S subunit contains two RNA molecules, 23 S and 5 S, and 30 different protein molecules. None of the proteins found in one subunit is also present in the other. There are very large numbers of ribosomes in bacterial cells, and the number increases with the growth rate. *Salmonella typhimurium* has more than 50,000 ribosomes per cell when growing rapidly, but fewer than 5000 per cell when growing slowly. We will discuss the role of ribosomes in protein synthesis in Chap. 12.

The genetic information of the procaryotic cell is contained in a DNA molecule that is not complexed with proteins to form a chromosome or separated from the cytoplasm by a membrane as in the eucaryotic cell. In *E. coli*, on which the most thorough studies have been done, the genetic information is carried by a single, double-stranded molecule (see Chap. 3) covalently joined to form a circle. The length of the molecule is approximately 1 mm, i.e., about 1000 times the length of the cell in which it is contained. It is obvious, therefore, that the DNA must be rather tightly packed in some manner. Electron micrographs of thin slices of bacterial cells show *nucleoids* or *nuclear areas* (Fig. 4-9). These are areas of greater electron density than the surrounding cytoplasm due to the presence of large numbers of phosphorus atoms that are heavier than the carbon, hydrogen, oxygen, and nitrogen atoms making up the bulk of the cytoplasm. In rapidly growing cells, the replication of DNA outstrips the physical separation of growing cells into daughter cells, and as many as four identical DNA molecules may be present in a single cell. Such rapid replication must require an orderly folding of the molecule. One proposed model for the organization of DNA within the cell

---

*S is the Svedberg unit and is the unit of measurement of the velocity of sedimentation in an ultracentrifuge. Both weight and shape affect the value of S, and S values for subunits are thus not additive.

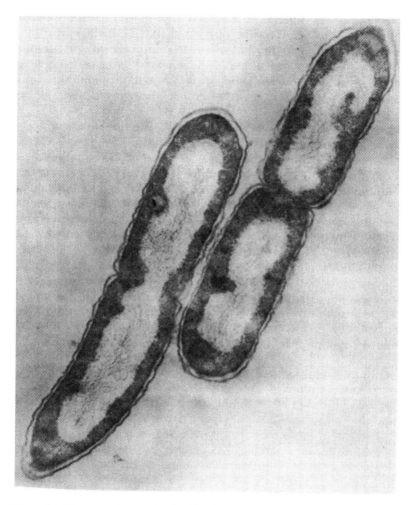

Figure 4-9 Electron micrograph of a thin slice of a Gram-negative bacterium, *Erwinia carotovora.* Both cells are in the process of dividing. In one, division is nearing completion and the nuclear (light) areas of the two daughter cells are well separated. In the other, the segregation of the nuclear areas has not been completed. Note the relatively thin cell wall of this Gram-negative bacterium. *(Courtesy of Edward A. Grula, Oklahoma State University.)*

involves the formation of a number of supercoiled loops attached to an RNA core. An enzyme called a *swivelase,* which can cut a supercoiled length of DNA to allow it to unwind for replication and then can reseal the nick, has been isolated from *E. coli.*

Additional structures found in procaryotic cells are probably dispensable; i.e., they are found in some cells but not in others and, although they may serve a useful function, they are not necessary for survival as are the structures described previously. Three types of internal structures in addition to those already described are found in certain bacterial cells. These are storage granules of several types, vesicles, and endospores. Of these, vesicles are the only structures that may be required by the organisms in which they are found.

The circumstances under which storage products accumulate in the bacterial cell are generally thought to be those in which a nutrient essential for balanced growth is lacking. Thus, when no nitrogen is available but other nutrients, such as carbon and phosphate, are present in excess, cells form polymeric products, which they store internally in granular bodies. The early formation of storage granules by cells growing on rich medium, which we have observed repeatedly in the laboratory, leads us to believe that a nutrient deficiency is not always a required stimulus for the formation of storage products. The most common forms of carbon source storage in procaryotes are polysaccharides, usually branched molecules resembling the *glycogen* stored by animal cells, and *poly-β-hydroxybutyrate* (PHB) (Fig. 4-10), a molecule that stains with fat-soluble dyes such as Sudan black. A few bacteria form a starchlike polymer. Excess inorganic phosphate is stored by many organisms in the polymeric form as *volutin granules*. PHB granules are often large enough to be clearly visible in the light microscope without staining. Volutin granules also are frequently visible without staining but can be stained with basic dyes such as toluidine blue. The reddish-purple color of the granules is the basis for the name metachromatic granules, which is often given to polyphosphate. Glycogen granules are quite small and can be seen only in the electron microscope. The polymerization of material to be stored inside the cell renders it osmotically inactive and allows the cell to accumulate large amounts of nutrients without increasing the internal pressure of the cell. Physical compactness of polymers is also advantageous. In Chap. 10 we shall discuss the uptake and storage of nutrients and their subsequent utilization in relation to the activated sludge process.

A somewhat different type of storage occurs in some cells that utilize reduced sulfur compounds as an energy source. Sulfides are generally oxidized to elemental sulfur by these cells, and in large cells such as the filamentous bacterium *Beggiatoa* or the purple sulfur bacteria, the sulfur is stored in granules inside the cell (Fig. 4-11). The sulfur is generally oxidized further. In this case, no special storage product is synthesized; sulfur is a normal intermediate in the pathway of the oxidation of sulfide to sulfate. Sulfur granules are plainly visible in the light microscope as very shiny particles in the cell (see Fig. 4-11).

Three types of vesicular bodies are found in certain procaryotic cells. These are bounded by single-layered protein membranes and are not considered to be exceptions to the rule that procaryotic cells have no organelles comparable to those in eucaryotic cells, i.e., bounded by unit membranes. The most frequent of these are the *gas vesicles*, hollow cylinders filled with gas and visible in the microscope as vacuoles, each composed of several vesicles.

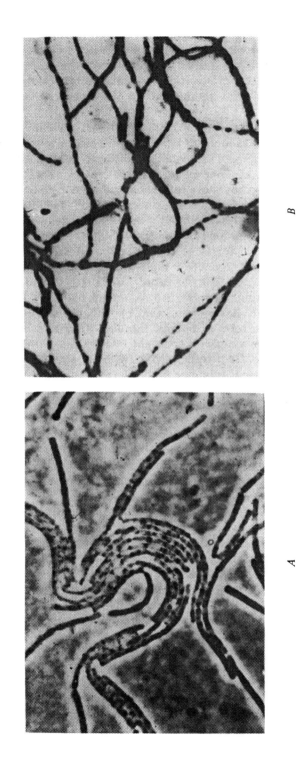

B

A

Figure 4-10 *Bacillus cereus* with numerous large granules of poly-β-hydroxybutyrate. (*A*) Cells growing on agar, unstained, phase contrast photomicrograph. (*B*) Cells stained with Sudan black B, a lipid stain that also stains poly-β-hydroxybutyrate. (*Courtesy of Lynn L. Gee, Oklahoma State University.*)

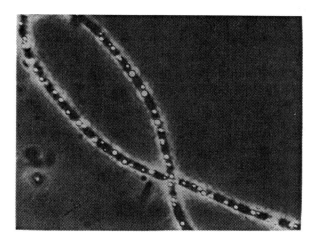

Figure 4-11 *Beggiatoa* with sulfur granules. The organism was obtained from sulfide-containing water (sulfur springs) in Platt National Park at Sulphur, Oklahoma. *(Courtesy of Lynn L. Gee, Oklahoma State University.)*

These vacuoles are flotation devices that occur in aquatic bacteria of a number of genera, including many blue-green bacteria. They apparently allow the cells to localize at depths where conditions are most favorable for growth. A somewhat similar organelle, the *Chlorobium* vesicle, is found only in the green sulfur bacteria. These are cylindrical vesicles located just beneath the cell membrane and containing the photosynthetic apparatus of the cell. In certain bacteria that use carbon dioxide as the carbon source, the enzyme responsible for carbon dioxide fixation (see Chap. 9) is located in bodies called *carboxysomes.*

Most bacteria do not form spores. Those that do belong almost exclusively to two genera of Gram-positive rods, *Bacillus* and *Clostridium.* Spores are much more resistant to radiation and desiccation than are vegetative cells and are remarkably resistant to heat, so that much of the difficulty experienced in heat sterilization of foods arises from the presence of spores. A single spore is formed for each cell, and the formation involves a complex series of events, resulting finally in the enclosure of a DNA molecule and a portion of the cytoplasm within a heavy wall, the multilayered spore coat. The spore may remain within the cell for a short time or may be liberated rapidly by disintegration of the vegetative cell. The spore is dormant until induced to germinate, usually by heat treatment, and spores may remain viable in the dormant state for many years. Spores are readily identifiable in wet mounts as highly refractile oval or spherical bodies with heavy walls. They are permeable to stains only after severe heating that damages the wall (Fig. 4-12).

Other variable components of bacterial cells are found outside the wall. These are capsules or slime layers, flagella, pili or fimbriae, and stalks, filaments, or other organs of attachment. We have mentioned the formation of capsules and slime layers by some eucaryotic organisms. The occurrence of these external layers is probably much more common among procaryotes.

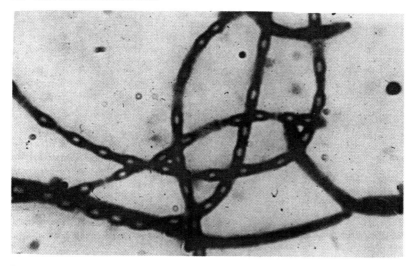

Figure 4-12 *Bacillus cereus* stained with crystal violet. Since the spores are impermeable to stains unless they have been heated drastically, they are visible as unstained oval areas in the cells. *(Courtesy of Lynn L. Gee, Oklahoma State University.)*

Most bacterial capsules and slimes are polysaccharides, either homopolysaccharides containing only one sugar or heteropolysaccharides containing several different monomers. A few bacteria form a polypeptide capsule containing D-glutamic acid. The capsules and slime layers of bacteria serve the same functions previously mentioned—protection from phagocytosis and from desiccation, at least temporarily. Capsular material binds cells together to form aggregates and may trap cells of other species as well. This may be important in floc formation (see Chap. 10). Capsules are visible in the microscope only against a dark background. They are best viewed in a wet mount with India ink (Fig. 4-13).

Many procaryotic species are permanently immotile; others are motile by means of flagella. A few types of bacteria possess a gliding motion that allows them to move over surfaces by an unknown mechanism; these cells are immotile in liquid. Two major patterns of distribution of flagella are found in bacteria, and these are of some taxonomic importance. *Polar* flagella are located only at the end of a rod-shaped cell (Fig. 4-14). These may be single or in tufts and may occur at one end or both. The arrangement of flagella in a tuft is called *lophotrichous*. In the *peritrichous* pattern of flagellation, flagella may occur anywhere on the cell, and they are often present in large numbers over the entire cell surface. Flagella arise within the membrane and pass through the wall. Bacterial flagella are thin tubular structures composed of spirally arranged protein molecules. Their diameter is much smaller than that of eucaryotic flagella or cilia, so they are not visible except in the electron microscope or in the light microscope after staining with a mordant that increases their diameter. Motility can be determined by observing young cells

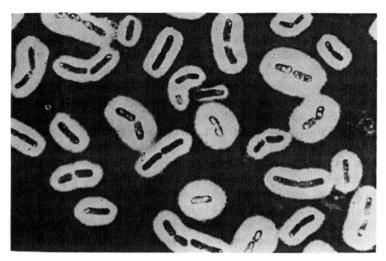

Figure 4-13 A capsule-forming *Bacillus* species suspended in India ink. *(Courtesy of Lynn L. Gee, Oklahoma State University.)*

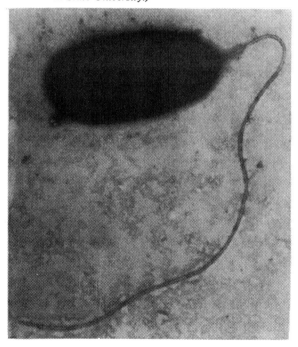

Figure 4-14 Electron micrograph of *Pseudomonas aeruginosa*, showing its single polar flagellum, ×58,000. *(Courtesy of Sue Pekrul and Brenda Mears, Oklahoma State University.)*

in a wet mount, and with practice, the type of flagellation often can be determined with considerable accuracy. Polarly flagellated cells move much more erratically than do peritrichously flagellated ones, having a tendency to somersault and tumble, whereas peritrichously flagellated cells are able to move in straight lines and reverse direction without turning. (Consider the difference in the motions of a shell rowed by a crew and a canoe propelled by a single paddle used only at the end of the canoe.) Flagella are advantageous to microorganisms under certain circumstances; they allow them to move toward higher concentrations of nutrients (*chemotaxis*) or away from toxic substances. Photosynthetic organisms exhibit *phototaxis*; exposed to light and dark areas, the cells accumulate in the lighted area by avoidance of the dark rather than by attraction toward the light.

Pili or fimbriae are found on some Gram-negative bacteria. These are somewhat similar to flagella in that they are also thin tubular structures made of protein molecules arranged spirally. However, they are shorter and thinner than flagella and straight, whereas flagella are wavy. Pili occur in large numbers, covering the cell surface with hairlike projections visible only by electron microscopy. The function of pili is not known, but they apparently aid the cell in adhering to surfaces, since they have "sticky" ends. Cells that produce pili often form a thin, tough pellicle of growth at the surface of a liquid culture, apparently held together by intertwining of the pili and attachment to adjacent cells. It has been proposed that pili may aid pathogenic bacteria in attaching to the surface of the host cells. A recent study of gliding motility showed pili at the end of the gliding cell (*Beggiatoa*) and suggested that pili might be involved in this type of motility by attaching to

Figure 4-15 A rosette formed by the stalked (prosthecate) bacterium *Caulobacter vibrioides.* *(Courtesy of Edward A. Grula, Oklahoma State University.)*

the surface in front of the cell and allowing the cell to pull itself along the surface. Certain special types of pili are formed by cells that carry extra genetic information in the form of plasmids, small DNA molecules that are concerned with the exchange of genetic information between cells. These pili occur in small numbers per cell and connect two mating bacterial cells.

A few types of bacteria form stalks or filaments by which they attach to solid surfaces or remain attached to each other (Fig. 4-15). These are generally continuous with the cell wall and membrane. Other cells produce filaments that bear daughter cells. Cells forming filaments and stalks are *prosthecate* organisms. The cells of some filamentous bacteria are enclosed in a sheath, a transparent tubular structure of polysaccharide, which is visible when empty in a phase contrast microscope. These sheaths, like some of the prosthecae formed by other bacteria, may terminate in a holdfast, an adhesive secretion that allows them to remain attached to a solid surface.

PROCARYOTIC MICROORGANISMS

The procaryotes may be divided into two major groups along historical lines. The first of these is the cyanobacteria, or blue-green bacteria, which were classified as algae until recently. The other group includes all the diverse organisms that traditionally have been considered to be bacteria.

Blue-green Bacteria

The blue-green bacteria are procaryotic, having no nucleus, mitochondria, chloroplasts, or other organelles typical of eucaryotic cells, and having a layer of peptidoglycan in the cell wall. However, their metabolism is quite similar to that of algae and plants. They are photosynthetic and they produce oxygen as do algae and plants.

The blue-green bacteria derive their color from the pigments associated with photosynthesis and, while most are actually blue-green, some have red pigments that impart a reddish or orange color to the cell. Blue-green algae may occur as single cells or as branched or unbranched filaments. They lack flagella, and motile species move by gliding on solid surfaces. Gas vacuoles are common in these organisms. Some species form a type of resting cell called an *akinete*; these are enlarged cells with heavy walls that are very resistant to desiccation. Other specialized cells called *heterocysts* are formed by some of the filamentous species (Fig. 4-16). These are rounded cells that contain no thylakoids (the layers of membrane that function in photosynthesis). It is believed that the heterocysts function in the fixation of nitrogen. Both unicellular and filamentous species frequently form gelatinous sheaths, which bind individual cells or filaments together in colonies or bundles.

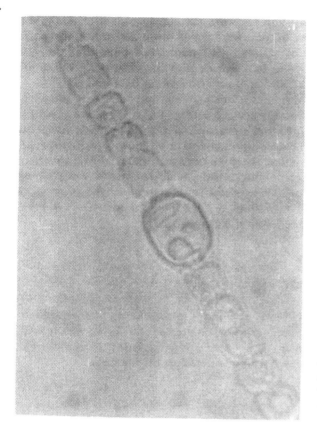

Figure 4-16 A portion of the filament of the cyanobacterium, *Anabena*, showing a heterocyst, the nitrogen-fixing cell. Other cells contain chlorophyll and are photosynthetic. *(Courtesy of Micheal Hunt, Oklahoma State University.)*

Bacteria

Some groups of microorganisms classified as bacteria are very similar in size and shape to blue-green bacteria and have the same type of gliding motility. They are nonphotosynthetic and may be descendants of blue-green bacteria that have lost the ability to form chlorophyll. Most bacteria, however, are unicellular organisms of much simpler morphology and smaller size than the majority of blue-green bacteria. Cells are spherical (cocci), rod-shaped, or spiral. Cocci may divide (1) in a single plane, forming chains; (2) in two planes, forming tetrads or larger sheets of cells; (3) in three planes, forming cubical packets; or (4) in any plane, forming irregular clusters. These cellular arrangements are important in the identification of bacteria since they are characteristic of the genus and species. Rods occur most often as single cells, but in some genera they may remain attached after division, forming chains. Since bacteria exhibit more variation in biochemical abilities and less variation in morphology than do other microorganisms, their biochemistry has been more thoroughly studied and is more important in differentiating genera and species. We will discuss the major groups of bacteria in Chap. 8.

THE VIRUSES

Viruses differ from cellular organisms in several very important ways:

1. They are much simpler in structure and composition. The two essential components of a virus are protein and nucleic acid and many contain no other compounds. Others are composed of protein and nucleic acid surrounded by a membrane of lipid and protein.
2. All cells contain both RNA and DNA, but a virus has only one type of nucleic acid. For viruses that contain RNA, the genetic information is carried by RNA rather than DNA. Cellular DNA is a double-stranded molecule that is circular in bacteria, and cellular RNA is single-stranded. Viral nucleic acids occur in several forms. DNA may be single- or double-stranded, linear or circular. RNA is almost always single-stranded, but a few viruses contain double-stranded RNA.
3. Viruses have no metabolic enzymes, use no nutrients, and produce no energy. They are obligate intracellular parasites and can reproduce only inside a living cell. The parasitism of viruses is unlike that of parasitic bacteria, protozoa, or fungi, which receive nutrients from the host cell. In a cell infected by a virus, the metabolism of the cell is diverted from the manufacture of new cell material to the manufacture of new viruses. Outside a cell, the virus is completely inert and inactive. Some viruses can even be crystallized. In short, they behave like large molecules of nucleoprotein.
4. Viruses do not reproduce as cells do by increasing in size and dividing. When a virus infects a cell, the protein and nucleic acid are separated; i.e., the virus no longer exists as an entity, and often only the nucleic acid enters the cell. Under the direction of the viral nucleic acid, the cell produces the viral proteins, the viral nucleic acid is replicated, and new viruses are assembled from the parts.

The architecture of all virus particles involves a protein shell surrounding and protecting the nucleic acid. This protein coat is made up of individual protein molecules of relatively small size. In some viruses all the protein molecules are identical; in others several types of protein occur. The protein molecules surrounding the nucleic acid are arranged in either helical or icosahedral (cubic) symmetry. Viruses with helical symmetry may be rigid or flexible rods, or the helical nucleoprotein may be a flexible ribbon wound into a sphere and wrapped in a membrane, as in influenza virus. The outer membrane of a virus is generally acquired from the cell as the virus is released. It is thus a typical cytoplasmic membrane, but it contains additional proteins made under the direction of the nucleic acid of the virus. Some viruses have a complex structure with a head containing the nucleic acid and a tail that attaches to the host cell.

The largest known animal virus is the smallpox virus, a brick-shaped virus with a thick fatty membrane measuring 200 nm in its longest dimension. This is slightly smaller than a small bacterial cell and is too small to be visible in an ordinary microscope. Most viruses are considerably smaller than the smallpox virus, although some filamentous plant viruses are much longer. All viruses are therefore visible only in the electron microscope. The size of a virus bears no relation to the size of the host it infects. Some of the larger and more complex viruses with both heads and tails infect bacteria whereas the human poliovirus is a tiny spherical particle only 27 nm in diameter.

Viruses are classified as plant, animal, or bacterial and are further classified on the basis of symmetry (helical or icosahedral), type of nucleic acid (RNA or DNA), and the presence or absence of a membrane. The plant and animal viruses are generally named for the disease they cause. Viruses are now the most important human pathogens, since bacterial diseases can be treated with a variety of antibiotics; these are not useful in treating viral diseases, because viruses depend upon the host cell for reproduction and it would be necessary to inhibit cellular processes to prevent virus multiplication. The prevention of viral infection by immunization and adequate sanitary procedures is therefore essential. The latter is particularly important for viruses for which no vaccine is available. In Chap. 14 we shall discuss further the problem of pathogenic microorganisms in the environment.

PROBLEMS

4.1 Both bacteria and viruses pathogenic to humans are excreted by infected individuals. Based on the differences between them, which would you expect to be able to survive longer in relatively unpolluted water? Why? Which might be able to multiply in a treatment process such as an oxidation pond? Explain.

4.2 At the beginning of this chapter, six functions that are indispensable to all organisms were listed. Compare the cellular structures or components involved in each of these functions in procaryotic and eucaryotic cells. It may be desirable to consult a text on microbiology such as Brock's *Biology of Microorganisms* (1979).

4.3 List ten structural features of microorganisms, procaryotic or eucaryotic, that are composed entirely or partly of protein.

4.4 Why are the organisms previously classified as blue-green algae now considered to be bacteria?

4.5 What function does each of the following perform for the cell?
- (*a*) Cell wall
- (*b*) Slime layer
- (*c*) Ribosome
- (*d*) Storage granules
- (*e*) Mitochondrion
- (*f*) Gas vesicle
- (*g*) Contractile vacuole

4.6 List the characteristics of viruses that distinguish them from all other microorganisms.

REFERENCES AND SUGGESTED READING

Adler, J. 1975. Chemotaxis in bacteria. *Annu. Rev. Biochem.* **44**:341–356.

Alexopoulos, C. J. 1962. Introductory mycology, 2d ed. John Wiley, New York.

Archibald, A. R. 1974. The structure, biosynthesis and function of teichoic acid. *Adv. Microb. Physiol.* **11**:53–95.

Bartnicki-Garcia, S. 1968. Cell wall chemistry, morphogenesis, and taxonomy of fungi. *Annu. Rev. Microbiol.* **22**:87–108.

Braun, V., and K. Hantke. 1974. Biochemistry of bacterial cell envelopes. *Annu. Rev. Biochem.* **43**:89–121.

Brock, T. D. 1979. Biology of microorganisms, 3d ed. Prentice-Hall, Englewood Cliffs, N.J.

Chapman-Andresen, C. 1971. Biology of the large amoebae. *Annu. Rev. Microbiol.* **25**:27–48.

Decad, G. M., and H. Nikaido. 1976. Outer membrane of Gram-negative bacteria. XII. Molecular-sieving function of cell wall. *J. Bacteriol.* **128**:325–336.

DiRienzo, J., K. Nakamura, and M. Inouye. 1978. The outer membrane proteins of Gram-negative bacteria: Biosynthesis, assembly and functions. *Annu. Rev. Biochem.* **47**:481–532.

Doetsch, R. N., and T. M. Cook. 1973. Introduction to bacteria and their ecobiology. University Park Press, Baltimore.

Douglas, John. 1975. Bacteriophages. Chapman and Hall, London.

Edmondson, W. T. (editor) 1959. Fresh water biology (Ward & Whipple), 2d ed. John Wiley, New York.

Ellwood, D. C., and D. W. Tempest. 1972. Effects of environment on bacterial wall content and composition. *Adv. Microb. Physiol.* **7**:83–117.

Fenner, F., B. R. McAuslan, C. A. Mims, J. Sambrook, and D. O. White. 1974. The biology of animal viruses, 2d ed. Academic Press, New York.

Fenner, F., and D. O. White. 1970. Medical virology. Academic Press, New York.

Fuller, R., and D. W. Lovelock (editors). 1976. Microbial ultrastructure. Academic Press, New York.

Gander, J. E. 1974. Fungal cell wall glycoproteins and peptidopolysaccharides. *Annu. Rev. Microbiol.* **28**:103–119.

Glauert, A. M., and M. J. Thornley. 1969. The topography of the bacterial cell wall. *Annu. Rev. Microbiol.* **23**:159–198.

Goldfine, H. 1972. Comparative aspects of bacterial lipids. *Adv. Microb. Physiol.* **8**:1–58.

Griffiths, A. J. 1970. Encystment in amoebae. *Adv. Microb. Physiol.* **4**:106–129.

Hawker, L. E., and A. H. Linton. 1971. Micro-organisms. Function, form and environment. American Elsevier, New York.

Hirsch, P. 1974. Budding bacteria. *Annu. Rev. Microbiol.* **28**:391–444.

Jahn, T. L., and F. F. Jahn. 1949. How to know the protozoa. Wm. C. Brown, Dubuque, Iowa.

Lang, N. J. 1968. The fine structure of blue-green algae. *Annu. Rev. Microbiol.* **22**:15–46.

Lascelles, J. 1968. The bacterial photosynthetic apparatus. *Adv. Microb. Physiol.* **2**:1–42.

Leedale, G. F. 1967. Euglenida/Euglenophyta. *Annu. Rev. Microbiol.* **21**:31–48.

Lemke, P. A. 1976. Viruses of eucaryotic microorganisms. *Annu. Rev. Microbiol.* **30**:105–145.

Lewin, J. C., and R. R. L. Guillard. 1963. Diatoms. *Annu. Rev. Microbiol.* **17**:373–414.

Luria, S. E., J. E. Darnell, Jr., D. Baltimore, and Allen Campbell. 1978. General virology, 3d ed. John Wiley, New York.

Murrell, W. G. 1967. The biochemistry of the bacterial endospore. *Adv. Microb. Physiol.* **1**:133–251.

Pfennig, N. 1977. Phototrophic green and purple bacteria: a comparative systematic survey. *Annu. Rev. Microbiol.* **31**:275–290.

Phaff, H. J. 1963. Cell wall of yeasts. *Annu. Rev. Microbiol.* **17**:15–30.

Pollock, M. R., and M. H. Richmond (editors). 1965. Function and structure in microorganisms, 15th Symposium of the Society for General Microbiology. Cambridge University Press, Cambridge, England.

Prescott, G. W. 1968. The algae: A review. Houghton Mifflin, Boston.
Prescott, G. W. 1970. How to know the freshwater algae. Wm. C. Brown, Dubuque, Iowa.
Razin, S. 1969. Structure and function in mycoplasma. *Annu. Rev. Microbiol.* 23:317–356.
Reaveley, D. A., and R. E. Burge. 1972. Walls and membranes in bacteria. *Adv. Microb. Physiol.* 7:1–81.
Rogers, H. J. 1965. The outer layers of bacteria: The biosynthesis of structure. *In* Function and structure in microorganisms. 15th Symposium of the Society for General Microbiology. Cambridge University Press, Cambridge, England.
Salton, M. R. J., and P. Owen. 1976. Bacterial membrane structure. *Annu. Rev. Microbiol.* 30:451–482.
Schmidt, J. M. 1971. Prosthecate bacteria. *Annu. Rev. Microbiol.* 25:93–110.
Seaman, G. R., and R. M. Reifel. 1963. Chemical composition and metabolism of protozoa. *Annu. Rev. Microbiol.* 17:451–472.
Shaw, N. 1975. Bacterial glycolipids and glycophospholipids. *Adv. Microb. Physiol.* 12:141–167.
Shively, J. M. 1974. Inclusion bodies of procaryotes. *Annu. Rev. Microbiol.* 28:167–187.
Silverman, M., and M. I. Simon. 1977. Bacterial flagella. *Annu. Rev. Microbiol.* 31:397–419.
Smith, R. W., and H. Koffler. 1971. Bacterial flagella. *Adv. Microb. Physiol.* 6:219–339.
Stanier, R. Y., E. A. Adelberg, and J. Ingraham. 1976. The microbial world, 4th ed. Prentice-Hall, Englewood Cliffs, N.J.
Stanier, R. Y., and G. Cohen-Bazire. 1977. Phototrophic procaryotes: The cyanobacteria. *Annu. Rev. Microbiol.* 31:225–274.
Stewart, W. D. P. (editor). 1974. Algal physiology and biochemistry. University of California Press, Berkeley.
Sutherland, I. W. 1972. Bacterial exopolysaccharides. *Adv. Microb. Physiol.* 8:143–213.

NUTRITION AND GROWTH CONDITIONS AS SELECTIVE AGENTS IN NATURAL POPULATIONS

The importance of microorganisms in the recycling of organic and inorganic matter has been emphasized in previous chapters. Microorganisms, particularly bacteria, are able to perform this role in nature because they exist in an almost infinite variety of species with different metabolic requirements and capabilities. It is believed by most microbiologists that all naturally occurring materials and all but a very few synthetic materials are subject to microbial attack.

The activities of microorganisms, from the human viewpoint, are either useful or harmful; indeed, the same activity may be useful or harmful depending on the environment in which it occurs. The unexpected growth of lactic acid bacteria such as *Streptococcus lactis* in milk irritates those who drink milk, but the growth of these same bacteria in milk under controlled conditions is the basis of important manufacturing processes in the dairy industry. The same organism that can cause gangrene in a puncture wound due to its ability to digest protein anaerobically is a valuable member of the mixed population in an anaerobic digester. These simple examples illustrate the need for an understanding of the growth habits of microorganisms by environmental microbiologists and environmental engineers who must deal with and often attempt to control the activities of microorganisms in their natural habitats.

THE MICROORGANISM AND ITS HABITAT

A natural habitat does not imply an environment unaffected by human activity, but rather one in which the species comprising the microbial population are those selected by interaction with the environment and with each other. Microorganisms, due to their small size, are readily dispersed over huge distances by air and water and by animate and inanimate agents. Thus, different types of microorganisms are constantly being introduced into a habitat and given the opportunity to proliferate in it. Physical and nutritional conditions select from the mixture of organisms those best adapted to that environment. Selection operates only upon differences among species, but even a small difference conferring a slight selective advantage can lead to relatively large variations in numbers of different species within a short time because of the extremely rapid rates of growth of which microorganisms are capable. Changes in conditions, whether imposed by external forces or brought about by the microorganisms themselves, can result in rapid changes in relative numbers of different species. Many environments are capable of supporting the growth of many different microbial species, and it is within such environments that the balance of species may shift frequently and rapidly due to relatively small changes in nutrient supply or physical conditions. Other environments are quite restrictive and are inhabited by only a few specialized types of microorganisms.

Knowledge of the nutritional and physical factors affecting microbial growth in general and the growth of individual species in particular is necessary to predict the types of organisms that may be expected to predominate in various habitats and to suggest measures that might be used to shift the population toward more desirable or useful types. While it would be ideal to study microorganisms under natural conditions, we can learn very little about their characteristics and activities without subjecting them to experimentation in laboratory conditions. Due to their small size, direct observation is not possible; nor is it possible by observing the entire population to determine which of the species comprising a mixed microbial population is responsible for any single activity. Each organism must be isolated and studied in pure culture so that its requirements and activities can be determined [see Brock (1966) for an excellent discussion of the role of pure cultures in the study of microbial ecology]. After the characteristics of the individual organisms have been determined, we are more apt to understand their behavior and interactions in the natural habitat. It is also desirable to work with mixed populations of microorganisms under controlled conditions in the laboratory. This allows us to study their interactions or simply to determine the collective properties of the mixture of species selected by the conditions imposed. In any case, a knowledge of the effects of various physical and nutritional factors on microbial growth is essential to any effort to understand or control the microbial inhabitants of a natural environment.

Both the physical and chemical characteristics of the environment influence microbial growth. Physical factors in general act as selective agents by determining the types of organisms that are able to grow and by influencing the growth rates of those organisms that can grow under the prevailing conditions. Chemical factors may or may not be selective. Some elements such as carbon and nitrogen, which are required in relatively large amounts and often can be used only in specific forms, may be very important in selecting predominant species. Other elements such as magnesium or iron, which are used in the same form by all microorganisms and are required in very small amounts, may exert little or no selective pressure.

A great variety of environmental factors that influence microbial growth can be enumerated. We shall emphasize in this chapter the factors most commonly encountered and those that are most effective in selecting the microbial species that can occupy a habitat, because these are the factors that might most feasibly be used in exerting technological control over microbial populations.

TEMPERATURE AND MICROBIAL GROWTH

Of the physical factors affecting microbial growth in any environment, one of the most influential in the selection of species is temperature. Microorganisms possess no means of controlling internal temperature, and the temperature within the cell is therefore determined by the external temperature. Each microorganism is able to grow only within a specific range of temperatures. This range may be quite narrow, e.g., less than 10°C for the human pathogen *Neisseria gonorrhoeae*. Organisms with such a restricted temperature range are called *stenothermal*. Most microorganisms, however, are *eurythermal*; that is, they are capable of growing over a range of 30 to 40°C. Fortunately, many human pathogens, having adapted to an environment—the human body— where the temperature is constant, are stenothermal and incapable of growing at temperatures more than a few degrees lower or higher than 37°C. This prevents their proliferation in surface waters or in such waste treatment facilities as oxidation ponds, activated sludge tanks, and trickling filters, except possibly during very hot weather.

The Cardinal Temperatures

Three temperature values, often called the *cardinal temperatures*, are commonly used to characterize the effect of temperature on the growth of a microbial species or strain. The maximum and minimum temperatures define the limits of the temperature range within which growth is possible; at lower or higher temperatures than these, no growth occurs. The *optimum* temperature is that at which growth is most rapid, i.e., at which the growth rate reaches its maximum value. The optimum temperature for most microorganisms is much closer to the maximum than to the minimum temperature.

It should be noted that the optimum temperature is based on the growth rate only and is not necessarily the temperature at which the maximum yield of cells is produced. The cardinal temperatures are considered to be characteristic of the organism, although they may be altered to some extent by other environmental factors such as pH, the concentrations of salts, or the available nutrients.

Growth at the minimum temperature is typically quite slow; the rate increases exponentially with increasing temperature, reaching a maximum at the optimum temperature and falling abruptly to zero at a few degrees above the optimum (see Fig. 5-1). For most microorganisms, the growth rate increases two- to threefold for each 10°C rise in temperature between the minimum and the optimum.

While any single microbial species is incapable of growth over a range of temperatures greater than about 40°C, different microorganisms can grow at temperatures from below 0°C to above 90°C. Based on their *optimum* growth temperatures, microorganisms are classified as *psychrophiles*, *mesophiles*, or *thermophiles*. Most microbial species are mesophilic, growing most rapidly at temperatures between 20 and 45°C. Psychrophiles have optimum temperatures below 20°C, whereas thermophiles grow most rapidly above 45°C.

Many microorganisms that have been reported to be psychrophilic do not fit the present definition of this group. Some microbiologists have considered any microorganism capable of multiplication at 0°C to be psychrophilic. Most of these organisms were actually mesophiles with broad temperature ranges that could grow only very slowly at 0°C. More recent studies have shown that *true psychrophiles* have maximum temperatures for growth of approximately 20°C and optimum temperatures of 15°C or below. Mesophiles that can grow at temperatures within the psychrophilic range have been called *facultative psychrophiles*, but a preferable term is *psychrotroph*. Since the temperature of most ocean waters is permanently well below 20°C, the removal of organic matter polluting these waters, e.g., oil spills, must be primarily dependent on

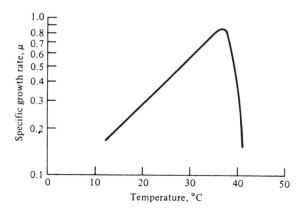

Figure 5-1 Hypothetical plot of the relationship between growth rate μ and temperature for a mesophilic bacterium with an optimum temperature for growth of 37°C and a Q_{10} value of 2. Actually Q_{10} values may not be constant over the entire temperature range.

the activity of psychrophilic microorganisms. Psychrotrophs, and possibly some psychrophiles, are important in the function of biological waste treatment facilities in cold climates, especially in winter. Psychrophilic microorganisms are also important nuisance organisms when they grow in refrigerated foods and other materials stored at low temperatures.

Growth at temperatures of 0°C and below occurs only where water exists in the liquid state, as in water with high concentrations of salts. Most bacteria that have been shown to grow at temperatures below 0°C in the laboratory are marine organisms. Both fungi and algae have been found in arctic regions, often growing in association as lichens. Algae, which are readily observed due to their pigmentation, are commonly seen on the surface of snow and under ice on oxidation ponds and reservoirs.

Among *thermophilic* microorganisms, those capable of growing at very high temperatures, i.e., above 60°C, are all procaryotes (see Table 5-1). Such extremely high temperatures are not found in many natural environments. Both blue-green bacteria and thermophilic species of eubacteria are commonly found in hot springs. Thermophilic bacteria also have been found in domestic hot water heaters and in heated industrial process waters. These process waters, when discharged into receiving streams, may increase water temperatures sufficiently to make a profound difference in the metabolic activities of the microbial population (e.g., see White et al., 1977).

It is important to realize that growth and survival are not necessarily affected in the same way by temperature. Temperatures above the maximum at which growth can occur are generally lethal and thus affect both growth and viability. Temperatures below the minimum at which growth is possible are not normally lethal and thus affect growth but not viability. Microorganisms may remain dormant without growing for considerable periods of time at low temperatures, and this fact is used to advantage by microbiologists since it offers a convenient method of storage of microorganisms. Cultures of mesophilic and thermophilic microorganisms are commonly stored at 4°C. Storage at very low temperatures in the frozen state is possible

Table 5-1 Approximate upper temperature limits for different microorganisms

Organism	Temperature, °C
Protozoa	45–50
Eucaryotic algae	56
Fungi	60
Photosynthetic bacteria (including cyanobacteria)	70–73
Bacteria	>99

Source: Brock, 1979.

with proper precautions regarding the suspending medium and rate of freezing. Some loss of viability occurs, but a fraction of the population survives.

Time is an important factor in considering the effect of temperature, particularly temperatures above the range of growth. Some effects of elevated temperatures are reversible, and a short exposure to an elevated temperature may not be lethal although longer exposure at that same temperature would be. The higher the temperature, of course, the less time is required for killing and the greater is the probability that irreversible damage will occur.

Molecular Basis of Effects of Temperature

The molecular basis of temperature limitation of growth is not understood. Two classes of molecules, *lipids and proteins*, have been implicated in various mechanisms proposed to explain the effects of temperature on growth and viability. Both types of molecules are affected in known ways by changes in temperature, and both perform essential functions in the cell.

Lipids and temperature Lipids containing fatty acids are essential structural components of membranes, both the cytoplasmic membrane common to all cells and the internal membranes of eucaryotes and some procaryotes. The melting points of such lipids increase with the degree of saturation of the fatty acids. The cytoplasmic membrane must maintain the proper balance of fluidity and structural integrity to allow control of the passage of molecules in and out of the cell while preventing the loss of essential cellular contents. The importance of the content of saturated fatty acids to organisms growing at different temperatures is evident when the composition of lipids in the membrane of a psychrophilic species is compared with that in closely related mesophilic and thermophilic species. There is a direct correlation between the degree of saturation of the fatty acids of the membrane and the temperature range within which the organism is able to grow. An individual microorganism can also respond to growth at different temperatures within its tolerable range by changing the degree of saturation of its fatty acids. *Escherichia coli*, for example, can vary the percentages of saturated and unsaturated fatty acids in its cytoplasmic membrane by almost threefold when grown at 10 and 43°C—the extremes of its temperature range. Higher organisms exhibit the same phenomenon. Beef fat from animals grown in cold climates contains a higher percentage of unsaturated fatty acids and has a lower melting point than does fat from animals raised in warm climates.

The effect of temperature on lipids is the basis of proposed mechanisms for both cessation of growth at low temperatures and death at high temperatures. At low temperatures, the fluidity of the membrane may be decreased sufficiently to prevent the functioning of transport systems, so that substrates cannot enter the cell rapidly enough to support even a low rate of growth. The transport of nutrients has been shown to vary with the growth temperature and degree of saturation of fatty acids in the membrane. At high temperatures, membrane lipids may melt, causing loss of the structural integrity of the membrane and leakage of the cell contents. It must be remembered that "low" and "high" temperatures are not the same for different organisms. Psychrophiles are killed by even short exposure to room temperature in the laboratory, and death is accompanied by leakage of macromolecules from the cell. Thermophiles, on the other hand, can remain viable at temperatures near the boiling point of water but cease to grow at minimum temperatures far above those that are lethal for psychrophiles.

Our present knowledge of membrane structure and function is not sufficient to explain these great differences in organisms growing at the two temperature extremes, but membrane function is undoubtedly important in determining the temperature at which an organism can grow or survive. Since bacterial membranes are the sites of the majority of the essential life processes of the cell, any perturbation of the membrane might be expected to affect at least one of these processes. It has been suggested (Brock, 1979) that the more complex internal membranes of the eucaryotes are more susceptible to loss of function with increasing temperature than is the cytoplasmic membrane. If so, this may explain the fact that eucaryotes have not been found at temperatures higher than 60°C. Procaryotes that are dependent on internal membrane systems, e.g., the photosynthetic bacteria, are not able to grow at temperatures higher than 70 to 75°C.

Proteins and temperature The other class of molecules involved in effects of temperature on growth and viability, *proteins*, perform much more varied functions in the cell than do lipids. Proteins, like lipids, are important structural components of membranes. Proteins are also structural components of the ribosome, along with RNA. Most of the many different protein molecules in the cell function as enzymes that catalyze the many reactions required for growth of the cell. For each of these roles, the specific protein involved must maintain a precise three-dimensional structure. Loss of function results from an alteration in the conformation of the molecule.

An increase in temperature affects proteins by causing thermal denaturation, an alteration of the functional spatial arrangement, which is usually irreversible. Elevated temperatures thus can inactivate many essential processes in the cell by inactivating the protein involved. It has been proposed that growth ceases when the temperature reaches the point at which the most heat-sensitive essential protein of the cell is denatured. Experimental evidence that this can occur is found in temperature-sensitive mutants. These are bacterial strains in which a single essential enzyme has been altered genetically so that it is extremely sensitive to temperature. Such a mutant of *Escherichia coli*, for example, may grow at the normal rate at a temperature of 25°C but may not be able to grow at all at 37°C, which is the normal optimum temperature for *E. coli*.

Thermal denaturation of proteins can cause loss of an essential enzyme function, alteration of membrane structure, or inactivation of the protein-synthesizing machinery due to alteration of ribosome conformation. In different organisms, the first function to be affected by increasing temperature may differ, but in any organism, growth must cease when one essential function ceases. In general, enzymes that perform identical functions in thermophiles and mesophiles are very similar in amino acid composition, molecular weight, and other physicochemical characteristics, but enzymes from thermophiles are functional at temperatures above those at which their mesophilic counterparts are totally inactivated. The basis for this difference in thermostability has not been determined.

In addition to the action of temperature in setting the upper and lower limits for growth of microorganisms, there is a continuous effect on the rate of growth at temperatures between the minimum and maximum. As temperature increases, the rate of each enzyme-catalyzed reaction in the cell increases and the rate of cell growth increases concomitantly. The Q_{10} value for enzyme activity—that is, the ratio of activities at temperatures 10° apart—is the same as that for growth, between 2 and 3. At the optimum temperature for any microorganism, some protein denaturation is probably occurring, but this undesirable effect is just offset by the increased activity of the undenatured enzyme molecules and perhaps by an increased rate of protein synthesis. A further small increase in temperature increases the rate of denaturation to a degree that cannot be balanced by increased activity, and the growth rate decreases sharply. Temperatures above the maximum for growth are lethal, and death is dependent on the temperature and time of exposure.

Selective Effect of Temperature

Since temperature determines the rate of growth as well as the limits of growth, it is apparent that temperature must be an important factor in the selection of predominant microbial species in any habitat. In a mixed population of microorganisms, an increase of several degrees in the temperature may eliminate some species because their maximum limit has been exceeded. If the species eliminated constituted a significant fraction of the total microbial population, other organisms whose growth rate had been increased by the temperature shift would be able to take advantage of the greater availability of nutrients and thus increase rapidly in numbers. In a habitat where frequent variations in temperature occur, e.g., day-night variation, one would expect to find few organisms that are incapable of growth at the highest temperatures reached.

The fact that increased temperature increases enzyme activity suggests that more rapid rates of substrate removal could be achieved if the temperature of biological treatment processes such as activated sludge could be increased and controlled. Such a modification of procedure presently is not feasible from an economic viewpoint but may become so in the future or might be so now in the treatment of wastes containing substrates that are metabolized very slowly at the low temperatures characteristic of activated sludge installations. In industrial fermentations, temperature is closely controlled at the optimum for production of the desired product. This optimum must be determined experimentally and may or may not be the same as the optimum temperature for growth of the organism.

pH AND MICROBIAL GROWTH

A second environmental factor that, like temperature, influences the growth rate and limits growth is the hydrogen ion concentration, i.e., the acidity or alkalinity, of the aqueous environment. This is most conveniently expressed as pH, the negative logarithm of the molar concentration of hydrogen ion, H^+.

Each microbial species is characterized by minimum and maximum pH values defining the limits of the range of pH values within which growth is possible. The optimum pH value for any species is that at which the growth rate is most rapid. The minimum and maximum values that limit growth usually differ by only three to four pH units. However, it must be remembered that a difference of four pH units represents a change in H^+ concentration of ten thousandfold, so that a "narrow" pH range actually involves a very broad range of H^+ concentrations. The optimum pH value for growth is usually approximately midway between the minimum and maximum.

While there are many exceptions, it is possible to make some general statements concerning the pH preferences of different types of microorganisms [see Altman and Dittmer (1966) for data tables]:

1. Most bacteria have pH optima near neutrality and minimum and maximum pH values for growth near 5 and 9, respectively.
2. Most fungi prefer an acid environment and have minimum pH values between 1 and 3 with an optimum pH near 5.
3. Most blue-green bacteria have pH optima higher than 7.
4. Most protozoa are able to grow in the pH range 5 to 8, with an optimum pH near 7.

Some bacteria, a few protozoa, and many fungi are capable of growing over surprisingly broad pH ranges. One species of the fungus *Penicillium* has been reported to have a minimum pH value for growth of 1.6 and a maximum of 11.1. The sulfur-oxidizing bacterium *Thiobacillus thiooxidans* grows most rapidly in the pH range 2.0 to 3.5 and has been reported to grow over the entire range of less than 0.5 to greater than 9.5, although more recent studies have placed its maximum pH at 6.0. A marine species, *T. neapolitanus*, grows over the range of 3.0 to 8.5. In contrast, a marine bacterium, *Nitrospina gracilis*, which oxidizes nitrite to nitrate, grows only between pH 7.0 and 8.0.

Although some human pathogens have a restricted pH range for growth, the adaptation to the constant environment of the body has apparently not occurred as frequently with pH as with temperature. *Streptococcus pneumoniae*, the cause of bacterial pneumonia, grows over a narrow pH range, 7.0 to 8.3, with an optimum of 7.8, but the pathogenic species of *Salmonella* and *Shigella* are capable of growth over the same broad range, pH 4.5 to 9.5, as are related, nonpathogenic organisms, such as *Enterobacter aerogenes*. Since the pH of many natural environments is near neutrality, pH is not often a factor in the survival of most human pathogens outside the body.

Two aspects of the relationship of microorganisms to pH deserve particular attention. First, changes in the pH of the environment are most likely to be brought about by microorganisms themselves. Second, the internal pH of the cell, unlike the temperature, is not determined solely by the environment. The microorganism can control the passage of ions, including the hydrogen ion, into and out of the cell, at least within limits. Therefore, the effects of pH on the cell may be indirect.

Alteration of pH by Microbial Activity

Many metabolic activities of microorganisms result in the formation of acidic or alkaline products. Organisms capable of growing in the absence of oxygen obtain energy through the fermentation of organic compounds, primarily carbohydrates. Sugars are converted to a variety of products, many of which are organic acids. These are released from the cell and can result in a decrease in external pH of sufficient magnitude to inhibit growth. The lactic acid bacteria, for example, which are used in the manufacture of dairy products such as buttermilk and sour cream, convert the lactose in milk to lactic acid and lower the pH to approximately 4.0 to 4.5, at which point they cease to grow. If other, more acid-tolerant organisms are present, they may continue to grow, and the predominant species thus change due to the lowering of the pH by metabolic products.

Organisms growing in an aerobic environment also produce acids. Oxidation of organic compounds may produce sufficient carbon dioxide to lower the pH values significantly if the aqueous environment is not sufficiently buffered. Bacteria growing at low oxygen tension (see below) may only partially oxidize substrates and often release acidic metabolic intermediates. Other types of stress may cause the release of incompletely metabolized acidic products (see Chap. 13). Even in the presence of sufficient oxygen during normal aerobic growth, some bacteria may release acidic organic products that can later be taken into the cell and further metabolized. In a mixed population of bacteria, the products released by one type of organism may, of course, be used by another species.

A few specialized types of bacteria produce large amounts of acid and a very low pH unsuited to the growth of other bacteria. Several species of *Thiobacillus* are capable of reducing the pH of their surroundings to less than 1.0 by oxidizing sulfur or reduced sulfur compounds such as hydrogen sulfide or thiosulfate to SO_4^{2-} (sulfuric acid). These organisms are responsible for crown corrosion in concrete sewers, where hydrogen sulfide is produced when sewage becomes anaerobic. They are also responsible for the low pH value of acid mine drainage waters, but their ability to oxidize sulfides is of value in certain mining operations where they are used to leach metals such as copper from low-grade ores. These organisms can also be used to release oil from oil shale. Their ability to form acid from sulfur makes them useful in dissolving rocky material, allowing easier extraction of the oil. The acetic acid bacterium *Acetobacter* is used in the commercial production of vinegar, in which ethanol is oxidized to acetic acid. These organisms can tolerate a pH value as low as 1.0.

Microorganisms also can cause an increase in the pH of their surroundings by the release of alkaline products or by removal of certain ions from the environment. The most common cause of increased pH is the metabolism of proteins, peptides, or amino acids. Deamination, i.e., removal of the amino group as NH_3, results in the formation of ammonium hydroxide when ammonia is released by the cell. Use of anions, e.g., the nitrate of $NaNO_3$, causes an increase in external pH values, whereas the use of cations, e.g., the ammonium ion of $(NH_4)_2SO_4$, causes a decrease in pH. Algae, which use inorganic carbon for the synthesis of cell material, i.e., as a carbon source, may increase the pH of an oxidation pond or a stream to 9 or above by the removal of large amounts of carbonate ion.

Obviously, when microorganisms are cultivated in the laboratory or in industrial processes, their propensity for altering the pH of their environment must be controlled to maintain the pH that is optimum for growth or for production of the desired product. This is accomplished in laboratory cultures by the addition of a buffer to the medium. The most commonly used buffer in synthetic media is a mixture of phosphate salts that has a maximum buffering capacity near neutrality. In cultures producing very large amounts of acid, powdered $CaCO_3$ is used. This is insoluble, and as acids are formed, they react with the $CaCO_3$ to produce insoluble calcium salts with the release of CO_2. In commercial installations, the pH is often controlled by automatic titration, i.e., continuous monitoring of pH changes with addition of acid or base as required.

In the biological treatment of domestic sewage, pH is not of concern since the pH value of sewage is near neutral and some components of sewage provide buffering action (e.g., amino acids are fairly good buffers). Some carriage waters also contain inorganic ions, such as carbonate, which resist lowering of the pH level by acidic metabolic products. In the treatment of many industrial wastes, however, it is necessary at least to provide initial adjustment of pH values to the neutral range if biological treatment is to be successful. This is a significant economic factor in the treatment of extremely acid or alkaline wastes. Combined treatment of wastes with differing pH values or a combination of an industrial waste with domestic sewage may, when feasible, circumvent or reduce the need for pH neutralization prior to treatment.

In anaerobic treatment of industrial wastes or sewage sludge, pH control is of perhaps even greater concern than in aerobic treatment, because the organisms needed for the success of the process have a more narrow range of pH tolerance. Therefore, the pH of anaerobic digesters must be monitored closely.

Molecular Basis of Effects of pH

The control of growth rate and viability by pH is no more completely understood than is the effect of temperature. An obvious explanation might lie in the effect of the pH on enzymatic activity, since each enzyme is active within only a specific and usually narrow pH range and displays maximum activity at an optimum pH. However, this explanation cannot account for all the effects of pH. While it is technically impossible to measure the internal pH value of a microbial cell, it seems reasonable to conclude that the internal pH cannot be identical with that of the environment, and it even seems likely that the pH may vary in localized areas within the cell. Determinations of the pH optima of enzymes have shown that the enzymes of cells capable of growing in extremely acid or alkaline environments do not necessarily have pH optima corresponding to the optimum pH for growth of the cell, and they may even be totally inactive at a pH at which growth occurs. It is not unusual to find, in cells that grow most rapidly at pH values near neutral, specific enzymes that are optimally active at a pH of 9 or above and other enzymes that are inactive or only slightly active at pH 9. The significance of such measurements is somewhat questionable, however, since determinations of the effect of pH on the activities of individual enzymes involve measurements with purified enzyme preparations. It is possible that the enzyme in its intracellular location may be bound to other cellular components that could alter the effect of pH on its activity. (The same precaution applies, of course, to measurements of temperature optima for purified enzymes.)

An additional consideration complicates the relationship between the cell and the internal and external concentrations of H^+ ions. The most recently proposed mechanism for the transport of a variety of ions and organic compounds across the bacterial cell membrane involves the establishment of a pH gradient across the membrane. The translocation of protons (H^+ ions) generates an electrochemical gradient, or proton-motive force, across the membrane, which drives the transport of other materials into the cell. Much evidence has accumulated recently in support of this theory—the *chemiosmotic model* of Mitchell (Hamilton, 1975, 1977). Experiments with membrane preparations have shown that the pH gradient across the membrane is affected by the pH of the aqueous environment, and this in turn affects the driving force for transport. The proton-motive force is also implicated in the energy-generating processes of cells that form ATP through electron transport (see Chap. 7), and the external pH may possibly affect the growth of bacteria by altering their ability to synthesize ATP.

Although the mechanism by which pH affects growth rate and viability is not known, it seems reasonable to conclude that the effect of pH on the transport of materials across the membrane is a very important factor and perhaps the determining factor in influencing growth. The effects of pH on transport are both direct and indirect. A direct effect is exerted through alteration of the pH gradient across the membrane. For compounds that are transported by binding to specific membrane proteins, pH may control the configuration and thus the activity of the binding protein. The external pH also determines the ionization state of nutrients required by the cell and of compounds that may be toxic to it. Cells are more readily permeable to nonionized than to ionized compounds. The ionization of a required nutrient may make it unavailable to the cell, whereas a toxic compound may become inhibitory only at a pH at which it is not ionized. Some required inorganic nutrients may become less available with a change in pH because of the formation of insoluble salts.

Selective Effects of pH

As in the case of temperature, pH determines whether any microbial species can proliferate in a particular environment and the rate at which it can reproduce. In contrast to temperature, however, the primary determinants of the pH are usually the microorganisms themselves. As we have seen, microorganisms can alter the pH of their environment through various metabolic activities, and these alterations are often disadvantageous to the microbial species that cause them. Some microorganisms are able to create environments in which few other microorganisms can survive, and if they themselves can survive the conditions they create, they may thus eliminate competitors. In many cases, however, the alterations in pH caused by a microorganism may encourage the predominance of a competitor. Bacteria may produce acidic products that decrease the pH value in an insufficiently buffered environment and allow fungi to become predominant.

The pH of the environment is not always the result of microbial activity, since many inorganic salts present in soil and natural waters may influence their pH, and the discharge of industrial wastes into receiving streams may cause highly acidic or alkaline conditions in localized areas. However, the ability of microorganisms to alter pH is the basis of significant interactions between species. Since pH affects growth rate, changes in pH, like changes in temperature, may cause drastic shifts in the relative numbers of different species in the population. Sakharova and Rabotnova (1976) found a number of effects of changes in the pH of the external medium on *Bacillus megaterium*, a common soil and water organism. Utilization of the carbon and energy source, efficiency of substrate utilization, synthesis of protein, synthesis of storage materials of different types, and release of metabolic products from the cell, as well as other aspects of cellular metabolism were drastically affected by changes in pH over the range within which the organism can grow,

4.6 to 9.6. All of these factors are very important in determining the ability of a microbial species to compete with other species in a mixed population. There is a great need for study of the effects of both pH and temperature and of the interactions between these two important environmental variables using mixed cultures of microorganisms under controlled conditions in the laboratory.

OXYGEN AND MICROBIAL GROWTH

Until Pasteur investigated the production of alcohol by yeasts, it was believed that life was possible only in the presence of air. For higher organisms this is correct, but many bacteria and some protozoa can grow only in the total absence of air (oxygen), while many other bacteria and some fungi and protozoa are capable of growth in either the presence or absence of oxygen. Algae, since they produce oxygen during photosynthesis, are aerobic organisms.

Organisms that require oxygen are classified as *strict or obligate aerobes*. Those that cannot grow in the presence of oxygen are *strict or obligate anaerobes*, and those that can grow with or without oxygen are *facultative anaerobes*. Some microorganisms, classified as *microaerophiles*, require low concentrations of oxygen (reduced oxygen tension) and do not grow either at atmospheric oxygen pressure or in the absence of oxygen. Such organisms may grow well in polluted waters where the oxygen concentration is low. One such organism, often found in polluted streams, is *Sphaerotilus natans*. Some microaerophilic bacteria are human and animal pathogens, e.g., *Listeria* and *Erysipelothrix*. The latter require elevated partial pressures of carbon dioxide for good growth.

Oxygen is required by aerobes for two purposes. The major requirement is as a terminal electron acceptor for the electron transport system necessary for the generation of energy (see Chap. 7). A very small amount of oxygen is required in certain enzymatic reactions. The oxidation of hydrocarbons, for example, requires the addition of molecular oxygen to the molecule. Synthesis of sterols and unsaturated fatty acids also involves reactions that require molecular oxygen. Organisms growing in the absence of oxygen cannot metabolize hydrocarbons and may require sterols or unsaturated fatty acids. Yeasts require oleic acid, an 18-carbon unsaturated fatty acid, when growing anaerobically, although they are able to synthesize it during aerobic growth. Other anaerobic organisms, however, are capable of synthesizing unsaturated fatty acids by a different series of reactions that do not involve oxygen.

Toxic Effects of Oxygen

Oxygen can be toxic even to strict aerobes when supplied at greater than atmospheric pressure (e.g., when the atmosphere in an incubator contains more than 20 percent oxygen by volume). Anaerobic organisms are usually quite sensitive to oxygen, although a few are aerotolerant. Strict anaerobes are killed by exposure to oxygen. This sensitivity may be extreme, as in the case of some methane-forming bacteria, which can be grown in the laboratory only by taking great care to remove all traces of oxygen from their environment. These organisms function well in the totally anaerobic environment of the anaerobic digester, where any traces of oxygen are removed by facultative anaerobes present in the mixed population, and in the anaerobic muds at the bottom of polluted ponds, swamps, and marshes.

The toxicity of oxygen results from the formation of toxic products in enzymatic reactions involving oxygen. Flavoproteins such as the amino acid oxidases form the toxic product hydrogen peroxide:

$$R-\underset{\underset{NH_2}{|}}{CH}-\overset{\overset{O}{\|}}{C}-OH + O_2 + H_2O \rightarrow R-\overset{\overset{O}{\|}}{C}-\overset{\overset{O}{\|}}{C}-OH + NH_3 + H_2O_2$$

Small amounts of even more toxic products are also formed in enzymatic reactions involving oxygen. The free radical form of oxygen, superoxide anion, O_2^-, and the products formed from superoxide, singlet oxygen, $^1O_2^*$, and hydroxyl free radical, $OH\cdot$, are extremely reactive and highly toxic.

Microorganisms that are able to grow in the presence of oxygen form enzymes that destroy these toxic products. All aerobic and aerotolerant microorganisms form the enzyme superoxide dismutase, which catalyzes the reduction of superoxide to hydrogen peroxide and thus prevents the formation of more toxic products:

$$2O_2^- + 2H^+ \rightarrow O_2 + H_2O_2$$

Aerobic organisms also form a second enzyme, catalase, which decomposes hydrogen peroxide:

$$2H_2O_2 \rightarrow O_2 + 2H_2O$$

A few organisms that are capable of growth in air, notably the lactic acid bacteria, form no catalase. These organisms are able to decompose hydrogen peroxide by a different mechanism in which the enzyme peroxidase catalyzes the reaction of hydrogen peroxide with any of several types of organic compounds. The hydrogen peroxide in this reaction is converted to water. A simple test to determine whether an organism forms catalase involves placing a drop of 3 percent hydrogen peroxide on a colony. The presence of catalase is indicated by vigorous bubbling as oxygen is released.

Selective Effects of Oxygen

Since some microorganisms exhibit an absolute requirement for oxygen, others require reduced oxygen tension, and still others die when exposed to oxygen, the availability of oxygen in the environment is an extremely important factor in selecting the organisms that inhabit that environment. Obligate aerobes can be maintained only in habitats that are not subjected to significant periods of anaerobiosis, whereas obligate anaerobes will be eliminated from any habitat in which they are exposed to oxygen unless they are able to persist in resistant forms such as spores. In habitats subject to frequent alternations between aerobic and anaerobic conditions, facultative anaerobes or organisms indifferent to the presence of oxygen, e.g., the lactic acid bacteria, have a selective advantage since they can continue to proliferate with or without oxygen.

Organisms with different oxygen requirements may grow in close proximity in environments not subject to frequent mixing. In muds at the margin of ponds and streams, aerobes and facultative anaerobes may flourish in the upper layer, effectively removing all oxygen so that anaerobes are able to grow in the deeper layers. Similar relationships may exist in deep ponds and lakes.

Oxygen Tension

The effects of oxygen tension, i.e., the dissolved oxygen concentration in the liquid medium, on aerobic organisms for which oxygen is not toxic have received considerable attention. The possible effects of major practical interest are those on the respiration (oxygen utilization) rate, biomass yield, cell composition, viability, utilization of soluble and colloidal carbon and energy sources, autodigestion, production of specific enzymes, and formation of metabolic products.

Extracellular products of facultatively anaerobic organisms can be greatly modified, quantitatively and qualitatively, by varying the oxygen tension. For example, *Enterobacter aerogenes*, using glucose under anaerobic conditions, forms few cells but considerable amounts of ethanol, formic acid, butanediol, acetic acid, acetoin, and carbon dioxide. Under slightly aerobic conditions, ethanol and formic acid are no longer made. Under greater oxygen tension, acetoin and butanediol are no longer made. Finally, at higher oxygen tension, acetic acid is no longer produced. The amount of soluble organic end products decreases, and greater amounts of carbon dioxide are produced as the system experiences a greater degree of aerobiosis. On the other hand, maximum production of acetic acid by the fungus *Aspergillus niger* requires significantly greater oxygen tension than that required for maximum growth.

The production of specific extracellular enzymes can be affected by the oxygen tension. In some aerobic organisms, increased aeration increases the production of certain extracellular enzymes, whereas in others, enzyme production is hampered by high oxygen concentration. Such aspects are extremely important to the breakdown or hydrolysis of macromolecules found in wastewater and to the aerobic as well as anaerobic digestion of excess biomass at treatment plants. Various types of proteinases, lipases, carbohydrases, and nucleic acid-degrading enzymes (nucleases) are needed. Also, the activity of these enzymes after they are synthesized can be affected by oxygen tension (and degree of agitation). Some enzymes are inhibited at high oxygen tension; thus, a high degree of aerobiosis may either help or hinder, depending on the species of organisms and types of substrates present. The fact that there are great numbers and types of macromolecular substrates to be degraded offers one of the explanations for conflicting reports on the effects of oxygen tension and mixing in bioreactors, since these effects may depend on the nature of the wastewater and other operational conditions affecting the chemical and ecological makeup of the biomass. This variability points out the need for environmental technologists to examine many experiments and make many observations on the behavior of the biomass under varying conditions before accepting generalizations or drawing conclusions.

Much of the work on the effects of oxygen tension has been concerned with respiration rate, organic carbon source utilization, and cell production. For the most part, in aerobic organisms capable of existing in anaerobic environments, i.e., facultative anaerobes of the type commonly found in aerobic biological treatment plants, the respiration rate (oxygen uptake rate) increases as the oxygen tension or agitation is increased. The amount of increased respiratory activity with increased aeration can vary with the species of microorganisms present and, for the same species, with the nature of the organic carbon source. However, maximum aerobic respiration can be observed at low oxygen tension values; such is the case with *E. aerogenes*, an organism commonly found in municipal sewage.

It should be made clear that in most cases studied, the increased respiration rate with increased DO tension involves very low DO levels. It generally has been found that above a certain "critical" DO concentration, increased levels of DO add little to aerobic respiration rates. The critical level is usually very low for dispersed as compared with flocculated cells. As a point of reference, the critical DO level is defined as the DO concentration at which the oxygen uptake rate of the cells is half of the maximum rate recorded for the system when DO is present in abundance. Thus, if a hyperbolic relationship were observed, the critical DO would be the equivalent (analytically but not necessarily physiologically) of the shape factor K_m of the Michaelis-Menten equation. The critical dissolved oxygen concentration has usually been found to be less than 0.1 mg/L. Above this level, the DO does not greatly affect the respiration rate. For flocculated heterogeneous microbial populations, the critical DO is generally found to be higher—0.5 mg/L. To provide a factor of safety, maintenance of a 2.0 mg/L concentration is generally recommended. Above critical DO levels, some increase in respiration and decrease in biomass yield have been observed in systems wherein both DO and vigor of mixing (i.e., mixing and velocity gradient) were increased (Rickard and Gaudy, 1968b). Holding mixing energy constant at a reasonably high value while varying the DO at concentrations above the critical range (1.4 to 7.1 mg/L) did not provide a noticeable change in the respiration rate or cell yield, but a slight increase in the total removal of soluble substrate was noticed.

It may be reasoned that as floc size increases, the critical DO increases because the oxygen molecule must penetrate farther into the biomass to reach interior cells and, assuming it is better to have these operate aerobically, it may be better to use higher DO. It could also be argued that no matter what the DO concentration in the medium, oxygen may never penetrate to the most interior cells unless the floc size is controlled to reasonable limits. The recent introduction of oxygenation systems using pure oxygen rather than air gives practical importance to such considerations. It has been argued that maintenance of supersaturated DO levels and decreased agitation or mixing in activated sludge processes foster the building of larger and more rapidly settling floc particles while providing penetration of sufficient oxygen to the interior of the floc. Selection of a system using either pure oxygen or air is often based entirely on consideration of space and cost, but there is also need to compare the two systems from a scientific and technological standpoint, and on this score there is considerable debate (Kalinske, 1976; Parker and Merrill, 1976). There may be an optimum floc size that could be controlled by agitation force with the system held at a specific DO by enhanced oxygen content in the aerating gas. Such possibilities await highly controlled experimental study, as does the question of whether it is a good idea to even attempt to maintain the interior population aerobic. The difficulty in measuring the DO at the interior of the floc represents an investigational challenge in itself. In any event, if it is possible to decrease the vigor of agitation and mixing by increasing DO tension, there is also a practical limit to the extent to which floc size can be allowed to increase, because it is essential to keep the contents mixed in the reactor. Furthermore, complete mixing is a hydraulic regime recommended in most process design models, and this requires considerable mixing force, especially if high biomass concentrations are used in the treatment reactor. The problems of providing absolute scientific proof in regard to the possible functional advantages of each means of providing oxygen are great, and some engineers take the attitude that only full-scale plant comparisons can provide a basis for deciding which type of system to use. Such an attitude is understandable, but it is somewhat reminiscent of Charles Lamb's instructive story—that is, it ought not be necessary to burn down the whole house to determine the best way to roast a pig. Clearly, more investigational work is warranted in this area, and it should be possible to obtain definitive and decisive answers at much lower experimental costs than those incurred by the taxpayers who support 75 percent of the cost of full-scale municipal plants.

Mixing and Aeration

The degree and vigor of mixing can exert profound effects on both the selection of species and the general behavior of the species thus selected. In general, mixing tends to homogenize the population in regard to the distribution of species within the system, because it tends to prevent localized differences in temperature and chemical composition (nutrients, dissolved oxygen, hydrogen ion, etc.) and it encourages microbial confrontations that might not occur were it not for this important hydraulic consideration. Mixing is often discussed concurrently with aeration because the vigor of mixing enhances the transfer of oxygen, which in turn may determine not only the types of organisms present (aerobic, anaerobic, or facultative) but also the rate of metabolism, the efficiency of substrate utilization, and the amount of microbial product, whether this product is the cells themselves or specific compounds produced by the cells.

Mixing also has ramifications to the separation processes following microbial growth in engineered systems. In some industrial processes, the aim is to expose as much cell surface to nutrient as possible; thus, one aims for as much dispersed growth (in contrast with flocculated or agglutinated cells) as possible. Vigorous agitation can be expected to enhance dispersion. However, when we consider a fluidized secondary biological wastewater treatment process (e.g., activated sludge), dispersed growth may be disadvantageous because current economic considerations dictate quiescent settling as the method of separating the mixed liquor and the cells. Increased agitation (mixing) can lead to smaller floc size and, unless the microbial flocs reassemble during the short time of travel from the agitated growth reactor to the quiescent settling tank, the efficiency of separation may be hampered. In addition, in such systems, the metabolic efficiency of the removal of soluble and colloidal organic matter may be related to the diversity of the population in the feeding community represented by the various types of microbial species within the floc particle. This microecosystem may be disrupted by shear forces. On the other hand, insufficient mixing can cause the floc size to become too large, militating against rapid settling as well as decreasing the compactability or thickening properties of the settled biomass. Also, as noted previously, if floc particles are too large, the passage of nutrients to the interior portions can be hampered, lowering the efficiency per unit biomass. Slowing of the transport of substrate and oxygen may cause widely differing growth conditions in parts of the floc community. If the floc particle is too large, the interior regions may become essentially anaerobic. If oxygen cannot penetrate, there is reason to surmise that larger substrate molecules have less chance of doing so. The lack of exogenous substrate as well as oxygen can lead to fermentative digestion of the cells, thus releasing organic matter that already had been extracted from the wastewater. This may be an important source of inefficiency or, on the other hand, the death of cells and release or formation of organic compounds not originally in the wastewater may enhance species diversity, which is the most valuable result of the use of heterogeneous rather than pure cultures in waste treatment systems. The above statements are made in a speculative vein, largely because one must be aware of the reasonable possible effects of mixing but also because there are few definitive experimental data upon which to base conclusions.

In addition to effects on the metabolic and physical behavior and species makeup of the biomass, mixing has another important ramification to the environmental technologist in the area of design of growth reactors and the analysis of metabolic reactions occurring therein. It will be shown in Chap. 6 that if the fluidized system of substrate and cells can be shown to be homogeneously mixed, mathematical manipulations and predictive equations for design can be simplified.

Most of the information in regard to mixing is tied to considerations of aeration, and there has been great controversy over the relative effects of DO concentration (oxygen tension) and agitation (mixing). The two have been considered together, and although efforts have been made by many to judge their separate effects, questions still remain and there is practical need to resolve them. One might ask whether vigor of agitation alone has any effect on the rate of metabolism. In an anaerobic system, one can avoid the complication of concomitant consideration of DO level. The general finding in industrial fermentations and in anaerobic digestion of sludge is that some mixing is beneficial in enhancing the rate of fermentation and/or gas production. The general conclusion has been that a stirred reactor gives better results than does a nonstirred reactor. There is little or no information allowing one to make quantitative conclusions regarding the increase in rate with increasing vigor of mixing. In general, in sludge digestion, mixing increases the rate of gas formation and allows higher loading of digesters by making better use of the reactor volume. There are fewer nonreacting portions, temperature is more nearly the same in all parts of the digester, and mixing facilitates release of the gaseous reaction products (largely carbon dioxide and methane). Mixing and aeration are important unit operations, and much valuable information on mass transfer concepts, equipment, methods of aeration, and mixing can be found in the literature (e.g., see Kalinske, 1976; Parker and Merrill, 1976). In regard to the effects of mixing on the chemical and ecological character of the biomass, there is evidence that an increase in the mixing energy can reduce the cell yield (proportion of the carbon source channeled to synthesis of biomass) and increase the amount of substrate respired in heterogeneous microbial populations; an increase in oxygen uptake has also been observed in pure culture systems. Increased agitation has been observed to decrease the proportion of filamentous forms in the biomass. Whether this is due to increased shearing of filaments of fungi, which are generally highly aerobic, or whether increased DO levels discourage the growth of filamentous bacteria that seem to grow better at lower DO levels, such as *Sphaerotilus natans*, cannot really be concluded on the basis of available data. Some studies made over a wide range of velocity (mixing) gradients show little or no increase in efficiency of removal of soluble carbon source and amounts of RNA, DNA, and protein synthesized. There is some evidence of decreased carbohydrate synthesis as agitation is increased.

NUTRITIONAL REQUIREMENTS FOR GROWTH

Knowledge of the exact chemical composition of a specific type of microbial cell would allow one to design an ideal nutrient solution for that organism containing all the required elements in the correct proportions. Such detailed information is seldom available, nor is it often necessary. Since all living cells, with a few exceptions such as the silica-containing diatoms, are composed of similar types of compounds, they have similar elemental compositions and therefore require the same elements in roughly the same relative amounts. Also, while a microorganism cannot reproduce if an essential element is totally unavailable, the composition of cells may vary somewhat, depending on the available nutrients and other factors such as temperature and pH.

Required Elements

The elemental composition of a typical bacterial cell, *Escherichia coli*, was shown in Table 3-1. Only four elements, carbon, oxygen, nitrogen, and hydrogen, make up more than 90 percent of the dry weight of the cell. These elements, plus phosphorus and sulfur, comprise the macromolecules of the cell. The remaining 4 percent of the dry weight of the cell includes a large number of elements—potassium, sodium, calcium, magnesium, chlorine, iron, manganese, cobalt, copper, boron, zinc, molybdenum, and others. Elements required in very small amounts are referred to as *trace elements*. It is difficult to demonstrate a requirement for them since they occur, in sufficient amounts to supply the cells' needs, as contaminants of other components of a synthetic medium. Certain elements are required in greater quantities by some cells than by others. Nitrogen constitutes a smaller proportion of fungal than of bacterial cells. Magnesium, as a constituent of chlorophyll, would be required in greater amounts by photosynthetic than by nonphotosynthetic organisms. Iron, which is a constituent of the cytochromes and of catalase and peroxidase, would generally be present in higher concentrations in aerobes than in anaerobes, although anaerobes also require iron as a cofactor for other enzymes. The trace elements in general, as well as potassium, calcium, and magnesium, serve as inorganic cofactors for, or constituents of, specific enzymes or organic cofactors. Certain cations such as zinc also may be involved in maintaining the structural integrity of the cell wall in Gram-negative bacteria.

Since the elemental requirements of the vast majority of living cells are qualitatively and quantitatively alike, the elemental composition of the environment is not a significant factor in the selection of microorganisms that may flourish in that environment except for the few microorganisms, e.g., diatoms, with special requirements. The chemical composition of the environment is, nonetheless, probably the most important factor in the selection of species in many habitats. This selectivity results from the fact that microorganisms vary greatly in their ability to obtain required nutrients from specific sources. Two closely related species that have exactly the same elemental composition and the same requirements for pH, temperature, and oxygen, may be unable to occupy the same habitat because of a single difference in nutritional requirements. The specific form in which a required element is available thus becomes a controlling factor in selecting species. In an environment containing an abundance of organic molecules of many types, which in the laboratory would be called a *complex medium*, a great variety of microbial species would be able to grow, and other factors such as growth rate at the prevailing pH and temperature would become of primary importance. However, many natural habitats, particularly surface waters, are not rich in a variety of organic compounds, so nutritional considerations become paramount in selecting the predominating species.

Not all required elements are equally important in determining predominating species. All elements present in the cell in minor proportions, i.e., phosphorus, sulfur, etc., can be utilized in the form of inorganic salts by almost all microorganisms. A few organisms require organic sulfur, but in general the same forms of these elements can support the growth of the vast majority of microbial species. Of the four elements comprising the bulk of the dry weight of the cell, i.e., carbon, oxygen, nitrogen, and hydrogen, only carbon and nitrogen are of selective importance. The hydrogen and oxygen of the cell are derived from water or from other compounds utilized by the cell. Thus, the major differences in nutritional requirements of microorganisms are based on the specific sources of carbon and nitrogen that they are able to utilize for the synthesis of cellular material.

Organic Growth Factor Requirements

The reproduction of a cell requires synthetic reactions of two general types. First, small organic molecules such as amino acids, purines, pyrimidines, sugars, vitamins, and many others must be synthesized from whatever starting materials are available to the cell, unless the required compounds can be obtained preformed from the surrounding liquid. Second, many of these small molecules must be assembled into the structural and functional macromolecules of the cell. All such reactions, syntheses of small molecules and their polymerizations, are catalyzed by specific enzymes. Only cells containing the genetic information that specifies the structure of a particular enzyme protein can carry out the reaction for which that enzyme is the specific catalyst. The ability of the cell to utilize specific compounds as starting materials for the synthesis of cellular material is also determined by the genetic information of the cell, since such starting materials must be partially metabolized and converted into intermediary metabolic products that serve as substrates for the initial reactions of the various biosynthetic pathways. More will be said of all these reactions later. The point to be stressed here is that each reaction in the cell is carried out by a specific enzyme, the structure of which is specified by the DNA of the cell. If that genetic information is absent or incorrect, the reaction cannot occur. Thus, it is the differences in genetic information possessed by different species of microorganisms that determine their metabolic abilities and their nutritional requirements for growth. Microorganisms, as well as higher organisms, that are not able to synthesize all of the small organic molecules required for proper function and for reproduction are nevertheless able to survive and reproduce if an external supply of the required compounds is available. A defect in the ability to polymerize small molecules cannot be compensated by external supply of the polymer, e.g., a protein, RNA, or DNA. Synthetic deficiencies leading to requirements for growth factors are therefore limited to those involving small molecules as products, i.e., the first type of synthetic reaction mentioned above.

Microorganisms exhibit a broad spectrum of synthetic abilities. At one extreme are the most self-sufficient organisms, the *prototrophs*. Given an assortment of inorganic salts containing the required elements and carbon dioxide or a single organic compound as starting material, these organisms can synthesize all of the hundreds of compounds that make up the cell. At the other extreme are those organisms that have very limited synthetic abilities. Such organisms require a large number of organic growth factors, i.e., amino acids, vitamins, purines, and pyrimidines. *Leuconostoc mesenteroides*, one of the lactic acid bacteria, is an example of these nutritionally exacting bacteria. It can synthesize proteins, for example, only if supplied with an external source of all the amino acids. Many human and animal pathogens have adapted to their parasitic mode of life by losing many synthetic abilities, and they thus require very complex media when cultured in the laboratory.

Microorganisms that cannot synthesize the compounds required for growth are classified as *auxotrophs*, and a specific required compound is called an *auxotrophic requirement*. These terms are most frequently used in referring to auxotrophic mutants that differ from the prototrophic parental strain in having lost the ability to synthesize a specific growth factor due to a genetic change, or mutation.

Carbon Source

The compound or compounds from which the carbon of the cellular material is derived is the *carbon source*. Prototrophs can utilize a *sole carbon source*, either an inorganic, i.e., carbon dioxide (autotrophic prototrophs), or an organic compound (heterotrophic prototrophs). It should be emphasized again that the ability of any species to use any specific carbon source depends on its genetic information. Some prototrophs, such as some species of the genus *Pseudomonas*, are capable of utilizing any one of approximately 100 different types of organic molecules as sole carbon source, whereas others are much more restricted in their carbon source utilization. For an organism such as *Leuconostoc*, the complex mixture of preformed growth factors that it needs may be considered collectively as the carbon source. For most microorganisms, a single carbon source can supply almost all of the cell carbon if one or a few auxotrophic requirements are available. Even organisms that utilize carbon dioxide as carbon source may require one or two vitamins, which contribute a negligible fraction of the carbon of the cell. A carbon source that can be utilized by the cell for the synthesis of most of the cellular material, with the remainder being supplied by preformed required compounds, may be considered as a *major carbon source.*

Since carbon is the essential element in all organic compounds and comprises 50 percent of the dry weight of the cell, the presence of a utilizable carbon source is a major determining factor in the ability of any microbial species to flourish in a given habitat. For microorganisms with specific auxotrophic requirements, the presence or absence of the required compounds in sufficient amounts to support growth is an even more restrictive environmental factor. Microorganisms with multiple growth requirements normally are found in abundance only where living or decaying animal or plant material is available to supply their needs. Those that obtain the required nutrients from living plant or animal cells are generally pathogenic, although some live in a symbiotic relationship with the host (rumen organisms, nitrogen-fixing bacteria in root nodules of legumes).

Nitrogen Source

Most of the organic molecules that make up the microbial cell contain nitrogen as well as carbon and, as in the case of carbon, the starting material from which the cellular nitrogen is derived is called the *nitrogen source.* Also, as with carbon, the nitrogen source may be organic or inorganic and may be a sole source or a major source of the nitrogen in the cell.

Nitrogen occurs in a greater variety of inorganic forms than does carbon, the most common being ammonia, NH_3, nitrate, NO_3^-, nitrite, NO_2^-, and nitrogen gas or dinitrogen, N_2. Ammonia is the most readily utilized of the inorganic forms of nitrogen. Its utilization requires no oxidation or reduction, since the nitrogen in cellular constituents such as amino acids, purines, and pyrimidines is at the same oxidation-reduction level as is ammonia; the valence of N is -3. The ability of an organism to utilize ammonia as a source of nitrogen, i.e., to *assimilate* ammonia, depends on its possession of one or more of the four enzymes that incorporate ammonia into organic compounds by reactions called *aminations*, the addition of an amino group to the molecule. The products of these four reactions, glutamic acid, glutamine, and asparagine, then transfer the amino group to other compounds in reactions called *transaminations.* An organism possessing the genetic information for synthesis of the enzymes involved in amination and transamination can utilize ammonia as a sole nitrogen source.

Assimilation of other inorganic forms of nitrogen requires additional enzymatic reactions. Nitrate, nitrite, or dinitrogen must be reduced to the level of ammonia and then utilized for amination. Only organisms possessing the necessary enzymes can utilize these compounds as sources of cellular nitrogen. Nitrate can be used as a nitrogen source by a variety of fungi and bacteria, but these are fewer in number than those that can utilize ammonia. Nitrate is commonly utilized by photosynthetic microorganisms. The assimilation of nitrate involves as a first step its reduction to nitrite by the enzyme nitrate reductase. Nitrite is then further reduced to ammonia. Nitrite can also be used as a nitrogen source by some microorganisms.

Use of atmospheric nitrogen Nitrogen gas, N_2, which makes up approximately 80 percent of the earth's atmosphere, is the most abundant form of nitrogen in nature. Its use as a source of nitrogen is called *nitrogen fixation*. It is somewhat surprising (and unfortunate from the point of view of food production) that after millions of years of evolution so few organisms have developed the ability to use this abundant form of nitrogen. Only procaryotes can fix nitrogen, and among this group the ability is limited to photosynthetic organisms (bacteria and blue-green bacteria) and to a few other genera or species. Among the nonphotosynthetic bacteria, nitrogen fixation is limited almost completely to the members of three families, Azotobacteraceae, Rhizobiaceae, and Bacillaceae. For the first two families, nitrogen fixation is the primary characteristic that distinguishes them from other families of bacteria. The two families are differentiated on the basis of their modes of life. The Azotobacteraceae are aerobic free-living organisms found in soil and surface water, whereas the Rhizobiacaeae live symbiotically in root nodules on leguminous plants such as soybeans and alfalfa. The Bacillaceae are not grouped together on the basis of their ability to fix nitrogen but rather because all the members of the family form spores. Two genera, *Bacillus* and *Clostridium*, include most of the species of the family, and both genera include some species capable of fixing nitrogen. However, nitrogen fixation is rare among species of *Bacillus*, occurring in only two of the 48 species and in these only under anaerobic conditions. *Clostridium* and *Bacillus* are differentiated on the basis of their relation to oxygen; *Bacillus* includes species that are obligate aerobes or facultative anaerobes, whereas the genus *Clostridium* includes only obligately anaerobic species. Several species of *Clostridium* are capable of fixing nitrogen and of these, *C. pasteurianum*, which is very active in nitrogen fixation, has been used in extensive studies of the mechanism by which bacteria reduce nitrogen to the level of ammonia, the form in which it is utilized for biosynthetic reactions.

There has been intense interest in the bacterial fixation of nitrogen. The industrial production of ammonia for fertilizer, using the Haber process for reduction of atmospheric nitrogen, accomplishes the same end result as does the biological process, but it requires high temperature and pressure and consumes large amounts of energy. It was hoped that an understanding of the biological process might suggest an alternate method of industrial production that would be less expensive, but these studies, while contributing a great deal to our understanding of the biological process, have not yet resulted in the development of a new and cheaper process for the industrial production of ammonia.

Since food production in most parts of the world, including our own country, is limited by the supply of fertilizer, particularly nitrogen, and since the most abundant source of nitrogen everywhere is atmospheric nitrogen, efforts to utilize this source biologically have become increasingly important as both the cost of energy and world population have increased. Biologists have concentrated their research efforts in attempts to develop a more widespread ability to use nitrogen biologically. Two possible avenues of approach are under study. The first involves attempts to increase the number of species that can fix atmospheric nitrogen by transferring the *nif* genes, i.e., the DNA that carries the information for synthesis of the enzymatic nitrogen-fixing system, from a nitrogen-fixing organism to others. It has been possible to achieve this in a few experiments, using as donor the *nif* genes of *Klebsiella pneumoniae*, a member of the family Enterobacteriaceae, to which *E. coli* also belongs. *K. pneumoniae*, like the nitrogen-fixing species of *Bacillus*, is a facultative anaerobe and can fix nitrogen only under anaerobic conditions.

An even more immediately useful accomplishment would be the adaptation of symbiotic nitrogen-fixing organisms to plants other than legumes. Several reports of additional symbiotic nitrogen-fixing relationships have recently appeared, but the data are not convincing. With continued research, however, human technology will probably achieve what natural evolution has not—a greater variety of biological nitrogen-fixing systems, whether symbiotic or simply free-living soil bacteria, which may increase the combined nitrogen content of the soil.

It is important to restate the warning given in Chap. 1 that humans, through rapid expansion of commercial fixation of nitrogen into ammonia and its addition to the soil, are already challenging the microorganisms as the biosphere's most prolific nitrogen fixers. This activity, while helping to increase the food supply, may be leading to environmental problems in balancing the nitrogen cycle. Regardless of the mode of accomplishment of nitrogen fixation, a sociotechnological decision must be made at some time in the not-too-distant future about the amount of nitrogen enrichment that can be permitted without causing an upset in the balance of the life-support system.

In regard to the aerobic treatment of industrial wastewaters that are deficient in nitrogen, one can envision a very useful application for nitrogen-fixing organisms. Finn and Tannahill (1973) have studied this possibility using an impure culture of *Azotobacter* to treat nitrogen-free synthetic wastes containing such carbon sources as carbohydrates, acids, and aromatics.

Requirements for organic nitrogen compounds Most organic growth factors required by microorganisms contain nitrogen as well as carbon and thus contribute to the nitrogen content of the cell. In many such microorganisms, the major source of cellular nitrogen may be inorganic, most often ammonia. These organisms are not deficient in either amination or transamination reactions but in a specific reaction or series of reactions leading to synthesis of the specific growth factor required. Such growth requirements can sometimes be satisfied by a precursor of the required compound if the reaction in which the organism is deficient leads to formation of that precursor, and if other enzymes that convert the precursor to the required product are made by the cell. If two or more such products have a common precursor, both requirements may be satisfied if that precursor is available and can be converted to both products. For example, homoserine is a precursor of three amino acids—threonine, methionine, and isoleucine. Inability to form homoserine could lead to requirements for both methionine and threonine but not for isoleucine, since threonine is a precursor of isoleucine. All three amino acids could be synthesized if the single precursor homoserine were available as an externally supplied organic growth factor. These possible substitutions for specific growth requirements are of greater theoretical than practical interest, since a microorganism in a natural environment is much more likely to encounter an exogeneous supply of amino acids than of their precursors as a result of the presence of protein from decaying animal or vegetable material.

Some microorganisms are unable to use any form of inorganic nitrogen for biosynthetic purposes. Such organisms require an organic nitrogen compound as nitrogen source and may be able to use any of several such compounds as a sole source of cellular nitrogen. Such organisms may be assumed to be deficient in the ability to aminate organic compounds, using ammonia as the source of amino nitrogen. Thus, all organic nitrogen in cellular material must originate by transamination reactions from the single organic nitrogen source.

Control of Synthetic Reactions

The fact that a microorganism is capable of reproducing in a minimal medium—that is, one containing only inorganic salts and a single source of carbon—does not indicate that organic growth factors will not be utilized if an exogenous supply is available. While some microorganisms cannot utilize organic growth factors, the number of such organisms is much smaller than it was once thought to be, and further study may reveal that all microorganisms are capable of using at least some preformed growth factors. The ability to conserve energy and nutrients by ceasing to synthesize compounds that are available in the environment is an important metabolic control mechanism possessed by many microorganisms and will be discussed at some length in Chap. 12. Bacteria such as *E. coli* can grow in minimal medium containing a single organic compound and inorganic salts of the required elements, since they synthesize all the components of the cell. If placed in a complex medium containing extracts of meat and yeast cells, which supplies all the small organic molecules normally synthesized by the organism, the biosynthetic enzymes for these molecules are not made, and preformed growth factors are utilized exclusively. Under these conditions growth is much more rapid, as might be expected, since it is necessary for the cell to only transport the molecules it requires and combine them into the various polymers that make up new cells. In studying the nutritional requirements of a specific microorganism, it is essential to distinguish between compounds that can be utilized for various purposes and those that are absolutely required for growth.

Energy Source

In addition to the required elements in utilizable forms and a supply of any organic growth factors that the cell cannot synthesize, a utilizable source of energy is essential for the maintenance of viability and for reproduction. The major requirement for energy in most microbial species is for biosynthetic reactions, both those involved in the synthesis of small organic molecules and those that form polymers such as proteins and nucleic acids. Energy is required also for transport of nutrients into the cell and for movement if the organism is motile. For many microorganisms, the same organic compound that is the carbon source also is the energy source. Organisms that can obtain carbon and energy from compounds such as amino acids or organic bases such as purines can often utilize a single such compound as a sole source of carbon, nitrogen, and energy and, in some cases, if the compound also contains sulfur or phosphorus, as a sole source of these elements. *Pseudomonas aeruginosa*, an aerobic prototroph, can utilize the amino acid histidine or the purine uric acid as a sole source of carbon, nitrogen, and energy in a medium containing inorganic salts that supply other elements. Organisms that are able to utilize single organic compounds to serve multiple nutritional needs probably have a selective advantage in many natural environments where the supply of organic compounds is limited.

Selective Effect of Nutrients

The selective effect of the chemical composition of a habitat is related to the variety of sources of carbon and nitrogen available and to the organic growth factors present. A habitat rich in organic matter, e.g., some soils, ponds polluted with human or animal wastes or with decaying vegetation, or a sewage treatment plant for domestic wastewaters, will support the growth of a variety of species of microorganisms. All of the nutrients required by most species should be available, and competition among species for nutritional requirements may have little influence on predominance. In a relatively nonpolluted surface water, the absence of a variety of organic growth factors may restrict the species that can proliferate to prototrophs or those with only minor growth factor requirements. In such cases, the ability to utilize the available forms of carbon and nitrogen may determine which organisms can reproduce.

One might expect that biological treatment facilities for industrial wastes would contain a very restricted population of microorganisms, since such wastes may contain no organic growth factors and the organic compounds available as carbon source may be utilizable by relatively few species of microorganisms. Often such wastes contain no nitrogen and must be supplemented with ammonium salts as nitrogen source for microbial growth. However, in such installations, microbial cells themselves may supply sufficient quantities of organic growth factors to support the growth of a variety of species. In the dense microbial population maintained in such treatment processes, cells may die and lyse or release cellular contents that can be utilized by other cells requiring specific growth factors.

The recycling of settled biological solids in the activated sludge process, the formation of thick layers of microbial growth on surfaces of trickling filter media, rotating disks, and similar devices, and the prolonged detention time in oxidation ponds all lead to retention of old cells in the system and to low ratios of substrate to biological solids. Both greater cell age and competition for limited supplies of nutrients by dense populations may be expected to result in a greater frequency of cell death than would occur in an actively growing population of cells. However, even in the latter type of system, e.g., in a "once-through" chemostat (see Chap. 6), we have found that many of the species comprising the heterogeneous population are unable to grow in pure culture in the medium used. It is known that two microorganisms are often able to grow in association in media that support the growth of neither alone, and in several such cases each overproduces and excretes an organic growth factor required by the other. Thus, the ability of a microbial species to establish itself in a natural habitat may be dependent not only on the physical conditions and the nutrients available but also on the other microbial species present. Such interactions between species are not limited, of course, to beneficial effects, since one species of microorganism may form a product that inhibits the growth of another.

For all microorganisms, the types of carbon source, nitrogen source, and energy source available in the environment are determining factors in selecting the species capable of proliferating in that environment. In Chap. 8, carbon and energy sources will be discussed further and used as the basis for a broad classification scheme for metabolic groups of microorganisms.

PROBLEMS

A synthetic waste containing the ingredients listed below is inoculated with a very small amount of activated sludge from a plant that treats municipal waste.

Glucose	2000 mg/L
$(NH_4)_2SO_4$	200 mg/L
$MgSO_4 \cdot 7H_2O$	100 mg/L
KH_2PO_4	3000 mg/L
Na_2HPO_4	6000 mg/L
$FeCl_3 \cdot 6H_2O$	1 mg/L
$CaCl_2 \cdot 2H_2O$	10 mg/L
Tap water	100 ml/L
Distilled water	To volume

Problems 5.1 through 5.5 refer to this synthetic waste.

5.1 The vessel containing the waste is placed in a water bath maintained at 25°C, and air is bubbled through the waste at a sufficient rate to maintain a dissolved oxygen concentration of at least 4 mg/L. Discuss the selective factors that will determine which of the organisms present in the inoculum will grow in the synthetic waste.

5.2 Assume that the biomass produced in this waste has the average composition given in Table 3-1.

(a) Which ingredient of the waste will limit the amount of biomass produced?

(b) Calculate the amount of biomass that can be produced (milligrams per liter dry weight).

(c) How much COD will be removed from the waste, assuming 50 percent of the glucose is used for synthesis and 50 percent for respiration?

(d) Calculate the amount of the limiting ingredient that would have to be added to allow removal of all the COD.

5.3 What two purposes are served by the potassium and sodium phosphates in the medium? Why are such high concentrations used?

5.4 If the temperature control of the water bath were adjusted to 40°C after the biomass had started to grow, what would be the effects on the organisms making up the biomass as the temperature rose?

5.5 Two species of bacteria, A and B, are isolated from a municipal activated sludge treatment plant. Neither will grow in the synthetic waste described above. Various additions are made to the synthetic waste and both bacteria are tested for ability to grow, with the results shown below.

	Growth response	
Addition to waste	Species A	Species B
Amino acid mixture	+	+
Glutamic acid	+	−
Aspartic acid	+	−
Histidine	+	+
None	−	−

Explain the nitrogen source requirement of each of these organisms.

5.6 Define obligate aerobe, obligate anaerobe, facultative anaerobe, and microaerophile. Discuss the effect of oxygen on each.

5.7 Discuss the functions of the enzymes that protect cells from the toxic effects of oxygen.

5.8 If a small amount of soil is placed in a flask, covered to a depth of approximately 2 in with a medium containing glucose, K_2HPO_4, and $MgSO_4$ in tap water, and left sitting in the dark in contact with air, a fairly heavy growth of bacteria will develop in 1 to 2 weeks. The predominant organisms are species of *Azotobacter* and *Clostridium*. What factors operate to select these organisms and allow both to grow?

5.9 If a bacterium were isolated from an oxidation pond in December when the water temperature was 10°C, what criteria would be used to classify it as a psychrophile or a psychrotroph? Why would this be important in predicting its usefulness in the year-round purification process?

5.10 Discuss the importance of mixing in a fluidized reactor, e.g., in an activated sludge process.

REFERENCES AND SUGGESTED READING

Altman, P. L., and D. S. Dittmer (editors). 1966. Experimental biology. Federation of American Societies for Experimental Biology, Bethesda, Md.

Arthur, H., and K. Watson. 1976. Thermal adaptation in yeast: Growth temperatures, membrane lipid, and cytochrome composition of psychrophilic, mesophilic, and thermophilic yeasts. *J. Bacteriol.* **128**:56–68.

Brill, W. J. 1977. Biological nitrogen fixation. *Sci. Am.* **236**:68–81.

Brock, T. D. 1966. Principles of microbial ecology. Prentice-Hall, Englewood Cliffs, N.J.

Brock, T. D. 1969. Microbial growth under extreme conditions. *In* Microbial growth, 19th Symposium of the Society for General Microbiology. Cambridge University Press, London.

Brock, T. D. 1979. Biology of microorganisms, 3d ed. Prentice-Hall, Englewood Cliffs, N.J.

Brock, T. D., and G. K. Darland. 1970. Limits of microbial existence: Temperature and pH. *Science* **169**:1316–1318.

Brown, C. M., D. S. Macdonald-Brown, and J. L. Meers. 1974. Physiological aspects of microbial inorganic nitrogen metabolism. *Adv. Microb. Physiol.* **11**:1–52.

Chapman, T. D., C. H. Matts, and E. H. Zander. 1976. Effect of high dissolved oxygen concentration in activated sludge. *J. Water Pollut. Control Fed.* **48**:24–86.

Cole, J. A. 1976. Microbial gas metabolism. *Adv. Microb. Physiol.* **14**:1–92.

Finn, R. K., and A. L. Tannahill. 1973. The "Azotopure" process for treating nitrogen-deficient wastes. *Biotechnol. Bioeng.* **15**:413–418.

Fridovich, I. 1975. Oxygen: Boon and bane. *Am. Sci.* **63**:54–59.

Hamilton, W. A. 1975. Energy coupling in microbial transport. *Adv. Microb. Physiol.* **12**:1–53.

Hamilton, W. A. 1977. Energy coupling in substrate and group translocation. *In* Microbial energetics, 27th Symposium of the Society for General Microbiology. Cambridge University Press, London.

Hawker, L. E., and A. H. Linton (editors). 1971. Micro-organisms. Function, form and environment. American Elsevier, New York.

Hughes, D. E., and J. W. T. Wimpenny. 1969. Oxygen metabolism by microorganisms. *Adv. Microb. Physiol.* **3**:197–232.

Hutner, S. H. 1972. Inorganic nutrition. *Annu. Rev. Microbiol.* **26**:313–346.

Inniss, W. E. 1975. Interaction of temperature and psychrophilic microorganisms. *Annu. Rev. Microbiol.* **29**:445–465.

Kalinske, A. A. 1976. Comparison of air and oxygen activated sludge systems. *J. Water Pollut. Control Fed.* **48**:2472–2485.

Mackenthun, K. M. 1969. The practice of water pollution biology. U.S. Department of the Interior, U.S. Government Printing Office, Washington, D.C.

Mainzer, S. E., and W. P. Hempfling. 1976. Effects of growth temperature on yield and maintenance during glucose-limited continuous culture of *Escherichia coli*. *J. Bacteriol.* **126**:251–256.

Morita, R. Y. 1975. Psychrophilic bacteria. *Bacteriol. Rev.* **39**:144–167.

Morris, J. G. 1975. The psysiology of obligate anaerobiosis. *Adv. Microb. Physiol.* **12**:169–246.

Morris, J. G. 1976. Oxygen and the obligate anaerobe. *J. Appl. Bacteriol.* **40**:229–244.

Parker, D. S., and M. S. Merrill. 1976. Oxygen and air activated sludge: another view. *J. Water Pollut. Control Fed.* **48**:2511–2528.

Reid, G. K. 1961. Ecology of inland waters and estuaries. Reinhold, New York.

Ricica, J. 1966. Techniques of continuous laboratory cultivations. *In* Theoretical and methodological basis of continuous culture of microorganisms, I. Malek and G. Fencl (editors). Academic Press, New York.

Rickard, M. D., and A. F. Gaudy, Jr. 1968a. Effect of oxygen tension on O_2 uptake and sludge yield in completely mixed heterogeneous populations. Proceedings, 23rd Industrial Waste Conference, Purdue University, pp. 883–893.

Rickard, M. D., and A. F. Gaudy, Jr. 1968b. Effect of mixing energy on sludge yield and cell composition. *J. Water Pollut. Control Fed.* **40**:R129–R144.

Sakharova, Z. V., and I. L. Rabotnova. 1976. Effects of pH on physiological and biochemical properties of a chemostatic culture of *Bacillus megaterium. Mikrobiologiya* **46**:15–21. (Available in English translation from Plenum Publishing Corporation, New York.)

Singleton, R. Jr., and R. E. Amelunxen. 1973. Proteins from thermophilic microorganisms. *Bacteriol. Rev.* **37**:320–342.

Skinner, K. J. 1976. Nitrogen fixation. *Chem. Eng. News,* **54**:22–35.

Stanier, R. Y., E. A. Adelberg, and J. Ingraham. 1976. The microbial world, 4th ed. Prentice-Hall, Englewood Cliffs, N.J.

White, J. P., D. P. Schwert, J. P. Ondrako, and L. L. Morgan. 1977. Factors affecting nitrification *in situ* in a heated stream. *Appl. Environ. Microbiol.* **33**:918–925.

Wimpenny, J. W. T. 1969. Oxygen and carbon dioxide as regulators of microbial growth and metabolism. In Microbial growth, 19th Symposium of the Society for General Microbiology. Cambridge University Press, London.

SIX

QUANTITATIVE DESCRIPTION OF GROWTH

In this chapter we shall be concerned primarily with the quantitative description of microbial growth. Growth results from the assimilation of nutrient materials, and one may consider it as a gain in size, or mass, of individual cells, or as an increase in the number of cells, i.e., an increase in the population. An increase in numbers occurs along with an increase in size but is not necessarily an immediate consequence of that increase. With heterogeneous microbial populations such as are found in water courses, soil, and biological treatment plants, it is quite difficult to enumerate the numbers of different species, or, for that matter, even total numbers of cells. In most industrial and environmental applications, growth is measured as a gain in mass even though the quality and activity of the biomass are dependent in large measure on the types and numbers of living cells present. The prime objective in this chapter is a pragmatic one—to formulate mathematical expressions that describe the rates of substrate utilization and biomass accumulation in a biological system. Such models may be useful to the fundamental understanding of the mechanisms resulting in growth. They form a basis for modern concepts of design and operation of biological wastewater treatment processes.

The work of environmental technologists and industrial microbiologists requires that they predict the rate and amount of microbial growth and often the metabolic processes related to growth, such as production of specific enzymes or other organic products. Kinetic description is needed in order to design processes and control their operation. The kinetic formulations and models become predictive equations and are as vital to the design and operation of biological engineering systems as are material stress equations to the design and analysis of engineering structures. However, it must be remembered that the environmental technologist is attempting to predict the behavior of living rather than inanimate engineering material.

Ideally, kinetic models and the equations devised for their description are truly quantitative manifestations of the function of specific mechanisms assumed in the models. More often than not, however, the mechanisms that the kinetic equation describes are not known or have not been fully delineated. We may recall that such was the case with the Michaelis-Menten equation. The mechanism involving the making and breaking of an enzyme–substrate complex was hypothesized as one that could be expressed by a hyperbolic relationship between the rate v and the substrate concentration S. Many years later, the existence of the complex was demonstrated experimentally. Thus, this hypothesis was a theory that proved to be correct.

All too often it is overlooked that what we choose to call "theoretical" is simply unproved hypothesis. This is a ready trap for applied scientists and engineers because they work under the pressure of urgent necessity to predict an outcome and its rate of attainment regardless of the state of their knowledge of the chemical, physical, or biological mechanism causing the outcome. To be sure, there is some soothing of professional conscience if one can say the equation fits a theory. On the other hand, we may ask: Why shouldn't it, when the chances are that the theory was devised to fit the equation? Thus, the two are not really married until the theory (hypothesis) invented to fit the kinetic expression of the experimental observation is itself shown to be correct by new experiments quite independent of the original one. This is the scientific approach, and it is as practical as it is scientific.

Similarity of kinetic expression does not give one license to conclude similarity of mechanistic action; recall that the rectangular hyperbola can be used to express enzyme kinetics and the kinetics of adsorption of gases on surfaces. While there are analogous aspects in the two processes, they are different mechanistic occurrences. It will be shown in this chapter that the relationship between the specific growth rate of microbial cells and the concentration of a required nutrient often can be expressed by an equation that plots a rectangular hyperbola, as can another relationship to be discussed later—that between the observed, or net, specific growth rate and the observed cell yield. It must be kept in mind, then, that kinetic expressions are vital for practical engineering reasons and useful as quantitative predictive expressions of proven mechanistic theory, but in themselves they do not provide definitive proof of any theory of mechanism.

Since the mechanisms that cause the observed kinetic behavior of a process may be unknown, can we have confidence in the kinetic expressions? We can, if their predictive value is proven useful repeatedly, but this confidence does not excuse a lack of continuing investigation into the mechanistic cause of the observed kinetic event. For example, over many years of recorded history, human beings were able to predict with extreme precision the observed, repetitive movements of the sun and the stars. Such predictions have provided us with the very means of kinetic expression, i.e., the units of time. Throughout most of this period, these learned individuals employed a hypothesis (theory), the correcting of which nearly cost Galileo his life. Thus, it is possible to formulate predictive expressions of real utility without really understanding the phenomenon our equations describe, provided we are skilled in experimentation or, as in the case of the ancients, good observers and able to describe quantitatively a well-run and much-repeated experiment. Since our state of knowledge of the basic mechanisms of microbial growth processes is incomplete, it is necessary to base the needed kinetic description on repeated experimentation, because the predictive equations are in large measure merely extrapolations of past events. If we hope to improve this situation, it is necessary to continue to seek mechanistic causation lest various biological hypotheses become erroneous dogma.

GENERAL RELATIONSHIPS AND METHODS OF MEASUREMENT

Taking an entirely practical approach, let us assume that an experiment is run in some sort of reaction vessel in the laboratory. Aerobic conditions are maintained by vigorous mixing and aeration. The reactor is seeded with a small inoculum of organisms from sewage or soil. The growth medium is natural wastewater containing some soluble organic compounds that the microorganisms can metabolize quite readily and some that are not metabolizable. Samples are taken at the initiation of the experiment, and the amount of soluble organic matter is measured. This can be done by filtering out the cells and running tests such as COD or TOC on the filtrate; in this example, COD is used. The dry weight of the cells retained on the filter is also measured. Periodically, as the experiment proceeds, other samples are taken and the concentrations of substrate S measured as COD, and biomass, or cells, X, are plotted versus time. The accumulated oxygen uptake is also measured and plotted.

A plot of the results of such an experiment is shown in Fig. 6-1. This type of result can be shown to happen again and again, with some variations to be discussed later. As the cells grow and respire, they utilize the organic food source. When the metabolically usable carbon source is exhausted, the cells stop growing; i.e., the biomass ceases to accumulate (see dashed line on the graph). Any further respiration may be called *endogenous*, because there is no longer a utilizable exogenous carbon source to respire, and the oxygen uptake proceeds at the expense of a loss in weight of the biomass. The portions of the curves to the left of the dashed line reflect events that occur in the purification, or substrate removal, phase, and events to the right occur in the endogenous, or autodigestive, phase. In a batch system such as the one shown in the graph, in which the initial biomass concentration X_0 is small in relation to the initial concentration of growth-limiting nutrient (the carbon source in this experiment), it is safe to assume that for practical purposes the

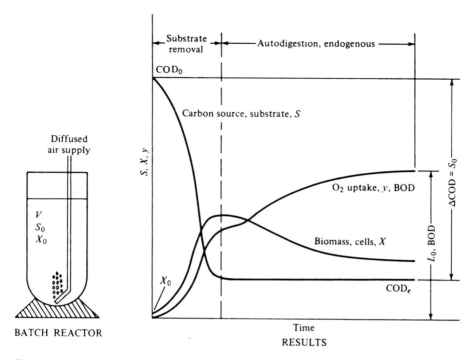

Figure 6-1 Batch reactor (*left*) and typical data plots of growth, respiration, and removal of soluble substrate during aerobic incubation. Note the distinction between ΔCOD and L_0.

purification and endogenous phases occur in such a sequential fashion. Reasons for this will be discussed later. First, it is important to discuss ways to measure the amount of metabolically available carbon source S_0, the concentration of biomass X, and the oxygen uptake.

Measurement of Usable Substrate (ΔCOD)

Referring to Fig. 6-1, it is seen that a significant concentration of COD remains unused (COD_e); i.e., not all the organic matter or COD in the sample was used by the microbial population. Whether the COD remaining is organic matter originally in the feed or organic compounds produced by the cells during growth is an important environmental consideration, but it is not germane to the present discussion. The important fact is that the amount of carbon source used by the organisms during growth was equal to ($COD_0 - COD_e$); this amount of carbon source removed is designated as ΔCOD in the figure. The ΔCOD is thus a measure of the amount of carbon source in the sample that was available as food material to the microorganisms; it is a measure of the strength of the waste.

The above definition of ΔCOD is not independent of the microbial population in the initial seed. It is possible that a different source of initial cell inoculum would give a higher or lower value of COD_e for the same COD_0, i.e., a different ΔCOD value. Thus, if ΔCOD is to be a valid measure of the amount of microbially usable carbon source, attention must be paid to two additional considerations. First, it must be stipulated that the seed population has been acclimated to the organic matter. Acclimation can be accomplished by growing a population through three or four daily transfers in the medium, that is, by establishing a prior growth history during which it is seen that any initial time lag in growth disappears after a few growth periods and that the residual COD prior to each transfer does not change to any great extent. Thus, the purpose of acclimation is to ensure that the cells in the original heterogeneous population have undergone biochemical acclimation and selection of the species best able to grow on as much of the carbon source in the sample as is possible. Second, one should try the whole operation several times, using various sources of initial seed population. It must be remembered that the "natural" sampling universe is immensely diverse. Soil, sewage, and lake or river water samples taken on different days may contain widely different types and numbers of cells. Also, it can be expected that there will be some variability in the nature and amount of the various carbon sources in the sample of wastewater. Thus, considerable amounts of data are needed in order to establish a dependable quantitative estimate of the concentration of microbial carbon source. Also, one should run the tests under favorable growth conditions, e.g., a temperature range of 20 to 25°C, neutral pH, and abundant air supply.

Instead of COD, TOC or TOD might have been used, or any good measure of total organic matter in the filtrate. Previously (see Chap. 3) we used these analyses as measures of organic matter, and now it is seen that they can be used as tools in a bioassay, a ΔCOD test (or ΔTOD or ΔTOC) to assess the amount of the organic matter that is available to an acclimated microbial population. Such a measure is central to the study of the behavior of microbial systems growing on mixtures of unknown compounds and to the successful design and operation of biological waste treatment processes. Other procedures available for measuring the amount of organic matter might include qualitative and quantitative analyses of the waste, which are indeed expensive and not particularly useful unless the microbial response to all of the compounds singly and in combination is known. Also, one may resort to measurement, not of the carbon source, but of the accumulated oxygen uptake during metabolism of the carbon source, i.e., to the biochemical oxygen demand. However, as shall be shown later, BOD is not a very dependable measurement of the organic loading S_0. The measurement of the loading to a bioreactor is as essential to its design as is the estimation of stress loadings to the design of a building or bridge.

The determination of ΔCOD as described above is by no means without flaws, but it has the great advantage of measuring the amount of biologically usable organic material in a mixture of unknown compounds rather than

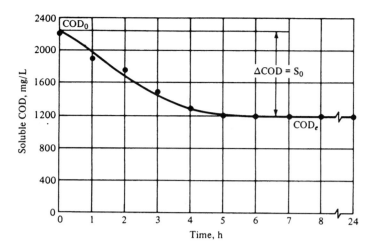

Figure 6-2 ΔCOD for a sample of dilute kraft pulp mill digester blow-down liquor. (*Data of D. Scott, Jr. and A. F. Gaudy, unpublished.*)

assessing the amounts of specific carbon sources present in a sample of water. It allows one to deal with S as a heterogeneous mixture of compounds in the same way one must deal with X as a heterogeneous mixture of microorganisms. X represents the concentration of a biomass, a sludge, or an average species, which we shall try to characterize as similar in many ways to a single pure culture; ΔCOD represents the heterogeneous counterpart for carbon source or substrate. If properly determined, ΔCOD represents the maximum potential substrate value of the sample. The amount of organic matter remaining, COD_e, may be rather high. For example, one might experience such a situation for a waste high in lignin compounds (see Fig. 6-2). This class of compound is not readily metabolized and would not be expected to contribute to the substrate, i.e., the loading to a biological treatment plant. Although lignin analyses were not run during the experiments shown in Fig. 6-2, one may assume with safety (based on past observations) that by far most of the COD_e consisted of lignin that was in the original sample.

Even when one runs an experiment using as carbon source a single compound that is known to be metabolized easily, there will be some residual COD, i.e., COD_e. In the experiment shown in Fig. 6-3, the substrate was glucose; COD_e was very low and only traces of glucose could be found. The COD_e in this case consisted of compounds produced by the biomass during the metabolic purification process. The COD_e may consist of small amounts of many different compounds. Some may be metabolic intermediates and/or end products; some may be bits of cellular structural components, nucleotides, peptides, and other contents from dead or dying cells. Thus, ΔCOD as a measure of S_0 makes no distinction as to whether the residual COD was due to nonmetabolizable organic matter in the original waste sample or to organic by-products of the purification mechanism.

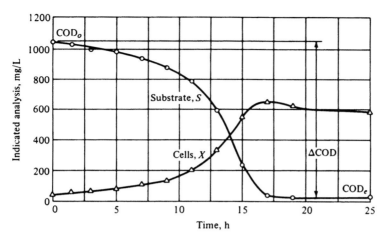

Figure 6-3 ΔCOD and biomass curves for a sample containing glucose as carbon source. Note the low COD_e compared with that in Fig. 6-2. (*Data of T. T. Chang and A. F. Gaudy, unpublished.*)

It should be apparent that repeated ΔCOD tests are needed on wastes that contain a mixture of metabolizable and nonmetabolizable compounds. Since the metabolizable fraction is not known, repeated runs are needed to establish that the value of COD_e is probably as low as it can be. From the results of one or two experiments, it is not possible to conclude whether something other than exhaustion of the usable carbon source caused the biomass to stop growing. If, in an experiment conducted with a compound such as glucose, as in Fig. 6-3, the residual COD leveled off at 300 mg/L rather than 30 mg/L, one would immediately question the resulting value of ΔCOD. It would be a valid result for that experiment, but that experiment, for one reason or another, would not be valid for measuring ΔCOD. If caution had been taken to provide reasonably good environmental conditions (DO supply, temperature, pH, nutrients, etc.), the reason for the high COD_e would in all probability be due to the nature of that particular biomass, and this is reason for running additional ΔCOD tests using different sources of seed organisms. Such a result could also be taken as an indication that the biomass at the time of the experiment was not sufficiently acclimated. Although it may have removed all of the glucose, in the environmental technologists' terms, it could not remove the carbonaceous organic compounds for which glucose was the source. Thus, the biomass was not totally acclimated, and the result was not consistent with previous experimental observations that indicated that results such as those shown in Fig. 6-3 could be expected for that source of carbon.

In the case of biological treatment plants, the nature and origin of the residual COD are of great importance because this material enters the environment, i.e., the natural water courses. Determination of its subsequent fate and its effect on the aqueous environment is vital to policies regarding technological control of the aqueous environment. The fact that there is some residual, nonmetabolized organic matter does not mean that this organic matter will not be metabolized in the natural reactors in the absence of the readily available carbon sources represented by the ΔCOD and in the presence of species of organisms that may not have been in the biomass that exerted the ΔCOD. The probability is that sooner or later the COD_ℓ will become a usable substrate, and under ordinary conditions the rate at which it is used will be rather slow—slow enough, in fact, that the receiving stream can accommodate this lighter organic loading within the aerobic decay leg of the carbon-oxygen cycle; that is, the material will be metabolized slowly enough so that natural reaeration can maintain aerobic conditions. The aim of the secondary biological treatment plant is to provide a situation wherein this posttreatment metabolism of residual organic matter occurs without undue stress on the environment. This does not imply that COD_ℓ, no matter what its source, cannot or should not be subjected to further reduction before it enters the aqueous environment. There are cases in which it is not desirable to let the natural reactors perform this task, and subsequent processing (physical-chemical or biological) may be used to reduce residual organic matter (see Chap. 2). Also, there may be special ways to operate biological treatment plants so as to maintain populations that can concurrently remove the more readily available and the difficultly metabolized organic matter, but these await development. Successful design and operation of the secondary plants dictated by law cannot await such developments, which indeed might not even be necessary if the secondary plants could remove close to 100 percent of the measured ΔCOD of the waste.

Measurement of Oxygen Uptake

Oxygen uptake (aerobic respiration) is an essential part of aerobic microbial growth, and it is recalled that this process is primarily responsible for the fact that the metabolism of 1 g of organic matter does not ordinarily lead to the synthesis of 1 g of new cells. Also, it is recalled that the need to maintain the natural aqueous environment in an aerobic condition is the major justification for requiring secondary treatment of wastewaters.

Oxygen utilization is such an important concern that the amount of oxygen used in the aerobic metabolism of a waste sample has been used for many years as a measure of the strength of a waste; that is, it has been used as a measure of the amount of the organic substrate S_0. It may be seen in Fig. 6-1 that the accumulated oxygen uptake or biochemical oxygen demand (BOD) of the waste is not equal to the ΔCOD, which is a measure of organic substrate S_0. Both BOD and ΔCOD are measured in terms of oxygen, but ΔCOD measures available carbon source and BOD measures the portion of the available carbon source that is respired.

We might ask: Can it be expected that these two parameters may ever be the same? If not, are they related by a constant proportionality factor? There is one way the measured oxygen uptake or BOD can approach the value measured for ΔCOD. In Fig. 6-1 it is seen that the net amount of biomass synthesized ΔX ($\Delta X = X_t - X_0$) decreases after the substrate removal phase; i.e., ΔX decreases in the endogenous or autodigestive phase. If the time t is long enough, it is not impossible for X_t to return to the value of X_0, in which case $\Delta X = 0$, and all of the organic matter in the waste that was usable as a microbial substrate, ΔCOD, has been biochemically oxidized. In this case, the ultimate value of the accumulated oxygen uptake curve (i.e., ultimate BOD, designated as L_0) would approach the value of ΔCOD. In such cases it can be said that the available carbon source was totally oxidized. It is this possibility that forms a conceptual basis for use of the so-called extended aeration modification of the activated sludge process.

Total oxidation in a closed system, such as the batch experiments shown in Figs. 6-1, 6-2, and 6-3, does not always occur, and when it does it is a slow process (t measured in many days). Thus, as a measure of the amount of organic substrate, BOD is not a very good parameter. If there were a constant relationship between BOD and ΔCOD (or ΔTOD or ΔTOC), BOD could be used as an assay for the amount of organic substrate. Unfortunately the ratio BOD/ΔCOD varies, largely because the degree of autodigestive decrease in biomass varies with the microbial population that is present. It is well to reemphasize here one of the differences between a pure culture seed and one that consists of a heterogeneous mixture of various types of microorganisms. If the biomass in the experiment shown in Fig. 6-1 consisted of a single microbial species, the decrease in X after removal of the substrate would be due only to oxidation of internal carbon in the living cells and oxidation by living cells of parts of cells that had died. The probability of X returning eventually to the level of X_0 would be very small indeed, because a single species cannot be expected to make all of the enzymes needed to break down all the various structural macromolecules it has synthesized. Surely, no individual cell can be expected to "endogenate" itself to nothingness. In heterogeneous populations, the autodigestive reactions that occur in the single-species population are operative but there is the additional presence of the food chain, which is active in the decay leg of the carbon-oxygen cycle (see Fig. 3-1). The chances of a large decrease in X and a resulting increase in BOD are greater, but there is no guarantee that in any randomly selected acclimated seed the predator–prey relationships will lead to the same amount of autodigestion each time the experiment is run. Thus, there is an unreliable relationship between ΔCOD and BOD. It was important to discuss the concept of BOD in relation to Fig. 6-1 because such discussion aids in understanding and interpreting results of the BOD test. The BOD test as well as the concept of BOD will be discussed more completely in Chap. 10, but certain aspects of BOD that are germane to the measurement of oxygen uptake require discussion in this chapter.

We shall describe three methods of measuring oxygen uptake; the standard BOD method, the open-reactor method, and the manometric technique.

Measurement of decrease in dissolved oxygen in closed reactors For the present, the standard method of measuring exertion of biochemical oxygen demand, i.e., the standard BOD test, is important because it provides a means of measuring oxygen utilization in very dilute systems. In Fig. 6-1, if it is assumed that both X_0 and S_0 are very low—so low that only a few milligrams of oxygen would be used during the entire experiment—there would be no need to aerate the liquid. One could saturate the fluid with oxygen and measure the initial dissolved oxygen concentration (DO_i), fill the reactor to the brim, and seal the entire body of reaction liquid from the atmosphere. The DO measured at some future time t, i.e., DO_t substracted from DO_i, would be the amount of oxygen used during the time period ($\Delta O_2 = DO_i - DO_t$). Replicate sealed reactors could be analyzed for DO at various other times, and in this way one could obtain data for plotting the accumulated oxygen uptake curve. The solubility of oxygen in water is low, generally in the range of 7 to 9 mg/L, depending on temperature, barometric pressure, and dissolved and suspended organic and inorganic solids concentrations.

Obviously, the system must be very dilute to use this method of measuring oxygen uptake, and such is the case with the standard BOD measurement in the closed system of the BOD bottle (see *Standard Methods*, APHA, 1976). If, on the other hand, X_0 and S_0 were high, the procedures described above could not be used to generate an entire oxygen utilization curve, but one could gain some idea of the instantaneous rate of oxygen utilization. For example, the mixed liquor could be oxygenated and the DO_i measured. If the biomass concentration X_0 were high, the DO would drop very quickly. If DO_t were measured within the short time before DO reached 0, the difference between DO_i and DO_t, i.e., ΔO_2 used during the short time interval Δt, could be calculated as a rate of oxygen uptake. In such an experiment, one usually finds that the DO concentration decreases linearly until just before zero DO is attained. The drop in DO during the small interval of time could be reported in any convenient time units—minutes, hours, days, etc. Naturally, caution must be exercised in extrapolating the short-term value to longer periods, i.e., assuming the rate that was measured would continue in a linear fashion for hours or days. Since the rate of oxygen uptake is dependent on the biomass concentration X, one could express it per unit of weight of biomass by dividing by the average concentration of X during the period Δt. When X is high, the change in X during the short time interval needed to utilize 4 to 6 mg/L of oxygen is very small. If such a rate measurement is made during the substrate removal phase, X will increase during Δt; if made during the endogenous phase (see Fig. 6-1), X will decrease during Δt due to endogenous respiration and autodigestion. Such a measurement of the specific rate of oxygen utilization in the endogenous phase can be useful in adjudging the vitality of the biomass; one can make a case for concluding that the higher the endogenous uptake per unit weight of biomass, the greater is the fraction of cells that are capable of metabolism.

More often than not, one deals with systems of substrate and biomass that are more concentrated than those for which the respiration curve can be obtained using a standard BOD procedure. One way to deal with such systems is to dilute them by whatever amount is necessary to allow use of the standard BOD technique. This procedure is unsatisfactory, although widely practiced, because the rate of respiration is affected by dilution (see Chap. 10). The methods described in the two following sections are more reliable for measuring the oxygen uptake in more concentrated systems.

Calculation of oxygen utilization from the DO profile in an open reactor In Fig. 6-1, the aeration vessel is neither completely filled with reaction fluid nor closed to the atmosphere. It is aerated by diffusing air through the submerged aeration tube. Aeration also could be accomplished by stirring or agitating the reaction fluid to enhance the transfer of oxygen from the atmosphere to the liquid. For an open reactor, the oxygen uptake of the cells can be estimated if the rate and kinetic mode of aeration of the reaction system are known. The reaction "system" refers to the mixed liquor and the vessel, because the reaeration rate is determined by both the geometry of the vessel and the chemical characteristics of the liquor.

The transfer of gaseous oxygen to the liquid phase is a physical process that follows a first-order, decreasing-rate kinetic pattern that can be stated adequately in mathematical terms. For example, if one filled a reactor with growth medium and did not add cells, or perhaps put cells in the reactor but added a disinfectant that killed them, thereby preventing any oxygen utilization due to metabolism, and if the liquid were stripped of all dissolved oxygen at the beginning of the aeration period, the DO concentration after turning on the air supply would increase along curves of the type shown in Fig. 6-4. The

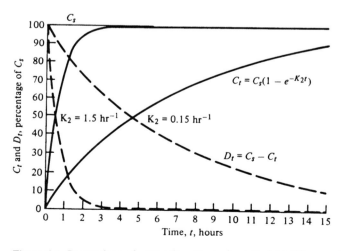

Figure 6-4 Comparison of reaeration curves for a tenfold difference in K_2.

DO concentration C_t increases at an ever-decreasing rate and finally approaches a value equal to the solubility of oxygen (C_s, the saturation concentration) under the conditions of the experiment. After C_s is attained, the DO does not increase further because the rate of its transfer into the liquid becomes equal to the rate at which it is transferred out again to the atmosphere; that is, an equilibrium is attained.

The DO saturation concentration C_s is dependent on the chemical composition of the medium. While the medium is largely water, the solubility of oxygen is not the same as in pure water. The DO deficit curve ($D_t = C_s - C_t$) is easily obtained from the DO data if one knows the value of C_s. The rate of change of both curves, dissolved oxygen C and deficit D, is the same and can be shown to be proportional to the DO deficit at any time t. Both curves follow the rate law of a monomolecular reaction, and the net driving force for transfer of oxygen from the gas phase to the liquid phase is the deficit from saturation. As the deficit is reduced, the driving force becomes smaller and the rate of transfer decreases. The rate of increase in DO concentration, dC/dt, is proportional to the deficit:

$$\frac{dC}{dt} = K_2(C_s - C_t) \qquad (6\text{-}1)$$

The proportionality constant K_2 is referred to as the *reaeration constant*. Considering the rate of change in DO, Eq. (6-1) can be rewritten in either of the following forms:

$$\frac{dC}{dt} = K_2C_s - K_2C_t \qquad (6\text{-}2)$$

or

$$\frac{\Delta C}{\Delta t} = K_2C_s - K_2C_t \qquad (6\text{-}3)$$

Equation (6-3) is readily applied to experimental data for determining K_2 and C_s. Its use is illustrated by the example shown in Fig. 6-5. The upper graph is a plot of measured DO versus time during an experiment such as the one shown in Fig. 6-4. The lower graph shows the use of a plot of Eq. (6-3) to calculate K_2. It is required to calculate values of dC/dt at various values of C_t, and numerical approximations of dC/dt, i.e., $\Delta C/\Delta t$, are obtained from the experimental data as illustrated. In the example shown (see upper graph), the value of $\Delta C/\Delta t$ at $t = 5.5$ h using a 1-h interval for calculation is 0.58 mg/L/hr at a C_t value of 4.3 mg/L. This point is shown in the lower plot (see darkened circle). After obtaining and plotting several values, the slope of the line is determined; from Eq. (6-3), this slope is equal to K_2. When C_t is equal to C_s, $\Delta C/\Delta t = 0$; i.e., the intercept on the C_t axis is the value of C_s.

Using the C_s and K_2 values thus obtained, the concentration of DO at any time C_t can be found in accordance with the integrated form of Eq. (6-1). If $C_t = 0$ at $t = 0$, then

$$C_t = C_s(1 - e^{-K_2t}) \qquad (6\text{-}4)$$

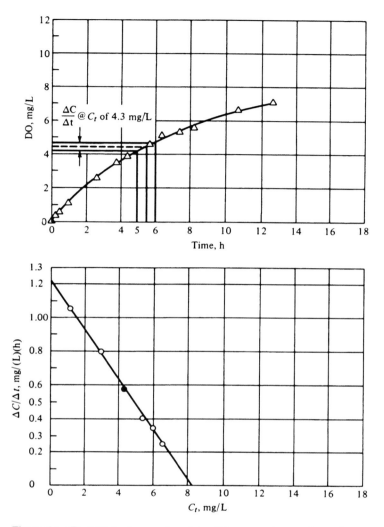

Figure 6-5 (*Top*) Plot of reaeration data; DO versus time to determine corresponding values of $\Delta C/\Delta t$ and C_t. (*Bottom*) Plot of $\Delta C/\Delta t$ versus C_t to determine the values of K_2 and C_s; see Eq. (6-3). $K_2 = 0.145 \text{ h}^{-1}$, $C_s = 8.3 \text{ mg/L}$. (*Data of M. P. Reddy and A. F. Gaudy, unpublished.*)

If there is an initial DO concentration C_0, then

$$C_t = C_s - (C_s - C_0)e^{-K_2 t} \tag{6-5}$$

Using these equations and the determined values of C_s and K_2, the calculated values of C_t at various times t can be computed and plotted to determine whether the calculated curve provides an adequate description of the experimental data.

Having thus estimated K_2 and C_s, which control the rate of input of dissolved oxygen to the system, it is possible to measure the microbial oxygen uptake, i.e., utilization of dissolved oxygen, if the amount of DO present at any time t during the experiment can be determined. This can be assessed by frequent measurement of the dissolved oxygen concentration. The method of estimating the oxygen uptake curve is illustrated with the aid of Fig. 6-6 and Table 6-1. The upper portion of the figure shows a plot of the dissolved oxygen concentration values during an experiment run like the one shown in Fig. 6-1, for which a "DO profile" was not shown. In this experiment the S_0 consisted of 60 mg/L of a 1:1 glucose–glutamic acid mixture, and X_0 was an

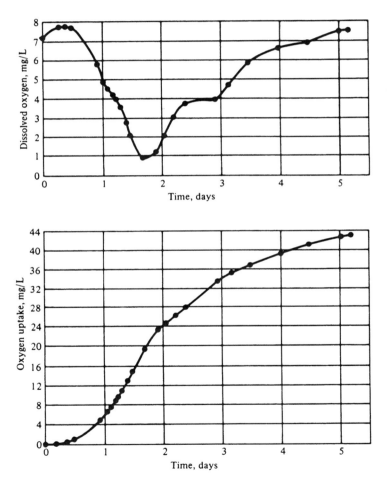

Figure 6-6 (*Top*) Dissolved oxygen profile developed in an open stirred reactor with $K_2 = 0.099\ h^{-1}$. The carbon source was a 1:1 mixture of glucose and glutamic acid; total concentration is 60 mg/L. The initial microbial population was obtained by seeding the reactor with 0.5 percent (*v/v*) settled fresh sewage. Temperature was 23°C. (*Bottom*) Oxygen uptake curve calculated from the DO profile shown above. (*K. M. Peil and A. F. Gaudy, unpublished.*)

Table 6-1 Calculation of oxygen uptake from open-jar reactors

(1) Hour	(2) D, mg/L	(3) K_2D, mg/(L × h)	(4) Δt, h	(5) $K_2D \Delta t$, mg/L	(6) DO, mg/L	(7) ΔDO, mg/L	(8) (5) − (7), mg/L	(9) O_2 uptake, mg/L
0.00	1.50	0.15	6.25	0.74	7.20	0.62	0.12	0
6.25	0.88	0.09	2.75	0.24	7.82	0.00	0.24	0.12
9.00	0.88	0.09	2.50	0.23	7.82	−0.09	0.32	0.36
11.50	0.97	0.10	10.50	2.04	7.73	−1.98	4.02	0.68
22.00	2.95	0.29	2.50	0.85	5.75	−0.95	1.80	4.70
24.50	3.90	0.39	2.00	0.80	4.80	−0.32	1.12	6.50
26.50	4.22	0.42	2.00	0.87	4.48	−0.31	1.18	7.62
28.50	4.53	0.45	1.25	0.57	4.17	−0.21	0.78	8.80
29.75	4.74	0.47	1.75	0.86	3.96	−0.42	1.28	9.58
31.50	5.16	0.51	2.00	1.11	3.54	−0.83	1.94	10.86
33.50	5.99	0.59	2.00	1.25	2.71	−0.63	1.88	12.80
35.50	6.62	0.66	4.75	3.41	2.08	−1.25	4.66	14.68
40.25	7.87	0.78	5.75	4.40	0.83	0.32	4.08	19.34
46.00	7.55	0.75	3.00	2.11	1.15	0.93	1.18	23.42
49.00	6.62	0.66	4.00	2.44	2.08	0.94	1.50	24.60
53.00	5.68	0.56	4.50	2.37	3.02	0.73	1.64	26.10
57.50	4.95	0.49	12.75	6.17	3.75	0.13	6.04	27.74
70.25	4.82	0.48	5.25	2.29	3.88	0.82	1.47	33.78
75.50	4.00	0.40	8.00	2.72	4.70	1.12	1.60	35.25
83.50	2.88	0.29	12.50	3.13	5.82	0.72	2.41	36.85
96.00	2.16	0.21	11.50	2.30	6.54	0.28	2.02	39.26
107.50	1.88	0.19	13.00	2.05	6.82	0.58	1.47	41.28
120.50	1.30	0.13	3.75	0.48	7.40	0.00	0.48	42.75
124.25	1.30	0.13			7.40			43.23
		$C_s = 8.70$ mg/L				$K_2 = 0.099$ h^{-1}		

The observed dissolved oxygen concentrations at various times throughout the open-jar experiment are shown in col. (6). The deficit was computed by subtracting the dissolved oxygen concentration from the value of C_s, i.e., 8.7, and these values are recorded in col. (2). Each deficit is multiplied by the K_2 value, 0.099 h^{-1}, and the product is entered in col. (3). The time interval is recorded in col. (4), and in col. (5) the average values during the time interval taken from col. (3), e.g., (0.15 + 0.09)/2, are multiplied by Δt to estimate the amount of oxygen that has been put into the system during the time interval [see col. (5)]. In col. (7) the change in the dissolved oxygen concentration is recorded. During the first time interval in the example shown, the DO rose 0.62 mg/L. However, during the interval, 0.74 mg/L of oxygen entered the system. The difference 0.12 is recorded in col. (8) as the oxygen uptake or BOD exerted during this time interval. Similar calculations can now be made for the next time interval, i.e., 6.25 to 9 h, and the oxygen uptake during this interval, i.e., 0.24 mg/L, is then added to that in the previous interval, and the accumulated oxygen uptake curve (BOD curve) is given in col. (9).

Source: Gaudy, 1975.

0.5 percent (v/v) inoculum of settled municipal sewage. The reaeration characteristics of the system were determined to be $K_2 = 0.099$ h^{-1} and $C_s = 8.7$ mg/L.

To make use of the above information in estimating the oxygen uptake curve, consider that the observed DO profile is the result of two opposing oxygen transfers. The first is the rate of transfer into the liquid, $dC/dt = K_2D$:

$$\frac{dC}{dt} = K_2(C_s - C_t) \tag{6-1}$$

or

$$\frac{\Delta C}{\Delta t} = K_2D_t \tag{6-6}$$

Therefore, the amount transferred into the medium ΔC in a small increment of time Δt is, from Eq. (6-6):

$$\Delta C = K_2D_t \, \Delta t \tag{6-7}$$

The opposing transfer, i.e., the transfer out of the liquid due to microbial oxygen utilization, in the same time interval is the oxygen uptake, ΔO_2 uptake. Thus, the change in DO along the profile:

$$\Delta DO = \Delta DO_{reaeration} - \Delta DO_{utilization}$$

$$\Delta DO = K_2D_t \, \Delta t - \Delta O_2 \text{ uptake} \tag{6-8}$$

Solving for ΔO_2 uptake and summing increments, the accumulated oxygen uptake curve is represented as follows:

$$\text{Accumulated } O_2 \text{ uptake} = \sum \Delta O_2 \text{ uptake} = \sum (K_2D_t \, \Delta t - \Delta DO) \tag{6-9}$$

In the experiment shown in Fig. 6-6, the DO profile first rose, indicating that reaeration was greater than oxygen utilization. The rate of oxygen utilization rapidly exceeded the rate of reaeration and the DO decreased. After attaining a short-lived minimum, recovery of the DO proceeded rapidly, but after about 2.5 days, recovery was slowed by increased oxygen utilization. Finally, after the third day, the system reached the initial level of dissolved oxygen.

In Table 6-1, the measured DO profile recorded in col. (6) is used to calculate the oxygen uptake curve in accordance with Eq. (6-9). In col. (5) the average amount of oxygen added to the liquid during the time interval is computed, and in col. (8) it is added algebraically to the change in DO during the time interval [col. (7)]. In col. (9) the ΔO_2 uptake for each time interval is accumulated to obtain the plotting points for the oxygen uptake curve shown in the lower portion of Fig. 6-6. There is a slight break or discontinuity in the oxygen uptake curve (a very slight plateau) corresponding to the period of rapid recovery after the low point in the DO profile. The significance of this phenomenon and reasons for it will be discussed more fully later.

It should be apparent that if one knows or can determine K_2 and C_s and has a graph such as this representing the best estimate of the oxygen uptake curve, it should be possible to calculate the DO profile in a reactor by solving Eq. (6-8) or (6-9) for ΔO_2 uptake (see Peil and Gaudy, 1975; Gaudy, 1975). Such an estimate is sometimes made for natural reactors such as rivers, for which t is converted to distance along the river from the point at which a waste is added. Determination of K_2 in rivers is not made as directly as in the method described above for large or small growth reactors. The rate of reaeration in rivers is determined largely by the degree of agitation of the water as the flow proceeds downstream. Several factors influence K_2. Among these are two of primary importance—the value of K_2 is increased by an increase in the velocity of flow and decreased by an increase in the depth of flow. Values of K_2 can change for various downstream locations along the path of flow, necessitating changing values of K_2. Various formulations for predicting K_2 are available in the literature (see O'Connor and Dobbins, 1958; Krenkel and Orlob, 1962; Thackston and Krenkel, 1966; Churchill et al., 1962; Isaacs and Gaudy, 1968; Tsiviglou et al., 1965; Nemerow, 1974; Streeter, 1926; Streeter and Phelps, 1925).

Whether this approach is used to estimate the DO in a natural reactor or to obtain the oxygen uptake curve in a laboratory reactor or a large-scale batch process, it is emphasized that the approach is sound but approximate. Values of K_2 and C_s can change even during the experiment in a small-scale reactor, because they are in some measure dependent on the chemical composition of the aerating liquor, and this may change during the experiment.

Manometric measurement of oxygen uptake The most widely used manometric apparatus for measuring the oxygen uptake of cells is the respirometer developed by Warburg in the mid-1920s. This instrument represents a modification of earlier devices introduced at the turn of the century. The Warburg apparatus was used in investigative work in the water pollution control field in the late 1940s and early 1950s for measuring the biochemical oxygen demand of wastewater samples. Today various applications of the principle are used in the laboratory and in on-line devices for monitoring the oxygen uptake at wastewater treatment plants.

The method is based on the fact that in a closed system.consisting of gaseous and liquid phases and operating at constant temperature and volume, any change in the amount of gas can be gauged by the accompanying change in pressure. In this technique, a known volume of reaction liquid containing cells and substrate is placed in a reactor of known volume, and the system is then closed to the atmosphere. The gas phase is bounded at one end by the gas–liquid interface in the flask and at the other by a suitable manometer fluid (see Fig. 6-7). Any deflection in the manometer is due to either a change in barometric pressure or a change in the amount of a constituent of the gas, e.g., oxygen in the air. Corrections for changes in barometric pressure are easily made by using another flask and manometer (without cells) as a barometer and correcting for the manometer deflection. The flasks are held at constant temperature by immersion in a water bath; they are shaken at a known speed and amplitude to agitate the liquid surface in the flask.

The cells grow, using the oxygen dissolved in the liquid, which causes oxygen from the gas phase to diffuse into the liquid. Aerobic metabolism

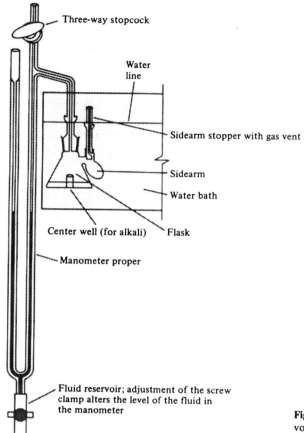

Figure 6-7 The Warburg constant-volume respirometer. (*Adapted from Umbreit et al., 1964.*)

produces large amounts of carbon dioxide, which escape to the gas phase but are taken up in an alkaline solution placed in the center well of the flask. Therefore, the change in pressure is due only to the amount of oxygen that has been transferred from the gas phase to the liquid phase.

It can be shown (Umbreit et al., 1964) that the value of oxygen uptake in microliters per millimeter change in manometer height h is:

$$\mu l \; O_2 \; \text{uptake} = h \left(\frac{V_g (273/T) + V_f m}{P_0} \right) \tag{6-10}$$

In Eq. (6-10), V_g is the volume of the gas phase, V_f is the volume of liquid, m is the solubility of oxygen in the reaction liquid, T is temperature in absolute degrees $(273 + C^\circ)$, and P_0 is standard pressure expressed in millimeters of manometer fluid. The value enclosed in parentheses is equal to the flask constant k, and several procedures for its determination are available

(Umbreit et al., 1964). The value of the oxygen uptake in microliters can be converted to milligrams per liter, as shown below:

$$\mu 1\, O_2 \left(\frac{32}{22.4} \frac{1}{V_f} \right) = mg/L\, O_2 \qquad (6\text{-}11)$$

The size of the Warburg vessels varies from 15 ml to as large as 150 ml. The larger sizes are generally used in water pollution control investigations. Reaction fluid volumes of 20 to 40 ml are not uncommon.

In addition to its use as a means of measuring oxygen uptake, the Warburg apparatus makes an excellent experimental reactor. It accommodates a number of flask and manometer setups such as shown in Fig. 6-7. Commonly, 18 flasks can be used at one time, although smaller and larger sizes are available. One may then use the apparatus for studies of bacterial kinetics, determining S, X, and oxygen uptake as well as other parameters by withdrawing flasks at various sampling times. Since one can make manometer readings to determine oxygen uptake very rapidly and easily, there is a continual check on the reproducibility of each "compartmentalized" sample. It is emphasized that since the oxygen measurement depends on interface transfer of oxygen, care should be taken that the upper limit of oxygen uptake recorded was in fact determined by the maximum capability of the biomass and was not due to a limiting rate of gas transfer from the gaseous phase to the liquid phase. One may determine the limiting transfer shaker rate by adjusting the shaker rate and the size of the oxygen-utilizing system. Simple procedures for testing this aspect as well as a detailed discussion of principles and special techniques are available in the literature (see Umbreit et al., 1964).

Measurement of Biomass

Whether one's interest lies in measuring the change in weight X of the biomass (preferred for most purposes) or the number of individuals N in the population constituting the biomass, there are various options in methods. Both direct and indirect means are available for either type of measurement. The choice of method is in most cases determined by the use to which the data are to be put and by the type of system being investigated. The various methods will be discussed with particular emphasis on the use of such measurements in environmental science and engineering applications.

Direct measurement of dry weight The dry weight of biomass per unit volume of reaction fluid is readily obtained by separating the cells from the fluid and then drying and weighing the cells in a tared container. The separation of the cells from the liquor may be accomplished by centrifugation or filtration. If the former is used, the pellet can be carefully washed from the centrifuge tube into a tared container with a few milliliters of distilled water. If filtered, the cells can be entrapped on a tared membrane filter. The increase in weight of the container after drying measures the dry weight of the biomass. In either case, one of the keys to accuracy is the relative weight of the container and the biomass in the sample. The ratio of weight of container to weight of sample should be as small as possible. Low-weight tare containers can be made from thin aluminum foil of the type commonly used in households. The filtration procedure is preferred because it offers less chance for handling error and loss of sample.

Dry weight measurements can be used to estimate numbers of cells. One can make a correlation curve between the weight of the biomass and the number of cells in the population by making a direct count of the number of cells in a known volume. This will be discussed later. Alternatively, cell numbers may be estimated by assuming an average value of dry weight for individual cells. For example, an average dry weight value for the bacterium *Escherichia coli* is 2×10^{-13} g. In heterogeneous microbial populations, there are many species and sizes and types (and weights) of cells. Also, the correlation will change from time to time as changes in species predominance occur, so any correlation of numbers in a population with weight of the biomass cannot be highly recommended.

Volumetric measurement The dry weight can be estimated from the packed volume in a graduated, small-bore centrifuge tube (e.g., a hematocrit tube). One can construct a standard curve relating milliliters of biomass to milligrams of biomass and then estimate other values of the latter from the volumetric reading at a specified gravitational force and length of centrifuging. At best, this method gives rough approximations of the weight, and the correlation itself may be affected by the concentrations of cells in the biomass and by the type of species present. One can also attempt to correlate the volume of the centrifuge pellet with numbers of cells, but this, too, is a rough approximation.

Measurement of various chemical components In certain cases, changes in weight of biomass can be estimated by direct measurement of changes in a constituent of the biomass, such as nitrogen, protein, or DNA. However, since the chemical composition of cells varies according to the species, environmental conditions, and rate of growth, such measurements, while often providing valuable information about the system, are not widely used as a sole quantitative measure of the weight of biomass produced during growth.

Measurement of turbidity A dispersed suspension of microorganisms, like any colloidal system, exhibits a Tyndall effect, and the light absorbed or scattered as a monochromatic beam passes through the suspension can be used as a relative measure of increase in the biomass. A spectrophotometer is used to measure the light transmitted through a bacterial suspension with a light path of approximately 16 to 20 mm. The transmittance T is related to the absorbance of light or optical density OD by the equation $OD = \log(1/T)$. In making such measurements, it is assumed that the increase in optical density is directly related to an increase in the number of cells and that the weight is proportional to the cell number. These assumptions are not entirely correct, because the shape of the cell can affect the passage of light, and mass can change without a change in the number of cells. Also, the absorbance is proportional to the number of cells only in rather dilute systems. However, the method is sufficiently accurate for most purposes when X is no greater than 300 mg/L and when cells are not flocculated. Cells can be dispersed by brief treatment in a blender and diluted to an appropriate concentration.

In addition to constructing a correlation curve between optical density and weight of the biomass, one can correlate optical density and the total number of cells or particles in the suspension. The color of the suspending fluid plays a significant role in selecting the optimum wavelength of mono-chromatic light to be used. One may use any of several commercial colorimeters or spectrophotometric devices. It should also be noted that turbidity that is not due to cell growth can detract from the validity of the correlation. Often such turbidity can be subtracted in the reagent blank correction. However, if such turbidity is subject to change, it is not easily corrected.

Even though there are a number of drawbacks to this method, which signal caution in applying it, the procedure offers a convenient way of supplementing biomass weight data. Figure 6-8 shows a correlation curve of absorbance and weight for a heterogeneous microbial population of sewage origin. Often, one is more interested in measuring the rate of increase in biomass concentration than the absolute amount. This permits the use of relative rather

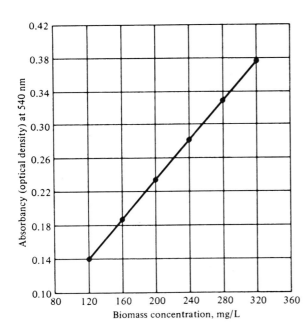

Figure 6-8 Correlation between absorbance and biomass concentration for a heterogeneous microbial population growing on a colorless medium.

than absolute values of biomass. The determination of the rate of growth requires many plotting points, and the use of optical density facilitates obtaining these data. A few weight determinations during the experiment can provide one with information with which to relate OD and weight (as shown in the figure).

Measurement of the number of particles Measurement of the number of particles in a given volume of sample can be correlated to the weight of the biomass in suspension. The particles may be counted electronically or by microscopic observation. Electronic devices can provide information on both the number and size distribution of the particles, but they are subject to various operational and interpretative difficulties when applied to heterogeneous populations in which there are many forms of cells. Direct microscopic observation of a small (representative) volume of sample in specially calibrated and ruled glass slides provides a means of estimating the number of particles (cells) in a unit volume of sample. These methods have some useful application in various types of investigations but are not uniquely suited to the estimation of the biomass size for the quantitative description of growth.

Measurement of the viable population In some types of experiment, measurement of the number of cells capable of replication can be correlated to changes in the weight of the biomass. The correlation is best when the cells are growing steadily from a small to a larger population. Also, in many instances the prime interest is assessing the viable population or the relation between viable numbers of cells and the weight of the biomass. Such a ratio is sometimes provided as a basis for estimating the *active fraction* of a heterogeneous biomass in river slimes and in biological treatment process reactors. Viability, defined within the boundary of the test to be described below, means the ability of cells to reproduce. However, viability is not necessarily related to the ability to use substrate. A biomass can assimilate considerable amounts of carbon source and use it for synthesis and respiration without replication. Thus, many cells incapable of replication may be "active" in the biomass.

Methods for determining numbers of viable cells depend on the ability of the organisms in a sample to replicate when transferred to a fresh source of nutrient. The source of nutrient may be the same as the one from which the sample was taken or it may be different. In the latter case, it is usually one containing as many known growth factors as possible; i.e., it consists of what may be called a well-supplemented diet. The aim here is to obtain an estimate of the maximum number of cells capable of replication in the sample. However, the ability of cells to replicate in a complete medium does not indicate whether the cells were replicating in the medium from which the sample was taken.

It should also be realized that there is no one type of growth medium, nor is there one set of conditions for incubation, that will permit growth of all possible types of microorganisms in an activated sludge. For example, different media and conditions are required by heterotrophic and autotrophic (e.g., nitrifying) bacteria. Algae, protozoa, and many fungi require special and differing methods of cultivation. The methods described here are useful primarily for the enumeration of aerobic or facultatively anaerobic chemoheterotrophic bacteria.

In general, there are two methods for enumeration. In the first, the organisms are deposited on or in a growth medium solidified by the presence of agar, and in the second, a dilution technique is used.

Colony counts using solid media Estimating the viable count by the *pour plate method* is perhaps most universal. A small volume of an appropriately diluted sample is placed in a petri dish. Usually, 1 ml of dilute sample is used. Approximately 15 ml of warm liquid medium containing agar is added, and the contents are carefully mixed. The cells are fixed in place as the agar solidifies and, after an appropriate incubation time (24 to 72 h), the individual cells have replicated to such an extent that visible colonies of cells are formed. The number of such colonies multiplied by the known dilution factor is the estimated viable count per unit volume of the original sample. The assumption is that each visible colony grew from a single cell; thus, if the cells are flocculated, they must be thoroughly dispersed before making the test. Microscopic observation should be used to determine whether the cells are dispersed. In a 3.5-in petri dish, there should be between 30 and 300 colonies.

The sample may be plated on the surface of the medium after the agar has solidified and been stored for approximately 48 h after pouring. A known volume may be spread (*spread plating*) over the entire surface area of the agar, or a micropipet may be used to plate a very small volume of dilute cell suspension (*spot plating*). These methods may have some advantage over pour plates because agar plates contaminated during the pouring of the agar can be discarded, and there should be less counting error because all colonies are in the same plane (the surface of the agar). Also, in the case of counting aerobic organisms, surface growth provides more aerobic conditions. We have found both spot and spread plating to be serviceable methods in environmental microbiological investigations (Gaudy et al., 1963). Another method that offers the same advantages at some increase in costs is the use of membrane filters. A sample is poured over a membrane filter, preferably one marked with grids to facilitate counting of the colonies which develop. The filter is then placed on an absorbent pad of such thickness that it will take up approximately 2 ml of nutrient solution. The nutrient held in the pad diffuses to the cells on the filter. The filter may also be placed on an agar plate. Since rather large volumes can be passed through the filter, the method can be used for dilute suspensions.

Null-point dilution in liquid medium The basis of this method is determination of the dilution factor for a sample that no longer provides a sufficient seed of microorganisms to permit growth in fresh liquid media. Any convenient quantitative measurement of growth may be used. For example, gas formation has been used in a standard test for coliform organisms (the MPN, or most probable number, determination, see Chap. 14). Development of turbidity may be used if one is interested in general growth characteristics. The viable count in the original sample is estimated using applicable probability theory (see Chap. 14). This method is not as accurate as the plating techniques unless a large number of replicate samples is used, but it is applicable to samples that contain very few microorganisms.

The membrane filter technique previously described is also applicable to small populations, since filtration effectively concentrates the cells in the sample.

Recommended methods It is seen from the above that many methods are available for measuring X. For most microbiological work, the most useful methods are dry weight determination using membrane filtration to separate cells from liquid, the turbidity measurement for correlation with weight or viable count, and the pour plate and surface plating methods for determining viable counts.

GROWTH IN BATCH SYSTEMS—THE GROWTH CURVE

In this section, the growth curve shown in Fig. 6-1 will be discussed in greater detail, with particular emphasis on its progress to the upper value of X. Such a growth curve is shown in Fig. 6-9. Graphs such as this appear in many microbiology texts and are used to describe the "rise and fall" of microbial populations. The curve is simply a generalized description of experimental observations. There may be other ways to compartmentalize the events, but this is a good starting point and one that facilitates quantitative description.

The kinetics as well as the cause for this course of events are related to properties of the microbial population and the environmental conditions in the system. Among the important environmental conditions are pH; temperature; dissolved oxygen concentration; amount and type of carbon source, nitrogen, phosphorus, and other nutrients; presence of inhibitory substances; and changes in all of these that may occur during growth of the population. Thus, the problems of devising ways to describe either the overall process or the recognizable portions of it are tremendous, and it is of great importance to determine, and if possible control, the environmental conditions in order to increase the useful application and interpretation of any mathematical description.

In the following discussion of the mathematics used to describe microbial growth, we shall use the symbol X to represent the size of the microbial population in units of dry weight (milligrams per liter) since this is the only practical measurement in work with heterogeneous populations.

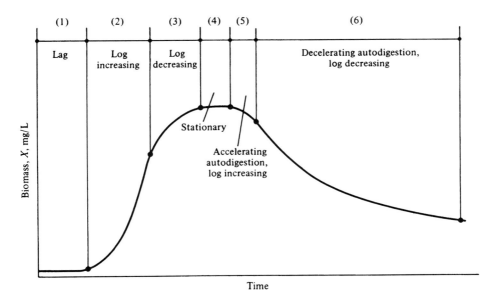

Figure 6-9 Distinguishable portions of biomass growth and decay curve.

The Lag Phase of Growth

A lag period may be considered a time of adjustment to a new environment. Cells obtained from sewage, soil, or a stock culture medium, wherein the environment was different from that into which they are now placed, can be expected to undergo a period of adjustment prior to assuming their maximum rate of growth in the new environment. Such a lag is not necessarily a dormant period. The cells may be taking in significant amounts of substrate, they may be synthesizing new enzymes, and they may be undergoing enlargement prior to division, but, in any event, they have not yet begun the orderly and steady replication characteristic of the succeeding phase.

The lag period may be divided into two parts. The first is one in which very little change in cell mass and no change in cell numbers take place; in the second the growth rate is increasing toward attainment of the rate characteristic of the exponential phase. The nature of the lag phase has been the object of study for many years, and more work is needed to describe its physiology. There is some work suggesting that during the latter stages of this phase, as the system approaches exponential growth, these "young" cells are more sensitive to changes in the growth environment.

Lag periods have obvious and significant ramifications to bioassay procedures. Consider the effect of a lag on a bioassay in which only one initial and one final measurement are taken, e.g., the oxygen uptake during the standard 5-day BOD test. Avoidance of the effects of a lag period is one of the reasons why acclimated seed is stipulated. One may also consider the change in environment when, for example, cells from an activated sludge process, which have been held in the absence of exogenous substrates and without aeration in the bottom of a settling tank, are returned to the aerobic environment of the aeration tank. Some portion of the residence time in the activated sludge tank may be given over to a lag prior to assuming the characteristic growth rate for the system. Fortunately, lag periods may be decreased by using a large inoculum of cells, which is the case in the activated sludge treatment process. A lag period such as that shown in Fig. 6-9 usually can be eliminated if the starting cells are taken from a system that has already attained the succeeding phase of growth.

The Exponential and Declining Phases of Growth

The exponential and declining growth phases are discussed together because there is often a close mathematical relationship between them, and together they form the substrate removal or purification phase depicted in Fig. 6-1. Referring to Fig. 6-9, the rate of change in X, i.e., the slope of the curve, dX/dt, increases in phase 2, passes through an inflection point, and thereafter decreases until the maximum concentration of X is obtained and the population enters the stationary phase in which $dX/dt = 0$.

There are many reasons that cells do not grow indefinitely, including exhaustion of nutrients, depletion of the dissolved oxygen supply, crowding, growth-induced changes in the chemical environment, and production of toxic substances. It can be appreciated that in wastewater treatment the aim is to control the environmental conditions so that the phenomenon limiting growth is exhaustion of the carbon source. Equations can be developed that follow the general sigmoidal ("S") shape of phases 2 and 3, but for the present phases 2 and 3 can be considered separately.

Often in phase 2, the rate of increase in X, dX/dt, is proportional to the concentration X in the system, and the proportionality factor can be shown to be constant. Designating the proportionality factor with the symbol μ:

$$\frac{dX}{dt} = \mu X \tag{6-12}$$

and μ is often called the *specific growth rate*, i.e., the growth rate per unit of biomass. The units of μ are time^{-1}. Upon integration, Eq. (6-12) may be written as:

$$X_t = X_0 e^{\mu t} \tag{6-13}$$

Equation (6-13) allows one to predict (calculate) the weight of cells at any time from a knowledge of the initial concentration X_0 and the specific growth rate μ.

According to these equations, the size of the biomass is increasing exponentially or logarithmically. The mode of kinetic expression of growth can be described as "first-order, increasing-rate," and this is recognized as the growth law of Malthus. Equation (6-13) can be put in conventional straight-line form for easy determination of μ by taking the logarithms of both sides:

$$\ln X_t = \ln X_0 + \mu t \tag{6-14}$$

Only when μ is constant can Eq. (6-14) plot a straight line; consequently, any straight-line portion of a semilogarithmic plot of growth data marks the extent of exponential growth. Figure 6-10 is a semilogarithmic plot of growth data. It is clearly seen that there is an exponential phase of growth. The value of the specific growth rate μ can be obtained from the data by solving Eq. (6-14) for μ:

$$\mu = \frac{\ln X_t - \ln X_0}{t} \tag{6-15}$$

If one chooses for a time interval the time required for X to double, t_d, then Eq. (6-15) can be written as follows:

$$\mu = \frac{\ln 2}{t_d} = \frac{0.693}{t_d} \tag{6-16}$$

Using Eq. (6-16), μ is easily calculated from the data.

The exponential growth phase may extend through several doubling times

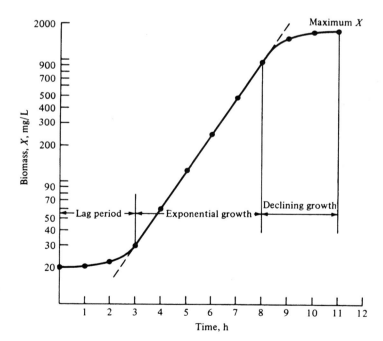

Figure 6-10 Semilogarithmic plot of a growth curve showing extent of exponential or logarithmic growth.

or it may not even last through one. But any straight-line portion, no matter what its extent, fits the mathematical definition of exponential (logarithmic) growth. It is important to emphasize that it is the mathematical description of an exponential phase of growth rather than a physiological definition with which we are concerned in describing the process. When cells are in an exponential growth phase, their doubling time is constant. If X is measured in number of cells, the time required to double the numbers is called the *generation time g*. The numerical values of g and t_d may or may not be equal, depending on the constancy of the relationship between mass and numbers of cells. It should be noted that determination of the existence and extent of an exponential phase as well as the value of its numerical descriptor μ are directly related to the quantity and quality of the growth data (plotting points) obtained. One is depending on the data to determine whether logarithmic growth occurs, so a rapidly performed measurement like that of optical density offers tremendous advantages. Caution is required in order to avoid assuming that a straight-line portion exists simply because a plot is made on semilogarithmic paper.

Figure 6-10 also shows a method of determining the extent of the lag phase. It begins at time 0 and ends when exponential growth begins, that is, where the growth curve becomes tangent to the straight line marking the exponential growth phase. Lag time can be expressed simply in time units, as shown on the figure, or with reference to the exponential phase, in doubling times by dividing the lag time by the value of t_d determined for exponential growth.

The declining phase of growth begins where the curve departs from a straight line. This point of departure is the inflection point shown in the arithmetic plot (see Fig. 6-9). From this point, the value of dX/dt begins to decrease. Sometimes the specific rate at which it is decreasing, μ', can be shown to be constant. This situation is mathematically analogous to the kinetic mode of expression by which the DO was approaching the saturation level in Fig. 6-4, and the same mathematical treatment is applicable. The maximum value of X replaces C_s, and μ' replaces K_2. Thus, both μ and μ' can represent first-order specific growth rate constants, but μ is the constant for an exponentially increasing rate, whereas μ' represents an exponentially decreasing rate. It is important to note that the maximum value of biomass X is not maintained for long periods of time. The biomass concentration may remain steady for a time, but it soon begins to decrease. It is also well to note that the inflection point separating phases 2 and 3 is not necessarily at the halfway point on the curve. It may lie above or below the halfway point.

Any of a number of environmental conditions may govern the extent of the exponential phase, but during batch growth, an exponential phase, as measured by practical experimental criteria, usually can be observed.

The Stationary Phase of Growth

As stated previously, the most common cause for cessation of growth is exhaustion of an essential nutrient, and in most biological treatment processes the aim is to make the carbon source the limiting nutrient. Thus, the peak in the growth curve usually coincides with the low point of the substrate removal curve, as shown in Fig. 6-1. Cells transferred to a new medium at the beginning of the stationary period often grow again in the exponetial phase without a lag just as they do when transferred from the exponential or declining phase. However, the longer they remain in the absence of substrate, the greater is the probability of breakdown of metabolic machinery, which occurs in the stationary phase.

If, after reaching the maximum concentration of X, all the cells died and did not disintegrate, X would remain constant and the stationary phase would be the terminal condition. In the normal course of events, this does not happen. The majority of cells remain alive; i.e., they respire. However, since the food supply is gone, they turn to internal carbon sources (endogenous substrates); the natural consequence is a decrease in the mass of X in the system. For example, certain enzyme proteins formerly used in metabolizing the original exogenous substrates are no longer needed; thus they may become substrates themselves and are oxidized to help the cell maintain itself in a living state. In mixed populations in this stage, there is the probability of species interaction and attack on one species by another. For example, one cell may elaborate specific enzymes that disintegrate cell walls and membranes of others (lysozymes), or viral agents (bacteriophage) may cause an infected cell to break open, thereby releasing its organic compounds, which can become external nutrients for the intact cells. Also, grazing populations of bacterial predators may consume part of the population. The time required for these processes to be set in motion and the rates at which they proceed vary, but in general for heterogeneous populations the stationary phase is rather short; the size of the population peaks and begins to decline very shortly thereafter.

Accelerating and Decelerating Phases of Autodigestion

These portions of the biomass curve are considered together for the same reasons that phases 2 and 3 were discussed together. Whereas phases 2 and 3 can be described as exponential accelerating and exponential decelerating growth, phases 5 and 6 can be mathematically characterized as exponentially accelerating "death" and exponentially decelerating "death." Special caution is needed in specifying the way in which X is measured. When cells are growing (phases 2 and 3), the correlation between viable cell count and biomass concentration as measures of X is often rather good, but this is generally not so during the autodigestion and endogenous phases (phases 5 and 6). Phase 5 may have a drastic effect on the viable count; in fact, in pure cultures the cells may sometimes undergo nearly total loss of viability (death) in an increasing exponential kinetic mode with little or no change in X measured as biomass. In many systems of natural populations with X based on weight measurement, phase 5 is rather short-lived and the size of the

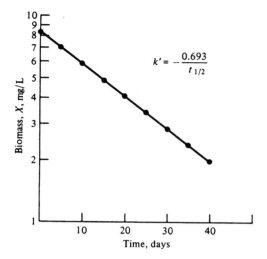

$$k' = -\frac{0.693}{t_{1/2}}$$

Figure 6-11 Decelerating autodigestion (logarithmically decreasing), phase 6 of Fig. 6-9.

biomass decreases at an ever-declining rate, as in phase 6. If the decrease in X is such that half of the biomass concentration is autodigested in a constant time interval, the following equations are applicable:

$$\frac{dX}{dt} = -k'X \tag{6-17}$$

$$X_t = X_0 e^{-k't} \tag{6-18}$$

$$k' = \frac{\ln X_0 - \ln X_t}{t} \tag{6-19}$$

$$k' = \frac{\ln 2}{t_{1/2}} = \frac{0.693}{t_{1/2}} \tag{6-20}$$

The value of k' can thus be determined in a manner analogous to the determination of μ. Rather than doubling time t_d, one measures the half time $t_{1/2}$, i.e., the time required for the biomass concentration to decrease by one-half (see Fig. 6-11). Equations (6-17) through (6-20) may be recognized as the same as those depicting the kinetics for radioactive decay. While one may be certain that radioactive decay will occur in accordance with these equations, the autodigestion of a biomass may not always follow such a kinetic expression. Again, the environmental technologist can only plot the data to determine whether the system under investigation does follow exponential decay laws, with full knowledge that there is no law saying that it must.

RELATIONSHIPS BETWEEN X AND S

Thus far we have dealt with simple ways to describe various phases of the growth and death of a biomass. Both growth and death are of vital importance because together they constitute the decay leg of the carbon-oxygen cycle. In an engineering sense, growth is synonymous with the purification phase in wastewater treatment, and the autodigestion phase is the same as aerobic sludge digestion. In this section we shall discuss some useful mathematical relationships between X and S; these will be used later in models to describe and control growth in continuous-flow reactors such as biological treatment plants.

The Yield Coefficient Y

Referring again to Fig. 6-1, the peak biomass concentration occurs at the end of the substrate removal phase, and the amount of biomass produced is significantly less than the amount of carbon source, ΔCOD, that was removed. The mass of cells produced per unit of substrate removed is called the *cell yield Y*:

$$Y = \frac{X_t - X_0}{S_0 - S_e} = \frac{\Delta X}{\Delta S} \tag{6-21}$$

The value of this fraction can vary because it is dependent on the nature of the substrate and the species present, as well as on the environmental conditions under which the cells metabolize the carbon source. With respect to any one experiment, one may ask whether Y is constant throughout the substrate removal period. Results of many experiments indicate that it is constant for pure cultures and also for heterogeneous populations.

Figure 6-12 shows the results of an experiment in which a mixed microbial population of sewage origin acclimated to glycerol was grown on glycerol. Figure 6-12C is particularly important because it shows that the ratio of biomass X synthesized to substrate removed S_r is constant throughout the substrate removal phase. One could have measured Y at any place along the curve in the substrate removal phase and found the same value. Thus we may write:

$$Y = \frac{dX/dt}{dS/dt} \tag{6-22}$$

Normally in batch experiments, Y is measured at the top of the growth curve as a matter of convenience.

One can have confidence in the cell yield value only when it is based on the measurement of both cells produced ΔX and substrate removed ΔS (or S_r). That is, it should not be assumed that all of the carbon source has been removed when growth ceases. Also, any measurement of Y after maximum growth is attained can lead to falsely low values of the yield coefficient. The value of the yield coefficient obtained during or at the end of the substrate removal period is called the *true cell yield Y_t*. Since any value measured after this, i.e., in the autodigestive phase, will be lower, it can be stipulated that Y_t is the maximum cell yield for that experiment. A yield value measured some

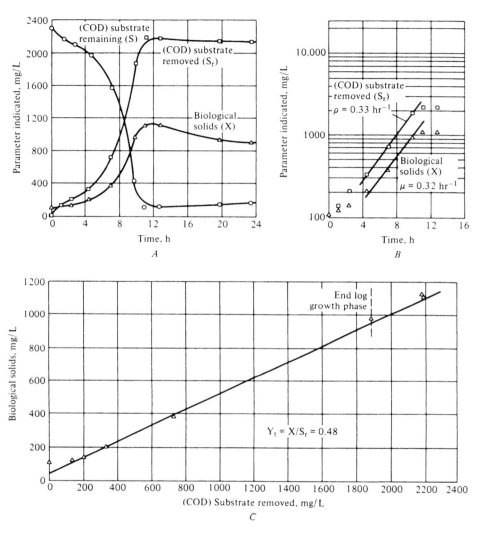

Figure 6-12 (*A*) Growth and substrate removal for heterogeneous population growing on glycerol. (*B*) Semilogarithmic plot of growth and substrate removal. (*C*) Biomass versus substrate removed, showing constancy of cell yield Y_t in exponential and declining growth phases. (*From Ramanathan and Gaudy, 1972.*)

time into the endogenous or autodigestive phase is not a valid yield measurement.

Many factors affect the numerical value of the cell yield Y_t. For example, whereas a lowering of temperature decreases the growth rate, it often causes an increase in Y_t. Also it can be shown, depending on the nature of the substrate, that cell yield can be increased by increasing the carbon/nitrogen ratio in the medium (see Chap. 10). However, in the normal case, the cell yield Y_t exhibits relative constancy for a given species grown on a given carbon source. For some species, e.g., *Escherichia coli*, the cell yield remains fairly constant for many compounds of the same class; e.g., cell yields on different carbohydrates are very similar.

Since the cell yield depends on the species present as well as on the nature of the carbon source, one must expect variation in Y_t values for natural populations even when they are grown on the same substrate under identical growth conditions. Table 6-2 shows a summary of Y_t values for heterogeneous populations of sewage origin grown on simple substrates in minimal salts media. All experiments were conducted under similar, closely defined conditions (temperature, aeration, mineral salts concentrations, pH, etc.). In using numerical descriptions of the growth of natural populations, one must always be cognizant of the fact that while the biomass possesses biochemical and biomechanical properties, it is an ever-changing ecosystem,

Table 6-2 Statistical summary of sludge yield values for heterogeneous populations of wastewater origin grown on various carbon sources

Carbon source	Average cell yield, %	Number of determinations	Range	Standard deviation	Coefficient of variance	95% Confidence limits
Arabinose	44.8	6	32–51	—	—	—
Cellobiose	50.3	4	34–61	—	—	—
Fructose	53.0	8	34–69	10.4	19.6	44.4–61.6
Galactose	51.9	24	36–76	11.3	21.8	47.1–56.7
Glucose	61.9	118	36–88	12.5	20.1	59.6–64.2
Glycerol	46.5	31	31–61	9.4	20.2	43.1–49.9
Lactose	47.1	12	30–61	7.5	15.9	42.4–51.8
Maltose	51.7	7	39–86	—	—	—
Mannose	52.2	8	36–70	13.4	25.6	41.1–63.3
Ribose	45.7	9	36–56	6.4	14.0	40.8–50.8
Sorbitol	50.6	39	35–59	6.0	11.8	48.7–58.5
Sorbose	52.0	12	23–73	16.7	32.1	41.4–62.6
Sucrose	53.1	12	33–77	14.1	26.5	44.2–62.0
Xylose	50.4	18	25–71	11.4	22.6	44.8–56.0
Acetic acid	41.2	6	26–53	—	—	—
Butyric acid	45.0	1	—	—	—	—
Citric acid	31.0	1	—	—	—	—
Glutamic acid	50.0	1	—	—	—	—
Lactic acid	29.0	1	—	—	—	—
Malonic acid	41.0	1	—	—	—	—
Propionic acid	38.0	2	36–40	—	—	—
Pyruvic acid	68.0	1	—	—	—	—

Cell yield is expressed as the precentage of the organic carbon source that is converted to cells, i.e., (dry weight of cells produced/weight of carbon source utilized) × 100.

Source: Ramanathan and Gaudy, 1972.

and this variation is often reflected in changes in the numerical values of the parameters used to describe it. In defining properties of this biomass, one can take this into account by using a range of values rather than seeking precise numerical amounts. Engineers are accustomed to the variability and approximate nature of properties of materials, and this biomaterial is subject to perhaps even more variability than is any other engineering material.

It is important not only to know that the cell yield "constant" is not a constant for heterogeneous biomasses but also to have some idea of the probable range of values to be expected. From the data in Table 6-2 one could conclude that in dealing with a predominantly carbohydrate carbon source in a natural wastewater, a Y_t value of approximately 0.5 would be a fairly good average; however, in making quantitative calculations, it would be wise to test the consequence of Y_t values ranging from 0.4 to 0.6. In placing such a range on the calculations, there is some surety that the analyst has bracketed the range of values of Y_t that might be expected.

Effect of S on μ

To this point, the specific growth rate μ has been defined by Eq. (6-12), and a relationship between the rate of substrate removal and the growth of biomass Y_t has been established [Eq. (6-22)]. It has been shown that the specific growth rate μ may be constant for a time during growth (exponential phase), but eventually it decreases until it becomes 0 at the top of the growth curve. If the carbon source has been removed when growth stops and the experimental environment is not deficient in anything but the carbon source, it may be concluded with some confidence that the decreasing growth rate was related to, or in any event may be correlated with, the decreasing concentration of carbon source S. Thus, we would conclude that at higher concentrations of substrate, the specific growth rate may not be related to S; i.e., substrate could be present at concentrations in excess of those that will affect μ, and only at lower concentrations is μ dependent on S. Concentrations of S that affect μ are found to vary with the species of microorganism and the nature of the carbon source.

If one ran a number of growth experiments using an acclimated culture and different concentrations of a specific carbon source and plotted the data, the family of curves obtained would be similar to one of those shown in Fig. 6-13. Figure 6-13A shows a case in which the slope of the exponential phase is the same at initial substrate concentrations S_0 of 1000 to 400 mg/L. Only at the two lowest concentrations does the slope of the exponential phase decrease, and then only slightly. Obviously, μ is dependent on substrate concentration only at very low concentrations of S.

In Fig. 6-13B, there are again distinct exponential phases for each initial substrate concentration, but there is considerably more spread in the curves. This type of result is most frequently observed for heterogeneous populations. Note that the spread of the curves increases as S_0 decreases; i.e., μ does not vary directly with S_0.

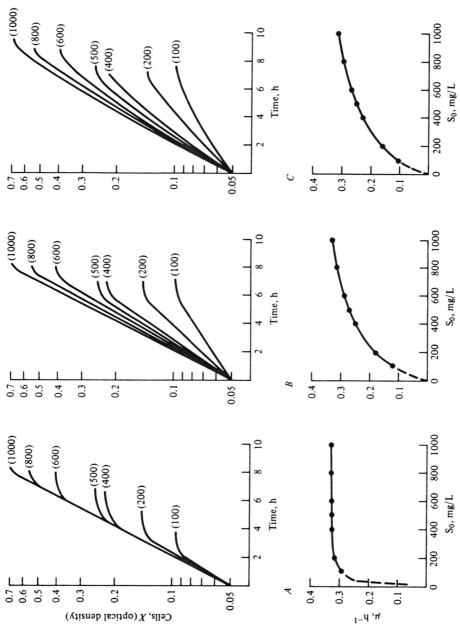

Figure 6-13 Examples of various types of batch growth curves that can be observed at different concentrations of substrate. Curves like the ones shown in *B* are usually observed for heterogeneous microbial populations. (*Top*) Growth curves at initial substrate concentrations shown in parentheses. (*Bottom*) Specific growth rate versus initial substrate concentration.

In Fig. 6-13C, all of the plots are curves; exponential growth is most closely approached at the highest concentration of substrate, but it is clear that to increase the straightening tendency as S_0 is increased would require an initial substrate concentration considerably higher than 1000 mg/L. That is, only at very high substrate concentrations could exponential growth be said to occur in this system. Even though exponential growth did not occur, the specific growth rate is related to substrate concentration and its dependence on it decreases as the substrate concentration is increased.

In all three cases, A, B, and C, the growth rate is dependent on the substrate concentration at some concentration. If the values of μ are calculated and plotted versus S_0, graphs of the types shown in the lower portion of Fig. 6-13 are obtained. The values of μ can be found using Eq. (6-16). For case C, however, there is no exponential phase, and the specific growth rate can only be estimated by determining the initial slope of the curve.

In the three cases there are similarities as well as differences. Cases A and B are similar in that exponential growth is developed in both; they differ in the degree of dependence of μ on S_0. Cases B and C both show considerable dependence of μ on S_0, but in case C, μ does not remain constant over a sufficient range of substrate concentrations to register an exponential phase. In system C, the gradual straightening out of the curves as the initial substrate concentration is increased indicates that at some S_0 higher than 1000 mg/L, exponential growth may occur.

In the early 1940s, Monod was beginning research on growth mechanisms that led to the discoveries in microbial genetics and metabolic control mechanisms for which he is most widely known. Understandably, one of his initial efforts involved the mathematical description of the course of microbial growth from the beginning to the stationary phase. For the pure cultures and substrates he studied (e.g., *Escherichia coli* and *Bacillus subtilis*; glucose and mannose), his results conformed more to case A than to B or C. In fact, he did not note divergence from a constant exponential value for μ until the S_0 concentration was somewhat lower than that used in the example of Fig. 6-13. The important aspect to note is that he did find a consistent but slight dependence of μ on S_0. Because μ was dependent on S only at very low concentrations, it was difficult to determine the nature of the effect. It can be seen from the lower graph of Fig. 6-13A that there are no points in the region that is of the most interest, i.e., below a μ value of 0.3 h^{-1}. At very low initial substrate concentrations, it is difficult to obtain enough plotting points to define a straight line on semilogarithmic paper (which is the analytical definition of exponential growth). However, as X approaches its maximum value in each of the growth curves in the upper graph, i.e., as it approaches a stationary phase, the curves break rather sharply. If one were to consider small increments, ΔX, along the breaking portion of any one of the curves, values of μ could be calculated as the slope of the tangent at the average X for any particular increment of time. The substrate concentration, being nearly exhausted in this declining stage of the curve, is difficult to measure.

However, Monod had previously concluded from other experiments that the cell yield was constant throughout the increasing and declining phases. Thus, from knowledge of the cell yield, it was possible to calculate the substrate concentration corresponding to the slope of the tangents to the growth curve at the average value of X in each increment examined. In this way he was able to get plotting points for the graph of μ versus the substrate concentration with which to fill out the curve at the lowest values of μ and S. His plot was not of μ versus S_0 but of μ versus S. In all cases, the data points depicted a sharply breaking curve similar to the one shown in the lower graph for case A.

The significant point here is that in the plots of μ versus S or S_0, μ approaches its upper value along a curve of ever-decreasing slope. The maximum value of μ is therefore approached as an asymptote. Monod noted that this curve might be described by the equation of a rectangular hyperbola. Such a curve can be used to describe the data obtained in many adsorption studies (the Langmuir equation) and in studies of the effect of substrate concentration on enzyme velocity (the Michaelis-Menten equation). The data can be plotted in any of the straight-line forms shown in Fig. 3-34. The asymptote μ_{max} and the shape factor, which can be designated as a saturation constant K_s, can be used in the hyperbolic equation to calculate the value of μ at any value of substrate concentration, and the accuracy with which the rectangular hyperbola describes the data can be assessed. The relationship can be expressed as follows:

$$\mu = \frac{\mu_{max}S}{K_s + S} \tag{6-23}$$

In Eq. (6-23), S rather than S_0 is used because Monod did not (could not) employ the initial substrate concentrations to obtain most of the data he used in forming the relationship; in his case μ changed very rapidly with changes in S at the lower concentrations. For cases B and C (and for case A, if one could measure exponential growth at low substrate concentrations), one could write an equation of the form

$$\mu = \frac{\mu_{max}S_0}{K_s + S_0} \tag{6-24}$$

It is emphasized that Monod would have determined μ using S_0, but the dependence of μ on S was so slight that it was not possible to detect differences in μ in the exponential phase. An examination of his data shows that exponential phases did develop at substrate concentrations lower than those needed to develop the maximum value of exponential μ, just as in the example shown in Fig. 6-13A. Thus Eq. (6-23) may be considered an approximation of Eq. (6-24). It is clear from Fig. 6-13A and B that μ is related to the substrate concentration but not so closely tied to it that there is a change in μ for every slight change in S. There is no doubt that growth occurs at the expense of a decrease in substrate concentration; therefore, during growth, for example at $S_0 = 200$ mg/L, the substrate concentration must have changed considerably during the 4 or 5 h that μ remained constant (see Fig. 6-13B). Thus, if one accepts the previous statement that most batch growth studies of this type yield data like those shown in Fig. 6-13B, Eq. (6-24) must be considered a more accurate representation of the data than is Eq. (6-23). This will be an important consideration when examining and using the descriptive theory of growth in continuous-flow reactors rather than in batch systems. Equation (6-23) is a more accurate descriptor of experimental data of the type shown in part C, since mathematically this equation states that for any change in S there is a change in μ. For each curve in case C, the slope continually decreases as S is decreased during growth.

The value of the shape factor K_s has an effect on the change in μ at different values of S or S_0. K_s, i.e., the value of S when $\mu = \frac{1}{2}\mu_{max}$, is much lower in case A than in case B. The effect of K_s on the shape of a plot of μ versus the substrate concentration (S_0 or S) is shown in Fig. 6-14. It is apparent that the higher the value of K_s, the flatter is the curve; thus, when K_s is 10 mg/L, a small difference in S (at very low values of S) evokes a rather large change in μ, whereas when $K_s = 150$ mg/L, a much larger change in S is required to change μ. In Monod's studies, the values of K_s were very low—ranging from 4 to 20 mg/L. Early attempts to verify Monod's findings

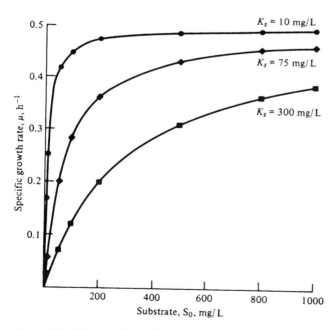

Figure 6-14 Effect of K_s on the relationship between specific growth rate μ and initial substrate concentration S_0. Assumed conditions: $\mu_{max} = 0.5$ h^{-1}; $K_s = 10$ (O), 75 ($\triangle$), and 300 ($\square$) mg/L. Calculations for plotting points were made using Eq. (6-24).

also indicated rather low values of K_s, which led to a general conclusion, bordering on dogma, that K_s was extremely low for most species of microorganisms growing on simple substrates.

Monod's aim at this time was to devise an equation to describe the growth curve (exclusive of the lag phase). Using Eqs. (6-12), (6-22), and (6-23), Monod was able to integrate the combined equations into an expression as follows:

$$\mu_{max}t = \frac{K_s + S_0 + (X_0/Y_t)}{S_0 + (X_0/Y_t)}\left(\ln\frac{X}{X_0}\right) - \frac{K_s}{S_0 + (X_0/Y_t)}\ln\left[\frac{S_0 + (X_0/Y_t) - (X/Y_t)}{S_0}\right]$$

(6-25)

For a given μ_{max}, K_s, and Y_t, one can use Eq. (6-25) to determine X at any time during growth when the initial concentration of cells X_0 and the initial concentration of substrate S_0 are known. It should be understood that Eq. (6-25), in accordance with Eq. (6-23), which was used in its derivation, predicts that there is no exponential phase unless the substrate concentration is in excess of that which will permit the system to grow at μ_{max}. If K_s is small, the amount of substrate needed to attain μ_{max} is also small, e.g., compare Fig. 6-13A and B. Thus, the smaller the value of K_s, the more closely the equation will be able to reproduce the curve of growth. One can test the accuracy with which the equation can describe the growth curve by deter-

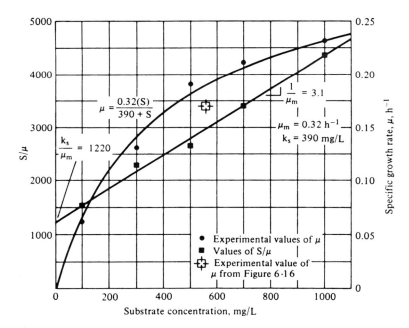

Figure 6-15 Relation between specific growth rate μ and initial substrate concentration for culture isolated from municipal sewage growing on glucose. (*From Gaudy et al., 1973*.)

mining K_s and μ_{max} for a system of cells and substrate, and then using Eq. (6-25) to compare the calculated and observed growth curves.

If one studies a system (species and carbon source) for which K_s is fairly high, one can, by selecting S_0 at some level below that needed to generate μ_{max}, test whether exponential (logarithmic) growth exists for this low value of S_0. Figure 6-15 shows plots of both μ versus S and S/μ versus S (reciprocal) for an organism growing on glucose minimal medium; the organism was isolated from municipal sewage. The value of K_s for this organism was rather high—390 mg/L. The curved line through the plot of μ versus S_0 was calculated from Eq. (6-24), and it is seen to fit the data fairly well. Figure 6-16 shows a plot of growth data using this same organism. The substrate was again glucose ($S_0 = 560$ mg/L). The dashed lines are drawn through the experimental data, whereas the solid lines are plotted from values calculated using Monod's Eq. (6-25). From the top graph it can be concluded that the

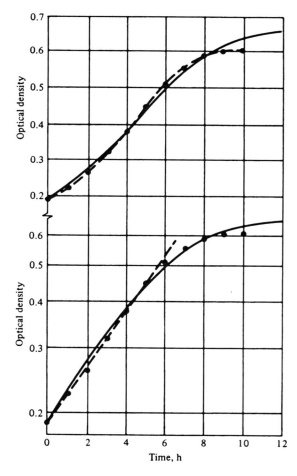

Figure 6-16 Comparison of observed growth curve (dashed lines and plotted points) with growth curve calculated from Eq. (6-25) (solid lines). The organism was the one used to generate the results shown in Fig. 6-15. (*From Gaudy et al., 1973.*)

equation provides a fairly good approximation of the growth curve. From the bottom graph (semilogarithmic plot) it is seen that, as mentioned before, the equation does not allow for an exponential phase at this initial substrate concentration. The S_0 of 560 mg/L was well below that required to permit μ to approach μ_{max} ($\mu_{max} = 0.32$, see Fig. 6-15). Thus, it should be clear that substrate does not need to be present in excess in order for exponential growth to occur in a batch system. It is evident from the semilogarithmic plot of Fig. 6-16 that exponential growth was attained.

Figure 6-17 shows the results of a growth experiment using a hetero-

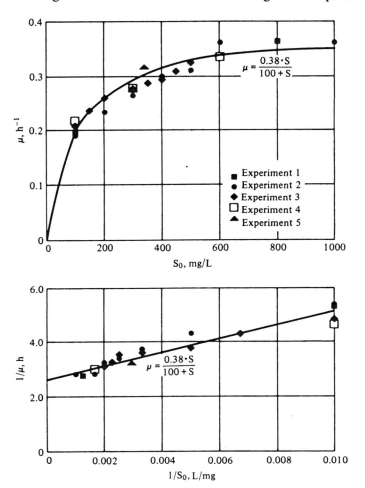

Figure 6-17 Variation of specific growth rate μ with initial substrate concentration S_0 for a heterogeneous population of sewage origin. Each plotted point was determined from growth experiments in which μ was obtained from a semilogarithmic plot. The hyperbolic curve (*upper portion*) was calculated from the equation shown using values of the constants K_s and μ_{max} obtained from the reciprocal plot of the same data (*lower portion*). (*From Gaudy et al., 1971.*)

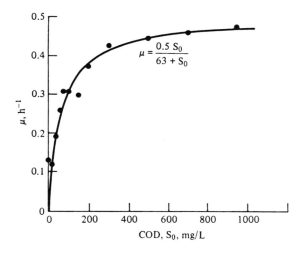

Figure 6-18 Monod plot for the soluble portion of municipal sewage. $\mu_{max} = 0.5\,h^{-1}$, $K_s = 63\,mg/L$. (*Adapted from Peil and Gaudy, 1971.*)

geneous microbial population of sewage origin growing on glucose. The general similarity of these results to those of Fig. 6-14 is evident.

Figure 6-18 shows the results of growth experiments using a heterogeneous population of sewage origin growing on the soluble portion of a municipal sewage. Again the curve is hyperbolic.

In view of the evidence available from experimental data, several conclusions regarding batch growth seem justified. First, from the best scientific approach, i.e., procurement of carefully obtained data and assessment of them in accordance with the mathematical description of exponential growth, one must conclude that exponential growth can occur at low substrate concentrations in batch processes. Second, substrate must be removed from solution during the phase of exponential growth. This is illustrated in Fig. 6-19; the substrate concentrations at various times during growth, measured using the COD test, are shown along with the growth curve in the top portion of the figure. Approximately 50 percent of the substrate was removed during the exponential phase. Figure 6-20 shows similar results for a system consisting of a mixed population of sewage origin. Thus, substrate concentration can change considerably while μ remains constant, and it is necessary to view Eq. (6-23) with some degree of respectful caution. Even though μ is not as responsive to S as the equation implies, the integrated form [Eq. (6-25)] can give a rather good approximation of the growth curve. The lack of immediate response of μ to a change in S can be looked upon as a looseness of coupling of μ to S or as a "slippage" in the coupling. One may also consider that the sluggishness with which μ responds to a change in substrate concentration is a kind of "kinetic inertia" that resists change. We shall return to this idea when we consider the concept of growth rate hysteresis in response to changes in S in continuous-growth reactors (see Chap. 13).

Like the cell yield Y_t, K_s and μ_{max} can be considered to be biomechanical

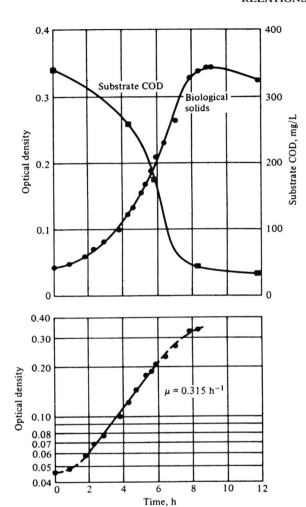

Figure 6-19 (*Top*) Growth and substrate removal at initial glucose concentration of 320 mg/L (340 mg/L COD). (*Bottom*) Semilogarithmic plot of optical density versus time for determination of the growth rate constant μ. (*From Gaudy et al., 1971.*)

properties of the biomass that can be used to describe the behavior of the biomass toward the carbon source. The numerical values of K_s and μ_{max} can change with temperature, nature of the carbon source, and other factors, just as can the value of Y_t. Also, like Y_t, the values of these kinetic parameters can change because of random changes in species predominance in heterogeneous populations. Thus, it is necessary to think of these two biomass "constants" in terms of a range of values rather than specific numerical values. For heterogeneous populations growing on various carbohydrates or municipal sewage at temperatures in the range of 18 to 25°C, μ_{max} values in excess of 0.4 to 0.6 h^{-1} are seldom observed. Using COD to measure concentration of the carbon source, values of K_s have been found to range between 50 and 200 mg/L, but values below 50 and above 200 mg/L are not uncommon.

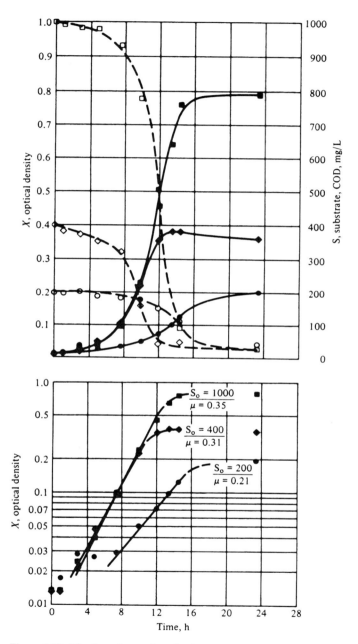

Figure 6-20 Kinetics of growth and substrate removal for heterogeneous microbial population of municipal sewage origin growing on glucose at three initial concentrations. (*From Gaudy et al., 1973.*)

When a sample of cells is taken from a very slowly growing system such as activated sludge and placed in fresh medium, the cells usually exhibit, for a while, the very slow growth rate characteristic of their previous growth history. Usually this period of slow growth represents only a small fraction of the substrate utilization period. Following the period, the growth rate increases and becomes more characteristic of the substrate concentration rather than of the previous environment. This may be looked upon as another manifestation of kinetic inertia. The early stage may even plot as a straight line on semilogarithmic paper, so that there may appear to be two exponential phases—a slow one that may consume considerable time but little substrate, and a faster one that consumes most of the substrate used prior to the inflection point separating the exponential and declining growth phases. At times the first phase does not approach a straight-line plot on semilogarithmic paper, and it is treated, as previously described, as a lag phase.

When the data plot two exponential growth phases, a decision must be made as to which provides the most valid information about the kinetic constants characteristic of the cells. It would seem that the second phase, representing the largest increase in biomass, is the one that should be used to determine K_s and μ_{max}, since it is more representative of the ultimate capability of the biomass under study and is thus more representative of the kinetic descriptors K_s and μ_{max}. During a period of slow exponential growth or lagging growth, much is happening in the biomass. In addition to occurrences previously described in discussing phase 1 of Fig. 6-9, selection of species can occur in heterogeneous populations. Thus, the cells that increase in numbers during a growth study are not necessarily the cells that predominated in the growth situation from which the seed was obtained. The abnormal growth curves sometimes obtained with cells from very slowly growing systems may be due to large proportions of nonviable cells in the seed or to the presence of cells that require organic growth factors available in a sludge but present in only small amounts in fresh medium. In the latter case, all of the viable cells may grow slowly using the low concentration of nutrients carried over into the fresh medium, and a second growth phase may occur when these nutrients are exhausted and cells able to grow without the nutrients begin to grow exponentially using the new carbon source that is present in a much higher concentration. Whatever the reason for the occurrence of two exponential phases, the second, most extensive phase does represent the specific growth rate that is attainable by the seed population at the given substrate concentration.

Description of microbial growth, particularly for the heterogeneous populations with which the environmental scientist and engineer must deal, is not the simple subject it is often made to appear in textbooks on microbiology and engineering. In some situations, we must arrive at a practical solution to the question of which exponential phase is the right one by defining the K_s and μ_{max} values for a biomass growing on a particular substrate as "constants" used to characterize a capability of the biomaterial in question, not necessarily its immediate behavior.

Relationships other than the rectangular hyperbola have been proposed to describe the effect of substrate concentration on growth rate. However, the Monod relationship provides the most generally satisfactory fit of the growth data (Gaudy et al., 1967; Ramanathan and Gaudy, 1969; Chiu et al., 1972b).

In general, the three biomass constants are sufficient biomechanical descriptors for characterizing growth in batch systems. However, when considering growth in continuous-flow reactors, it is necessary under certain conditions of growth to introduce another kinetic "constant" for describing the growth of the biomass at slow specific growth rates. This latter constant allows one to consider the possibility that a certain amount of exogenous substrate cannot be used to increase the biomass concentration but must be expended to provide the cells with energy and replacement carbon to maintain the biomass. The concept is an old and reasonable one analogous to the caloric maintenance requirement for growing and mature animals. However, for an organism as small as a microbial cell, the maintenance requirement would be expected to be extremely low; thus it could not be expected to be manifested quantitatively unless the population was extremely large and the time required for doubling the population was very long, i.e., the growth rate very small. There is no evidence for a maintenance requirement during the normal batch growth study starting with low initial cell inocula. As the concentration of the biomass increases, the cell yield is the same all along the growth curve; hence, over the range of growth rates and cell concentrations involved, a maintenance requirement is not demonstrable and need not be considered in kinetic descriptions of growth in batch systems.

CONTINUOUS CULTURE SYSTEMS

There are many types of reactors for growing microorganisms under continuous feeding conditions. Only one type will be considered here, that is, an entirely fluidized reactor system similar to the batch growth reactor shown in Fig. 6-1 except that it is modified as shown in Fig. 6-21 for continuous addition of substrate at concentration S_i and flow rate F. Also, there is continuous egress of mixed liquor at the same flow rate F. The reactor volume is designated as V, and the substrate and cell concentrations in the reactor are designated as S and X, respectively.

In assessing growth under conditions of continuous inflow and outflow of liquid, we assume that each drop of incoming medium is mixed instantaneously and completely in the reaction liquid volume. Thus the inflowing substrate concentration is immediately diluted by a factor F/V as it enters the reaction vessel. The ratio F/V is called the *dilution rate* or *flow rate* per unit volume and is designated by the symbol D. Thus, D is the reciprocal of the mean hydraulic retention time $\bar{t}$ for the reactor, V/F, i.e., the time required to

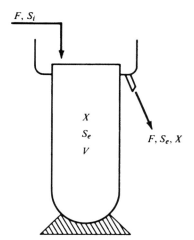

Figure 6-21 Completely mixed continuous-flow growth reactor.

replace the volume of aerating liquid:

$$\frac{F}{V} = D = \frac{1}{t}$$ (6-26)

As a consequence of complete mixing, the concentration of any reactant is the same in one part of the vessel as it is in another. This assumption also leads to the conclusion that the reactant concentrations S and X in the outflow are exactly the same as they are in the reaction liquid. Since V is the constant volume of aerating fluid, each increment of volume of inflow causes precisely the same volume of outflow. These assumptions will be examined analytically, but first it is desirable to make another assumption. Before becoming too involved with the mathematics of continuous culture, it is well to construct some physical picture of how such a system might behave.

Let it be supposed that at the start of operation the reactor was filled to the overflow volume V with a medium containing salts, buffer, and some small concentration of cells X but no carbon source. At the start of operation, a medium containing carbon source at concentration S_i is pumped into the reactor at flow rate F. In response to the inflowing substrate, X begins to increase. Some cells are continually lost from the reactor at a unit flow rate of D; i.e., cells leave the reactor at a rate DX mg/L/h. The cells grow in the reactor at a specific growth rate determined by the substrate concentration in the reactor; i.e., μ is governed by the hyperbolic relationship with S [Eq. (6-23)]. As X increases, the level of S decreases proportionally; the cell yield is constant. The decrease in S causes μ to decrease, thus decreasing the rate of increase (dX/dt) in X. It seems reasonable that a situation would eventually develop wherein the increase in X due to the inflowing substrate would be just counterbalanced by the loss of cells from the reactor; that is, X would attain some constant level, resulting in dX/dt approaching 0. The condition wherein $dX/dt = 0$ can be defined as a *steady state* with respect to X. One may hypothesize that if cell concentration X is steady, the cells must be growing at a steady specific growth rate μ. The cells are in an exponential phase of growth, since, if they were not, X would be either increasing or decreasing. Also, one can surmise that the value of μ must be related in some way to the hydraulic flow rate F, since it controls the loss of cells from the reactor and thus the rate at which they must be being replaced by growth in order to keep X constant. The above statements are developed mathematically below.

From the time pumping of the medium is initiated, the rate of increase in the concentration of cells in the reactor can be described using Eq. (6-12). The rate of decrease in X, or the cell debit, is DX, as previously mentioned. Thus, the rate of change in X is:

$$\frac{dX}{dt} = \mu X - DX \tag{6-27}$$

When the steady state is attained and $dX/dt = 0$,

$$\mu = D = \frac{F}{V} \tag{6-28}$$

Equation (6-28) has much engineering significance; it implies that the specific growth rate of the biomass can be hydraulically controlled. This is important because the hydraulic character of the reactor system can be controlled rather easily. In addition to hypothesizing about the behavior of this growth system, one may be tempted to ask whether, if one is able to hold X at a constant level, the cells that are so grown will be at some constant physiological level or state with respect to biochemical composition and metabolic capability, e.g., carbohydrate, protein, lipid, and nucleic acid content, type and amount of enzymes, etc.

The general line of reasoning and the mathematical expression of it described above were developed by Monod in much more elegant fashion in his second bench-mark paper on microbial growth in 1950. He recognized that his theory of steady state growth could provide microbial kineticists and physiologists with useful tools. Indeed, his own major use of continuous culture was to harvest large amounts of cells with like physiological characteristics for his studies on microbial genetics. In this same year, Novick and Szilard independently arrived at the same conclusions regarding continuous culture and derived equations similar to those of Monod. In 1956, Herbert et al. recast and expanded upon the concepts and equations. Over the past two decades, many workers have made contributions to the development of the kinetic theory of continuous culture. We shall make use of these hypotheses and demonstrated principles in attempts to describe growth in continuous-flow reactors. Principles will be used in a pragmatic way because they appear in many cases to provide ways and means to obtain predictive equations with which to control microbial growth and thus biological treatment processes in engineered and natural reactors. Where the experimental observations are at odds with the hypotheses and the equations developed by assuming that any hypothesis is correct, then there is need to refine or change the hypothesis. From an engineering standpoint, there is also a need to deal with these divergences and judge the adequacy of the predictive value even in cases where the theory is imprecise.

Hydraulic Mixing Regimes

Before developing equations for predicting growth and substrate removal, it is essential to justify the hydraulic assumption that makes Eq. (6-28) physically possible. That assumption is that the mixed liquor (X and S) is completely mixed. It is vital to the successful application of the predictive equations that the boundary conditions assumed in their development be physically realized or approximated to a satisfactory degree.

The two extremes of hydraulic mixing regimes in reactors are plug flow and complete mixing. In actual practice, many reactors, especially those designed before the most recent decade, lie somewhere between the two extremes.

In either type of reactor, the mean hydraulic retention time t is equal to V/F. In a plug flow or pipe flow reactor, all of the suspended particles, e.g., cells, and the molecules in solution, e.g., substrate, have the same residence time in the reactor—V/F. There is no forward or backward mixing as the material moves from inflow to outflow. It is indeed difficult to design a reactor that can provide perfect plug flow. Kinetically, it is analogous to a batch system with time interchangeably used with distance from the inlet to the outlet of the reactor. Plug flow is approached in long runs of pipes in which the cross-sectional area for flow is very small in comparison with the length of the pipe. It is only approximately approached in receiving streams and rivers. Plug flow was formerly considered to occur in rectangular activated sludge reactors, but this has been found to be an erroneous assumption. Plug flow can be approached in settling tanks when efforts are made to maintain the liquid under quiescent conditions.

Since plug flow means there is no or little forward or backward mixing as the material flows from inflow to outflow port, it is easily disturbed by mixing. Indeed, if the entire volume of liquor is vigorously mixed, a condition of "instantaneous" and complete mixing occurs. In this case, not all particles or molecules have the same residence time; some have extremely short residence times, whereas others have extremely long residence times.

Test for complete mixing Since the kinetic equations used to describe growth in continuous culture assume that the reactor is completely mixed, it is necessary that one test any reactor to which these equations are to be applied to determine whether it is completely mixed. Such tests are easily performed.

If at time 0, one began pumping clear water into a completely mixed reactor filled to the overflow exit with suspended solids, e.g., dead organisms

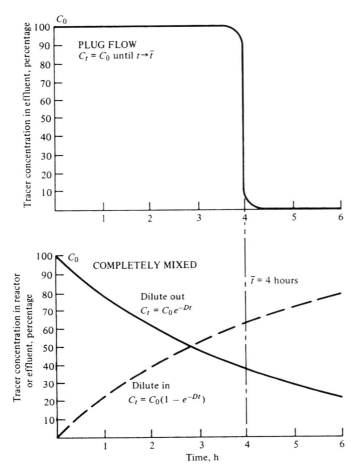

Figure 6-22 Comparison of dilute-out curves for plug flow and completely mixed reactors for assumed $D = 0.25\ \text{h}^{-1}$.

or a clay suspension, the first drop of fluid would replace some of the tracer material, and its concentration C in the reactor would be reduced in proportion to the unit flow rate D, i.e., by an increment $-DC$. The reason for using dead cells or inert suspended solids as a tracer is simply to assure that the only change in their concentration is due to their dilution by inflowing clear water. The second drop of inflowing fluid would reduce the concentration remaining in the reactor by the same factor D. Thus, the following equation may be written:

$$\frac{dC}{dt} = -DC \tag{6-29}$$

Equation (6-29) may be integrated to give

$$C_t = C_0 e^{-Dt} \tag{6-30}$$

Equation (6-30) describes the washout of the tracer material. Conversely, if the reactor had been filled with clear water and the tracer pumped into the tank beginning at time 0, the dilute-in curve would be given as follows:

$$C_t = C_0(1 - e^{-Dt}) \tag{6-31}$$

To illustrate the difference between plug flow and completely mixed reactors, Fig. 6-22 shows the type of results one would obtain in washout tests. The graph represents a plot of the concentration of the tracer substance measured in the reactor effluent at various times after beginning to pump in clear liquid containing no tracer. The sigmoidal curve at the time equal to $\bar{t}$ in the plug flow system shows that there is a small amount of forward and backward mixing, but in general the concentration of a tracer in the effluent stays the same until all of the fluid in the reactor is replaced by clear fluid at time $\bar{t}$. The bottom graph, as well as Eq. (6-30), shows that in the completely mixed system, the concentration of marker substance approaches 0 asymptotically (first order, decreasing rate).

In a laboratory apparatus, complete mixing is rather easily obtained by providing vigorous aeration. Figure 6-23 shows experimental dilute-in curves and theoretical curves calculated using Eq. (6-31) for a small laboratory reactor with aerating liquid volume V of 2.5 L. The reactor is simply a 4-in diameter test tube with a circular launder made from a section of 6-in diameter glass pipe fused to a 4-in diameter tube. The inflowing medium free-falls directly into the center of the circular surface. The liquid egress is around the periphery of the top edge of the vessel; the liquid is collected in the circular launder and is channeled to the effluent pipe (see Fig. 6-21). Aeration and mixing are by compressed air diffused into the liquid at the bottom of the tube. It is seen from Fig. 6-23 that complete mixing is approached rather closely and that elaborate inflow and outflow structures are not needed provided the mixing is vigorous enough. If one obtains such agreement of actual and theoretical curves, there is assurance that the reactor is completely mixed.

In the example shown, a soluble dye was used; thus it can be concluded that the reactor was completely mixed with respect to soluble substrate. When making dilute-in curves using a dye, it is important to be sure that the dye is not adsorbed on the walls of the feed reservoir, feed line tubing, or walls of the reactor. Also, it should be remembered that it is possible for the reactor to be mixed completely with respect to soluble substrate but not with respect to suspended particles, e.g., X. Thus, it is desirable to perform the same experiment with a particulate tracer such as dead cells or clay particles.

More vigorous mixing might be needed to provide complete mixing with respect to heavier solids, e.g., flocculated microorganisms that will settle to the bottom under quiescent conditions. After one has initiated continuous

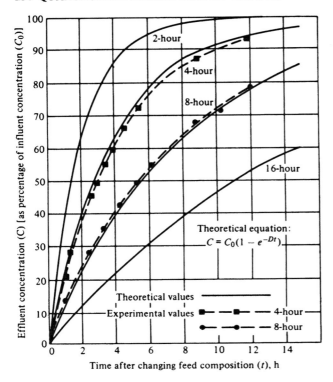

Figure 6-23 Theoretical and experimental reverse washout curves for a completely mixed aeration tank. (*From Komolrit and Gaudy, 1966.*)

growth, it is wise to check periodically to determine whether the reactor remains completely mixed. This can be done during continuous growth by determining whether the S and X sampled directly from within the reactor are the same as those sampled in the effluent. Usually a comparison of optical density readings will suffice to check on complete mixing with respect to X. Generally when a reactor is not completely mixed with respect to X, the reaction vessel accumulates cells; i.e., X in the reactor becomes higher than X in the effluent. When the reactor becomes incompletely mixed with respect to soluble substrate, short circuiting is usually the problem. Problems can usually be corrected by increasing the vigor of mixing, i.e., increasing the air flow rate, rather than by making elaborate changes in the mode of inflow and egress. Studies in our laboratories with various types of reactors and inflow and outflow arrangements indicate that the simple reactor setup shown in Fig. 6-21 provides complete mixing.

Once-through Systems

Well-established principles for writing material balances for the rates of change in the concentrations of biomass X and substrate S will be used in developing equations to predict the kinetic behavior of a continuous-flow reactor system under steady state conditions. A chemical equation is in essence a mass balance, so although the term *mass balance* is unique to the manufacturing processes, the principle upon which it is based is not foreign to those who have made use of the law of conservation of mass, i.e., those who have balanced a chemical equation. If one follows a few simple rules, writing mass balances provides excellent "bookkeeping" procedures with which to work through step-by-step accounting of the raw material (input) and products (outputs) of the system. The term *system* includes each process within the boundary established for the system by the analyst.

Figure 6-24 is a flow diagram for the continuous-flow, completely mixed reactor system shown in Fig. 6-21. The boundary of the system is defined by the broken lines (envelope). The reactor is represented by the box. It is not entirely a black-box approach because, in order to describe the biochemical changes that affect the mass rate of change with respect to X and S, one needs to establish a rate equation that is an adequate description of the process; that is, one needs a model. The kinetic model of Monod will be used to describe changes due to growth. It is important to emphasize that the mass balance does not yield the model; the model must be inserted into the mass balance. The mass balance provides equations that show how the growth model is related to operational and control parameters in describing the performance of the bioreactor system.

The mass balance for X is given as follows:

$$\begin{array}{cccc} \text{Mass rate of} \\ \text{change in } X \end{array} = \begin{array}{c} (+)\text{ change due} \\ \text{to inflow} \end{array} \quad \begin{array}{c} (+)\text{ change due} \\ \text{to growth} \end{array} \quad \begin{array}{c} (-)\text{ change due} \\ \text{to outflow} \end{array}$$

$$V\frac{dX}{dt} \quad = \quad 0 \quad + \quad \mu XV \quad - \quad FX \tag{6-32}$$

$$V\frac{dX}{dt} \quad = \quad \left(\frac{\mu_{max}S}{K_s + S}\right)XV \quad - \quad FX \tag{6-33}$$

The units are mass · time^{-1} (i.e., the product of volume and rate of change in concentration). Since in this system there were no inflowing cells, the value of X will increase due to growth and decrease due to outflow. If one divides Eq. (6-32) by V and lets dX/dt approach 0, Eq. (6-28) is obtained. This equation has already been derived using the same principles but without calling it a

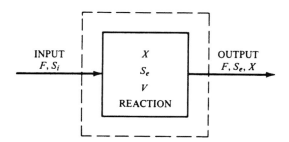

Figure 6-24 Flow diagram for a completely mixed once-through reactor. Dashed lines show the boundary of the system for which the mass balance is to be written.

mass balance. If the Monod relationship or model is now substituted for μ, Eq. (6-33) results. If one were dealing with a batch process, FX would be 0.

If Eqs. (6-12) and (6-22) are combined and integrated, and the resulting equation is combined with Eq. (6-33), Eq. (6-25) results. This is the equation previously used to estimate the time-dependent course of batch growth in Fig. 6-16. However if, instead of dealing with the time-dependent state, one lets $dX/dt = 0$ in Eq. (6-33), the following equation results:

$$\mu = D = \mu_{max}\left(\frac{S}{K_s + S}\right) \tag{6-34}$$

Solving for the steady state concentration of substrate S:

$$S = S_e = K_s\left(\frac{D}{\mu_{max} - D}\right) \tag{6-35}$$

Note that as a consequence of complete mixing, the concentrations of soluble substrate in the reactor S and in the reactor effluent S_e are the same.

A mass balance for S may be written as follows:

Mass rate of change in S	=	(+) change due to inflow	(−) change due to outflow	(−) change due to growth

$$V\frac{dS}{dt} \quad = \quad FS_i \quad - \quad FS_e \quad - \quad \frac{\mu X}{Y_t}V \tag{6-36}$$

or

$$V\frac{dS}{dt} \quad = \quad FS_i \quad - \quad FS_e \quad - \quad \frac{X}{Y_t}\left(\frac{S_e}{K_s + S_e}\right)V \tag{6-37}$$

In Eq. (6-36), the first two terms need no explanation. In the third term, the rate of substrate consumption is that due to growth, i.e., μX, which is related to the amount of substrate by the factor $1/Y_t$. The Monod equation may be substituted for μ as shown in Eq. (6-37). However, if one divides Eq. (6-36) by V and lets dS/dt approach 0, the term D appears in all terms. $D = \mu$ in the steady state; thus:

$$X = Y_t(S_i - S_e) \tag{6-38}$$

Equations (6-35) and (6-38) provide ways to solve for the two unknowns on the flow diagram, i.e., and S and X in the reaction vessel and effluent at various dilution rates. Figure 6-25 shows how S_e and X vary with dilution rate for cells grown at an S_i of 1000 mg/L with $\mu_{max} = 0.5\ h^{-1}$, $Y_t = 0.6$, and values of K_s of 10, 75, and 300 mg/L. If K_s is very low, as was observed by Monod and some of the early workers in continuous culture, the cells dilute out at a D close to the value of μ_{max}, and there is little change in S_e and X over a wide range of values of D (or μ). But at higher values of K_s, the substrate concentration in the effluent S_e begins to rise at progressively lower values of D. This is a consequence of the effect of the numerical value of K_s on the relationship between μ and S_e; e.g., compare Figs. 6-25 and 6-14. One can

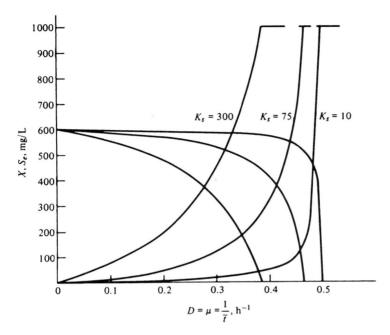

Figure 6-25 Effect of K_s on the dilute-out patterns for X and S_e as dilution rate is increased. Assumed conditions are: $S_i = 1000$ mg/L, $Y_t = 0.6$, $\mu_{max} = 0.5$ h^{-1}, $K_s = 10, 75, 300$ mg/L.

determine the dilute-out value of D by substituting S_i for S in Eq. (6-34) and solving for D. This value is called the *critical dilution rate* D_c. The critical dilution rates for the increasing values of K_s used in the example are 0.495, 0.465, and 0.385 h^{-1}.

Examination of Eq. (6-38) shows that it is a straightforward statement that one could possibly have surmised intuitively without benefit of a mass balance. Equation (6-35) reveals two characteristics of the theory of continuous culture that are worth pointing out since they do not entirely check with experimental evidence over a wide range of operational conditions.

Equation (6-35) indicates that the steady state level of substrate in the effluent S_e is independent of S_i. Also, X does not appear in the equation. This is not necessarily observed for systems growing at rather slow rates using ΔCOD as a means of measuring S_i. There is a tendency for S_e to increase somewhat for higher S_i values (which may be dependent on analytical methods of measuring S); also, there is some evidence that higher values of X can lead to decreases in the level of S_e. Thus, under certain conditions, generally very low specific growth rates and high biomass concentrations, the simple model equations may not predict the precise level of effluent substrate for the system. Nonetheless for many applications these equations are entirely adequate.

It must always be borne in mind that the model equations are attempts to simplify the quantitative description of a very complex process, and a model such as this need not apply universally over the entire range of operational conditions in order for it to have fundamental value and useful application. Equations (6-35) and (6-38) hold true for many pure culture systems, and one may ask whether they can be used to describe the growth of heterogeneous microbial populations. In view of the many changes in predominance of species and the resultant variability of the biological constants μ_{max}, K_s, and Y_t, the question of attainment of a steady state is of foremost interest. Figure 6-26 shows a plot of X and S_e for a natural microbial population of sewage origin growing with $S_i = 3000$ mg/L of glucose at a dilution rate of 0.33 h^{-1}. The calculated COD of 3000 mg/L of glucose is 3200 mg/L (i.e., $3000 \times 192/180$). The average S_i (COD) during the 2-week run was 3140 mg/L. S_e measured by the anthrone test (filtrate carbohydrate) was very low—averaging 17 mg/L—and a steady state was rather closely approached. The amount of substrate measured by the COD test was less steady, but a steady state was approximated with an average value of 113 mg/L COD. In most studies reported on growth of pure cultures, the substrate is measured by the

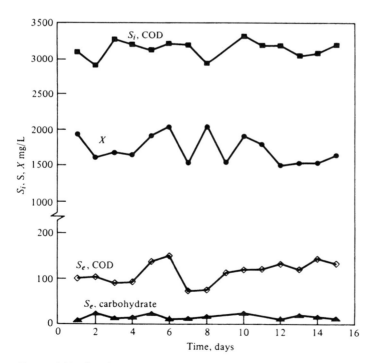

Figure 6-26 "Steady state" values of X and S_e for feed values S_i shown in a once-through reactor operating at $D = 0.33$ h^{-1}. (*Adapted from Ramanathan and Gaudy, 1969.*)

most specific test possible for the substrate; i.e., it is not measured by a nonspecific test such as one using the chemical oxygen demand analysis. Thus, it cannot be said that the variation in S as measured by COD is unique to heterogeneous populations. The nature of the unidentified COD (96 mg/L) was not determined in this study. The biomass concentration was much less steady than was substrate concentration, probably because of variations in cell yield accompanying shifts in species predominance. The average biomass concentration was 1730 mg/L. The variations in X are not attributable to simply sampling error, since, in studies at other dilution rates, samples taken at hourly intervals plot in general along lines connecting the daily sample values. Thus, there is natural cycling in X. In any event, one may conclude that the system approached "pseudo" steady state in X.

When a number of such runs is made at various dilution rates and the average values of X and S for each are plotted, results of the type shown in Fig. 6-27 can be obtained. The solid lines through the data points describe curves similar to those calculated in accord with Eqs. (6-35) and (6-38) and plotted in Fig. 6-25. However, as D is increased, the washout curves exhibit a *tailoff* not accounted for by the equations. The tailoff effect has been noted in pure culture studies as well as in the studies shown here using heterogeneous populations. For pure cultures, the effect has been attributed to various causes, e.g., mutations to higher μ_{max} values and the selection of these

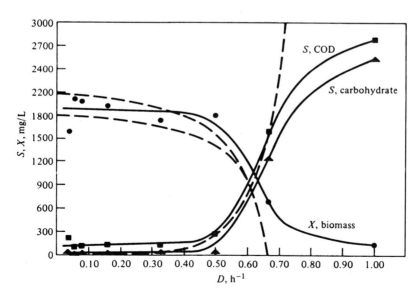

Figure 6-27 Dilute-out curves in S and X for increasing values of D for heterogeneous microbial populations growing on glucose. $S_i = 3000$ mg/L. Dashed lines show dilute-out pattern calculated from Eqs. (6-35) and (6-38) using experimentally determined values of μ_{max}, K_s, and Y_t. (*Adapted from Ramanathan and Gaudy, 1969.*)

mutants at the high dilution rates. The tailoff has also been attributed to wall growth and its faster sloughing at higher flow rates through the reactor. With heterogeneous populations there is evidence for selection of species exhibiting higher μ_{max} values (Gaudy et al., 1967; Ramanathan and Gaudy, 1969; Chiu et al., 1972a). It is reasonable to surmise that higher flow rates would tend to wash out slower growing cells. Also plotted on Fig. 6-27 in dashed lines are the values of X and S_e calculated from the model equations using μ_{max} and K_s values obtained from batch growth studies of the type shown in Fig. 6-12 through 6-19, and using as inoculum cells taken from the continuous-flow reactor during growth at the various values of D. Two dotted lines are shown for X. The lower line represents X calculated using cell yield values obtained from batch studies, whereas the higher values of X represent those obtained using cell yield values calculated from the continuous-flow reactor data. In any event, it can be seen that until the dilute-out curves are well on the way to washout, the "theory" of continuous culture provides a satisfactory tool for predicting behavior of the system with respect to X and S_e under steady feeding and environmental conditions.

In addition to the effects of S_i and X on the concentration of S and the tailoff effect, another possible weakness in the predictive value of the theory as originally given is often found. Whereas the tailoff in S_e and X occurs at very high values of dilution rate, there is often a decrease in biomass concentration, i.e., in observed yield, $Y_0 = X/(S_i - S_e)$, at very low dilution rates. The decrease in the observed cell yield at slow growth rates is conveniently, and with some logic, attributable to the effects of cell maintenance and/or endogenous respiration.

Cell Maintenance and Endogenous Respiration

The cell maintenance theory is basically a statement of well-known dietary facts about larger heterotrophic life applied to microorganisms. It is generally accepted that, in addition to receiving a balanced diet, animals must receive a certain minimum number of calories to maintain their weight at a particular level. For a given size and age of animal, a certain amount of food intake is needed to maintain the status quo, i.e., to repair the wear and tear of the structural and metabolic machinery of the cell. If the feeding rate falls below this amount, weight decreases, and if it rises above the required level, weight increases. One may ask, Should not this maintenance requirement exist for microorganisms as well? As mentioned before, there is little direct evidence for its existence in batch growth studies of the type previously described, but the dropoff in X as S_i is held constant and D is decreased in once-through continuous-flow completely mixed growth reactors (chemostats) can be interpreted as evidence of the existence of a requirement for use of some of the carbon and energy source for maintenance of the cells in a viable condition.

In discussing Fig. 6-1, a practical definition of endogenous metabolism was used; it was respiration of internal rather than exogenous carbon source resulting in a decrease in the microbial biomass. If one considers, however, that this process of aerobic decay can also proceed in the presence of an exogenous source of carbon and energy, one can envision a tie or a relationship between endogenous respiration and use of the exogenous energy source for maintenance. From the standpoint of a quantitative effect on the observed cell yield at decreasing values of D or μ, one can consider that the observed specific biomass accumulation rate (specific growth rate) represents a net specific growth rate μ_n, resulting from concurrent synthesis and endogenous oxidation of some of the synthesized materials not only for growth but also to maintain the viability of the cells that have already been synthesized. From the standpoint of estimating the effect on the amount of X produced for a given amount of carbon source fed, it may make little difference whether the respired carbon is taken from "deep" storage or short-term storage products of the cell, or whether or not the energy used for maintenance is derived directly from the exogenous carbon source, resulting in a reduction in the amount that can be channeled into synthesis. The net effect is the same in either case. Both cell maintenance and endogenous respiration in the presence of exogenous substrates are concepts about which much has been written, and investigations into their physiological and mechanistic causes continue to receive attention. These are extremely important concepts about which more knowledge is needed. Some of the references cited at the end of this section should be consulted for more detailed information (also see Chap. 10). The main interest here is in devising some way to incorporate the observed experimental effect into the predictive equations for growth because, as will be brought out later, this concept has important practical ramifications to the function of biological wastewater treatment plants as well as other processes in the environment.

The rate of respiration for maintenance (or endogenous respiration) can change because of changes in environmental conditions, such as temperature, and in natural populations because of changes in the predominance of species. Also, the factor for maintenance or endogenous respiration may in fact be a variable that depends on the specific growth rate. Nevertheless, the effect of this phenomenon can be considered quantitatively by assuming it to be a kinetic "constant." Once such an assumption is made, a wise approach is to determine experimentally whether the assumption is generally satisfactory for quantitative description and to determine the range of values one might expect; that is, the maintenance or endogenous effect can be handled like the other three "constants," μ_{max}, K_s, and Y_t.

Since the decrease in biomass has been equated to maintenance energy requirements and endogenous metabolism, it may be aptly called the *specific decay rate* and will be designated by k_d. It will be considered as a factor that decreases the rate of change in the amount of biomass in a growing system.

Thus, the material balance equation for X is revised as follows:

$$\begin{array}{c}\text{Mass rate of} \\ \text{change in } X\end{array} = \begin{array}{c}(+)\text{ rate of change} \\ \text{due to growth}\end{array} \quad \begin{array}{c}(-)\text{ rate of change} \\ \text{due to decay}\end{array} \quad \begin{array}{c}(-)\text{ rate of change} \\ \text{due to outflow}\end{array}$$

$$V\frac{dX}{dt} \quad = \quad V\mu X \quad - \quad Vk_d X \quad - \quad FX \qquad (6\text{-}39)$$

Performing the same operations on this mass balance equation as were performed on the previous ones, the following equation is obtained:

$$\mu = D + k_d \qquad (6\text{-}40)$$

The specific growth rate μ is no longer equal to D; it is still controlled hydraulically but its magnitude is affected by the new biomass constant. The mass balance equation for substrate is written exactly as before:

$$V\frac{dS}{dt} = FS_i - FS_e - V\frac{\mu X}{Y_t} \qquad (6\text{-}36)$$

Note that no term for S used for maintenance is included, since it has already been included in the mass balance equation for X. Operating on this equation in the same manner as we did on Eq. (6-36):

$$X = \frac{Y_t D(S_i - S_e)}{\mu} \qquad (6\text{-}41)$$

Substituting Eq. (6-40) for μ:

$$X = \frac{Y_t D(S_i - S_e)}{D + k_d} \qquad (6\text{-}42)$$

Substituting the Monod relationship [Eq. (6-23)] for μ in Eq. (6-40), the value of S in the steady state is obtained:

$$S_e = \frac{K_s(D + k_d)}{\mu_{max} - (D + k_d)} \qquad (6\text{-}43)$$

It is interesting to compare the theoretical (calculated) effect of k_d on the behavior of X and S at various dilutions, i.e., to compare Eqs. (6-42) and (6-43) with Eqs. (6-38) and (6-35). Figure 6-28 shows plots of X and S_e for cells with the following kinetic constants: μ_{max}, $0.5\,h^{-1}$; K_s, 75 mg/L; Y_t, 0.6; and k_d, 0 or $0.005\,h^{-1}$. The inclusion of k_d makes little difference in the value of S_e; it results in only slightly higher values throughout the entire range of dilution rates. For example, at $D = 0.4\,h^{-1}$, $S_e = 300$ mg/L without k_d and 320 mg/L when k_d is included in the predictive equations. Also for dilution rates in the range 0.1 to D_c, there is little effect on X. However, at very low dilution rates, an effect on X is readily apparent. For example, at a value of $D = 0.05\,h^{-1}$ ($\bar{t} = 20$ h), X without inclusion of k_d is 595 mg/L, whereas it is 540 mg/L when k_d is included in the predictive equation; that is a relatively small difference. However, at $D = 0.01\,h^{-1}$ ($\bar{t} = 100$ h), without k_d, X approaches 600 mg/L, whereas it is only 399 mg/L when k_d is included. This is a

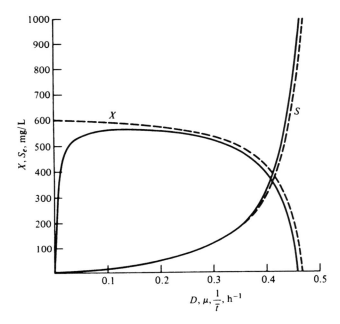

Figure 6-28 Effect of maintenance energy or biomass decay factor k_d on S and X at various dilution rates in a once-through reactor. Assumed conditions: $\mu_{max} = 0.5\ h^{-1}$, $K_s = 75\ mg/L$, $Y_t = 0.6$, $k_d = 0.005\ h^{-1}$ (0.12 day^{-1}), $S_i = 1000\ mg/L$. Dashed line, $k_d = 0$, from Fig. 6-25; solid line, $k_d = 0.005\ h^{-1}$.

significant decrease. If one were to calculate the cell yield using Eq. (6-38) at this low dilution rate, it would amount to only two-thirds of the maximum or true cell yield Y_t characteristic of this particular biomass.

The difference between the true cell yield Y_t and the cell yields observed at different values of D (the observed cell yield Y_o) is dependent on the dilution rate as well as on the numerical value of k_d (the higher the k_d, the greater is the difference at a given dilution rate). The observed cell yield Y_o is not a true biological constant. It is not characteristic of the biomass, but rather of the biomass and the dilution rate.

A numerical value for k_d can be determined from experimental values of Y_o obtained at various dilution rates D. In developing an equation for plotting the experimental data, it is essential to consider the consequences of Eq. (6-40); solving the equation for D yields:

$$D = \mu - k_d \qquad (6-44)$$

Without considering k_d, $D = \mu$, but now D represents the net specific growth rate, which may be designated as μ_n. It is the specific rate at which the biomass can accumulate due to the combined effect of growth (μ) and the autodigestion or maintenance requirement (k_d). Thus, the following identity can be stated:

$$\mu_n = D = \frac{1}{t} = \mu - k_d \qquad (6\text{-}45)$$

Since k_d is a constant, both μ_n and μ can be said to be hydraulically controlled by the dilution rate or the mean hydraulic retention time V/F. The relationship of Y_0 to Y_t, μ_n, and k_d can be derived easily by noting that the observed yield is the ratio of X in the reactor to the amount of substrate that has been removed:

$$Y_o = \frac{X}{S_i - S_e} \qquad (6\text{-}46)$$

The equation is the same as Eq. (6-38) for Y_t neglecting k_d. Substituting the value of X according to Eq. (6-42):

$$Y_0 = \frac{Y_t D(S_i - S_e)}{k_d + D} \frac{1}{S_i - S_e}$$

Then

$$Y_o = \frac{Y_t D}{k_d + D} \qquad (6\text{-}47)$$

Substituting Eq. (6-45) for D in Eq. (6-47):

$$Y_o = \frac{Y_t \mu_n}{k_d + \mu_n} \qquad (6\text{-}48)$$

Equations (6-47) and (6-48) are of the same form as the Michaelis-Menten equation, the Langmuir adsorption isotherm, and the Monod equation; that is, the relationship between the observed cell yield and the net specific growth rate can be described by a rectangular hyperbola. The reader is again reminded that similarity of kinetic expression is not reason to assume similarity of the mechanistic cause for the descriptive kinetic of the phenomenon being considered.

A plot of Y_o versus μ_n calculated from the values of X and ΔS for the system plotted in Fig. 6-28 ($k_d = 0.005$ h^{-1}) is shown in Fig. 6-29A along with curves for $k_d = 0.05$, 0.02, and 0.0005 h^{-1}. The hyperbolic nature of the plots is apparent, as is the effect of the numerical value of k_d on the shape of the curve. Y_t is approached asymptotically as D (or μ_n) increases. Figure 6-29B is a linearized plot of Eq. (6-48) obtained by taking the reciprocal of the equation (note the similarity to the linear plots of the Michaelis-Menten and Monod equations):

$$\frac{1}{Y_o} = \frac{1}{Y_t} + \frac{1}{D}\left(\frac{k_d}{Y_t}\right) \qquad (6\text{-}49)$$

or

$$\frac{1}{Y_o} = \frac{1}{Y_t} + \frac{1}{\mu_n}\left(\frac{k_d}{Y_t}\right) \qquad (6\text{-}50)$$

Equations (6-49) and (6-50) provide a convenient way to determine the values of Y_t and k_d for experimental values of Y_o and μ_n. The Y_t value determined using the plot and these equations should be checked against Y_t values determined from batch experiments using cells taken from the reactor during growth at various values of μ_n. Sometimes, but not always, the Y_t values obtained in batch (Y_{t_B}) may be somewhat lower than the values obtained from the ordinate intercept of a maintenance plot such as that shown in Fig. 6-29B.

Figure 6-29C shows another way to obtain numerical values of k_d and Y_t from a plot of data obtained from continuous-flow reactor studies. It is readily seen that Eq. (6-45) for μ_n can be written as follows:

$$\mu_n = \frac{dX}{dt}\left(\frac{1}{X}\right) - k_d \tag{6-51}$$

Substituting Eq. (6-22) for dX/dt:

$$\mu_n = Y_t\left(\frac{dS}{dt}\right)\frac{1}{X} - k_d \tag{6-52}$$

The term $(dS/dt)(1/X)$, the rate of substrate utilization per unit concentration of biomass, is defined as the *specific substrate utilization rate U*:

$$U = \frac{dS}{dt}\frac{1}{X} \tag{6-53}$$

Equation (6-52) can now be rewritten as follows:

$$\mu_n = Y_t U - k_d \tag{6-54}$$

This is the equation plotted in Fig. 6-29C. It should also be clear that the ratio of the net or observed specific growth rate μ_n to the specific growth rate μ is equal to the ratio of the observed cell yield Y_o to the true biomass yield Y_t; i.e.,

$$\frac{\mu_n}{\mu} = \frac{Y_o}{Y_t} \tag{6-55}$$

This relationship can be derived by substituting the value of k_d from Eq. (6-45) into Eq. (6-48).

The major evidence for the existence of a maintenance energy requirement is the decrease in observed cell yield as μ_n is decreased in continuous culture systems. This is not necessarily proof that the "theory" of maintenance energy is correct. However, the cell maintenance theory does provide a convenient way to explain the experimental observation of the decrease in observed cell yield at low specific growth rates. Also, a word of caution is needed regarding the treatment of k_d as a cellular or biomass constant. There is really no guarantee that k_d itself does not vary with the specific growth rate μ. We will return to the subject of cell maintenance when cell recycle is considered, because the return of cells to the reactor provides an effective

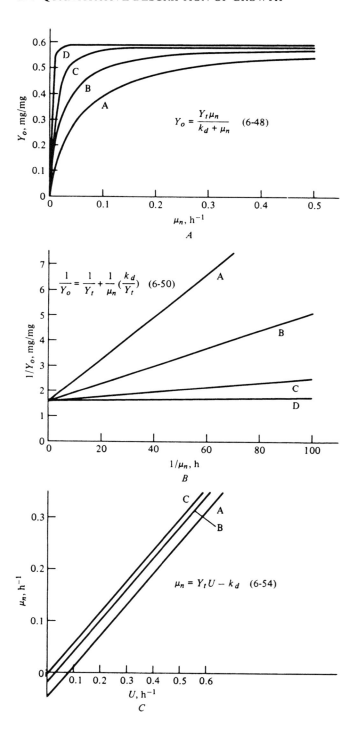

Table 6-3 Summary of predictive equations in accordance with the theory of continuous culture after Monod, Novick and Szilard, and Herbert ($k_d = 0$) and similar equations including a cell maintenance constant ($k_d \neq 0$)

	$k_d = 0$		$k_d \neq 0$	
Specific growth rate μ and net specific growth rate μ_n, time^{-1}	$\mu = D$	(6-28)	$\mu = D + k_d$ and $\mu_n = \mu - k_d = D$	(6-40) (6-45)
Reactor and effluent substrate concentration, mass $\cdot$ volume^{-1}	$S_e = \dfrac{K_s D}{\mu_{max} - D}$	(6-35)	$S_e = \dfrac{K_s(D + k_d)}{\mu_{max} - (D + k_d)}$	(6-43)
Biomass cell concentration, mass $\cdot$ volume^{-1}	$X = Y_t(S_i - S_e)$	(6-38)	$X = \dfrac{Y_t D(S_i - S_e)}{D + k_d}$	(6-42)

means of reducing the specific growth rate, and at lower specific growth rates the maintenance effect is significant. That is, as μ approaches k_d, μ_n approaches 0; thus, Y_o approaches 0. It has been seen that for once-through systems one can reduce μ (or μ_n) only by decreasing the dilution rate D. Generally, for heterogeneous populations, there is little decrease in Y_o at dilution rates greater than $\frac{1}{24}$ h^{-1}.

The predictive formulations for X and S for once-through systems are summarized in Table 6-3. The cell output, or production rate of cells, can be expressed in units of concentration, i.e., DX (mass)(volume)$^{-1}$(time)$^{-1}$, or as a mass rate of cell production, i.e., FX (mass)(time)$^{-1}$.

Cell Recycle Systems

Cells can be recycled to the growth reactor by passing the mixed liquor exiting the reactor through a separation operation as shown in the flow diagram in Fig. 6-30. The separator may be a sedimentation tank, flotation device, centrifuge, or filter. This flow diagram is like that for the activated sludge process, which was shown in Fig. 2-4. In the case of activated sludge, the separation operation traditionally has been carried out in settling tanks

Figure 6-29 Various forms of equations expressing the concept of cell maintenance: (A) rectangular hyperbola; (B) linear form, reciprocal of form in part (A); (C) linear form with μ given as $Y_t U$. Curves labeled A, B, C, and D on the graphs are for k_d values of 0.05, 0.02, 0.005, and 0.0005 h^{-1}, respectively. Other biokinetic constants used in the calculations were $Y_t = 0.6$, $\mu_{max} = 0.5$ h^{-1}, $K_s = 75$ mg/L.

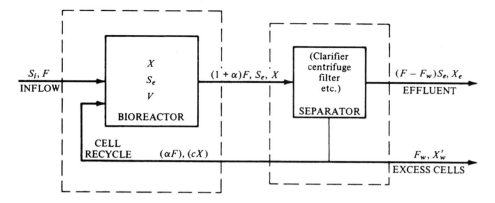

Figure 6-30 Flow diagram for a completely mixed reactor with cell feedback. Dashed lines show the boundary of the system for which the mass balances are written. $F_R = \alpha F$; $X_R = cX$.

(clarifiers), but for other continuous growth processes, any of the separation devices cited may be used. Settling is less costly, but it is effective only for flocculated cells. If the cells remain as individuals and do not clump together into floc particles, settling alone cannot be used because the wet weight of an individual bacterial cell is approximately 10^{-12} g, and such biocolloids will not settle. Regardless of the means of separation, the end result is one stream of fairly clarified effluent and one of thickened or concentrated cells. In the flow diagram, a portion of the thickened cells, usually in the underflow, is returned to the reactor. This feedback of cells has very important effects on the kinetic behavior of the system, because, as will be shown later, cell feedback lowers the specific growth rate μ.

Recycle systems using c as a control parameter The equations to be developed first are those according to Herbert (1956). It can be seen from the flow diagram that recycle offers, in addition to dilution rate D, two other variables with which to control the system. These are the hydraulic recycle ratio α, which is the ratio of the cell recycle flow F_R to the feed flow F,

$$\alpha = \frac{F_R}{F} \tag{6-56}$$

and the recycle cell concentration factor c, which is the ratio of cell concentrations in the recycle X_R and in the reactor X:

$$c = \frac{X_R}{X} \tag{6-57}$$

Practicable values of α are less than unity, and values of c are greater than unity.

The predictive equations in X and S_e are again developed by writing material balances around the process. In Fig. 6-30, there are two boundary envelopes, one around the growth process (the bioreactor) and one around the cell separator. It should be emphasized that the equations are to be derived for describing the biological process of growth and substrate utilization, not cell separation. The latter is a separate unit operation and should be treated as such. Some engineers hold that the envelope should include both the growth reactor and the separator; the biological treatment plant is considered as a joint unit process–unit operation. However, the growth equations predict X and the soluble substrate S_e. There is no model to predict the concentration of cells in the effluent X_e when simple sedimentation is the separation operation. It is well known that in the clarifier overflow there is, in addition to S_e, some small amount of suspended matter, i.e., some cell concentration X_e, due to the fact that the settling tank is not 100 percent efficient. Both S_e and X_e constitute organic pollutants. The only thing that the engineer can do in regard to X_e is to try to hold it as low as possible.

Generally, when one is designing a system to meet a specific effluent requirement, it is good practice to assume a reasonable value for X_e (often this value is stipulated as an effluent requirement) and then control S_e by manipulation of the control variables D, α, and c, so that the overall organic loading in the clarifier overflow does not exceed the effluent requirement. Since S_e is the only effluent characteristic that can be controlled and predicted by the model equations, the material balance equation is correctly written around the bioreactor. Also, if one were to write a total mass balance that included both the reactor and the clarifier, the recycle line would not be cut by the boundary envelope (see Fig. 6-30), and the resulting equations would not include the control parameters α and c; thus, they could not be used to control the process. Therefore, while it may simplify the equations somewhat to write balances around the whole process, it is wrong to do so from both the microbiological and the engineering points of view. It is important that the environmental technologist be left free to solve individual but related problems. From a purely microbiological point of view, the challenge is to understand the mechanisms and factors controlling autoflocculation of the biocolloids in the reactor; from an engineering standpoint, the challenge is to control or artificially enhance flocculation or to devise better means than simple sedimentation for separating the cells and the purified liquid. Mass balance equations for various conditions of operation are developed below.

Case a: $k_d = 0$, $S_R = S_e$ Assuming k_d is negligible and S in the recycle is equal to S_e, the mass balance for X is:

$$V\frac{dX}{dt} = (\alpha F)(cX) + V\mu X - (1 + \alpha)XF \qquad (6\text{-}58)$$

inflow   growth   outflow
(recycle)

The mass balance for S is:

$$V\frac{dS}{dt} = \underset{\text{inflow}}{FS_i} + \underset{\text{recycle}}{\alpha FS_e} - \underset{\text{utilization}}{\frac{\mu XV}{Y_t}} - \underset{\text{outflow}}{F(1+\alpha)S_e} \qquad (6\text{-}59)$$

Making the same assumptions and proceeding the same way as for the once-through system, the mass balance in X yields the following equation:

$$\mu = D(1+\alpha-\alpha c) \qquad (6\text{-}60)$$

It is seen that μ is no longer equal to D but is decreased by the recycle factor $(1+\alpha-\alpha c)$, i.e., by the combined effects of the recycle flow ratio and the cell concentration ratio. As before, the Monod relationship [Eq. (6-23)] is substituted for μ and the equation is solved for S_e:

$$S_e = \frac{\mu K_s}{\mu_{max} - \mu}$$

Substituting the value of μ from Eq. (6-60):

$$S_e = \frac{K_s D(1+\alpha-\alpha c)}{\mu_{max} - D(1+\alpha-\alpha c)} \qquad (6\text{-}61)$$

Solving the mass balance for substrate for X:

$$X = \frac{Y_t D(S_i - S_e)}{\mu} \qquad (6\text{-}62)$$

and

$$X = \frac{Y_t D(S_i - S_e)}{D(1+\alpha-\alpha c)} \qquad (6\text{-}63)$$

or

$$X = \frac{Y_t(S_i - S_e)}{1+\alpha-\alpha c} \qquad (6\text{-}64)$$

Equations (6-60), (6-61), and (6-64) are very much like their counterparts for the once-through systems [Eqs. (6-28), (6-35), and (6-38)] except for the factor $(1+\alpha-\alpha c)$. It should be pointed out that there is a limiting relationship between α and c. While these are selectable parameters for controlling μ and thus S_e and X, there is a practical limit to the freedom of selection. For example, consider the consequences of selecting $\alpha = 0.25$ and $c = 5$. In this case $(1+\alpha-\alpha c) = 0$ and the equations are not applicable. In like manner, the term can never be a negative quantity. From the example above, if c were to be greater than 5, α would have to be less than 0.25, and if α were greater than 0.25, c would have to be decreased to keep the term from going to 0 and negating the practical use of the equations.

In activated sludge processes, values of α usually range between 0.25 and 0.50. The concentration factor has not been consciously considered as a control parameter in engineering design by most designers, but one can estimate an average value of c by considering how much settling and compaction of biomass one can generally expect (or hope for) in the settling tank used as the cell separator. In general, if the value of X entering the clarifier is neither unusually low nor high, i.e., if it ranges between 2000 and 3000 mg/L, a well-flocculated and settling biomass can be expected to compact itself four- or fivefold in the clarifier. Thus, usual values of c will be slightly less or greater than 4; values of X_R will generally range between 8000 and 15,000 mg/L. It should also be apparent that there is a practical limit on the control one can exert on μ by manipulating α and c. One can control μ by all three selectable parameters, D (i.e., $1/\bar{t}$), α, and c, when designing or sizing the reactor. But after making the reactor a particular volume, thus fixing D for any particular value of feed flow F, only the parameters α and c remain available as selectable parameters in the operation of the reactor. The ability to operate the reactor to control microbial growth is of equal or perhaps even greater importance than is the design of the reactor.

These ways and means of controlling growth are, of course, not available during operation if the pumps, conduits, and other facilities do not have the capacity and/or flexibility to allow the exercise of judgment and selection. When dealing with laboratory reactors, if the experimenter finds that the final process setup is not satisfactory, he or she can make another reactor, change the pumps, or use different size tubing without too much loss of time and money. However, when millions of dollars are spent to contruct large-scale reactor systems that are not operationally flexible, much of the financial resource is lost and may not be replaceable if changes in the design are needed. There is, of course, the important question of accountability to be considered. For these reasons, it is wise to allow for operational flexibility in the design and construction of the process system. This may take such forms as the use of oversized conduits and variable flow pumps, or provision for adding flocculating chemicals to the reactor when needed.

Case b: $k_d \neq 0$, $S_R = S_e$ Including k_d and assuming that S in the recycle is equal to S in the effluent, the mass balance for X is:

$$V\frac{dX}{dt} = \underset{\substack{\text{inflow} \\ \text{(recycle)}}}{(\alpha F)(cX)} + \underset{\text{growth}}{V\mu X} - \underset{\text{decay}}{Vk_dX} - \underset{\text{outflow}}{(1+\alpha)FX} \qquad (6\text{-}65)$$

The mass balance for S is:

$$V\frac{dS}{dt} = \underset{\text{inflow}}{FS_i} + \underset{\substack{\text{inflow} \\ \text{(recycle)}}}{\alpha FS_e} - \underset{\text{utilization}}{\frac{\mu XV}{Y_t}} - \underset{\text{outflow}}{(1+\alpha)FS_e} \qquad (6\text{-}66)$$

Solving these equations as before:

$$\mu = D(1+\alpha-\alpha c) + k_d \qquad (6\text{-}67)$$

$$S_e = \frac{K_s[D(1+\alpha-\alpha c)+k_d]}{\mu_{max}-[D(1+\alpha-\alpha c)+k_d]} \qquad (6\text{-}68)$$

$$X = \frac{Y_t D(S_i - S_e)}{D(1 + \alpha - \alpha c) + k_d} \tag{6-69}$$

The specific growth rate μ is controlled by four parameters [Eq. (6-67)]: k_d, a property of the biomass, and D, α, and c, which are control parameters selected by the designer and operator of the process.

Although not readily apparent in the final equations, μ is also controlled by the Monod equation [Eq. (6-23)], which was used in deriving the model equations. This relationship between μ and S_e is an essential ingredient in the "theory" of continuous culture. The biological constants K_s and μ_{max} affect the value of S_e [Eq. (6-23)], S_e affects the value of X, and X affects the value of μ [Eqs. (6-57), (6-68), and (6-69)]. Thus, all of the factors are interrelated in the simultaneous solution of the mass balance equations in X and S.

Up to this point, we have examined two cases with the difference being the inclusion or omission of the effect of the maintenance or autodigestion constant k_d. In both cases, it has been assumed that S_e in the recycle was the same as S_e in the effluent. This assumption is a reasonable one, but over a wide range of dilution rates, as will be seen later, S_e is so low that it can be neglected. Also, although not designed to remove soluble organic substrate, the separator may actually remove a small amount; thus, S_e may be smaller in the recycle line than in the effluent. For example, in an activated sludge process, it is not unreasonable to surmise that small amounts of soluble substrate will be removed in the clarifier or settling tank. If actively metabolizing cells growing in the aeration tank are put into the clarifier, they will continue to be metabolically active in the separator. One can reason that, once in the clarifier in the absence of active aeration, they will cease using substrate, but this is more hope than reality. Many of the species constituting an activated sludge are capable of anaerobic metabolism. This fact can cause operational problems, e.g., gasification and flotation of previously settled sludge (rising sludge), but the facultative nature of many of the organisms in activated sludge is not without its merits. One can imagine the consequences of retaining strictly aerobic organisms for 1 to 3 h in an oxygen-deficient environment. The cells may die and there would be little reason for recycling them. The aim, of course, is to remove the substrate in the aeration tank; thus there should be little substrate to be oxidized in the clarifier.

For practical purposes, one might write the mass balances to describe the process (over the practical range of dilution rates) on the assumption that S_e in the recycle is negligible. Mass balances and model equations for these cases are summarized below.

Case c: $k_d = 0$, $S_R = 0$ Neglecting k_d and assuming that S in the recycle is 0, the mass balance for X is the same as Eq. (6-58), and for substrate the same as Eq. (6-59) except that the recycle term is 0. The resulting equations for μ, X, and S_e are:

$$\mu = D(1 + \alpha - \alpha c) \tag{6-60}$$

$$X = \frac{Y_t D[S_i - (1 + \alpha)S_e]}{D(1 + \alpha - \alpha c)} \tag{6-70}$$

$$S_e = \frac{K_s D(1 + \alpha - \alpha c)}{\mu_{max} - D(1 + \alpha - \alpha c)} \tag{6-61}$$

The only difference between these equations and those for case a is the substrate removal term in the numerator of the equation for X [compare Eqs. (6-63) and (6-70)].

Case d: $k_d \neq 0$, $S_R = 0$

$$\mu = D(1 + \alpha - \alpha c) + k_d \tag{6-67}$$

$$X = \frac{Y_t D[S_i - (1 + \alpha)S_e]}{D(1 + \alpha - \alpha c) + k_d} \tag{6-71}$$

$$S_e = \frac{K_s[D(1 + \alpha - \alpha c) + k_d]}{\mu_{max} - [D(1 + \alpha - \alpha c) + k_d]} \tag{6-68}$$

Comparison of cases a, b, c, and d To this point, four sets of conditions have been examined for recycle systems using the Monod model and Herbert's original treatment (case a) of selectable, controllable variables D, α, and c. The original model was adjusted for maintenance by the inclusion of k_d and for operational conditions when S in the recycle could be considered to be negligible.

Mathematical models of processes or systems are useful because they permit one to assess the effect of changes in various selectable parameters, biomechanical constants, and inputs on the resulting outputs, e.g., S_e and X. For present purposes, it is interesting to compare the effects of the different assumptions made in cases a, b, c, and d. Figure 6-31 shows the effect on X and S_e of various values of D for all four cases. Over a wide range of dilution rates, there is little difference in the curves for X whether one includes or omits consideration of the amount of substrate S in the recycle. There is even less difference in S_e, because the inflowing substrate concentration has no effect on S_e according to Eqs. (6-61) and (6-68).

The value of S_i used in this example was 1000 mg/L. However, common sense tells one that as D is increased to the dilute-out value, S_e can never become equal to S_i if the recycle flow is assumed to have no substrate. The fact that it has been assumed that there is no substrate in the recycle should allow S_i to rise only to a value of $S_i[F/(F + \alpha F)]$, as seen in the figure for cases c and d. The equations for X were different for all four cases, and the effect of this difference on the numerical values of X has little numerical significance for the values of the biological constants and the engineering constants used in the example. Inclusion or omission of k_d has the same general effect on X as it has for the once-through system. The effect of recycle can be appreciated by comparing the graphs for recycle cases a, b, c, and d with the dilute-out

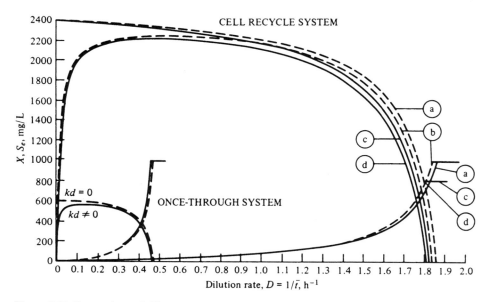

Figure 6-31 Comparison of dilute-out curves for cases a, b, c, and d in cell recycle systems and once-through systems with and without considering cell maintenance. Case a: $k_d = 0$, $S_R = S_e$. Case b: $k_d \neq 0$, $S_R = S_e$. Case c: $k_d = 0$, $S_R = 0$. Case d: $k_d \neq 0$, $S_R = 0$. Assumed values: biokinetic constants: $\mu_{max} = 0.5 \, h^{-1}$, $K_s = 75 \, mg/L$, $k_d = 0.005 \, h^{-1}$, $Y_t = 0.6$. Engineering parameters: $\alpha = 0.25$, $c = 4$.

curves for a once-through system (Fig. 6-28) replotted in the lower left portion of Fig. 6-31. The same biomass kinetic constants were used for both plots, so the effect is due solely to recycle, i.e., to operational control, not to any change in the microbial population. It is still true in the recycle system that cells will be washed out when μ approaches μ_{max}. However, in the recycle system, μ is no longer equal to D as it was in the once-through system. If one calculates μ, for example, at $D = 1.5$, which is three times higher than μ_{max}, it will be found that the actual μ for the system is much less than μ_{max}. Also, one can calculate the dilution rate at which the washout of cells occurs in the same manner as was done for the once-through system. It is apparent that cell recycle allows operation at higher dilution rates (lower mean hydraulic retention times) because cell recycle depresses the value of μ. It is also apparent that S_e approaches 0 asymptotically as μ is decreased. Thus, lower specific growth rates give lower values of S_e. Hence, from a standpoint of biological treatment of wastewaters, cell recycle has obvious advantages. Indeed, it has other advantages that will be made apparent later.

Experimental behavior of cell recycle systems Model equations are useful only when they relate to experimental observations by providing an adequate prediction of results. Results from rather extensive tests to determine whether

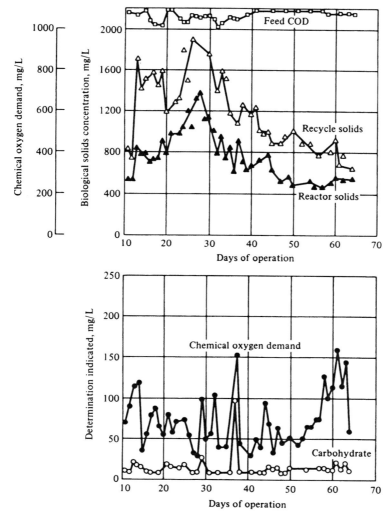

Figure 6-32 "Steady state" growth of heterogeneous microbial populations in a cell recycle system operating in accord with the model of Herbert. $S_i = 1000$ mg/L of glucose, $D = 0.33$ h^{-1}, $\alpha = 0.25$, $c = 1.5$. (*From Ramanathan and Gaudy, 1969.*)

laboratory systems could be run and results predicted in accordance with the foregoing recycle equations using heterogeneous microbial populations indicate that the equations are valid, but the results permit only a qualified endorsement of their use for designing or predicting the behavior of activated sludge systems processing soluble organic matter (Ramanathan and Gaudy, 1969, 1971).

Figure 6-32 shows an example of the type of results obtained with heterogeneous microbial populations in continuous growth reactors using the parameters of Herbert's model. After the first 20 days, c (the ratio of recycle solids to reactor solids) was maintained fairly constant at the value selected for this particular run ($c = 1.5$). However, the biomass concentration was obviously not in a steady state condition. The values for S_e (see lower graph) as measured by carbohydrate (anthrone test) were fairly steady. There was considerable fluctuation in S_e as measured by soluble COD in the effluent, but it is the unsteadiness in X that causes concern about the utilization of Herbert's model equations for the description of the growth of heterogeneous populations. As explained for the once-through system, fluctuations in X can be attributed to changes in the value of cell yield Y_t due largely to changes in the predominating species. Once can imagine the operational difficulty encountered in trying to keep c constant. Each time a change is noted in X, there must be a corresponding change in X_R in order to satisfy the boundary conditions stated in the derivation of the model equations. Also, the requirement to hold c as a system constant prevents the self-correcting feature, previously described in once-through systems, from returning the system to a steady state. For example, let it be assumed that during operation a slight increase in X is noted. In the once-through system the increase in X would cause a slight depression in the substrate concentration S_e in the reactor, which in turn would cause a decrease in μ. The decrease in μ would then cause a decrease in X, returning it to the previous value. Consider now the sequence of events and consequences of a slight increase in X in the recycle system. To keep c constant, there would have to be an increase in X_R. Rather than causing a quick return to a lower value of X, the higher X_R increases X still more, driving it further from the previous value.

Obviously, when the real situation in nature departs more than very slightly from one of the assumed conditions in the model (the assumption that a steady state in X will exist), the theoretical equations may need some adjustment in order to be applicable. Alternatively, some change in the operation of the process might make the assumed conditions hold true. Even with the variation from the steady state shown in Fig. 6-32, when the data for a number of similar experiments at various dilution rates are averaged and compared with values predicted using the model equations, there is fairly good agreement between predicted and average observed steady state results (Ramanathan and Gaudy, 1969). Thus, one can validly conclude that these equations can be used to design and predict the performance of systems using heterogeneous populations. However, the observed unsteadiness in X and the failure of the model to provide a way to correct it, coupled with the operational difficulty of maintaining c as a system constant, would seem to dictate a need to modify this system and the model equations for use with heterogeneous microbial populations. Thus, while the kinetic equations of cases a, b, c, and d are very useful in many industrial fermentation applications using pure cultures or mixtures of specific species, some modification is called for in modeling for the environmental application of biological wastewater treatment.

Recycle systems using X_R as a control parameter The realization that c is not a satisfactory control parameter surely does not warrant discarding the theory of continuous culture, especially when it has been shown, even with the problems thus far cited, to provide a reasonable prediction of the observed data. If c is not a satisfactory control parameter, it is necessary to consider another control. The recycle cell concentration X_R is the most logical choice. If this change in the model is made, two practical consequences must be considered. First, what changes, if any, would this cause in the model equations? Second, can X_R be held as a selectable control parameter in the operation of the system with more or less ease (or difficulty) than c? The ultimate question, depending on the answers to the first two, is: Can the equations based on constant X_R be used to design and predict the behavior of a biological treatment process such as activated sludge?

One approaches the derivation of the equations in the same manner as for cases a, b, c, and d using c as a control parameter, with the exception that X_R in the fraction X_R/X, rather than c, is inserted into the balance. The recycled cell mass is no longer $(\alpha F)(cX)$ but becomes $(\alpha F)X_R$; that is, X_R is no longer dependent on X. This leads to very different equations for X and S_e when one proceeds as before through the simultaneous solution of the mass balances in X and S. Also, it makes the recycle cells totally independent of the growth system, although they are assumed to have the same values of μ_{max}, K_s, Y_t, and k_d. These considerations require further discussion.

Table 6-4 shows the equations for cases a, b, c, and d, using X_R as a control parameter. For purposes of comparison, the equations using c as a control parameter are also shown. One notes that holding X_R independent of X complicates the solution of the system of simultaneous equations, leading to a quadratic form. While these equations are somewhat more complex than those using c as a control parameter, they do permit one to see the full web of interrelations among the factors that affect X and S_e. Now S_e is affected by S_i.

Comparison of models using c and X_R as control parameters For reasons described below, the equations for case d (including k_d and neglecting S in the recycle) [Eqs. (6-75), (6-80), and (6-81)] appear to be the ones most useful for the design and analysis of systems with heterogeneous populations, e.g., activated sludge processes. Thus, in order to compare the washout behavior of this model with that of Herbert, only case d is plotted in Fig. 6-33. In this figure, curves labeled a and b are reproductions of those for the once-through system and case d using c as a system constant from Fig. 6-31. Curves labeled c are those calculated using X_R as a control parameter [Eqs. (6-80) and (6-81)] with X_R at 10,000 mg/L. Perhaps the most striking difference between curves b and c is that curve c shows greater resistance to washout as the dilution rate is increased. This apparent increased stability at high dilution rates arises due to the assumption of constant X_R. The use of X_R as a system constant constitutes a rather basic difference in approach compared with that using c as an operational parameter. The mathematical consequences of carrying the fraction X_R/X along in the derivation of the equations with X_R as a constant

Table 6-4 Comparison of steady state kinetic equations according to the models of Herbert and of Gaudy and coworkers for predicting behavior of continuous-flow recycle systems

$c = \dfrac{X_R}{X}$ as control parameter (Herbert)		X_R as control parameter (Gaudy)	

Case a

$k_d = 0$

$$\mu = D(1 + \alpha - \alpha c) \tag{6-60}$$

$$S_e = \frac{K_s D(1 + \alpha - \alpha c)}{\mu_{max} - D(1 + \alpha - \alpha c)} \tag{6-61}$$

$$X = \frac{Y_t D(S_i - S_e)}{D(1 + \alpha - \alpha c)} \tag{6-63}$$

(Gaudy)

$$\mu = D\left(1 + \alpha - \alpha \frac{X_R}{X}\right) \tag{6-72}$$

$$S_e = \frac{-b \pm \sqrt{b^2 - 4ac}}{2a} \tag{6-73}$$

$$a = \frac{\mu_{max}}{1 + \alpha} - D$$

$$b = D(S_i - K_s) - \frac{\mu_{max}}{1 + \alpha}\left(S_i + \frac{\alpha X_R}{Y_t}\right)$$

$$c = K_s D S_i$$

$$X = \frac{Y_t(S_i - S_e) + \alpha X_R}{1 + \alpha} \tag{6-74}$$

Case b

$k_d \neq 0$

$S_R = S_e$

$$\mu = D(1 + \alpha - \alpha c) + k_d \tag{6-67}$$

$$S_e = \frac{K_s[D(1 + \alpha - \alpha c) + k_d]}{\mu_{max} - [D(1 + \alpha - \alpha c) + k_d]} \tag{6-68}$$

$$X = \frac{Y_t D(S_i - S_e)}{D(1 + \alpha - \alpha c) + k_d} \tag{6-69}$$

$$\mu = D\left(1 + \alpha - \alpha \frac{X_R}{X}\right) + k_d \tag{6-75}$$

$$S_e = \frac{-b \pm \sqrt{b^2 - 4ac}}{2a} \tag{6-76}$$

$$a = \frac{\mu_{max}}{1 + \alpha} - D - \frac{k_d}{1 + \alpha}$$

$$b = D(S_i - K_s) - \frac{\mu_{max}}{1 + \alpha}\left(S_i + \frac{\alpha X_R}{Y_t}\right) + \frac{k_d}{1 + \alpha}(S_i - K_s)$$

$$c = K_s D S_i + \frac{K_s k_d}{1 + \alpha} S_i$$

Case c

$k_d = 0$

$$\mu = D(1 + \alpha - \alpha c) \tag{6-60}$$

$$\mu = D\left(1 + \alpha - \alpha \frac{X_R}{X}\right) \tag{6-72}$$

$S_R = 0$

$$S_e = \frac{K_s D(1 + \alpha - \alpha c)}{\mu_{max} - D(1 + \alpha - \alpha c)} \tag{6-61}$$

$$S_e = \frac{-b \pm \sqrt{b^2 - 4ac}}{2a} \tag{6-78}$$

$$X = \frac{Y_t D[S_i - (1+\alpha)S_e]}{D(1 + \alpha - \alpha c)} \tag{6-70}$$

$$a = \mu_{max} - (1+\alpha)D$$

$$b = D[S_i - (1+\alpha)K_s] - \frac{\mu_{max}}{1+\alpha}\left(S_i + \frac{\alpha X_R}{Y_t}\right)$$

$$c = K_s D S_i$$

$$X = \frac{Y_t(S_i - S_e) + \alpha X_R}{1 + \alpha + (k_d/D)} \tag{6-77}$$

$$X = \frac{Y_t[S_i - (1+\alpha)S_e] + \alpha X_R}{1+\alpha} \tag{6-79}$$

Case d

$k_d \neq 0$

$$\mu = D(1 + \alpha - \alpha c) + k_d \tag{6-67}$$

$$\mu = D\left(1 + \alpha - \alpha \frac{X_R}{X}\right) + k_d \tag{6-75}$$

$S_R = 0$

$$S_e = \frac{K_s[D(1 + \alpha - \alpha c) + k_d]}{\mu_{max} - [D(1 + \alpha - \alpha c) + k_d]} \tag{6-68}$$

$$S_e = \frac{-b \pm \sqrt{b^2 - 4ac}}{2a} \tag{6-80}$$

$$X = \frac{Y_t D[S_i - (1+\alpha)S_e]}{D(1 + \alpha - \alpha c) + k_d} \tag{6-71}$$

$$a = \mu_{max} - (1+\alpha)D - k_d$$

$$b = D[S_i - (1+\alpha)K_s] - \frac{\mu_{max}}{1+\alpha}\left(S_i + \frac{\alpha X_R}{Y_t}\right) + k_d\left(\frac{S_i}{1+\alpha} - K_s\right)$$

$$c = K_s D S_i + \frac{k_d}{1+\alpha} K_s S_i$$

$$X = \frac{Y_t[S_i - (1+\alpha)S_e] + \alpha X_R}{1 + \alpha + (k_d/D)} \tag{6-81}$$

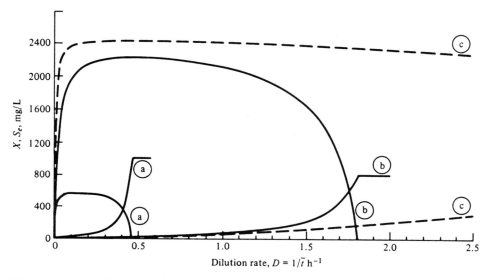

Figure 6-33 Comparison of dilute-out curves for (*a*) once-through systems, (*b*) recycle systems according to Herbert (with $k_d \neq 0$, $S_R = 0$), and (*c*) recycle system according to Gaudy and coworkers (with $k_d \neq 0$, $S_R = 0$). Assumed values: biokinetic constants: $\mu_{max} = 0.5\ h^{-1}$, $K_s = 75\ mg/L$, $k_d = 0.005\ h^{-1}$, $Y_t = 0.6$. Engineering parameters: (*b*) $\alpha = 0.25$, $c = 4$, (*c*) $\alpha = 0.25$, $X_R = 10,000\ mg/L$.

rather than *c* are readily apparent from a comparison of Eqs. (6-67), (6-68), and (6-71) with Eqs. (6-75), (6-80), and (6-81). However, is is well to consider the physical and biological significance of such a decision.

When one uses *c* as a constant, X_R is related to and made dependent on the value of *X*; but if X_R is the operational constant, *X* is related to and made largely dependent on the value of X_R at all but extremely high values of S_i. Since X_R is a constant, the lowest value to which *X* can be calculated is that which would exist when there is no contribution to *X* from growth. In Herbert's model, this washout value is quite naturally 0. However, if X_R is constant, the lowest value to which *X* can fall is that concentration corresponding to simple dilution with the waste flow. That is, *X* can fall to a concentration of $[\alpha/(1 + \alpha)]X_R$; in the example given here, the dilute-out value of *X* is 2000 mg/L, i.e., $10,000 \times 0.25/1.25$. Even at the high *D* of 2.5 h^{-1}, this value of *X* has not been approached in curve *c*, whereas the washout value for curve *b* is slightly greater than 1.8 h^{-1}. In a physical sense, the assumption is that one can obtain recycle solids from a source of cells that is not necessarily associated with the growth reactor. This assumption may not be entirely satisfactory from a theoretical point of view in regard to making models to describe the upper boundary conditions for growth in a system. However, from a practical engineering point of view, one would never contemplate using values of *D* so high (or $\bar{t}$ so low) that the system itself would not be able to produce the cells needed for X_R. Quite to the contrary, unless one wishes to use the system to produce cells, the aim is to make the cells grow slowly in order to produce as few cells in excess of those needed for recycling as is feasible. Production of large quantities of excess sludge is objectionable, because this sludge (at least for the foreseeable future) represents a waste product of the treatment process, and the mass rate of cell production increases as D increases (up to values where cell dilute-out becomes very rapid). Very high values of D are undesirable for other reasons as well; the most important of these involves consideration of shock loadings to the system, which will be discussed in Chap. 13.

It is apparent from Fig. 6-33 that for the reasonably attained values of X_R and c chosen in the example, both models yield similar behavior of S_e over a wide range of D values (i.e., below $D = 1.0\,h^{-1}$ or above $\bar{t} = 1\,h$). The curve for S_e can be lowered or raised by changing the value of X_R or c. Increasing either X_R or c will give lower values of S_e and higher values of X. There is, of course, a practical limit to the values of X_R or c that one can choose; X_R values greater than 15,000 mg/L are not easily obtained without the addition of some sort of sludge-thickening process. Since one's ability to operate the system is dependent on the physical ability to obtain the recycle cell concentration selected, it seems well to fix X_R rather than c by design. What is a practical upper value of c? As previously noted for $\alpha = 0.25$, c cannot be greater than 5 because this would make $\mu = 0$, which is the lower boundary of Herbert's growth model for cell recycle systems. However, one can calculate values of X and S_e for values of c between 4 and 5. If, for the conditions assumed in Fig. 6-33, c were 4.75 rather than 4, the curve for S_e would be even lower than the one shown, and the curve for X would yield so much higher values that this value of c would be impractical. For example, at $D = 0.5\,h^{-1}$, the calculated values of S_e and X are 5.86 and 8215 mg/L, respectively. To obtain this value of X, maintaining c at 4.75 requires an X_R value of nearly 39,000 mg/L, which is outside the realm of practical possibility. If c were 4.25, X would be 2967 mg/L and X_R would be 12,600 mg/L, which is close to the upper limit generally attainable without special thickening procedures.

It is also of interest to compare the models as D is decreased. At high D values, the lowest possible X value for curve c was 2000 mg/L, but as D is lowered, the biomass curve goes through the 2000 mg/L level. This behavior has considerable significance. As with the high D end of the curve, attainment of $X = 2000$ mg/L signifies a dilution rate at which there is no net growth of cells. At the high D values, the flow-through of liquid is so fast that washout is approached, but at the low values of D, growth is so slow that it can be balanced by the cell maintenance or cell decay rate k_d:

$$\mu = D\left(1 + \alpha - \alpha\frac{X_R}{X}\right) + k_d \tag{6-75}$$

$$\mu = \mu_n + k_d \tag{6-45}$$

As μ_n approaches 0, μ approaches k_d.

The dilution rate at which $\mu = k_d$ can be calculated using Eqs. (6-23) and (6-80). The value of S_e corresponding to $\mu = k_d$ (0.005 day^{-1} in this example) is obtained from Eq. (6-23), and this value is substituted into Eq. (6-80), which is then solved for D. For the biomass constants, operational conditions, and S_i of this problem, $D = 0.0167$ h^{-1}, i.e., $\bar{t} = 60$ h. Thus, an extended period of aeration is needed in order to operate a system at $\mu_n = 0$, i.e., a condition of no excess biological sludge production due to growth of organisms in the reactor.

The model also allows a negative specific growth rate. Below the value of D that yields $\mu = k_d$, the term $\alpha(X_R/X)$ in Eq. (6-75) becomes greater than $(1 + \alpha)$ because of the lowering of X in accordance with Eq. (6-81). Thus, the model allows the system to function both as an aerobic sludge digester for cells extraneous to the system and as a growth reactor and autodigester of cells. The boundaries of calculational possibilities cover the gamut of aerobic metabolic applications from aerobic digestion of sludge, to extended aeration, to high rate activated sludge. Such unified applicability of the theory of continuous culture is not readily possible using c as a control parameter. Again, considering Eq. (6-67), the only way μ can equal k_d (i.e., the μ_n term approach 0) is to let D approach 0, because c and α are constants. Thus, the Herbert model is not applicable to a situation in which $\mu \to 0$ unless $\bar{t} \to \infty$. The biological possibility and likelihood of total decay of previously synthesized cellular material is a rather controversial subject and will be discussed later. The point emphasized here is that if such a mode of operation is possible, the model using X_R as a system control parameter provides a ready quantitative estimate of the kinetic relationships that predict no excess sludge production, whereas the model using c as a system constant does not readily facilitate this.

Dependence of S_e on S_i It is important to analyze another difference between the model equations that arise as a consequence of using X_R rather than c as an operational engineering parameter. The theory of continuous culture predicts S_e independently of S_i, e.g., Eq. (6-35) or (6-43) for once-through systems and Eq. (6-68) for cell recycle systems. However, when X_R is used as a control parameter, S_e is dependent on S_i [see Eq. (6-80)].

There are two major concerns in considering relationships between S_i and S_e. The first involves the measurement procedure for S as well as the nature of the biomass. The general kinetic concepts of continuous culture were developed on the assumption that S represents a specific concentration of a specific nutrient being fed to the system, and that the biomass constants are specific for a single species being grown. Such items of practical importance to environmental engineering science as leakage of organic matter from the cells, production and elaboration of intermediates and end products, and protozoan and crustacean excrement are not considered in the theory of continuous culture. Of necessity, we try to measure organic pollution with an "umbrella type" of measure such as COD, TOC, TOD, ΔCOD, ΔTOD, or

ΔTOC. Also, we attempt to measure organic substrates by the BOD test, which was previously mentioned and is yet to be discussed. When one feeds a compound known to be readily metabolized and uses COD_i as a measure of S_i and COD_e as a measure of S_e, it is often found that COD_e increases for high values of S_i at constant values of D or μ. If one feeds a single compound as carbon and energy source, usually only traces of the original compound can be found in the effluent, provided the value of μ or D is not so high as to put the system in the area of rapid washout.

We have noted over a period of many years of gathering data that during the growth of heterogeneous microbial populations on a readily available carbon source, there is a tendency, especially in once-through reactors operated at the same value of D, for the percentage removal of COD to remain constant at approximately 95 percent over a fairly wide range of values of S_i. At rather low values of S_i there is a tendency for the percentage of COD removal to decrease. These findings appear to hold true provided that the carbon source and nothing else, neither nutrients nor environmental conditions, limits growth and metabolism, and provided the pH and temperature are held fairly constant. This means that $COD_e(S_e)$ rises approximately in a linear fashion as $COD_i(S_i)$ is increased. Usually only traces of the original compound can be found in the filtrate. Thus, one may expect to find slightly different results depending on the measurement used to assess S. Whether the increase in S_e with increased S_i is due to the nature of the method used to measure S or to the heterogeneity of the microbial biomass, or both, cannot yet be stated with certainty. Some microbial kineticists have challenged the theory of continuous culture on the matter of constancy of S_e with increasing S_i. One possible way in which S_e could increase with S_i at a given specific growth rate, μ or D, would be if the value of K_s were not always constant but increased with increasing S_i values (see Fig. 6-25). Both Contois (1959) and Fujimoto (1963) have presented such an argument, while others argue that deviations from "theory" are due to mixing effects (e.g., see Herbert et al., 1956; Tempest, 1968). Grady et al. (1972) found that for the pure cultures they examined (*Aerobacter aerogenes* and *Escherichia coli*) growing on glucose in a once-through reactor at S_i values of 500, 1000, and 1500 mg/L, the S_e values measured by a specific test for glucose were constant over the range of S_i values, in agreement with theory. However, for S_e measured as COD_e, there was an increase in S_e for increasing values of S_i. When heterogeneous microbial populations were grown in the same manner, there was an increase in S_e regardless of the test used to assess S_e (Grady and Williams, 1975). Radhakrishnan and SinhaRay (1974), in studies with *Bacillus cereus* growing on phenol in a once-through reactor, observed that values of S_e rose linearly (i.e., the percentage removal was constant) for the three values of S_i used, 200, 400, and 800 mg/L of phenol, in opposition to the theory of continuous culture. Phenol does present a somewhat different case than a substrate such as glucose because it is an inhibitor as well as a substrate (see Chap. 10).

From the data available, the kinetic mode of increase of COD_e with increasing COD_i cannot be stated, because massive amounts of data are needed before trends can be expressed in the quantitative manner needed for engineering prediction with heterogeneous populations. Usually small amounts of data lead to either a straight-line or a first-order relationship, which may be more characteristic of the mode of thinking of engineering scientists than of the kinetic mode of the data. About all that can be said at present is that the work in our own laboratories, coupled with the results cited above and the more recent results of Grady and Williams (1975), indicate that with heterogeneous populations and readily metabolized carbon sources, one can expect COD_e to increase with S_i. If the percentage removal of COD remains fairly constant over a wide range of S_i values, the doubling of S_i would be expected to lead to a doubling of S_e.

A rise in the amount of soluble organic matter in a reactor effluent with increasing concentrations of S_i is rather important in view of the increasing need for fine-tuning the control of the quality of effluents from biological treatment plants. It may prove far more cost-effective to invest monies and effort in gaining understanding for the purpose of controlling and reducing COD_e, or TOC_e, than to let the process tune itself, as it were, and to rely on expensive add-ons, e.g., activated carbon or other tertiary treatment processes, to meet future effluent standards of allowable organic content. Also, the nature and environmental fate of this effluent substrate need further consideration. Such considerations relate to organic assimilation capacity in the natural water courses receiving treatment plant effluents and also to socio-technological decisions in regard to the desirability of using this assimilation capacity (see Chap. 10).

The second important consideration here is some analysis of the reasons for the independence of S_e and S_i in continuous culture theory, and the dependence of S_e on S_i in the practical modification of the theory that uses X_R as an operational parameter. In a once-through system, S_e does not rise with S_i because cell yield is "constant" and X rises as S_i increases; S_e is affected only by factors that affect μ. The recycle equations operate in the same way because c is a constant, and X_R must vary proportionally with X; that is, X_R is not independent of X, and S_e is dependent on the values of the biological and engineering constants and is thus fixed once the selectable constants are fixed. However, for the equations using X_R as an independent control parameter, X_R is no longer required to keep pace with the increase in X due to growth on an increased S_i, and this is reflected in an increase in S_e.

It is interesting to examine the changes in S_e with S_i for the model using constant X_R. Figure 6-34 shows the relationship among S_e, X, and S_i, and compares the behavior of S_e and X with X_R and c as operational parameters. For constant X_R (upper portion of the figure), S_e increases at an ever-decreasing rate as X_R increases, whereas X increases linearly with S_i. For the operational conditions noted in the legend, S_e appears to be approaching a constant value asymptotically. The values in parentheses along the plot of X

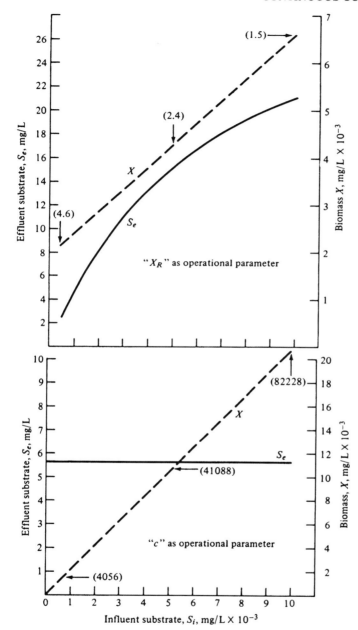

Figure 6-34 (*Top*) Effect of S_i on steady state values of X and S_e for the model according to Gaudy and coworkers, using X_R as an engineering control parameter. See Eq. (6-81) for X and Eq. (6-80) for S_e. (*Bottom*) Effect of S_i on steady state values of X and S_e for the model according to Herbert, using c as an engineering control parameter. See Eq. (6-71) for X and Eq. (6-68) for S_e. Assumed values: $X_R = 10,000$ mg/L, $c = 4$, $\alpha = 0.25$, $\mu_{max} = 0.5$ h^{-1} (12 day^{-1}), $K_s = 75$ mg/L, $Y_t = 0.6$, $k_d = 0.005$ h^{-1} (0.12 day^{-1}).

are the values of c at S_i values of 500, 5000, and 10,000 mg/L. With X_R constant, the value of c decreases as S_i increases. That is, the proportion of X contributed by X_R diminishes and the system functions more and more like a once-through system that, in accord with the theory of continuous culture, would approach a constant value of S_e with respect to increases in S_i.

In the lower portion of the figure (c constant), S_e remains constant at any value of S_i, and X increases linearly. In this case, X_R varies and the values in parentheses are those for X_R at S_i values of 500, 5000, and 10,000 mg/L. The values of X_R needed to keep c constant soon would become so high as to be unattainable by ordinary means of separation and thickening of the cells. Thus, if S_e were to be made independent of S_i by operating with c constant, one would soon have difficulty trying to increase X_R sufficiently to keep c constant and keep X at the level demanded by the equations.

From the upper graph of S_e, it is seen that at S_i concentrations lower than 2000 mg/L, S_e varies nearly linearly with S_i. While the equations using constant X_R offer the additional assurance that the model tends to agree with various experimental observations using colligative measures of S_e and S_i, there is, of course, no guarantee that S_e really does increase with S_i. Much more experimental data are needed to assure the generality of this observation. Also, it should be clear that the increase in S_e *as* S_i increases, which is predicted by Eq. (6-80), has no real basis in biological mechanism. It comes about only as a consequence of holding X_R constant as an operational parameter; i.e., it is a mathematical manipulation that has the effect of retaining S_i in the numerator of the equation for S_e. Furthermore, the increase in S_e for the higher values of S_i predicted by Eq. (6-80) is not inconsistent with the theory of continuous culture which states that S_e remains constant because μ remains constant [see Eq. (6-67)]. However, when one holds X_R constant, rather than c, X remains in the denominator. As always, an increase in S_i causes an increase in X, which in effect reduces c, which is no longer used as a control parameter. The corresponding decrease in the ratio X_R/X thus increases μ for increased values of S_i. The compensating increase in μ with S_i leads to an increase in S_e so that the relationship between S_e and μ that was used in deriving the predictive equations, i.e., the Monod relationship [Eq. (6-23)], still holds true (Gaudy and Manickam, 1978).

Any mechanistic relationship describing an increase in the amount of soluble organic matter derived from the original substrate in the effluent as the amount of feed substrate increases requires both experimental and theoretical investigation. From a practical engineering standpoint, the equations using constant X_R tend to predict or to calculate S_e in the right direction with respect to various experimental observations of residual COD as S_i increases. At the current stage of development of kinetic concepts for predicting the performance of heterogeneous populations using heterogeneous substrates, the fact that the prediction moves in this direction seems to offer another argument for using the system and model equations that provide engineering control over the biochemical dosage X_R.

Some practical considerations for design and operation with X_R as control parameter If one is to use any particular model in design and analysis, it is essential to sound engineering practice that one do everything possible to accommodate physically the boundary conditions laid down in the model. That is, if one assumes a completely mixed reactor in devising prediction equations, a completely mixed reactor should be designed. In like manner, if X_R is to be a constant, some provision for making it constant should be included in the design. Certain types of separators can be made to deliver desired cell concentrations. However, if one wishes to use a gravity clarifier (settling basin), there is no easy way to guarantee that the underflow will be at a constant concentration, and certainly no way to achieve the specific concentration selected by the analyst or designer. In the future there may be changes in the design of clarifiers, or they may be replaced by other separation operations. One way to retain use of the clarifier while obtaining constant X_R is to take some of the cells in the clarifier underflow, measure their concentration, and then bring the entire required volume of cells (sludge) to the desired concentration (by thickening or by diluting) in an aerated recycle sludge dosing tank. The recycle flow diagram of Fig. 6-30 could be modified as shown in Fig. 6-35. Daily, the recycle cell concentration can be made up to the desired X_R and continually pumped to the reactor.

To some who may be familiar with current practices in the design of wastewater purification control plants, the idea of controlling the recycle sludge concentration is objectionable because it is not now being done, and to do so would require an extra cost for construction and operation of the dosing tank. Some may feel that there is no need to control the dosage of a commodity that is gratuitously generated in the course of operating the

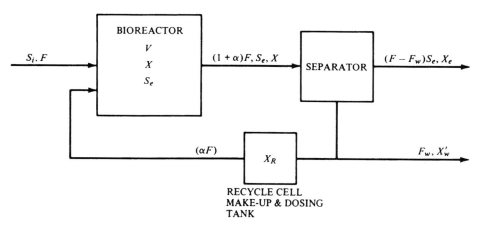

Figure 6-35 Flow diagram of Fig. 6-30 modified by addition of a recycle cell makeup and sludge dosing tank for maintaining X_R at selected concentrations. X_R rather than c is the control parameter.

system. Also, some may believe that it is operationally difficult to control X_R. These are important engineering aspects to be discussed.

If, instead of a biological treatment process, we were dealing with a chemical treatment process for purifying wastewater, there would, of course, be no question of making an expenditure for chemical dosing and storing equipment. It would be obvious that the chemical dosage should be controlled because it would be an accepted fact that the dosage controls the quality of the effluent. Also, the chemicals used are an expensive item of operational cost. In regard to the biological treatment plant, one can be convinced easily that the "chemical" dosage (the recycled cells) to this biochemical process exerts a controlling influence on the quality of the effluent by calculating values of S_e from Eq. (6-80) for different values of X_R (see Fig. 6-36). It is well worthwhile to control this biochemical dosage. It is easily understood why the sludge, which is a biochemical reactant vital to the functional success of this system, is not valued when one considers that the amount of sludge in excess of that needed for recycle represents a major waste product of the waste treatment plant, and a major disposal problem. Nonetheless, the sludge that is dosed to the aeration tank is responsible for the success of the process, and in accordance with the descriptive equations, it is an important control parameter.

As a consequence of metering the supply of X_R from an aerated dosing tank, it might be expected that any soluble substrate in the clarifier underflow would be reduced by the extra treatment provided in the dosing tank. For this reason, S_e in the recycle can generally be considered to be 0 regardless of its

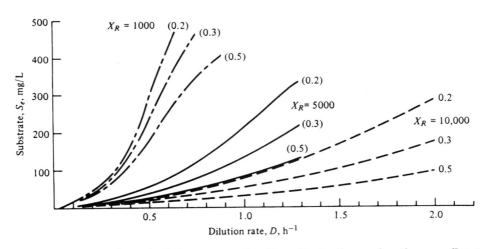

Figure 6-36 Effect of recycle sludge concentration X_R and hydraulic recycle ratio α on effluent substrate concentration S_e at various dilution rates D. Assumed conditions: $S_i = 1000$ mg/L. $\mu_{max} = 0.5 \, h^{-1}$, $K_s = 75$ mg/L, $Y_t = 0.6$, $k_d = 0.12 \, day^{-1}$. Values of α are given in parentheses; S_e was calculated from eq. (6-80).

concentration in the clarifier underflow. Thus, the equations for cases c and d are more applicable than those for cases a and b (Table 6-4).

Studies in laboratory pilot plants to determine the predictive value of the equations and the ease or difficulty of operation of a system using X_R as one of the control parameters have provided results indicating the applicability of the approach (Gaudy and Srinivasaraghavan, 1974; Srinivasaraghavan and Gaudy, 1975; Saleh and Gaudy, 1978; Manickam and Gaudy, 1978). Operationally, it is quite easy to maintain a preselected value of X_R, especially if the value is somewhat lower than the concentration of compacted sludge in the clarifier underflow. Values of X_R of approximately 10,000 mg/L can usually be obtained by diluting the portion of clarifier underflow to be recycled. Thus, selection of values of X_R of this magnitude would usually obviate the need for thickening equipment. It should be pointed out, however, that many treatment plants use thickening processes to prepare excess sludge for treatment and disposal. Therefore if, in the field, X_R were selected higher than the concentration of the underflow, thickened sludge for recycle could be obtained without too much additional cost.

The concentration of sludge is easily estimated by measuring its optical density if samples are appropriately diluted and the biomass dispersed by blending prior to measuring the transmittance. The recycled sludge to be dosed to the aeration tank can be made up each 24 h. During the ensuing 24-h pumping period, there is little autodigestion and X_R remains fairly constant. Thus, from an operations standpoint, the use of constant X_R does prevent the problems encountered when attempting to operate with constant c as a control parameter.

Figure 6-37 shows typical results during operation of a pilot plant at constant X_R. In this run, X_R was selected as 10,000 mg/L. The steadiness in X and S is apparent. The quality of the effluent can be measured in a number of ways. In this figure, soluble COD S_e, cells in the clarifier overflow X_e, and total (unfiltered) COD S_t are shown. In addition, the BOD of the clarifier effluent and the concentration of carbohydrate (anthrone test) in solution were also measured but not as often as those analyses that are plotted. Soluble carbohydrate values were much lower (approximately 75 percent) than soluble COD values. Soluble COD is defined as organic matter capable of passing through a 0.45 μm (nominal pore size) membrane filter. Values of BOD (5-day incubation periods) of the unfiltered clarifier effluent were very low—in the range of 10 to 15 mg/L. The actual residual COD values were plotted, and actual COD values for the strength of the carbon source (glucose) are shown for S_i and S_e. These could have been corrected to ΔCOD values for S_i and to ΔCOD minus COD_e values as a measure of effluent quality S_e, but this was not done in order to simplify presentation of the experimental results. However, it is important to realize that residual soluble COD is not equivalent to S_e. To determine S_e, one must substract the residual COD observed from the determined ΔCOD (S_i) of the wastewater. The residual soluble COD plotted in Fig. 6-37 is significantly higher than the actual S_e. One may make a

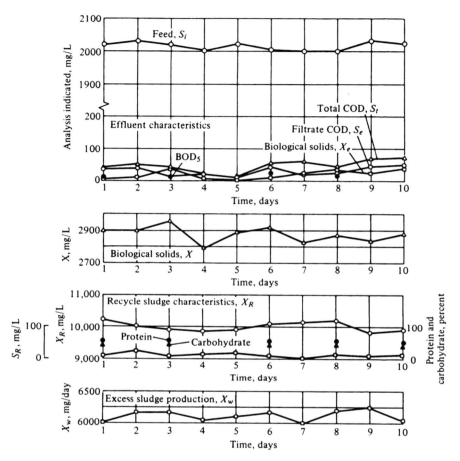

Figure 6-37 Operational characteristics for an activated sludge process pilot plant with constant X_R. (*From Srinivasaraghavan and Gaudy, 1975.*)

fairly conservative estimate of the base-line value of soluble COD_e by noting that during the run shown in Fig. 6-37, the soluble COD in the sludge return aeration tank (aerator No. 2 in Fig. 6-33) was on average 14 mg/L. For this run, 14 mg/L COD is perhaps the absolute minimum COD_e, because a small amount of a residual soluble COD in the effluent (an average of 31 mg/L in this run) combined with 10,000 mg/L of cells for 24 h provided for the "squeezing out" of 17 mg/L COD. Thus, by substracting 14 mg/L from the average 31 mg/L in the effluent, a closer but still conservative estimate of S_e would be 17 mg/L. Also, the low soluble COD concentration observed in the recycle substantiates the assumption made in the derivation of the equation that substrate in the recycle can be neglected.

After making a number of such studies at various specific growth rates and determining values of the biomechanic constants μ_{max}, K_s, Y_t, and k_d, one can calculate (i.e., predict) the steady state values of X and S_e that would exist at the selected values of the engineering control parameters D, α, and X_R. When this is done, the equations do predict observed values rather closely. The equations of cases c and d predict nearly the same values of S_e and X. However, the equations for case d, which include k_d, provide significantly better predictions of excess sludge, especially for low values of μ_n. The excess, or waste, sludge produced each day during the run shown in Fig. 6-37 is given in the lower graph. The excess sludge or the net accumulation of biomass in the system each day is an important consequence of growth and substrate removal because, in the field, the disposal of this excess growth represents a considerable problem. One can predict the amount of excess sludge using the model equations.

Since case d is most applicable, it will be used in deriving prediction equations. The equations can be obtained by writing a mass balance equation in X bounding the cell separator or clarifier (see Fig. 6-30). The mass balance for X is :

$$V\frac{dX}{dt} = (+)\,\text{inflow} \quad \begin{array}{c}(-)\,\text{cells in effluent} \\ \text{(overflow)}\end{array} \quad \begin{array}{c}(-)\ \text{cells in waste} \\ \text{line (underflow)}\end{array} \quad \begin{array}{c}(-)\,\text{cells in} \\ \text{recycle}\end{array}$$

$$V\frac{dX}{dt} = (1+\alpha)FX \quad - \quad (F - F_w)X_e \quad - \quad F_w X'_w \quad - \quad \alpha F X_R$$

$$(6\text{-}82)$$

Since it is assumed that no reaction involving a change in X in the clarifier takes place, the mass balance is a simple inventory of the biomass entering and exiting the clarifier. Also, since the clarifier does not provide perfect separation of mixed liquor, the term for X in the effluent or overflow, X_e, is included. It is well known that in practice the clarifier usually functions as an accumulator of settled biomass, i.e., the settled cells are not immediately removed; but by expanding the unit time to a day, one can assume that a steady state is approached, i.e., any accumulation is relieved daily. Thus, dividing Eq. (6-82) by V and letting dX/dt approach 0:

$$\frac{1}{V}[(F - F_w)X_e + F_w X'_w] = \frac{F}{V}[(1 + \alpha)X - \alpha X_R] \qquad (6\text{-}83)$$

The term in brackets on the left side of Eq. (6-83) represents both the rate of biomass wastage from the underflow $(F_w X'_w)$ and that inadvertently wasted in the overflow $(F - F_w)X_e$. Designating this term as the excess sludge and representing it by X_w:

$$X_w = VXD\left(1 + \alpha - \alpha\frac{X_R}{X}\right) \qquad (6\text{-}84)$$

Also, substitution of Eqs. (6-45) and (6-75) for μ_n leads to:

$$X_w = VX\mu_n \qquad (6\text{-}85)$$

and
$$X_w = \frac{VX}{\Theta_c} \qquad (6\text{-}86)$$

In Eq. (6-86), Θ_c is defined as the mean cell residence time or sludge (cell) age, and is the reciprocal of μ_n.

Equation (6-84), (6-85), or (6-86) can be used to predict the excess sludge for any of the cases examined. In the case of Herbert's approach, c replaces X_R/X and, since μ_n rather than μ appears in the equations, they hold true whether or not k_d is neglected. X_w will not be the same for all cases. Differences occur because of differences in the value of the predicted cell concentration X. In general, when the steady state equations in X, S_e, and X_w for case d [Eqs. (6-80), (6-81), and (6-84)] are used to calculate these values under specific experimental conditions, the predicted values are rather close to those experimentally observed during operation in continuous culture. Thus, the equations appear to be useful for both analysis and design of recycle systems using heterogeneous populations, e.g., activated sludge. More detailed description of the utility and use of this model can be found in the literature (Ramanathan and Gaudy, 1969, 1971; Gaudy and Srinivasaraghavan, 1974; Srinivasaraghavan and Gaudy, 1975; Gaudy et al., 1977; Gaudy and Kincannon, 1977a, 1977b; Manickam and Gaudy, 1978; Saleh and Gaudy, 1978). Bonotan and Yang (1976) have reported considerable success in using the model in pilot plant studies on an industrial waste.

Obtaining values of the biokinetic constants for use in any of the models is not a major problem. However, there is not yet a sufficient library of values from past experiments so that the range of probable values for various types of waste (carbon sources), temperature, pH, and other factors can be treated as a "shelf item" or handbook item. The values of the biokinetic constants should be determined by the environmental technologist during the treatability study and functional design stage, which for the most part has been woefully neglected in past practice of the profession. Also, the biomechanical properties should be checked periodically during the life of the process. The model equations can be useful in designing operational strategies as well as in the initial design of the process. The biomass constants for this living engineering material must be expected to exhibit much more variability than do the characteristic constants used to describe the functional behavior of inanimate engineering materials such as steel, concrete, and asphalt. Although the challenge is greater in the environmental field, the stakes are higher. In the future there will be as much or more opportunity for environmental professionals in operation as there is currently in design. The control of product quality in biological treatment facilities is a more complicated and sophisticated technological problem than it is for some of the other applied biological processes such as the making of spirits or the production of penicillin. The major product, reusable water, is much more important than those cited, and in time its monetary value will be more in line with its importance and will provide economic justification for the use of more sophisticated operational skills in its production.

It should be clear that interest in the kinetic constants will continue during operation. The Monod constants, μ_{max}, K_s, and the true cell yield Y_t, can be estimated from batch data, as previously described. They may also be calculated using data from a continuous-flow pilot reactor and the kinetic model equations recommended here. The value of the maintenance coefficient k_d is the most costly to determine because several continuous-flow runs must be made to obtain data for making plots such as those shown in Fig. 6-29. Values for k_d reported in the literature range from 0.025 to 0.48 day^{-1} for heterogeneous populations and somewhat higher for some pure cultures (see Chap. 10). How constant k_d will be for a given type of substrate and given environmental conditions over a range of net specific growth rates is not really known. However, Fig. 6-38 shows a maintenance plot for data obtained over a 4-year period, using the same carbon and nitrogen sources (glucose and NH_4^+), essentially constant temperature (20 to 22°C), constant pH (6.8 to 7.2), and natural populations obtained from a municipal treatment plant at various times during that period. The data show a rather high degree of constancy over the 4-year period.

For operation under relatively constant conditions, Eqs. (6-80), (6-81), and (6-84) predict relatively accurate values of S_e, X, and X_w for heterogeneous microbial populations. The growth system can be adequately described using four biochemical properties of the biomass, μ_{max}, K_s, Y_t, and k_d, and the systems can be controlled with respect to X, S_e, and X_w using three selectable engineering parameters, D (or $1/\bar{t}$), α, and X_R. Together these parameters determine μ_n or $1/\Theta_c$.

Although Eqs. (6-80), (6-81), and (6-84) can be used to design an activated sludge process (Gaudy et al., 1977), they are ideally suited for analysis of the system to predict the behavior of X, S_e, and X_w for various values of the biological constants and engineering parameters. When designing a process, the required S_e value is known or can be calculated from given or calculatable values for allowable organic loadings to the receiving stream. The approach is to calculate the μ_n corresponding to the required S_e, determine values of the biokinetic constants, then select the engineering parameters α and X_R, and determine the required volume of the reactor (or $\bar{t}$, V/F). The required net specific growth rate is given by Eq. (6-87) [from Eqs. (6-23) and (6-45)]:

$$\mu_n = \mu_{max}\frac{S_e}{K_s + S_e} - k_d \tag{6-87}$$

The mass balance equations for biomass and substrate concentrations can be solved simultaneously and expressed as shown in Eqs. (6-88) and (6-89):

$$X = \frac{Y_t[S_i - (1+\alpha)S_e]}{1 + k_d/\mu_n} + \alpha X_R \tag{6-88}$$

$$V = \frac{Y_t F[S_i - (1+\alpha)S_e] + X_R F}{k_d X} - \frac{(1+\alpha)F}{k_d} \tag{6-89}$$

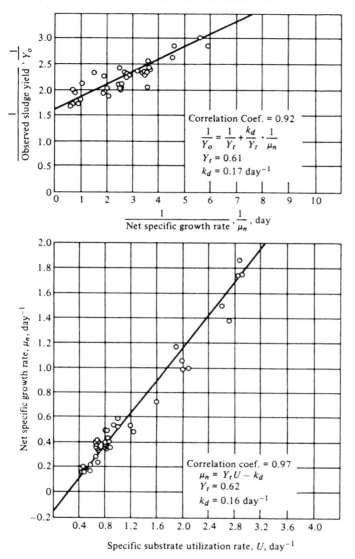

Figure 6-38 Plots of maintenance energy equations to determine true yield Y_t and maintenance coefficient k_d. (*Top*) Eq. (6-50). (*Bottom*) Eq. (6-54). (From Srinivasaraghavan and Gaudy, 1975, with additional unpublished data of M. Saleh.)

Substitution into Eq. (6-88) of the μ_n value calculated from Eq. (6-87) gives the value of X needed for substitution into Eq. (6-89). One can, with repetitive solutions of this equation and using a reasonably attainable range of values of α and X_R, arrive at the most desirable reactor volume that gives the required μ_n, i.e., S_e. For a more detailed discussion and stepwise solution of a numerical problem in design, see Gaudy and Kincannon, 1977a.

We have not dealt with any particular species of microorganisms but instead with a complicated mixture that has been treated more or less as a "species" representing the entire class of aerobic heterotrophs. However, it should be realized that this biomass includes species capable of anaerobic growth, both those that can substitute bound oxygen for molecular oxygen (the denitrifiers) and those that can grow without either (fermentative species) as well as the aerobic autotrophic species (nitrifiers). Whether we look at this biomass as a "species" of microorganisms, as an engineering material, or as a biochemical with which to control the biological purification of a wastewater, it should be realized that it is subject to change. These changes come about in two ways. First, because the biomass consists of many different species of living microorganisms, it is subject to the internal stresses resulting from the Darwinian forces of survival of the fittest and the dynamics of the food chain. Second, the various species of living cells of which the biomass is composed are subject, individually and collectively, to changes in response to variations in the external environment. Thus, for engineering purposes, the biomass is treated as an average species at the expense of some variability in the kinetic constants used to describe it.

It is important to note that the general approach and predictive equations given in this chapter can be applied to anaerobic systems, e.g., anaerobic treatment of strong wastes, anaerobic sludge digestion, and denitrification processes. They are also applicable to aerobic autotrophic growth, e.g., nitrification. The growth conditions and the values of the biokinetic constants will be quite different, but the same general growth model can be used to design the process and to predict system performance. Depending on the type of metabolism involved in the process, the diversity of species and the complexity of relationships between species may be greater or less than those for the aerobic treatment of organic wastes.

The predictive equations were formulated on the premise that a steady state would be maintained; i.e., $dX/dt \to 0$ and $dS/dt \to 0$. The available data indicate that in the main this assumption can be accepted as valid for heterogeneous populations. In biological treatment processes, maintenance of the steady state with respect to S_e is obviously more important than maintaining X in the steady state. In general, if S_i, F, temperature, and pH are held relatively constant, and X_R rather than c is used as a system control parameter, steadiness in X, S_e, and X_w will result. It must be remembered, however, that in the treatment plant F and S_i vary hourly, temperature generally is not controlled, and pH is seldom subject to control. We may properly ask: How good is the model and the system it describes when these environmental factors change? Solving this problem takes various forms. Can the equations be integrated in a convenient form for the transient rather than the steady state? If so, will they predict the actual response? Will the response be such that unacceptable amounts of effluent COD (S_e) will leak into the receiving stream before the system attains a new steady state? Will a new steady state be attained? Regardless of the equations and the responses they may or may not predict, how will the system actually respond? What magnitude of change or what types of change will engender a reasonably successful response and which will cause serious breakdown in the efficiency of the process? What combinations of environmental shock cause serious disruption of substrate utilization? For environmental shocks beyond those the system can successfully combat, what can be done in an engineering way to enhance the steadiness of effluent quality—e.g., equalization basins for leveling out of F and S_i, or add-on units to clean up the effluent after receiving excessive environmental stresses? Are there changes in operational procedures that could be used to combat the shock? Can the onset of a shock be assessed so there is advance warning?

The list of questions of a fundamental environmental microbiological nature and of an engineering nature can go on and on. More than two decades ago, a friend advised against studying biological treatment of waste because, he said, we already know about all we are going to know about it. Like the country boys in the song about Kansas City, "They have gone about as fer as they can go"! The point of the story is that, then as now, some have! But others have not! And it is the latter who will effect the needed technological control of the life-support system we often call the *environment*. Some of the questions just listed will be addressed in subsequent chapters.

PROBLEMS

6.1 (*a*) Calculate values for μ_{max} and K_s from the following pilot plant data for a system operating using Gaudy's model, with X_R as a control parameter.

$D = 0.08 \text{ h}^{-1}$	$D = 0.324 \text{ h}^{-1}$
$\alpha = 0.25$	$\alpha = 0.25$
$S_i = 2360 \text{ mg/L}$	$S_i = 2210 \text{ mg/L}$
$S_e = 10 \text{ mg/L}$	$S_e = 50 \text{ mg/L}$
$Y_t = 0.6$	$Y_t = 0.6$
$k_d = 0.005 \text{ h}^{-1}$	$k_d = 0.005 \text{ h}^{-1}$
$X_R = 7500 \text{ mg/L}$	$X_R = 7500 \text{ mg/L}$
$X = 2500 \text{ mg/L}$	$X = 2500 \text{ mg/L}$

(*b*) Suppose k_d was not previously known and you neglected it, i.e., you assumed $k_d = 0$, what effect would this have on your estimates of K_s and μ_{max}?

6.2 Assume you have an activated sludge pilot plant with an effective volume V in the reactor of 1.5 m³.

(*a*) If you operate the pilot plant under the following conditions, how much excess sludge X_w will you generate each day?

$\mu_{max} = 12 \text{ day}^{-1}$	$\alpha = 0.25$
$K_s = 125 \text{ mg/L}$	$X_R = 10{,}000 \text{ mg/L}$
$Y_t = 0.6$	$D = 0.125 \text{ h}^{-1}$
$k_d = 0.1 \text{ day}^{-1}$	$S_i = 500 \text{ mg/L}$

(*b*) Assume you now decide to run the process as an extended aeration system, i.e., no generation of excess sludge. At what detention time $\bar{t}$ should you run the pilot plant?

(*c*) Suppose after a time during operation as an extended aeration process, the k_d for the system changes to $0.13 \, \text{day}^{-1}$. If you wished to continue operation as an extended aeration process with α and X_R remaining the same as in parts (a) and (b), at what value of detention time $\bar{t}$ must you run the pilot plant?

(*d*) Suppose in part (*c*) you did not immediately realize that k_d had changed to $0.13 \, \text{day}^{-1}$. How would you first know that it had changed? What would be the steady state biomass concentration X if you did not change α or $\bar{t}$?

6.3 Assume you are going to operate an activated sludge pilot plant according to the model equations of Herbert. Use the following values for the biological constants and engineering parameters in your calculations.

$$\begin{aligned}
\mu_{\text{max}} &= 12 \, \text{day}^{-1} & \alpha &= 0.25 \\
K_s &= 75 \, \text{mg/L} & c &= 4.5 \\
Y_t &= 0.65 & D &= 3.0 \, \text{day}^{-1} \\
k_d &= 0.1 \, \text{day}^{-1} & V &= 2.0 \, \text{m}^3
\end{aligned}$$

(*a*) Calculate S_e and X for $S_i = 500 \, \text{mg/L}$ for $k_d = 0$, $k_d \neq 0$; and $S_R = 0$, $S_R \neq 0$.

(*b*) How does the assumption that $k_d \neq 0$ affect the sludge production?

(*c*) Increase S_i to $3000 \, \text{mg/L}$ and compare the required concentrations for X_R at this feed level with those for $S_i = 500 \, \text{mg/L}$. Discuss the practical feasibility of using c rather than X_R as an engineering control parameter.

6.4 (*a*) Calculate the effluent substrate concentration S_e using the equations according to Gaudy's model and the following information:

$$\begin{aligned}
\mu_{\text{max}} &= 10 \, \text{day}^{-1} & \alpha &= 0.25 \\
K_s &= 100 \, \text{mg/L} & X_R &= 10{,}000 \, \text{mg/L} \\
Y_t &= 0.6 & D &= 3 \, \text{day}^{-1} \\
k_d &= 0.1 \, \text{day}^{-1} & S_i &= 1500 \, \text{mg/L}
\end{aligned}$$

(*b*) Increase X_R to $15{,}000 \, \text{mg/L}$. To what extent will this improve the effluent quality (with respect to S_e)?

(*c*) Could you produce the same effluent quality as in part (*b*) by holding X_R at $10{,}000 \, \text{mg/L}$ and changing the value of α? What would α need to be to attain this level of S_e? What does this do to the actual $\bar{t}$ in the reactor, $\bar{t}_1$?

$$\bar{t}_1 = \frac{V}{F}\left(\frac{1}{1+\alpha}\right)$$

Would it seem best to attain the S_e of part (*b*) by changing X_R to $15{,}000 \, \text{mg/L}$ or by changing α?

6.5 The data in the accompanying table were obtained during a batch aerobic growth experiment on a nonstrippable substrate.

Time, h	Soluble COD, mg/L	Mixed liquor COD, mg/L
0	510	511
1	476	496
2	170	363
3	80	324
4	40	307
5	18	298
6	10	294

(*a*) Calculate the cell yield, assuming the empirical formula for microorganisms according to Porges and his coworkers is applicable (see Ch. 3).

(*b*) How much BOD has been exerted at 3 h?

6.6 Assume you have run a batch growth study using five concentrations of substrate S_0 and have generated the growth data shown in the accompanying table. Each system was started using an initial inoculation $X_0 = 10$ mg/L of cells previously acclimated to the substrate.

(a) Determine values of μ_{max} and K_s (if the data are adequate for such a determination; i.e., never assume a particular relationship exists but use your data to determine whether it does!).

(b) The substrate concentration in flask 5 was 150 mg/L at 10 h; calculate the cell yield.

(c) Calculate the cell concentration X at 3 h for $S_0 = 500$ mg/L and $X_0 = 5$ mg/L. How much substrate will remain in solution at this time?

Biomass concentration X in mg/L at time and S_0 indicated

Time, h	$S_0 = 100$	200	400	600	1000
0	10	10	10	10	10
1	13	14	15	15	16
2	16	20	22	24	25
4	27	38	50	56	62
6	45	74	110	131	153
8	50	90	200	308	380
10	50	90	210	300	520
12	50	—	—	—	—

6.7 (a) Determine the volume of the activated sludge tank required and the excess sludge production using Gaudy's model (X_R as a control parameter) for a wastewater flow $F = 1$ mgd (million gal/day) under the following conditions:

$$\mu_{max} = 10 \text{ day}^{-1} \qquad k_d = 0.1 \text{ day}^{-1}$$
$$K_s = 75 \text{ mg/L} \qquad \alpha = 0.25$$
$$Y_t = 0.6 \qquad X_R = 10,000 \text{ mg/L}$$

Organic loading: $\quad S_i = 500$ mg/L BOD_5 (5-day BOD)
Effluent standard: $\quad$ total $BOD_5 = 20$ mg/L
$\qquad\qquad\qquad\qquad$ suspended solids $X = 20$ mg/L

(b) Describe what calculations you might make to determine the optimum α and X_R combination to use.

(c) If you were designing an activated sludge process, would you make provision for the use of one α and one X_R or a range of these parameters? Give reasons for your decision.

(d) With reference to the plant designed in part (a), calculate the effect on S_e if the inflow F was reduced to 0.5 mgd but the sludge recycle flow F_R remained the same as in part (a). Calculate the effect on S_e if F was changed to 1.5 mgd and F_R remained the same as in part (a).

6.8 The accompanying data were obtained during batch growth studies. The values listed under the heading Δt represent the duration of the straight-line portion of a semilogarithmic plot of suspended solids (biomass) concentration X versus t, and the initial and final values for X represent those over the stated time interval for growth on the given S_0. Determine whether there is a relationship between the exponential specific growth rate μ and the initial substrate concentration. If the Monod equation looks like a reasonable possibility, determine the numerical values of μ_{max} and K_s, and show that the equation describes the experimental data adequately.

Growth data

S_0, mg/L	X_0, mg/L	X_t, mg/L	Δt, h
100	19	63	7.5
200	20	101	7.0
300	18	167	8.25
400	17	112	6.5
500	17	300	9.25
600	16	207	8.0
800	15	270	8.5
1000	21	759	10.25

6.9 The accompanying graph shows the results of a batch experiment in which the initial soluble COD of the substrate was 1000 mg/L. As the heterogeneous microbial population grew during the experiment, a small sample was diluted and the concentration was measured (see biological solids concentration X). Also, the oxygen uptake was measured during the experiment.

    (*a*) What is the numerical value of ΔCOD? What does ΔCOD represent?

    (*b*) What is the biochemical oxygen demand BOD at the end of the experiment?

    (*c*) Where does the endogenous phase begin in this graph?

    (*d*) Note the shape of the oxygen uptake curve. Can you suggest a possible reason why the oxygen uptake curve should develop a pause between phases. It is not an experimental error, nor is it due to nitrification.

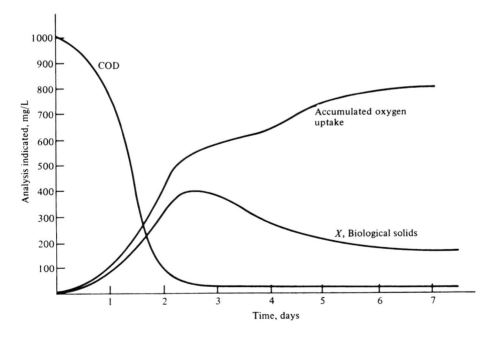

6.10 The accompanying data list shows average steady state results obtained during a pilot plant study on a biodegradable wastewater. S_i and S_e values are given in terms of soluble COD, and you have previously determined that the waste is highly biodegradable; i.e., there is little or no nonbiodegradable COD_i.

S_i, mg/L	S_e, mg/L	X_e, mg/L	X, mg/L	Excess sludge X_R, mg/L	F_w, L/day	X_w/F_w, mg/L/day	V, L	F, L/day
500	4	10	1129	5000	0.26	14,000	8.0	24.0
1000	6	18	1283	5000	0.60	13,500	8.0	24.0
2000	11	26	1593	5000	1.25	13,800	8.0	24.0
500	7	21	1142	5000	0.48	15,600	8.0	48.0
1000	13	30	1297	5000	1.36	12,100	8.0	48.0
2000	24	46	1608	5000	2.78	12,400	8.0	48.0
2000	41	88	1109	2500	2.43	13,400	8.0	48.0

From these data calculate the biological constants μ_{max}, K_s, Y_t, and k_d. Also find the volume of the reactor that is required to treat a waste with $S_i = 1000$ mg/L and a flow $F = 1$ mgd. Set α at 0.25 and X_R at 10,000 mg/L. The local effluent standard requires that the total COD of the effluent can be 30 mg/L, and the suspended solids concentration cannot exceed 15 mg/L. Size the aerator for a flow of 1 mgd. Also predict the amount of excess sludge in dry weight per day. State the basis for and comment on any assumptions or judgments you need to make in completing this design assignment.

6.11 The accompanying sketch shows a plot of the calculated substrate concentration S_e and biomass concentration X in a once-through reactor at different dilution rates D, from slow ones, e.g., $\frac{1}{48}$ h^{-1}, to very rapid ones at which cells and substrate dilute out of the system. Neglect k_d and assume that the values for μ_{max}, K_s, and Y_t are 0.5, 100, and 0.6, respectively.

(a) Calculate and plot the dilute-out curves for X and S using these constants; then calculate and plot the dilute-out curves for other values given in parts (b), (c) and (d).

(b) Change the cell yield Y_t to 7.5, and show graphically the effect of higher yield.

(c) Change μ_{max} to 0.7 h^{-1} and graph the effect on the dilute-out curves.

(d) Change K_s to 450 mg/L and show graphically the effect of increased K_s.

(e) Calculate the critical dilution rate D_c for parts (a), (b), (c), and (d).

(f) Compare and discuss the curves, i.e., the effect the values of the constants have on the curves.

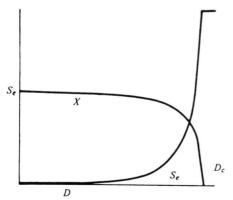

Dilute-out curve for a completely mixed once-through growth reactor

6.12 Calculate the net specific growth rate μ_n and sludge age Θ_c for the data shown on the pilot plant flow diagram.

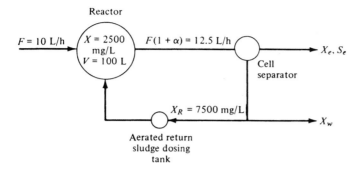

Reactor

$F = 10$ L/h $\quad X = 2500$ mg/L $\quad V = 100$ L $\quad F(1 + \alpha) = 12.5$ L/h $\quad$ Cell separator $\quad X_e, S_e$

$X_R = 7500$ mg/L $\quad X_w$

Aerated return sludge dosing tank

REFERENCES AND SUGGESTED READING

American Public Health Association. 1976. Standard methods for the examination of water and wastewater, 14th ed. APHA, Washington, D.C.

Bonotan, F. M., and P. Y. Yang. 1976. The application of constant recycle solids concentration in activated sludge. *Biotechnol. Bioeng.* **18**:145–165.

Chiu, S. Y., L. T. Fan, I. C. Kao, and L. E. Erickson. 1972a. Kinetic behavior of mixed populations of activated sludge. *Biotechnol. Bioeng.* **14**:179–199.

Chiu, S. Y., L. E. Erickson, L. T. Fan, and I. C. Kao. 1972b. Kinetic model identification in mixed populations using continuous culture data. *Biotechnol. Bioeng.* **14**:207–231.

Churchill, M. A., H. L. Elmore, and R. A. Buckingham. 1962. Prediction of stream reaeration rates. *J. Sanitary Eng. Div. ASCE* **88**:1–46.

Contois, D. E. 1959. Kinetics of bacterial growth: Relationship between population density and specific growth rate in continuous culture. *J. Gen. Microbiol.* **21**:40–50.

Fujimoto, Y. 1963. Kinetics of microbial growth and substrate consumption. *J. Theor. Biol.* **5**:171–191.

Gaudy, A. F. Jr. 1975. Prediction of assimilation capacity in receiving streams. *Water and Sewage Works*, Part I **122**:62–65, Part II **122**:78–79.

Gaudy, A. F., Jr., F. Abu-Niaaj, and E. T. Gaudy. 1963. Statistical study of the spot plate technique for viable cell counts. *Appl. Microbiol.* **11**:305–309.

Gaudy, A. F. Jr., and E. T. Gaudy. 1971. Biological concepts for design and operation of the activated sludge process. U.S. Environmental Protection Agency Report No. 17090 FQJ 09/71. U.S. Government Printing Office, Washington, D.C.

Gaudy, A. F. Jr., and E. T. Gaudy. 1972. ΔCOD gets nod over BOD test. *Ind. Water Eng.* **9**:30–34.

Gaudy, A. F. Jr., and D. F. Kincannon. 1977a. Functional design of activated sludge processes. *Water and Sewage Works* **124**:76–81.

Gaudy, A. F. Jr., and D. F. Kincannon. 1977b. Comparison of design models for activated sludge. *Water and Sewage Works* **123**:66–70.

Gaudy, A. F. Jr., and T. Manickam. 1978. Some effects of influent COD on effluent COD. Proceedings, National Conference on Environmental Engineering, pp. 397–404. American Society of Civil Engineers, New York.

Gaudy, A. F. Jr., A. Obayashi, and E. T. Gaudy. 1971. Control of growth rate by initial substrate concentration at values below maximum rate. *Appl. Microbiol.* **22**:1041–1047.

Gaudy, A. F. Jr., and M. Ramanathan. 1971. Variability in cell yield for heterogeneous microbial populations of sewage origin grown on glucose. *Biotechnol. Bioeng.* **13**:113–123.

Gaudy, A. F. Jr., M. Ramanathan, and B. S. Rao. 1967. Kinetic behavior of heterogeneous populations in completely mixed reactors. *Biotechnol. Bioeng.* 9:387–411.

Gaudy, A. F. Jr., and R. Srinivasaraghavan. 1974. Experimental studies on kinetic model for design and operation of activated sludge processes. *Biotechnol. Bioeng.* 16:723–738.

Gaudy, A. F. Jr., R. Srinivasaraghavan, and M. Saleh. 1977. Conceptual model for design and operation of activated sludge processes. *J. Environ. Eng. Div. ASCE* 103:71–84.

Gaudy, A. F. Jr., P. Y. Yang, R. Bustamante, and E. T. Gaudy. 1973. Exponential growth in systems limited by substrate concentration. *Biotechnology. Bioeng.* 15:589–596.

Grady, C. P. L. Jr., L. J. Harlow, and R. R. Riesing. 1972. Effects of growth rate and influent substrate concentration on effluent quality from chemostats containing bacteria in pure and mixed culture. *Biotechnol. Bioeng.* 14:391–410.

Grady, C. P. L. Jr., and D. R. Williams. 1975. Effects of influent substrate concentration on the kinetics of natural microbial populations in continuous culture. *Water Res.* 9:171–180.

Herbert, D. 1956. A theoretical analysis of continuous culture systems. *Soc. Chem. Ind.* Monograph No. 12:21–53.

Herbert, D., R. Elsworth, and R. C. Telling. 1956. The continuous culture of bacteria—a theoretical and experimental study. *J. Gen. Microbiol.* 14:601–622.

Hiser, L. L., and A. W. Busch. 1964. An eight-hour biological oxygen demand test using mass culture aeration and COD. *J. Water Pollut. Control Fed.* 36:505–516.

Isaacs, W. P., and A. F. Gaudy, Jr. 1968. Atmospheric oxygenation in a simulated stream. *J. Sanitary Eng. Div. ASCE* 94:319–344.

Komolrit, K., and A. F. Gaudy, Jr. 1966. Biochemical response of continuous flow activated sludge processes to qualitative shock loadings. *J. Water Pollut. Control Fed.* 38:85–101.

Krenkel, P. A., and G. T. Orlob. 1962. Turbulent diffusion and the reaeration coefficient. *J. Sanitary Eng. Div. ASCE* 88:53–83.

Manickam, T., and A. F. Gaudy, Jr. 1978. Response to qualitative-quantitative shock of activated sludge operated at constant sludge concentration, Proceedings 33d Purdue Industrial Waste Conference. Purdue University, Lafayette, Ind.

Monod, J. 1942. Recherches sur la croissance des cultures bacteriennes. Hermann et Sie, Paris.

Monod, J. 1950. La technique de culture continue. Theorie et applications. *Ann. Inst. Pasteur* 79:390–410.

Nemerow, N. L. 1974. Scientific stream pollution analysis. McGraw-Hill, New York.

Novick, A., and L. Szilard. 1950. Description of the chemostat. *Science* 112:715–716.

O'Connor, D. E., and W. E. Dobbins. 1958. The mechanics of reaeration of natural streams. *Trans. ASCE* 123:641–684.

Peil, K. M., and A. F. Gaudy, Jr. 1971. Kinetic constants for aerobic growth of microbial populations selected with various single compounds and with municipal wastes as substrates. *Appl. Microbiol.* 21:253–256.

Peil, K. M., and A. F. Gaudy, Jr. 1975. A rational approach for predicting the DO profile in receiving waters. *Biotechnol. Bioeng.* 17:69–84.

Radhakrishnan, I., and A. K. SinhaRay. 1974. Activated sludge studies with phenol bacteria. *J. Water Pollut. Control Fed.* 46:2393–2418.

Ramanathan, M., and A. F. Gaudy, Jr. 1969. Effect of high substrate concentration and cell feedback on kinetic behavior of heterogeneous populations in completely mixed systems. *Biotechnol. Bioeng.* 11:207–237.

Ramanathan, M., and A. F. Gaudy, Jr. 1971. Steady state model for activated sludge with constant recycle sludge concentration. *Biotechnol. Bioeng.* 13:125–145.

Ramanathan, M., and A. F. Gaudy, Jr. 1972. Studies on sludge yield in aerobic systems. *J. Water Pollut. Control Fed.* 44:441–450.

Saleh, M. M., and A. F. Gaudy, Jr. 1978. Shock load response of activated sludge operated at constant recycle sludge concentration. *J. Water Pollut. Control Fed.* 50:764–774.

Srinivasaraghavan, R., and A. F. Gaudy, Jr. 1975. Operational performance of an activated sludge process with constant sludge feedback. *J. Water Pollut. Control Fed.* 47:1946–1960.

Streeter, H. W. 1926. The rate of atmospheric reaeration of sewage polluted streams. Reprint 1063, Public Health Reports. U.S. Government Printing Office, Washington, D.C.

Streeter, H. W., and E. B. Phelps. 1925. A study of the pollution and natural purification of the Ohio River. Bulletin No. 146, U.S. Public Health Service. U.S. Government Printing Office, Washington, D.C.

Symons, J. M., R. E. McKinney, and H. H. Hassis. 1960. A procedure for determination of the biological treatability of industrial wastes. *J. Water Pollut. Control Fed.* 32:841–852.

Tempest, D. W. 1968. The continuous culture of mirroorganisms. *Methods Microbiol.* 2:259–279.

Thackston, E. L., and P. A. Krenkel. 1966. Longitudinal mixing and reaeration in natural streams, Technical Report No. 7. Sanitary and Water Resources Engineering, Department of Civil Engineering, Vanderbilt University, Nashville, Tenn.

Tsiviglou, E. C., R. L. O'Connell, C. M. Walter, P. J. Codsil, and G. S. Logsdon. 1965. Tracer measurements of atmospheric reaeration—laboratory studies. *J. Water Pollut. Control Fed.* 37:1343–1362.

Umbreit, W. W., R. H. Burris, and J. F. Stauffer. 1964. Manometric techniques, 5th ed. Burgess, Minneapolis.

ENERGY GENERATION AND UTILIZATION IN BIOLOGICAL SYSTEMS

The ultimate source of energy for all life on earth is the sun, although only photosynthetic organisms—the green plants, algae, cyanobacteria, and a few other genera of bacteria—can benefit directly from radiant energy. The organisms that carry out "plant-type" photosynthesis convert carbon dioxide to organic compounds that serve as a source of chemical energy for most other living organisms and, in the process, produce oxygen. Thus, the two essential ingredients of the energy-yielding systems of animals and many microorganisms, oxygen and an organic energy source, are products of photosynthesis, i.e., synthesis powered by light.

In Chap. 1 we examined several cycles of vital importance to the maintenance of our life-support system. In this chapter we shall see that the carbon cycle is intimately related to the generation of energy in biological systems. In carbon dioxide, carbon is in its most oxidized state, and the energy required to reduce it to the level of carbohydrate is furnished by sunlight. That energy is released when carbon is again oxidized to carbon dioxide by organisms that utilize chemical energy. The two possible sources of energy for living organisms, radiant energy and chemical energy, are thus not independent since, were it not for the continual input of energy from the sun, the chemical energy available on earth, at least in forms usable by humans, would be rapidly exhausted. The degradative leg of the carbon cycle is equally important since the concentration of carbon dioxide in the atmosphere is very low; the entire amount available in the earth's atmosphere has been estimated to be sufficient to sustain photosynthesis at the present rate for only about 2 years if oxidation of food material to carbon dioxide ceased.

All living organisms utilize a common form of energy storage. Whatever the energy source, energy derived from that source is stored as chemical energy in the form of ATP, adenosine triphosphate (see Fig. 3-40). A major portion of this chapter will be devoted to the mechanisms by which ATP is formed. Three such mechanisms exist—*substrate level phosphorylation, oxidative phosphorylation*, and *photophosphorylation*. We shall examine each of these as they are utilized by microorganisms, but first a brief discussion of the thermodynamic principles involved in the release and utilization of energy is necessary for understanding the role played by ATP in the living cell. For a more complete discussion of thermodynamics, the reader should consult one of the references listed at the end of the chapter.

THERMODYNAMIC PRINCIPLES

Energy may be defined as the capacity for doing work. Thermodynamics deals with energy in its various forms, thermal, chemical, and electric, and the laws governing energy transformations in chemical or physical processes.

The work for which microorganisms require energy is related primarily to the production of new cells. A minor requirement for some microorganisms is the mechanical work of movement, the amoeboid or gliding movement of the cell or the movement of flagella or cilia, but except where motion is necessary for capturing food, it is a luxury that can be dispensed with by many motile microorganisms without affecting viability or reproduction. Work is also required to transport materials through the cytoplasmic membrane or, in eucaryotes, through membranes of organelles or from one part of the cell to another. These are minor requirements, however, compared with the work of biosynthesis. Synthetic reactions proceed only when energy is supplied from an external source. In nonbiological systems, energy is often supplied in the form of heat, but biological reactions must occur at the relatively low temperatures compatible with life. As we shall see later, the major requirement for energy in the microbial cell is for polymerization reactions, i.e., the formation of proteins from amino acids, nucleic acids from nucleoside triphosphates, etc. The work of chemical bond formation is performed through the input of chemical energy derived from other, energy-yielding reactions. The compound that mediates this transfer of energy is ATP.

The laws of thermodynamics allow us to predict which reactions will produce energy (exergonic reactions) and which will consume energy (endergonic reactions) and to calculate the amounts of energy produced or consumed. The three laws of thermodynamics may be stated simply as follows:

1. Energy can be neither created nor destroyed; i.e., the total amount of energy in nature is constant.
2. The total amount of entropy in nature is constantly increasing; i.e., spontaneous reactions tend toward increased randomness (entropy).

3. At absolute zero (0 K or $-273.16°C$) there is no molecular motion, i.e., entropy is zero.

Free Energy

We have defined energy as the capacity for doing work, but it is the actual amount of work that can be accomplished by the energy input to any system that is important. Consideration of the first two laws of thermodynamics led Gibbs and Helmholtz to formulate an expression that defines useful energy, that is, the energy available to do work. When a chemical reaction occurs under isothermal conditions, as is true of biological reactions, part of the energy released is lost because of the tendency of entropy to increase, and the remainder is useful energy, or free energy. The change in free energy ΔG was given as Eq. (3-1) in Chap. 3:

$$\Delta G = \Delta H - T \Delta S$$

where ΔH is the change in enthalpy (or internal energy), T is the absolute temperature at which the reaction takes place, and ΔS is the change in entropy. The units of ΔG are calories per mole. A calorie is defined as the amount of heat energy required to raise the temperature of one gram of pure water from 14.5 to 15.5°C at atmospheric pressure. (The symbol ΔF has exactly the same meaning as ΔG.) The standard free-energy change $\Delta G°$ is defined as the value of ΔG for the reaction under standard conditions, when all reactants are present at initial concentrations of 1 mol/L.

Only reactions for which ΔG has a negative value proceed spontaneously, since ΔG is a measure of the change in the amount of useful energy in the system. If the useful energy is to increase, there must be an inflow of energy from an external source; i.e., the reaction will not occur spontaneously. Thus, the direction of change in the amount of free energy determines whether a reaction can occur, and the magnitude of the change determines the degree to which the reaction will proceed toward completion.

Measurement of Free-Energy Change

The relationship between ΔG and the extent to which a reaction proceeds is an important one and provides one means of measuring ΔG. A reversible reaction

$$A + B \rightleftharpoons C + D$$

will proceed from left to right until equilibrium is reached. At equilibrium, $\Delta G = 0$ and entropy is at its maximum value. The equilibrium constant for the reaction, K_{eq}, is the ratio of the product of the concentrations of the reaction products to the product of the concentrations of the reactants at equilibrium:

$$K_{eq} = \frac{[C] \times [D]}{[A] \times [B]} \tag{7-1}$$

The standard free-energy change of the reaction is related to the equilibrium constant because it is the change in free energy that drives the reaction:

$$\Delta G^\circ = -RT \ln K_{eq} \qquad (7\text{-}2)$$

where R is the gas constant [1.987 cal/(mol)(deg)]. In biological systems, molar concentrations of reactants are not expected and the actual free-energy change ΔG is related to the standard free-energy change by the equation

$$\Delta G = \Delta G^\circ + RT \ln K_{eq} \qquad (7\text{-}3)$$

The value of ΔG under physiological conditions may be quite different from ΔG°, but since the exact conditions under which reactions proceed inside the cell are impossible to determine, values of ΔG° are useful as standard reference points in comparing the free-energy changes of different reactions and in determining whether a reaction is thermodynamically possible.

The fact that a large negative value of free-energy change is associated with a reaction has no influence on the kinetics of the reaction, i.e., the rate at which the reaction proceeds. Such a reaction may proceed almost to completion but at so slow a rate that for practical purposes it may be considered not to occur at all. In nonbiological systems, heat or catalysts such as finely divided metals may be used to increase the rate of reaction. In biological systems, enzymes perform the catalytic function of activating the reactant molecules to increase the reaction rate. Catalysts, whether biological or nonbiological, cannot influence the equilibrium of a reaction nor can they cause a reaction that would not occur spontaneously, albeit at an infinitely slow rate, to take place. The laws of thermodynamics determine whether a reaction can occur and the extent to which it will proceed but do not predict the rate of the reaction, whereas catalysts, including enzymes, can influence only the rate.

If sufficiently accurate methods for determining the concentrations of reactants and products are available, a reaction may be allowed to proceed until equilibrium is attained, i.e., until no further reaction occurs, with or without a catalyst, and the value of K_{eq} is determined, from which the standard free-energy change can be calculated using Eq. (7-2). If the reaction is carried out at pH 7.0, the standard free-energy change is represented by the symbol $\Delta G'$. It is customary when using a pH other than 7.0 for biological reactions to specify the pH at which the determination was made. It should be noted that the value of K_{eq} is theoretically based on activities rather than concentrations, but these are generally assumed to be equivalent.

Other methods for determining the value of ΔG° are available (Casey, 1962; Mahler and Cordes, 1966). One is of particular interest because it is based on the energy released during oxidative reactions, and this type of reaction is the source of energy for the living cell. Oxidations in biological systems very seldom involve the direct participation of oxygen but are accomplished instead by the removal of electrons. The transfer of electrons generates an electric current, which can be measured as electric energy. The equivalence of electric and chemical energy is stated by the first law of thermodynamics and may be expressed by the equation:

$$\Delta G^\circ = -nF \, \Delta E_0 \qquad (7\text{-}4)$$

where n is the number of electrons involved in the reaction, and F is the faraday (96,500 coulombs, i.e., the number required to reduce or oxidize 1 mol). ΔE_0 is the difference in electrode potentials (in volts) between two reacting systems, oxidized reactant–reduced product and reduced reactant–oxidized product. E_0 is the oxidation-reduction (redox) potential of a system in which the activities of oxidized and reduced forms are equal, measured with reference to the standard hydrogen electrode, which is arbitrarily assigned a redox value of 0. The subscript m (E_m) is sometimes used to indicate that the measurement was made using equal concentrations of oxidized and reduced forms. E_0 measured at pH 7 is designated by the symbol E_0'. Values of E_0 for biologically important redox systems have been measured and are useful in determining the potential of one system for reducing another. A system with a larger negative value will reduce (transfer electrons to) any system with a less negative value or a positive value of E_0, and the amount of free energy released by the reaction is proportional to the difference in the E_0 values. Larger differences in E_0, then, indicate a greater tendency to proceed toward completion of the reaction, i.e., higher values of K_{eq}, since, from Eqs. (7-2) and (7-4), $\Delta G = -nF \, \Delta E_0 = -RT \ln K_{eq}$.

COUPLED REACTIONS

We have discussed thermodynamic principles as applied to individual, isolated reactions, and determinations of thermodynamic constants must be made with such reactions. However, in biological systems, reactions never proceed in isolation. As we shall see in succeeding chapters, reactions in the living cell are consecutive, proceeding from an initial substrate to a final product or products, and such a series of consecutive reactions constitutes a metabolic pathway. Two consecutive reactions, or a series of reactions, are considered to be coupled since the product of the first reaction is a reactant in the following reaction. A thermodynamically unfavorable reaction may therefore be "pulled" by a following reaction that removes the product and thus displaces the equilibrium to the right. In coupled reactions in which A is converted to B, B to C, and C to D,

$$A \rightarrow B \rightarrow C \rightarrow D$$

the reaction $B \rightarrow C$ may proceed to only a slight extent or not at all in isolation, but the removal of C by its conversion to D in the coupled reaction allows the conversion of B to C to proceed to completion. This is simply another way of saying that if a series of reactions is to occur spontaneously, it is only necessary that the overall ΔG, the algebraic sum of the ΔG values for the individual reactions, be negative.

It is not necessary that reactions occur as parts of the same reaction sequence or metabolic pathway in order to be coupled. In fact, the entire concept of the utilization of energy released from the energy source in catabolic reactions to drive the energy-requiring anabolic, or biosynthetic, reactions is based on a different type of coupling of reactions. This type involves reactions that utilize a common "carrier" molecule. The most important of such carrier molecules are the pyridine nucleotides, NAD (see Fig. 3-40) and NADP, which couple oxidative and reductive reactions by acting as common electron carriers, and ATP, which acts as a phosphate carrier and transfers energy from exergonic to endergonic reactions. These molecules are coenzymes; that is, they participate in large numbers of different enzymatic reactions and thus link otherwise unrelated reactions so that thermodynamically unfavorable reactions become thermodynamically possible due to an overall drop in the amount of free energy.

ATP and the Transfer of Energy

ATP is ideally suited for its role in the transfer of energy for two reasons. First, it is a so-called energy-rich compound. It contains three phosphate groups in a linear arrangement; the first is joined to ribose by an ester linkage, and the bonds connecting the phosphates to each other are anhydride bonds, that is, bonds formed by removing water from two molecules of phosphoric acid,

$$HO-\underset{\underset{OH}{|}}{\overset{\overset{O}{\|}}{P}}-OH + HO-\underset{\underset{OH}{|}}{\overset{\overset{O}{\|}}{P}}-OH \xrightarrow{\overset{H_2O}{\diagup}} HO-\underset{\underset{OH}{|}}{\overset{\overset{O}{\|}}{P}}-O-\underset{\underset{OH}{|}}{\overset{\overset{O}{\|}}{P}}-OH \qquad (7\text{-}5)$$

Acid anhydrides are unstable, highly reactive compounds, and these bonds are often called *high-energy* bonds and are represented by the symbol $\sim$. ATP can be represented in biochemical shorthand notation as

$$\text{Adenine–ribose–P} \sim \text{P} \sim \text{P}$$

Hydrolysis of either of the high-energy bonds is accompanied by the release of a large amount of energy, which can be used to form a phosphorylated compound of lower energy. In other words, ATP has a tendency to donate phosphate to other compounds [see Mahler and Cordes (1966) for a complete explanation of the high-energy bond]. The second reason for the role of ATP as the cell's "mobile power supply" (Casey, 1962) is the ability of ATP to accept phosphate groups from other phosphorylated compounds of higher energy level to form ATP. The intermediate position of ATP between the energy-rich and energy-poor compounds makes it the ideal phosphate carrier. Table 7-1 shows the free energies of hydrolysis of a number of biologically

Table 7-1 Free energies of hydrolysis of some biologically important phosphate compounds

Compound	$\Delta G'$, kcal/mol
Phosphoenolpyruvate	−14.8
1,3-Diphosphoglycerate	−11.8
Acetylphosphate	−10.1
ATP	−7.3
Glucose 1-phosphate	−5.0
Fructose 6-phosphate	−3.8
Glucose 6-phosphate	−3.3
3-Phosphoglycerate	−2.4
Glycerol 1-phosphate	−2.2

Source: Data from Lehninger, 1971.

important phosphorylated molecules. The free energy of hydrolysis of ATP is approximately 7000 cal/mol at pH 7.0 under standard conditions,* and this is arbitrarily defined as the dividing line between high- and low-energy bonds. The phosphate esters, such as glucose 6-phosphate, are relatively stable compounds with a much lower tendency to donate their phosphate groups, whereas other phosphorylated metabolic intermediates have higher phosphate transfer potentials and are able to donate phosphate to ADP to form ATP. The importance of this gradient of phosphate transfer potential will become apparent in our later discussion of metabolic pathways. It should be mentioned here that the values in Table 7-1 as well as other values of $\Delta G°$ or $\Delta G'$ are subject to some degree of uncertainty. Values obtained by different methods (or by different experimenters) vary by several hundred calories. Also, as mentioned previously, these values are certainly not representative of the actual free-energy change under physiological conditions. Various estimates of ΔG for ATP hydrolysis in the cell indicate that the value is probably closer to 12,000 cal/mol. The other values given are, of course, in need of the same sort of adjustment.

In Chap. 3, we used a hypothetical example to illustrate the function of ATP in coupling exergonic and endergonic reactions. Using the values of $\Delta G'$ in Table 7-1, we can now examine two actual reactions coupled through ATP as a common intermediate. Acetylphosphate can be formed as a product of glucose metabolism, with the energy required for its formation being supplied

---

*The free energy of hydrolysis is a convenient measure of the useful energy released when a bond is broken, e.g., in the reaction $ATP + H_2O \rightarrow ADP + H_3PO_4(Pi)$. Actual hydrolysis of the bond would result in wasting the energy as heat, but free energies of hydrolysis are useful for comparison.

by the oxidative decarboxylation of pyruvic acid (see Chap. 9). The energy thus stored can be transferred through ATP and used in the formation of α-glycerol phosphate, which is required for the synthesis of certain lipids. Since acetylphosphate has a higher free energy of hydrolysis than does ATP, it can transfer phosphate to ADP to form ATP.

$$\text{Acetylphosphate} + H_2O \rightarrow \text{acetate} + H_3PO_4 \quad \Delta G' = -10.1 \text{ kcal/mol}$$

$$ADP + H_3PO_4 \rightarrow ATP + H_2O \quad \Delta G' = +7.3 \text{ kcal/mol}$$

Sum: Acetylphosphate + ADP →.acetate + ATP $\Delta G' = -2.8 \text{ kcal/mol}$

The phosphate group can then be transferred from ATP to glycerol, since the free energy of hydrolysis of ATP is greater than that of α-glycerol phosphate.

$$\text{Glycerol} + H_3PO_4 \rightarrow \alpha\text{-glycerol phosphate} + H_2O \quad \Delta G' = +2.2 \text{ kcal/mol}$$

$$ATP + H_2O \rightarrow ADP + H_3PO_4 \quad \Delta G' = -7.3 \text{ kcal/mol}$$

Sum: Glycerol + ATP → α-glycerol phosphate + ADP $\Delta G' = -5.1 \text{ kcal/mol}$

The free-energy change for the overall reaction is the algebraic sum of the individual values of $\Delta G'$, $-2.8 + (-5.1) = -7.9 \text{ kcal/mol}$:

Acetylphosphate + glycerol → acetate + α-glycerol phosphate

$$\Delta G' = -7.9 \text{ kcal/mol}$$

Acetylphosphate and glycerol do not, of course, react directly, but the energy captured in the energy-rich bond of acetylphosphate drives the endergonic reaction of synthesis of α-glycerol phosphate by coupling the two reactions through ATP, which serves as a cofactor for both reactions. Such coupled reactions are often represented as follows:

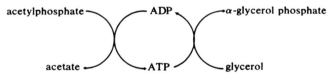

FORMATION OF ATP

As stated previously, the three mechanisms by which ATP is formed are substrate level phosphorylation, oxidative phosphorylation, and photophosphorylation. These are associated with different modes of life. *Substrate level phosphorylation* occurs in all organisms that utilize organic compounds as energy source. However, in aerobic organisms it contributes only a minor fraction of the total ATP synthesized, whereas in almost all anaerobic organisms, i.e., in organisms whose metabolism is fermentative, substrate level phosphorylation is the only known mechanism for ATP formation.

Oxidative phosphorylation occurs in organisms that have respiratory metabolism, either aerobic or anaerobic. The oxidation of reduced electron carriers involves passage of the electrons through an organized electron transport system and finally to a terminal electron acceptor. The formation of ATP is coupled to the transfer of electrons. *Photophosphorylation* is quite similar to oxidative phosphorylation in that formation of ATP is coupled to the transfer of electrons through an electron transport system. Both oxidative phosphorylation and photophosphorylation are placed in the single category of *electron transport phosphorylation* by some authors. In this chapter we shall be concerned primarily with the reactions and cellular structures specifically involved in ATP synthesis. In later chapters, we will consider the relationship of these ATP-forming reactions to the overall metabolic activity of the cell.

As we have stated previously, energy is released in oxidation-reduction reactions and, in biological systems, oxidation occurs through the removal of electrons. When an organic compound is oxidized, the reaction usually involves the simultaneous loss of two protons, or H^+ ions, so that the oxidation actually is a dehydrogenation, i.e., the removal from the molecule being oxidized of two atoms of hydrogen, 2H; the convenient shorthand notation −2H for such a reaction is often used. Conversely, reduction usually involves the addition of 2H. For example, the reversible reaction in which acetaldehyde is reduced to ethanol is often written as

$$\underset{\text{acetaldehyde}}{CH_3—\overset{\overset{\displaystyle O}{\|}}{CH}} \underset{-2H}{\overset{+2H}{\rightleftharpoons}} \underset{\text{ethanol}}{CH_3CH_2OH} \tag{7-6}$$

It is understood that the two hydrogens removed from the organic substrate are used to reduce a primary electron acceptor, i.e., an *electron carrier* such as NAD. As discussed above, one of the important ways in which reactions can be coupled is through the reduction and oxidation of such electron carriers, using a variety of organic or inorganic compounds as donors and acceptors of electrons. The importance of coupled oxidation-reduction reactions is twofold. Many such reactions, as we shall see later, are essential for the generation of ATP. Of equal importance is the fact that reduced electron carriers must be reoxidized if the metabolic activity of the cell is to continue. The accepting of 2H by NAD in an oxidative reaction is a temporary expedient—a permanent acceptor for the hydrogen must be found so that NAD can continue to be available to accept additional hydrogen. This is a particularly important problem for the organism in fermentation, where no net oxidation of the energy source can occur (see Chap. 11).

In all energy-yielding metabolism, then, electron carriers are reduced and must be reoxidized by passing electrons to a *terminal electron acceptor*. The difference between fermentative and respiratory metabolism lies in the means by which this is accomplished. In fermentation, an *organic* compound, usually a metabolic product formed from the energy source, serves as the terminal electron acceptor. In respiration, an *inorganic* compound is the terminal electron acceptor. In aerobic respiration, the terminal electron acceptor is oxygen, whereas in anaerobic respiration it is most often nitrate or sulfate ion. Some of the reduced NAD may be reoxidized by acting as an electron donor in reductive steps of biosynthetic pathways, either directly or indirectly. We will discuss this in later chapters, but it is important to realize that this use of reduced electron carriers in cells with respiratory metabolism represents an energy cost to the cell, since the electrons so used are not available for electron transport and the coupled synthesis of ATP.

Substrate Level Phosphorylation

In very early studies of the steps involved in the fermentation of glucose to alcohol and carbon dioxide by yeast, it was discovered that glucose is phosphorylated to form the ester glucose 6-phosphate before being further metabolized. Complete unraveling of the pathway revealed that almost all the intermediate compounds formed in converting glucose to alcohol and carbon dioxide or to lactic acid (the product formed by glucose fermentation in muscle) are phosphorylated. Glucose is converted through a series of enzymatic reactions to two molecules of pyruvic acid, which is the first non-phosphorylated intermediate in the pathway. This pathway, glycolysis, or the Embden-Meyerhof-Parnas (EMP) pathway, is used by many different types of cells. Pyruvic acid is further metabolized in a number of different ways to form products characteristic of the organism. Two such fermentations are those mentioned above, the alcoholic fermentation of yeast and the lactic acid fermentation of muscle (in the absence of oxygen) or of lactic acid bacteria such as *Streptococcus lactis*, which is used in the dairy industry. Aerobically, pyruvic acid is completely oxidized to carbon dioxide and water.

We will examine the reactions of glycolysis in some detail in Chap. 9. At present, we are concerned with the reactions that allow the organism metabolizing glucose to store a part of the energy contained in glucose as ATP. This is accomplished by the conversion of glucose to an energy-rich phosphorylated intermediate that is capable of donating its phosphate group to ADP to form ATP. We have discussed the fact that there is a gradient of phosphate transfer potential, which was shown in Table 7-1. This means that if ATP is to be formed during the metabolism of glucose, an intermediate having a higher free energy of hydrolysis than that of ATP must be formed. Most of the reactions of glycolysis are preliminary reactions—phosphorylation by ATP, rearrangements, cleavage, and dehydration—which are secondary to the major purpose of the pathway, the creation of an energy-rich intermediate with high phosphate transfer potential. Other pathways for the metabolism of glucose and other compounds have a similar theme but utilize different reactions. All, however, are designed to achieve the synthesis of

ATP by the transfer of a phosphate group from a phosphorylated inter-mediate of the pathway to ADP. This mechanism of formation of ATP is called *substrate level phosphorylation.*

The conversion of glucose to two molecules of pyruvic acid involves 10 enzymatic reactions, only two of which are important in energy generation. In the first, the energy released by oxidation of an aldehyde to an acid is used to drive the energy-requiring addition of inorganic phosphate to the molecule [Eq. (7-7)]. The result is a mixed anhydride of a carboxylic acid and phos-phoric acid, which is highly reactive (i.e., unstable) and has a high phosphate transfer potential.

The initial reactant is glyceraldehyde 3-phosphate, a product of a series of four reactions in which glucose is phosphorylated and cleaved. The following reaction then occurs:

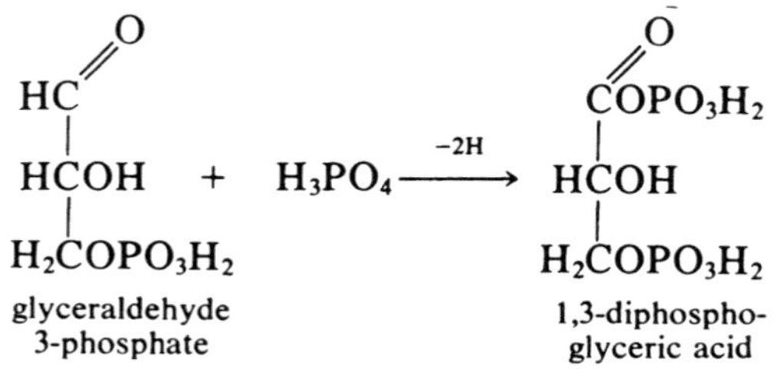

$$
\begin{array}{c}
\overset{\displaystyle O}{\overset{\|}{HC}} \\
| \\
HCOH \\
| \\
H_2COPO_3H_2 \\
\text{glyceraldehyde} \\
\text{3-phosphate}
\end{array}
+ \ H_3PO_4 \xrightarrow{\ -2H\ }
\begin{array}{c}
\overset{\displaystyle \bar{O}}{\overset{\|}{COPO_3H_2}} \\
| \\
HCOH \\
| \\
H_2COPO_3H_2 \\
\text{1,3-diphospho-} \\
\text{glyceric acid}
\end{array}
\tag{7-7}
$$

The two phosphate groups of 1,3-diphosphoglyceric acid are very different in potential as phosphate donors. The phosphate on carbon-1 is readily lost, with the release of a large amount of energy, whereas that on carbon-3 is in an ester linkage, a much more stable, less energy-rich bond. The free energy of hydrolysis of the anhydride phosphate, $\Delta G'$, is $-11,800$ cal/mol, while that of the ester phosphate is only -2400 cal/mol. Since the free energy required to add phosphate to ADP is approximately 7000 cal/mol, it is possible to form ATP using the anhydride phosphate but not using the ester phosphate.

In the next reaction in the same sequence, ATP is formed:

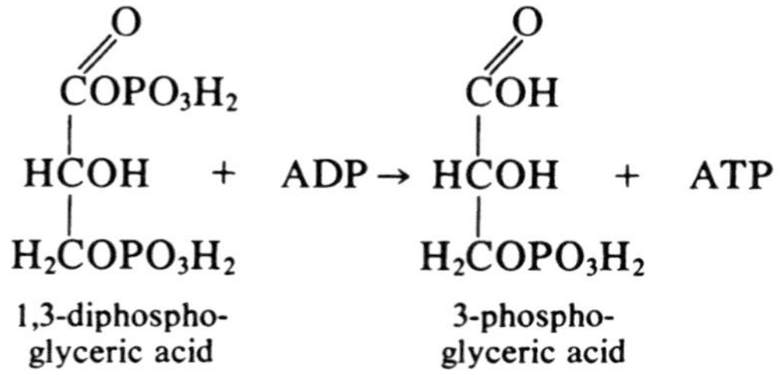

$$
\begin{array}{c}
\overset{\displaystyle O}{\overset{\|}{COPO_3H_2}} \\
| \\
HCOH \\
| \\
H_2COPO_3H_2 \\
\text{1,3-diphospho-} \\
\text{glyceric acid}
\end{array}
+ \ ADP \rightarrow
\begin{array}{c}
\overset{\displaystyle O}{\overset{\|}{COH}} \\
| \\
HCOH \\
| \\
H_2COPO_3H_2 \\
\text{3-phospho-} \\
\text{glyceric acid}
\end{array}
+ \ ATP
\tag{7-8}
$$

In this sequence of two reactions, the oxidation of glyceraldehyde 3-phosphate has been coupled to ATP synthesis. The overall change in the amount of free energy for the coupled reactions is very small, indicating that essentially all of the energy produced in oxidizing 3-phosphoglyceraldehyde to 3-phosphoglyceric acid was stored as high-energy phosphate in ATP, where it is available to do work for the cell.

The second reaction that results in ATP synthesis in glycolysis is a unique type of reaction that can be considered as an internal oxidation-reduction reaction. No actual reduction or oxidation of the molecule occurs; rather, a molecule of water is removed. This results in the destabilization of the phosphate group due to shifting of electrons within the molecule. The compound formed, phosphoenolpyruvate, has a free energy of hydrolysis of $-14,800$ cal/mol and is therefore an energy-rich phosphate donor, capable of phosphorylating ADP.

$$
\begin{array}{c}
\overset{O}{\overset{\|}{C}OH} \\
| \\
H\overset{|}{C}OPO_3H_2 \\
| \\
H_2\overset{|}{C}OH
\end{array}
\xrightarrow{\quad H_2O \quad}
\begin{array}{c}
\overset{O}{\overset{\|}{C}OH} \\
| \\
\overset{|}{C}OPO_3H_2 \\
\| \\
CH_2
\end{array}
\qquad (7\text{-}9)
$$

2-phospho-
glyceric acid

phosphoenol-
pyruvic acid

$$
\begin{array}{c}
\overset{O}{\overset{\diagup\!\!\diagup}{C}OH} \\
| \\
\overset{|}{C}OPO_3H_2 \\
\| \\
CH_2
\end{array}
+ \; ADP \rightarrow
\begin{array}{c}
\overset{O}{\overset{\diagup\!\!\diagup}{C}OH} \\
| \\
C{=\!\!=}O \\
| \\
CH_3
\end{array}
+ \; ATP
\qquad (7\text{-}10)
$$

phosphoenol-
pyruvic acid

pyruvic acid

Again two reactions have been coupled through a common intermediate—the product of one being the reactant in the other—resulting in the capture of energy in the form of ATP.

There are other substrate level phosphorylations, some of which we will examine in later chapters. All have the same common features—no electron transport is involved; no oxygen is involved, so that the reactions are anaerobic; and all proceed through the transfer of phosphate from an energy-rich phosphorylated derivative of the substrate to ADP (or another nucleoside diphosphate).

Efficiency of energy conservation In the glycolytic pathway for catabolism of glucose, approximately 28 percent of the energy released when glucose is fermented to two molecules of lactic acid is conserved in the high-energy bonds of ATP. As we shall see later, in the lactic acid fermentation two molecules of ATP are formed per molecule of glucose fermented. The difference in free energy between 1 mol of glucose and 2 mol of lactic acid is 52,000 cal. The free energy of hydrolysis of 2 mol of ATP is 14,600 cal. Thus, the energy conserved may be calculated as $(14,600/52,000) \times 100 = 28$ percent. In other words, the change in free energy for the overall pathway is:

$$\text{Glucose} + 2H_3PO_4 + 2ADP \rightarrow 2 \text{ lactic acid} + 2ATP + 2H_2O$$
$$\Delta G' = -37,400 \text{ cal/mol} \qquad (7\text{-}11)$$

Of the 52,000 cal of free energy produced in the glycolytic conversion of 1 mol of glucose to 2 mol of lactic acid, only 14,600 cal is captured by the organism, while 37,400 cal is wasted, i.e., dissipated as heat. The lactic acid fermentation thus achieves an efficiency of energy conversion of only 28 percent. Its efficiency based on the total free energy available from glucose is much lower, however. The total combustion (oxidation) of glucose to carbon dioxide and water releases 686,000 cal as free energy. That is:

$$\text{Glucose} + 6O_2 \rightarrow 6CO_2 + 6H_2O \qquad \Delta G° = -686,000 \text{ cal/mol} \qquad (7\text{-}12)$$

Fermentation, therefore, is highly inefficient in releasing the available free energy of glucose, leaving most of it in the product, lactic acid. The efficiency of the lactic acid fermentation in utilizing the energy *potentially available* from glucose is only 2 percent ($14,600/686,000) \times 100 = 2$ percent. This very low efficiency of utilization of an energy source is typical of fermentation (some fermentations are only half as efficient) and explains the fact that anaerobic organisms dependent on fermentation for energy require very large amounts of energy source for growth.

ATP-limited growth It has been shown with many different anaerobic organisms carrying out different types of fermentations that ATP formation is the factor that limits growth. If a fermentative anaerobe is supplied with all of the amino acids, vitamins, sugars, organic bases, and other requirements to synthesize new cells, so that its energy source can be used exclusively for the production of ATP, the amount of cells produced is exactly proportional to the amount of ATP formed. This growth yield is Y_{ATP}, i.e., yield based on ATP formed, and its value is 10 mg of cell dry weight per millimole of ATP. *Streptococcus lactis*, using glucose as energy source in a very rich medium containing all the required organic growth factors, will produce 20 mg (dry weight) of cells per millimole (180 mg) of glucose used, since it is able to synthesize 2 mol of ATP per mole of glucose fermented. A closely related species, *Leuconostoc mesenteroides*, ferments glucose by a different pathway that produces lactic acid, ethanol, and carbon dioxide in equimolar amounts as the end products and yields only 1 mol of ATP per mole of glucose fermented. It is capable of producing only 10 mg (dry weight) of cells per millimole of glucose fermented. In both of these organisms, as in other fermentative species under similar conditions, the glucose provided is used solely as an energy source, and none of the carbon from glucose is incorporated into new cellular material. All of the cellular carbon is supplied by organic growth factors.

Oxidative Phosphorylation

Organisms able to reoxidize reduced NAD or other electron carriers through electron transport are generally able to oxidize glucose completely, thus releasing all of the energy potentially available from the molecule. It is not necessary for these organisms to use partially oxidized products of glucose metabolism as electron acceptors, thereby leaving much of the potentially available free energy in the end products of the pathway. Rather, they are capable of using an external, inorganic electron acceptor. Glucose is degraded completely to carbon dioxide, and the electrons removed in oxidative reactions are transferred from the primary electron acceptor to the electron transport system and used finally to reduce the terminal electron acceptor, O_2, or another oxidized inorganic ion. The oxidation of glucose can still be considered an anaerobic process since oxygen is not involved directly and oxidation occurs by dehydrogenation. This anaerobic (non-oxygen-requiring) conversion of glucose to carbon dioxide is, however, dependent on the availability of NAD to accept electrons, and the degradative pathway is therefore coupled to the electron transport system through the oxidation and reduction of the common electron carrier NAD. This relationship may be represented for aerobic organisms using O_2 as terminal electron acceptor by the overall reaction shown in Fig. 7-1. It is interesting to note that energy is obtained not only from the oxidation of glucose but also from the oxidation of six molecules of water (Wald, 1966). In fact, half of the ATP formed by

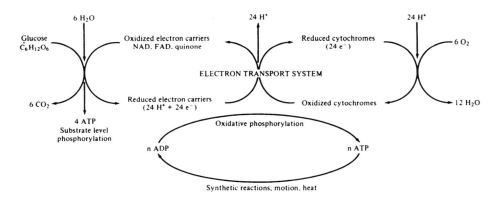

Figure 7-1 Complete oxidation of glucose. The net equation for the reactions is

$$Glucose + 6O_2 \rightarrow 6H_2O + 6CO_2$$

Glucose is metabolized through the Embden-Meyerhof-Parnas pathway and the Krebs (TCA) cycle (see Chap. 9). In three reactions in these pathways, water is added, two molecules per molecule of glucose. Half of the electrons used to generate ATP through oxidative phosphorylation coupled to electron transport are therefore derived from glucose and half from the added water. The number of molecules of ATP produced per molecule of glucose, n, is 32 in mitochondria but has not been determined in bacteria.

oxidative phosphorylation in most cells is not the result of oxidation of the energy source. The overall reaction is quantitatively the same for all organisms that use these pathways of glucose catabolism if the amount of ATP formed is not specified. In eucaryotic cells, the amount of ATP formed can be predicted; $n = 32$. The total amount of ATP synthesized in the above process is 36 mol of ATP per mole of glucose, since 2 mol of ATP per mole of glucose is formed by substrate level phosphorylation during glycolysis and 2 mol by substrate level phosphorylation in the Krebs cycle (see Chap. 9). For procaryotic cells, the net synthesis of ATP cannot be predicted. This difference is the result of the dissimilar electron transport systems in the two types of cells. All eucaryotic cells have electron transport systems neatly arranged within a specialized organelle, the mitochondrion, and the components of their electron transport chains are similar, if not identical. In procaryotes, the electron transport system is embedded in the cytoplasmic membrane, and there is no uniformity of components among different genera. However, the principles involved in electron transport and its function in ATP synthesis are the same regardless of the exact details of the pathway of electron flow.

In our discussion of thermodynamics, we referred to the equivalence of chemical and electric energy demanded by the first law and illustrated this by expressing the free-energy change of a reaction as

$$\Delta G^\circ = -nF \, \Delta E_0 \qquad (7\text{-}4)$$

Since the free energy required for the synthesis of ATP from ADP and inorganic phosphate under standard conditions (ΔG°) is approximately 7000 cal/mol, any oxidation-reduction reaction that makes available an amount of free energy in excess of 7000 cal can potentially be coupled to the formation of ATP. Using Eq. (7-4), we can calculate the value of ΔE_0 that would supply 7000 cal of free energy:

$$-7000 = -2(96,500) \, \Delta E_0$$

$$\Delta E_0 = \frac{7000 \text{ cal/mol} \times 4.18 \text{ CV/cal}}{2 \times 96,500 \text{ C/mol}} = 0.15 \text{ V}$$

The conversion factors are: 1 coulomb-volt = 1 joule; 4.18 joules = 1 cal; or the relationship, 1 faraday = 23,086 cal may be used. If the value of ΔG for ATP formation under physiological conditions is used, i.e., 10,000 to 12,000 rather than 7000, the difference in potential required for ATP synthesis may be as high as 0.26 V. We shall return to this required difference in redox potentials as we examine the sequence of electron transfers in electron transport systems.

Electron transport in mitochondria In eucaryotic cells, the mitochondrion contains a variety of enzymes, the components of the electron transport chain and the ATP-synthesizing mechanism, the nature of which is not yet known. The enzymes vary with the specific type of cell but usually include enzymes involved in oxidation-reduction reactions (dehydrogenases) for various in-termediates of carbohydrate and lipid metabolism and the enzymes of the Krebs (citric acid) cycle. Components of electron transport systems include flavoproteins, quinones, cytochromes, and iron-sulfur proteins.

Flavoproteins are a heterogeneous group of proteins with one charac-teristic in common—the use of a flavin coenzyme, flavin adenine dinucleotide (FAD) (see Fig. 3-40) or flavin mononucleotide (FMN), as the electron carrier. Both FMN and FAD are derivatives of riboflavin (vitamin B_2). Some flavo-proteins also contain metals, iron or molybdenum, sulfide, and heme (Fe) groups. They thus have characteristics of several different electron carriers—simple flavoproteins, iron-sulfur proteins, and cytochromes (heme Fe)—and may function as "a miniature electron-transport system on a single protein" (White, Handler, and Smith, 1973). Some such flavoproteins may transfer hydrogen and electrons directly to oxygen, forming hydrogen peroxide, and are called *oxidases*. Flavoproteins such as nitrate reductase, which contains iron, sulfur, and molybdenum, accept electrons from other iron-sulfur pro-teins or from cytochromes and transfer them to nitrate to form nitrite. Other flavoproteins serve as dehydrogenases, accepting electrons and transferring them to other electron carriers. The flavoproteins of the mitochondria do not transfer electrons directly to oxygen but function as dehydrogenases. The flavoproteins specifically involved in electron transport in mitochondria ac-cept electrons (and protons) from NADH, from the citric acid cycle inter-mediate, succinate, or from flavoproteins that oxidize other substrates, and transfer electrons to other carriers of the electron transport system, quinones or cytochromes.

Quinones, the most recently discovered components of the electron transport chain, are a family of small molecules with the structure shown in Fig. 7-2. The length of the side chain, i.e., the value of *n*, varies from 6 to 10 depending on the cell in which the compound is found. The redox potential of ubiquinone is very close to that of cytochrome b, but it is probable that ubiquinone transfers electrons from flavoproteins to cytochromes (see below).

The *cytochromes* are proteins containing iron in the form of a heme group like that in hemoglobin or catalase (see Fig. 7-3). The heme iron functions in the transfer of electrons since it can be reversibly oxidized and reduced:

$$Fe^{3+} \xrightleftharpoons[-e^-]{+e^-} Fe^{2+} \qquad (7\text{-}13)$$

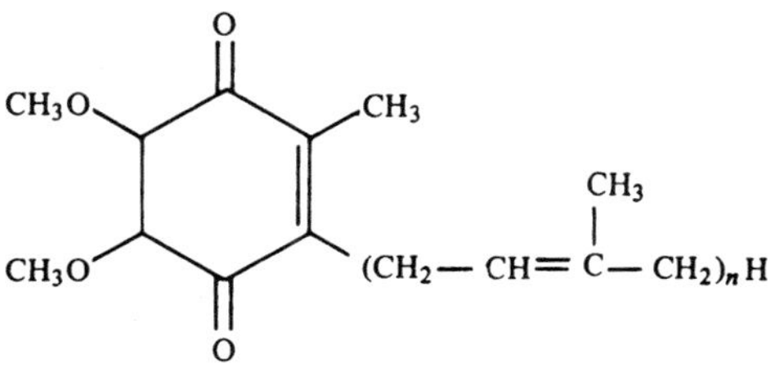

Figure 7-2 Ubiquinone (coenzyme Q) oxi-dized form. $n = 6$ to 10, differing with the organism.

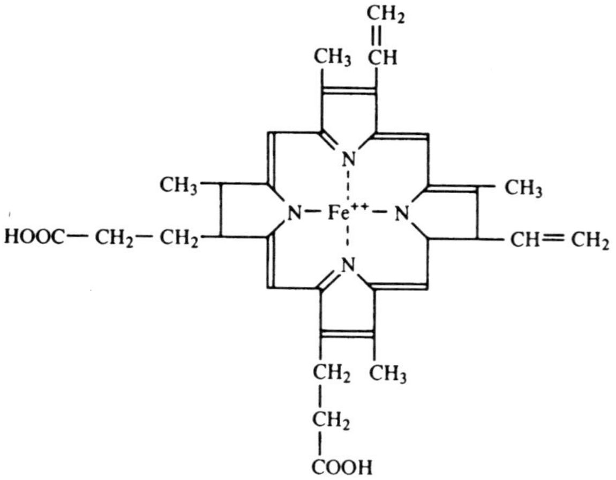

$$
\begin{array}{c}
CH_2 \\
\| \\
CH_3 \ CH
\end{array}
$$

CH₃—

HOOC—CH₂—CH₂—

N—Fe⁺⁺—N

CH₃

CH=CH₂

CH₂ CH₃
|
CH₂
|
COOH

Figure 7-3 The structure of the heme that serves as the prosthetic group in the cytochromes.

Cytochromes are readily detected spectroscopically and are classified on the basis of their absorption spectra and certain structural differences (Bartsch, 1968). In mitochondria, cytochromes of three types, a, b, and c, participate in electron transport. Additional types of cytochromes are found in bacteria.

Small proteins containing iron (nonheme) complexed with sulfur, in the same arrangement found in certain flavoproteins, also function in electron transport systems, particularly in bacteria. These are called *iron-sulfur proteins*. A number of such proteins have been found and, while some function in known ways in such processes as nitrogen fixation and photosynthesis, the functions of others are not understood. Some of these proteins, e.g., ferredoxin, have very low redox potentials ($E_0' = -0.49$ V) and have been called *high-energy* electron carriers for this reason (Benemann and Valentine, 1971).

The exact order of the electron transport components in mitochondria has been the subject of intensive investigation in many laboratories for years, as has the mechanism by which electron transport is coupled to ATP synthesis. A scheme that appears to represent at least the major pathway for mitochondrial electron transport is shown in Fig. 7-4. The greatest uncertainty lies in the relative positions of cytochrome b and ubiquinone (coenzyme Q). It is

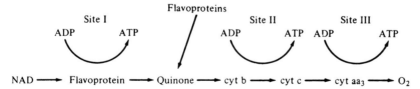

Figure 7-4 The mitochondrial electron transport chain for transfer of electrons from NAD to the terminal electron acceptor O_2. Flavoproteins reduced by organic substrates such as succinate or fatty acids may transfer electrons to the chain at the quinone level. Other components such as iron-sulfur proteins may be involved.

believed that a number of different flavoproteins may transfer electrons directly to ubiquinone, while others may transfer electrons to cytochrome b.

Efficiency of energy conservation Table 7-2 shows E_0' values for some of the biologically important redox systems. Using the value of $\Delta E_0'$ calculated above for the synthesis of ATP, we can determine that in the mitochondrial electron transport system there are three possible sites for ATP synthesis. For the oxidation of reduced NAD by flavoprotein, the amount of free energy released can be calculated as

$$\Delta G° = \frac{-nF\,\Delta E_0'}{4.18} = \frac{-2(96,500)(0.32 - 0.06)}{4.18} = -12,005 \text{ cal/mol}$$

Table 7-2 Redox potentials of biological systems

System	E_0', V
H^+/H_2	-0.42
NAD/NADH$_2$	-0.32
Flavoprotein$_{ox}$/flavoprotein$_{red}$	-0.06
Ubiquinone/dihydroubiquinone	$+0.10$
Cyt b$_{ox}$/cyt b$_{red}$	$+0.05$
Cyt c$_{ox}$/cyt c$_{red}$	$+0.22$
Cyt a$_{ox}$/cyt a$_{red}$	$+0.29$
Nitrate/nitrite	$+0.42$
O_2/H_2O	$+0.82$

Similarly, the amount of free energy available from the oxidation of cytochrome b by cytochrome c is approximately 8000 cal/mol and that from the oxidation of cytochrome a by oxygen is approximately 24,000 cal/mol. So, for each pair of electrons passing through electron transport from NADH$_2$ to O$_2$ in the mitochondrion, three molecules of ATP are synthesized. The difference in potential between NADH$_2$ and O$_2$ is 1.14 V, corresponding to a total free-energy decrease of 52.6 kcal/mol. The energy conserved in ATP is (3×7300) or 21,900 cal/mol, amounting to an efficiency of energy conservation of 42 percent, or 68 percent if a value of 12,000 cal/mol is used for the free energy of hydrolysis of ATP. (It should be obvious that calculations such as these contain a large element of uncertainty.)

P/O ratio While NAD, flavoprotein, and quinones accept not only electrons but protons as well, the cytochromes transfer only electrons. The protons are released by ubiquinone when electrons are transferred to cytochrome b, but in the final step, the reduction of O$_2$ to H$_2$O by the terminal cytochrome in the mitochondrial system (cytochrome oxidase), both protons and electrons are involved. The reaction may be written as

$$2H^+ + 2e^- + \tfrac{1}{2}O_2 \rightarrow H_2O \qquad (7\text{-}14)$$

We can summarize the mitochondrial electron transport system by stating that for each two hydrogen molecules transferred from reduced NAD to oxygen, three molecules of ATP are synthesized. This is commonly spoken of as the P/O ratio; i.e., for each atom of oxygen reduced, three molecules of inorganic phosphate are esterified, yielding a P/O ratio of 3.0 for oxidative phosphorylation coupled to electron transport in mitochondria. The P/O ratio is lower for electrons that enter the electron transport chain at the level of flavoprotein or at other points. As we shall see later, one of the dehydrogenases of the Krebs cycle, succinic dehydrogenase, uses a flavin cofactor FAD, rather than NAD, as electron carrier, and the electrons removed in the oxidation of succinic acid enter the electron transport system at the flavoprotein level, yielding a P/O ratio of 2.0 rather than 3.0. The two molecules of $NADH_2$ formed in glycolysis also yield two rather than three molecules of ATP. $NADH_2$ formed in the cytoplasm cannot enter the mitochondrion, and the electrons are transferred to another molecule that enters the mitochondrion and is oxidized by a flavoprotein (Stryer, 1975).

ATP synthesis and its control It was assumed for a number of years that oxidative phosphorylation, like substrate level phosphorylation, involved the formation of an energy-rich phosphorylated compound capable of transferring phosphate to ADP, with the energy for formation of the hypothetical intermediate $X \sim P$ being derived from the oxidation-reduction reactions of electron transport. A vigorous search for a compound with the functions ascribed to $X \sim P$ has been unsuccessful, and other mechanisms have been proposed to explain the coupling of ATP synthesis to electron transport (Boyer et al., 1977). However, there is no conclusive evidence in support of any proposed mechanism, although the chemiosmotic model mentioned in relation to substrate transport in Chap. 5 is favored by several investigators.

The two processes, electron transport and oxidative phosphorylation, are tightly coupled in mitochondria. This means that one cannot proceed without the other. If ADP or inorganic phosphate is not available for use in oxidative phosphorylation, electron transport ceases. If the electron carriers, particularly $NADH_2$, cannot be oxidized by transferring electrons to the electron transport system, all reactions requiring NAD cease and the catabolism of the energy source cannot proceed. The situation is somewhat more complicated than this, but essentially the tight coupling of oxidative phosphorylation and electron transport controls the rate of ATP synthesis so that ATP is not produced in much larger amounts than needed for biosynthetic reactions or other energy-requiring processes. The synthesis of materials that can be stored and used at a later time as an energy source is one useful means of recycling ATP to allow oxidation of the energy source to continue in cells that are not growing. Humans and other animals, of course, store excess carbon and energy source as fat in adipose tissue, and the synthesis of fat is controlled largely by the amount of ATP accumulated in the mitochondria. In warm-blooded animals, much of the energy source is used to generate heat; i.e., the free energy released is not conserved to do useful work but is used to maintain a constant internal temperature.

Electron transport in bacteria Electron transport and oxidative phosphorylation in bacteria differ in several important ways from the same processes in eucaryotic cells. Since bacteria have no mitochondria (being themselves approximately the same size as a single mitochondrion), their electron transport systems with associated enzymes and ATP-synthesizing apparatus are located in the cytoplasmic membrane. Much of the progress in research on electron transport and oxidative phosphorylation in mitochondria has been possible because the entire system can be separated from the rest of the cell, as isolated mitochondria, for study of these specific reactions. It has not been possible to do similar studies with bacteria, and the problems caused by the inability to study the electron transport and oxidative phosphorylation systems in isolation have delayed progress in understanding bacterial respiration and energy metabolism.

The great variety of components found in bacterial electron transport systems has also complicated the study of such systems, because no single organism or group of organisms can be chosen as a representative model for such studies. Electron transport systems in bacteria appear to be much more complex than the mitochondrial system. Branching is common, with electrons entering the electron transport chain at different levels and with more than one terminal oxidase, i.e., more than one electron carrier capable of transferring electrons to oxygen. The situation is further complicated by the ability of some bacteria to use alternate final electron acceptors in the absence of oxygen and the need to determine what differences exist between electron transport pathways to oxygen and to other electron acceptors, such as nitrate, in bacteria that can utilize both, e.g., the denitrifying bacteria. Similar problems arise in the case of bacteria that can utilize a variety of electron acceptors at different oxidation levels, e.g., the sulfate-reducing bacteria, which can use as final electron acceptor sulfate, trithionate, thiosulfate, and sulfite as well as fumarate (LeGall and Postgate, 1973).

The types of electron carriers found in bacteria are in general the same as those found in mitochondria; i.e., both eucaryotes and procaryotes utilize flavoproteins, quinones, cytochromes, and iron-sulfur proteins. However, there are probably few bacterial species with an electron transport chain exactly the same as that in mitochondria. Different bacterial species contain different combinations of electron transport carriers and even a single bacterial species may form different carriers or different relative amounts of the same carriers when grown under different conditions. The greatest variation occurs in the cytochromes. Whereas mitochondria contain cytochromes of the a, b, and c types, some bacterial species apparently may have cytochromes of only one or two of these types and may have several forms of the same cytochrome, e.g., cytochromes c_4 and c_5, or cytochromes a_1, a_2, and a_3. In addition, bacteria often have two cytochromes not found in mitochondria, cytochromes o and d, which act as terminal oxidases, transferring electrons to oxygen.

Despite the great variability in electron transport systems in bacteria and the technical difficulties involved in studying them, much progress has been made with a number of bacterial species. We will cite a few of these as examples to illustrate the differences between these and mitochondrial systems and the variability within bacterial systems. Additional information is available in several recent reviews (Bartsch, 1968; Haddock and Jones, 1977; Horio and Kamen, 1970; Kamen and Horio, 1970; White and Sinclair, 1971).

Figure 7-5 shows an electron transport system proposed for *Azotobacter*, which illustrates the branched pathways using multiple terminal oxidases, which appear to be common in bacteria (Cole, 1976).

Although P/O ratios have not been determined accurately in bacteria, it is proposed that the pathway to cytochrome a_1 or o contains three sites for energy conservation, i.e., three sites at which ATP synthesis is coupled to the transfer of electrons, whereas the pathway to cytochrome d contains only two such sites. Figure 7-6 shows a proposed pathway for electron transport to oxygen in *Escherichia coli*, which contains at least nine different cytochromes and two different quinones (Haddock and Jones, 1977).

Several advantages may result from having alternate pathways of electron

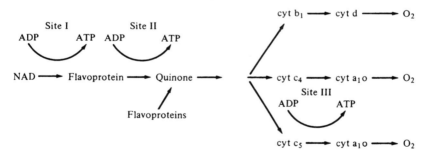

Figure 7-5 Possible branched path of electron flow in *Azotobacter*, with proposed sites of ATP synthesis. (*Adapted from Cole, 1976.*)

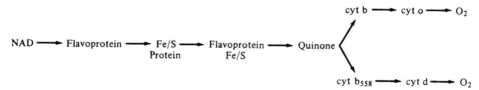

Figure 7-6 Possible branched pathway of electron transport in *Escherichia coli*. (*Adapted from Haddock and Jones, 1977.*) Sites of ATP synthesis are uncertain, but there are probably two.

transport. One may be the ability to continue to function in the presence of inhibitors that completely block electron transport in mitochondria. Cytochromes a_1 and o, like the cytochrome oxidase of mitochondria, are inhibited by low concentrations of cyanide, for example, whereas cytochrome d is relatively insensitive to cyanide. Cytochrome a_2, found in some bacteria, is also resistant to cyanide. Differences in affinity for oxygen may also be important to bacteria, since they are frequently exposed to environments with very different oxygen tensions. It is known that large differences in cytochrome content result from growth with high or low aeration rates. A species of *Achromobacter* grown at low aeration rates forms less cytochrome o than when grown at high aeration rates. In the two electron transport systems shown above for *Azotobacter* and *E. coli*, the pathway to oxygen through cytochrome d is used at low oxygen tensions and that through cytochrome o, which has a lower affinity for oxygen, at high oxygen tensions. The ability to switch from one pathway of electron transport to another, depending on environmental conditions, may have considerable survival value for bacteria.

Anaerobic electron transport Although the mitochondrial electron transport system can utilize only oxygen as terminal electron acceptor, bacteria are more versatile. As we have mentioned previously, some bacteria that use oxygen as terminal electron acceptor when it is available are able to use nitrate or possibly one of the more reduced forms of nitrogen—nitrite, NO_2^-, nitrous oxide, N_2O, or nitric oxide, NO—as an alternative terminal electron acceptor in the absence of oxygen. Figure 7-7 has two proposed schemes for electron transport systems using NO_3^- as terminal electron acceptor. Organisms that can use NO_3^- in this way are found in several different genera, and their activities are ecologically and economically important. We will discuss them in more detail below.

Other terminal electron acceptors used by some bacteria are carbon dioxide and sulfate and some of its reduced products. Organisms using these are specialized groups that cannot use either oxygen or nitrate as terminal electron acceptor and are strict anaerobes, usually killed by exposure to oxygen.

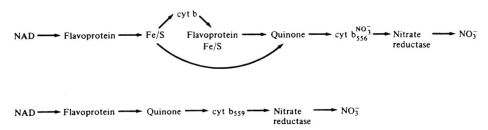

Figure 7-7 Two proposed schemes for flow of electrons from NAD to the terminal electron acceptor NO_3^-. (*Top*) Proposed for *Escherichia coli*. (*Adapted from Haddock and Jones, 1977.*) (*Bottom*) Proposed for *Klebsiella aerogenes*. (*Adapted from Stouthamer, 1976.*)

The sulfate-reducing bacteria belong to two genera, *Desulfovibrio*, a Gram-negative non-spore-forming curved rod, and *Desulfotomaculum*, a Gram-negative spore-forming rod. The group contains psychrophilic and thermophilic as well as mesophilic species and both fresh- and salt-water species, so they are widespread in nature. All use organic compounds as carbon and energy source, but some can oxidize hydrogen as energy source. Electrons removed from hydrogen or an organic energy source are used to generate ATP through oxidative phosphorylation coupled to electron transport. Electron carriers are similar to those found in aerobic organisms; b, c, and d type cytochromes, quinones, flavoproteins, and three high-energy iron-sulfur proteins, ferredoxin, flavodoxin, and rubredoxin, have been found in species of *Desulfovibrio*. The sequence of carriers involved in electron transport to sulfate is not known. A recent review by LeGall and Postgate (1973) offers an excellent discussion of the physiology of these bacteria and their ecological and economic importance. They are responsible for odors in marine or brackish water, muds, and other environments with high sulfate content, since they reduce sulfate to hydrogen sulfide, which is not only unpleasant but also highly toxic to both plant and animal life. They have been implicated in the corrosion of metals, although the mechanism of corrosion is not understood. Their ability to use sulfate as terminal electron acceptor under anaerobic conditions forming elemental sulfur, hydrogen sulfide, or metallic sulfides is important in many environments and constitutes one reductive portion of the natural sulfur cycle, the other portion being carried out by organisms that use sulfate as a source of sulfur, reducing it to the sulfhydryl, –SH, level (assimilatory reduction). Organisms that use carbon dioxide as terminal electron acceptor also presumably generate ATP by oxidative phosphorylation coupled to electron transport, although they have not been shown to contain cytochromes and the electron carriers have not been identified. These bacteria (the methane bacteria) are difficult to study in the laboratory because of their extreme sensitivity to oxygen and the difficulty of isolating and maintaining pure cultures. Their ability to use hydrogen as a source of electrons for the reduction of carbon dioxide to methane is important in recycling carbon to the atmosphere from anaerobic environments, since methane is highly volatile and is much less soluble in water than is carbon dioxide. The role of these bacteria in the anaerobic digestion of wastes will be discussed in Chap. 11.

ATP yields and uncoupling of ATP formation and use The ATP yield for oxidative phosphorylation coupled to electron transport using terminal electron acceptors such as nitrate, sulfate, or carbon dioxide is not known, for the same reasons that P/O ratios for aerobic respiration have not been determined with certainty. It is believed that electron transport to nitrate yields less ATP than that to oxygen, because less growth is obtained for the same organism using NO_3^- as electron acceptor as compared with that using oxygen (Stouthamer, 1976). The ATP yield for electron transport to sulfate or carbon dioxide is also assumed to be low since these organisms grow rather slowly. The tight coupling between electron transport and oxidative phosphorylation found in mitochondria has not been shown in bacteria (Forrest and Walker, 1971), but experiments of the type that demonstrate such coupling in mitochondria are not possible with bacteria. In anaerobic fermentative bacteria, the production of ATP limits growth, as shown by the constant Y_{ATP} for fermentation. Aerobic bacteria, however, are probably able to produce ATP in greater amounts than required for the biosynthetic reactions associated with growth, but even that statement cannot be made with certainty because of the many problems involved in making such determinations. Since respiration, i.e., utilization of oxygen, may proceed as rapidly in cells that are unable to grow because of the absence of an essential nutrient such as nitrogen as it does in growing cells, bacteria certainly have means of uncoupling growth and respiration. If ATP synthesis and respiration, i.e., electron transport, are tightly coupled, then there must be mechanisms by which the ATP synthesized can be utilized to supply ADP for use in oxidative phosphorylation. Some such mechanisms are known.

Many bacteria are capable of synthesizing one or more polymers that serve as a reserve of carbon and/or energy (Dawes and Senior, 1973). Glycogen-like polymers require ATP for synthesis, and the storage of glycogen thus serves both to regenerate ADP and to allow the cell to store carbon source in an environment where carbon and energy are available but nitrogen, for example, is not. This process, called *oxidative assimilation*, can be used to advantage in the treatment of nitrogen-deficient industrial wastes, as we shall see in Chap. 10. The polymeric form of inorganic phosphate, polyphosphate or volutin, which is stored by many cells, may serve as an energy store since the phosphate–phosphate anhydride bonds are of the same type and energy content as those in ATP. Whether it is used as a means of storing energy or simply as a store of inorganic phosphate, it also serves as a mechanism for recycling ATP that is not required for growth. Poly-β-hydroxybutyrate does not require ATP directly in its synthesis, but it does require $NADH_2$ and acetyl-CoA, which would otherwise generate ATP. It has also been proposed that bacteria may simply waste energy that cannot be used for biosynthetic reactions by hydrolyzing ATP, releasing the energy as heat. Another possible means of partial uncoupling of respiration and biosynthesis would be the use of alternate electron transport pathways in which there are fewer sites of oxidative phosphorylation.

Photophosphorylation

The synthesis of ATP in photosynthetic organisms, like that in aerobic chemotrophs, is coupled to electron transport, and, as in the case of oxidative phosphorylation, the mechanism of coupling is unknown. For all photosynthetic organisms except one group of bacteria, carbon dioxide serves as the sole source of cellular carbon, and the synthesis of cell material from carbon dioxide requires ATP and "reducing power," i.e., hydrogen (electrons) in the form of $NADPH_2$. The electron transport systems of these organisms supply both.

Oxygenic and anoxygenic photosynthesis There are two types of photosynthesis that are related to the source of reducing power used by the organism. In *oxygenic photosynthesis*, water serves as the source of reducing power (electron donor) and oxygen is produced as a by-product. The equation for the *photolysis of water* is

$$2H_2O \rightarrow O_2 + 4H^+ + 4e^- \tag{7-15}$$

Oxygenic photosynthesis is carried out by green plants, algae, and cyanobacteria (blue-green algae).

Bacterial photosynthesis is *anoxygenic*, i.e., no oxygen is produced; indeed, bacterial photosynthesis occurs only in the absence of oxygen. Those species of bacteria that are obligately photosynthetic are strict anaerobes. In bacterial photosynthesis, reducing power is supplied by reduced inorganic compounds or, in one group of bacteria, by organic compounds such as acetate or malate. Many photosynthetic bacteria are able to use molecular hydrogen, H_2, as electron donor. Reduced sulfur compounds, or elemental sulfur, are used by the green and purple bacteria; e.g.,

$$H_2S \rightarrow S^0 + 2H^+ + 2e \tag{7-16}$$

Function of photosynthetic pigments Oxygenic photosynthesis involves two reaction centers for the capture of light, photosystems I and II, whereas only photosystem I is found in bacteria. Photosystem II is responsible for the photolysis of water, which does not occur in bacteria.

The initial step in photosynthesis is the absorption of light by the antenna pigments of the photosystem. This assembly of light-harvesting pigments includes chlorophylls and, in most organisms, carotenoids and phycobilins, the numbers and types of pigments varying with the organism. These pigments are complexed with proteins and are located in the membranes of the chloroplast in eucaryotes or in the invaginated membranes of the procaryotes. Each photosystem contains a special type of chlorophyll, the reaction center chlorophyll, which is involved directly in the initiation of electron transport. Light energy absorbed by the light-harvesting pigments is transferred to the reaction center chlorophyll, which loses an electron to become positively charged. The electron ejected from the excited reaction center chlorophyll has a very high energy level. It reduces an electron carrier and is then transferred through the electron transport system, generating ATP by coupled phosphorylation of ADP.

Cyclic and Noncyclic Photophosphorylation In organisms that carry out oxygenic photosynthesis, two types of photophosphorylation may occur (see Fig. 7-8). *Cyclic photophosphorylation* generates ATP but does not provide

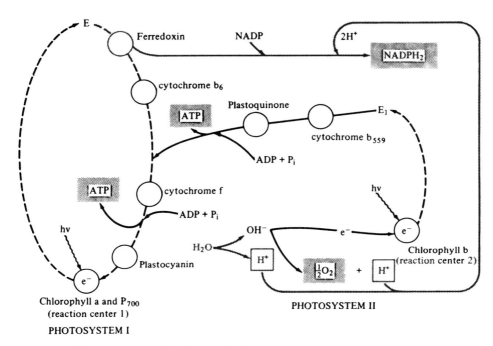

Figure 7-8 Electron transport in photosynthesis.

reducing power and requires only photosystem I. In *noncyclic photophosphorylation*, both photosystems I and II are required and both ATP and reducing power, in the form of $NADPH_2$, are provided.

In cyclic photophosphorylation, as the name implies, the electron lost from the reaction center chlorophyll passes through the electron transport chain and returns to the chlorophyll from which it originally came.

When electrons ejected from the reaction center chlorophyll are used to reduce NADP to $NADPH_2$ (noncyclic electron flow), they must be replaced by electrons from the hydrogen donor, water. The reaction center chlorophyll of photosystem II is excited by the transfer of energy from its associated antenna pigments and loses an electron, which is transferred through electron carriers to the reaction center chlorophyll of photosystem I. The chlorophyll of photosystem II now requires an electron to replace that lost to photosystem I, and this is supplied by an electron from water.

In this noncyclic photophosphorylation, the direction of flow of electrons is the reverse of that in oxidative phosphorylation, where electrons flow downhill from the more electronegative $NADH_2$ ($E_0' = -0.32$) through successively less negative electron carriers and finally to oxygen to form water ($E_0' = +0.82$). Electron flow in photosynthesis actually results in the reduction of NADP by water. This is thermodynamically impossible without the input of energy, and that energy is furnished by light, which boosts the redox potential of the electrons ejected from the reaction center chlorophyll so that the flow of electrons is actually downhill (in terms of energy) from reaction center I to NADP, from reaction center II to reaction center I, and from water to reaction center II, which after losing an electron has an E_0' value of $+1.0$ and can therefore be reduced by water ($E_0' = +0.82$). ATP synthesis is coupled to the transfer of electrons between photosystem II and photosystem I; at least part of the same electron transport chain functions in cyclic photophosphorylation, so that ATP is synthesized by both cyclic and noncyclic electron transport. However, it is probable that cyclic photophosphorylation is of little importance in organisms that carry out oxygenic photosynthesis and functions only as a shunt mechanism when NADP is unavailable (Lehninger, 1975).

Bacterial photosynthesis involves cyclic electron transport and photophosphorylation. Bacteria have only one type of reaction center, photosystem I. It is still a matter of some controversy whether bacteria are able to carry out noncyclic photophosphorylation. If this occurred, it would require a mechanism different from that in oxygenic photophosphorylation, since the photosystem responsible for the photolysis of water is absent in bacteria. It is tempting to conclude, based on the obvious similarity of Eqs. (7-15) and (7-16), that the mechanism of generation of ATP and $NADPH_2$ is the same in all photosynthetic organisms, the only difference being the identity of the electron donor. However, an alternate possibility exists and is supported by several investigators (and some rather convincing evidence). This requires the reversal of electron flow by energy derived from ATP.

Reverse electron transport It is possible to reverse the flow of electrons through a series of electron carriers by the application of energy from an external source (Peck, 1968). As we stated in our earlier discussion of thermodynamics, a reaction with a positive free-energy change cannot proceed spontaneously but must be driven by an input of external energy. Thus, in the normal course of events, electrons flow from $NADH_2$ ($E_0' = -0.32$) to cytochrome c ($E_0' = +0.22$). The difference in potential is sufficient to generate ATP. If ATP is hydrolyzed and releases its free energy, and this energy can be coupled to the electron transport system, electrons can be made to flow uphill; i.e., cytochrome c can transfer electrons to NAD or NADP to form the reduced pyridine nucleotide. This has been demonstrated experimentally and is called *reverse electron transport*. Therefore, if a sufficient supply of ATP is available, it is not necessary to utilize light energy directly in the generation of $NADPH_2$. Since the supply of ATP available to a photosynthetic bacterium from cyclic photophosphorylation should be limited only by the availability of light, it should be possible to generate $NADPH_2$ by reverse electron transport powered by the ATP generated by light-dependent cyclic photophosphorylation. If hydrogen ($E_0' = -0.42$) serves as the electron donor, no thermodynamic problem exists, since the H_2/H^+ couple is sufficiently electronegative to reduce NADP. However, other electron donors such as thiosulfate, which can be used by some photosynthetic bacteria, have less negative redox potentials and cannot reduce NADP without input of energy from ATP.

This problem is shared by the aerobic organisms that obtain energy from the oxidation of inorganic compounds and use carbon dioxide as the sole source of cellular carbon. In these nonphotosynthetic organisms, e.g., in *Nitrobacter*, which can use nitrite as energy source (electron donor), light energy cannot be used to generate reducing power for the reduction of carbon dioxide. Electrons removed from nitrite ($E_0' = +0.42$) cannot be used to reduce NADP ($E_0' = -0.32$) but can reduce other electron carriers with more positive redox potentials and generate ATP, which then can be used to reverse electron transport for the reduction of NADP.

It appears reasonable that all bacteria that use carbon dioxide as carbon source, whether the energy source is light or an inorganic chemical compound, use ATP to reverse electron transport for the reduction of NADP. ATP would be synthesized by reactions coupled to cyclic photophosphorylation in photosynthetic bacteria and to oxidative phosphorylation in organisms that oxidize inorganic energy sources.

ENERGY GENERATION AND THE ENVIRONMENT

As we stated at the beginning of this chapter, the processes by which microorganisms obtain and store energy are intimately related to the carbon-oxygen cycle and are therefore a vital factor in our life-support system. The algae and the blue-green bacteria use light as energy source and help to supply other organisms, including humans, with oxygen and organic matter, which enters the food chain and can serve as a source of chemical energy. Microorganisms that use chemical energy affect the environment in a variety of ways, depending on the specific energy source used and the way it is utilized.

Reactions related to energy generation probably have a greater direct effect on the environment than do reactions related to growth, because energy-yielding reactions create by-products of metabolism that are released into the environment. Nutrients required for growth are used consumptively; i.e., they are converted to cellular material so that the product of their consumption is new biomass. The use of the energy source is nonconsumptive. The energy source is converted to metabolic products, which are released or excreted by the cell. Some of these products are important in the recycling of elements; for example, carbon dioxide is the product of energy generation in aerobes using organic energy sources, and its production and release are essential to the maintenance of the carbon cycle. If the energy source is inorganic, the products of its oxidation may have a significant influence on the environment. We have mentioned one example of this—the formation of sulfuric acid as a result of the use of hydrogen sulfide as an energy source.

Oxidation of the energy source, as we have pointed out, requires that the

H^+ and electrons removed be transferred to an electron acceptor, and this also creates metabolic by-products that are released into the environment. In organisms using oxygen as the terminal electron acceptor, the product is water, and the energy metabolism of these organisms affects the environment primarily by the consumption of oxygen (BOD exertion). Use of sulfate as terminal electron acceptor results in the formation of hydrogen sulfide, which is odorous and toxic. Organisms that use organic electron acceptors affect the environment by converting the energy source to a variety of organic compounds. These affect other organisms, e.g., by lowering the pH or by serving as carbon or energy sources for other species.

Utilization of a chemical energy source, whether organic or inorganic, thus leads to release from the cell of both the reduced electron acceptor and the products of metabolism of the energy source. These conversions of energy source and electron acceptor to metabolic by-products are important in the carbon-oxygen cycle and in the cycles of other elements. Nitrogen is a unique element in that, in different inorganic forms, it serves as a major requirement for growth, as an energy source, and as an electron acceptor. An examination of some of the interconversions of nitrogen compounds by different organisms that use the various forms of the element for different purposes will illustrate the importance of energy-generating reactions to the nitrogen cycle and the effects of these reactions on the environment.

Energy and the Nitrogen Cycle

Bacteria that can use nitrate as an alternate electron acceptor generally do so only in the absence of oxygen or at very low oxygen tensions, although a few bacteria apparently can utilize oxygen and nitrate simultaneously. In most bacteria, the enzyme nitrate reductase, which transfers electrons to nitrate, is not formed in aerobic environments and, if already present in the cell, is inhibited by oxygen. The reduction of nitrate to nitrogen gas, which results in loss of nitrogen from soil and water, is called *denitrification*. Many bacteria that can reduce nitrate are not denitrifiers, since they reduce nitrate only to nitrite. Other bacteria may then reduce nitrite further, since nitrite is also utilizable by some bacterial species as a terminal electron acceptor. Nitrous oxide, N_2O, which is formed as a product of reduction of either nitrate or nitrite by some species, is also gaseous and escapes to the atmosphere. Under acid conditions, nitrite may be decomposed nonbiologically to form nitric oxide, NO, another gaseous product.

Denitrification is a process of considerable ecological importance with both beneficial and nuisance effects, depending on the circumstances under which it occurs. As an important part of the nitrogen cycle, denitrification is necessary for recycling nitrogen to the atmosphere. Denitrification, alone or in combination with nitrification, is also important in the removal of nitrogen from water, where the presence of inorganic nitrogen compounds in excessive amounts may cause several types of problems.

Inorganic nitrogen compounds may be introduced into water in a number of ways: (1) pollution by domestic wastes and by many industrial wastes; (2) runoff, particularly from fertilized land; (3) fixation of nitrogen by aquatic microorganisms, primarily cyanobacteria; and (4) rainfall, which washes gaseous nitrogen compounds from the air. Some of the nitrogen that reaches surface waters is in the organic form, and nitrogen fixation also produces organic nitrogen compounds, the nitrogen-containing components of the organisms that utilize N_2 as their nitrogen source. Organic nitrogen compounds are decomposed by bacteria and fungi, and at least part of the nitrogen is generally released as ammonia. Nitrogen-fixing microorganisms that are not consumed by large organisms die and are decomposed by other microorganisms, often in the bottom muds, again releasing ammonia. Ammonia, then, can originate from the decomposition of organic matter in water or can be introduced into the water by runoff or rainfall or as a component of domestic or industrial wastes.

Other inorganic nitrogen compounds, primarily nitrate, are also introduced into surface waters by rainfall, by runoff, and in waste streams. As we shall discuss later, nitrogen compounds more reduced than nitrate are oxidized to nitrate by bacteria that utilize these compounds as energy source. Such bacteria are strict aerobes, using NH_4^+, for example, as energy source and transferring the electrons removed from NH_4^+ to oxygen as terminal electron acceptor through electron transport coupled to ATP synthesis by oxidative phosphorylation. In water containing dissolved oxygen, NH_4^+ may be converted to nitrite and then to nitrate by bacterial oxidation. Consumption of oxygen for this purpose is the basis of the nitrogenous BOD. NH_4^+, it should be remembered, is also utilizable as a nitrogen source by most microorganisms—bacteria, fungi, and algae—so that there are two competing processes for utilization of NH_4^+, i.e., conversion to organic cell material (*assimilation*) and oxidation to nitrite and nitrate (*nitrification*). Since the organisms that oxidize ammonia can use carbon dioxide as sole carbon source, while other microorganisms that can only assimilate ammonia require organic carbon sources, the availability of organic material and other required nutrients may determine which process proceeds most rapidly.

The further transformations of inorganic nitrogen compounds also involve competing processes and depend on other factors. Nitrate, and to a lesser extent nitrite, can serve as nitrogen source for aquatic plants, algae, most fungi, and a number of bacterial species. As in the case of ammonia, the assimilation of nitrate requires that other nutrients, including carbon and energy sources, be available. The inorganic nitrogen assimilated in the form of nitrate, as in the case of ammonia, remains in the system as organic nitrogen and is subject to recycling through ammonia. Only nitrogen that is converted to gaseous products by denitrification is recycled as atmospheric nitrogen. All microorganisms that are capable of using nitrate or nitrite as terminal electron acceptor can also use oxygen and will use nitrate or nitrite only at low oxygen tension or in the absence of oxygen. These are also, with few exceptions, organisms that require organic carbon and energy sources. The requisite conditions for denitrification, then, are low dissolved oxygen concentration or anaerobiosis and the presence of required nutrients. It is possible, however, that denitrification may proceed to a considerable degree in the absence of nutrients other than an energy source, i.e., an oxidizable compound that furnishes electrons for the reduction of nitrate. As we have mentioned, bacteria can uncouple respiration and synthesis and can therefore utilize an energy source in the absence of growth. For denitrification under these conditions to be significant, a large population of denitrifying bacteria would be required initially since no increase in numbers could occur.

While low concentrations of usable nitrogen compounds are necessary to maintain the food chain in surface waters, ammonia, nitrate, and nitrite may cause undesirable effects if present in higher concentrations. If other required minerals are present, e.g., phosphorus, excess nitrate may lead to the growth of aquatic plants, and nitrate or ammonia may support "blooms" of algae. The oxidation of ammonia by nitrifying bacteria may deplete the dissolved oxygen supply, leading to fish kills and other harmful results. Ammonia itself is toxic to many aquatic organisms in concentrations of less than 1.0 mg/L. Nitrite in drinking water causes methemoglobinemia in infants ("blue babies") by combining with hemoglobin so that it cannot carry oxygen. This is a potentially fatal reaction. Nitrate in drinking water may be converted to nitrite by bacteria in the infant's intestine and produce the same effects.

Nitrite may also react with various amines to form nitrosamines, which are carcinogenic agents. Nitrate ingested in food or water may be used as electron acceptor by bacteria in the intestine, and the nitrite formed may then form nitrosamines.

The Environmental Protection Agency (EPA) has placed limits on the concentrations of all these ions in various types of water. The maximum allowable concentrations are shown in Table 7-3.

Table 7-3 Maximum allowable concentrations of inorganic nitrogen compounds

	Fresh water	Marine water	Public water supply
NH_3	0.02 mg/L	0.4 mg/L	0.5 mg/L
NO_3^-	100 mg/L NO_3^- + NO_2^-	—	10 mg/L
NO_2^-	10 mg/L	—	1 mg/L

As we pointed out in Chap. 2, biological treatment processes are engineered systems designed to carry out, on shore, reactions that would otherwise occur in the receiving water. All the transformations of inorganic nitrogen that occur in polluted waters can and do occur in various forms of biological waste treatment. In the biological treatment of wastes containing nitrogen in excess of the C/N ratio needed for synthesis of cellular material, nitrification converts excess NH_4^+ to nitrate if μ_n is sufficiently low. In secondary settling tanks under quiescent conditions, the supply of dissolved oxygen may be exhausted, and if denitrifying organisms are present in sufficient numbers, the formation of nitrogen gas may cause flotation of the sludge, i.e., a rising sludge. This leads to loss of solids in the effluent, one of the harmful effects of denitrification.

Rising sludge is a nuisance in places other than settling tanks. Sludges formed at the bottom of lakes, ponds, and other bodies of water may also rise if denitrification occurs when the oxygen supply is exhausted due to microbial activity.

Enforcement of the EPA standards for concentrations of inorganic nitrogen compounds in water may make it necessary to use denitrification as a means of removing excess nitrate from effluents before discharge. Nitrogen in excess of that needed for growth of the biomass in an aerobic treatment plant can be converted to NO_3^- by nitrifying bacteria. Then, by separate anaerobic processing, the NO_3^- is converted to nitrogen gas by denitrifying bacteria. However, since the organic substrates were removed in the preceding aerobic treatment, it is usually necessary to add a carbon and energy source. Since nitrogen does not form insoluble compounds, it is not readily removed by physical-chemical methods, and biological removal seems to be the most feasible method available. Excess nitrogen in the form of ammonia can be stripped from solution under alkaline conditions.

An economically important aspect of denitrification is its occurrence in soil. Nitrate is an excellent source of nitrogen for plants and has been used in fertilizers for many years. However, it is rapidly leached from the soil— accounting for much of the nitrate content of waters receiving runoff. If organic matter is available and if the oxygen supply is depleted because of waterlogging or microbial activity, a very significant amount of nitrate added as fertilizer can also be lost by the activity of denitrifiers in the soil.

In summary, it should be apparent that the maintenance of the nitrogen cycle, which was shown in Fig. 1-3, is dependent on the energy-generating metabolism of two groups of bacteria, the nitrifiers and the denitrifiers. Once inorganic nitrogen has been assimilated into organic matter at the reduction level of ammonia, it must be oxidized to NO_2^- or NO_3^- before being reduced to a gaseous form that can return to the atmosphere. The nitrifying bacteria use NH_4^+ and NO_2^- as energy source, oxidizing them to NO_3^-. The denitrifying bacteria, using organic compounds as energy source, pass the electrons to NO_3^- as terminal electron acceptor and form N_2, which returns to the atmosphere. While these reactions may have undesirable or harmful ecological effects, depending on the location in which they occur, they are as necessary to our life-support system as is the fixation of nitrogen, which is vital for our food supply.

PROBLEMS

7-1 $\Delta G_0'$ for the hydrolysis of glucose 6-phosphate at pH 7.0 and 25°C is -3.3 kcal/mol (see Table 7-1):

$$\text{Glucose 6-phosphate} + H_2O \rightleftharpoons \text{glucose} + H_3PO_4 \qquad \Delta G_0' = -3.3 \text{ kcal/mol}$$

Calculate the equilibrium constant K_{eq} for the reaction.

7-2 The enzyme glucose phosphate isomerase catalyzes the reversible conversion of glucose 6-phosphate to fructose 6-phosphate:

$$\text{Glucose 6-phosphate} \rightleftharpoons \text{fructose 6-phosphate}$$

If the enzyme is added to a 30-mmol solution of glucose 6-phosphate at pH 7.0 and 25°C, and the reaction is allowed to proceed to equilibrium, it is found that two-thirds of the glucose 6-phosphate has been converted to fructose 6-phosphate. Calculate K_{eq} and $\Delta G_0'$ for the reaction.

7-3 Using the redox potentials in Table 7-2, calculate the free-energy change involved in the oxidation of cytochrome b by nitrate. Is this sufficient to allow the formation of ATP if the transfer of electrons can be coupled to phosphorylation?

7-4 If an organism using nitrate as terminal electron acceptor synthesizes 2 mol of ATP per mole of $NADH_2$, what is its efficiency of energy conservation?

7-5 What groups of bacteria must use reverse electron transport? Explain why it is necessary.

7-6 What are the possible advantages to bacteria of alternative electron transport pathways to oxygen?

7-7 An anaerobe that ferments glucose by a pathway that forms 2.5 mol of ATP (net) per mole of glucose is inoculated into a medium containing 2000 mg/L of glucose and sufficient peptone and yeast extract to supply its needs for biosynthesis. Calculate the maximum amount of cells (dry weight) that could be produced if all the glucose were used as energy source.

7-8 Discuss the advantages and disadvantages of designing and operating an activated sludge aerator to deliver a "nitrified" effluent, i.e., one in which all inorganic nitrogen is in the form of nitrate.

7-9 List all the interconversions of the various forms of nitrogen (inorganic and organic) that are brought about by microorganisms. Divide these into three groups: (a) those that occur only under aerobic conditions; (b) those that occur only under anaerobic conditions; and (c) those that may occur either aerobically or anaerobically.

7-10 What are the three mechanisms by which microorganisms form ATP? Under what conditions does each mechanism function?

7-11 Why is it thermodynamically impossible for electrons from water to reduce NADP without an input of energy? Explain how this is accomplished.

7-12 List the differences between the two groups of photosynthetic microorganisms, i.e., those with plant-type and those with bacterial photosynthesis.

REFERENCES AND SUGGESTED READING

Bartsch, R. G. 1968. Bacterial cytochromes. *Annu. Rev. Microbiol.* **22**: 181–200.

Baum, S. J. 1978. Introduction to organic and biological chemistry, 2d ed. Macmillan, New York.

Benemann, J. R., and R. C. Valentine. 1971. High-energy electrons in bacteria. *Adv. Microb. Physiol.* **5**: 135–172.

Boyer, P. D., B. Chance, L. Ernster, P. Mitchell, E. Racker, and E. C. Slater. 1977. Oxidative phosphorylation and photophosphorylation. *Annu. Rev. Biochem.* **46**: 955–1026.

Brock, T. D. 1979. Biology of microorganisms, 3d ed. Prentice-Hall, Englewood Cliffs, N.J.

Casey, E. J. 1962. Biophysics. Reinhold, New York.

Cole, J. A. 1976. Microbial gas metabolism. *Adv. Microb. Physiol.* 14: 1–92.

Dawes, E. A., and P. J. Senior. 1973. The role and regulation of energy reserve polymers in microorganisms. *Adv. Microb. Physiol.* 10: 135–266.

Delwiche, C. C., and B. A. Bryan. 1976. Denitrification. *Annu. Rev. Microbiol.* 30: 241–262.

Doelle, H. W. 1975. Bacterial metabolism, 2d ed. Academic Press, New York.

Environmental Protection Agency. 1971. Water quality criteria data book. Vol. 2, Inorganic chemical pollution of freshwater, 18010 DPV 07/71. EPA, Washington, D.C.

Forrest, W. W., and D. J. Walker. 1971. The generation and utilization of energy during growth. *Adv. Microb. Physiol.* 5: 213–274.

Gest, H. 1972. Energy conversion and generation of reducing power in bacterial photosynthesis. *Adv. Microb. Physiol.* 7: 243–282.

Haddock, B. A., and W. A. Hamilton (editors). 1977. Microbial energetics. 27th Symposium of the Society for General Microbiology. Cambridge University Press, London.

Haddock, B. A., and C. W. Jones. 1977. Bacterial respiration. *Bacteriol. Rev.* 41: 47–99.

Harrison, D. E. F. 1976. The regulation of respiration in growing bacteria. *Adv. Microb. Physiol.* 14: 243–313.

Horio, T., and M. D. Kamen. 1970. Bacterial cytochromes. II. Functional aspects. *Annu. Rev. Microbiol.* 24: 399–428.

Kamen, M. D., and T. Horio. 1970. Bacterial cytochromes. I. Structural aspects. *Annu. Rev. Biochem.* 39: 673–700.

Kaplan, N. O., and E. P. Kennedy (editors). 1966. Current aspects of biochemical energetics. Academic Press, New York.

LeGall, J., and J. R. Postgate. 1973. The physiology of sulfate-reducing bacteria. *Adv. Microb. Physiol.* 10: 81–133.

Lehninger, A. L. 1971. Bioenergetics, 2d ed. W. A. Benjamin, Menlo Park, Calif.

Lehninger, A. L. 1975. Biochemistry, 2d ed. Worth, New York.

Mahler, H. R., and E. H. Cordes. 1966. Biological chemistry. Harper & Row, New York.

McCarty, P. L., and R. T. Haug. 1971. Nitrogen removal from wastewaters by biological nitrification and denitrification. *In* Microbial aspects of pollution, G. Sykes and F. A. Skinner (editors). Academic Press, New York.

Morris, J. G. 1975. The physiology of obligate anaerobiosis. *Adv. Microb. Physiol.* 12: 169–246.

Parson, W. W. 1974. Bacterial photosynthesis. *Annu. Rev. Microbiol.* 28: 41–59.

Peck, H. D. Jr. 1968. Energy-coupling mechanisms in chemolithotrophic bacteria. *Annu. Rev. Microbiol.* 22: 489–518.

Reid, G. K. 1961. Ecology of inland waters and estuaries. Reinhold, New York.

Stanier, R. Y., E. A. Adelberg, and J. Ingraham. 1976. The microbial world, 4th ed. Prentice-Hall, Englewood Cliffs, N.J.

Stouthamer, A. H. 1976. Biochemistry and genetics of nitrate reductase. *Adv. Microb. Physiol.* 14: 315–375.

Stryer, L. 1975. Biochemistry. W. H. Freeman, San Francisco.

Suzuki, I. 1974. Mechanisms of inorganic oxidation and energy coupling. *Annu. Rev. Microbiol.* 28: 85–101.

Thauer, R. K., K. Jungermann, and K. Decker. 1977. Energy conservation in chemotrophic anaerobic bacteria. *Bacteriol. Rev.* 41: 100–180.

Wagner, R. H. 1974. Environment and man, 2d ed. W. W. Norton, New York.

Wald, G. 1966. On the nature of cellular respiration. *In* Current aspects of biochemical energetics, N. O. Kaplan and E. P. Kennedy (editors). Academic Press, New York.

Walker, D. A. 1970. Photosynthesis. *Annu. Rev. Biochem.* 39: 389–428.

White, A., P. Handler, and E. L. Smith. 1973. Principles of biochemistry, 5th ed. McGraw-Hill, New York.

White, D. C., and P. R. Sinclair. 1971. Branched electron-transport systems in bacteria. *Adv. Microb. Physiol.* 5: 173–211.

EIGHT

METABOLIC CLASSIFICATION OF MICROORGANISMS

It is desirable in any science to utilize classification wherever possible to enhance learning and communication by organizing information in some logical manner. Just as the chemist classifies certain molecules as carbohydrates and others as proteins, living organisms must be grouped according to common characteristics that we hope will indicate relationships based on common origins. Taxonomy is the branch of biological science that seeks to discover relationships between living organisms so that a natural system for their classification can be devised.

For the higher organisms, the plants and animals, the following graded series or hierarchy of groupings is used:

Kingdom
Phylum (or division)
Class
Order
Family
Genus
Species

Attempts to apply this hierarchy to microorganisms generally lead to many difficulties, especially in the case of bacteria. In most groupings of bacteria, the highest category recognized is the family. Orders are recognized for some groups of bacteria, and in others genera are not even grouped into families. In one family, genera have been grouped into tribes. In short, classification of bacteria uses no completely consistent scheme. We shall use the terms *order*, *family*, *genus*, and *species* where appropriate. The significance of these terms as applied to microorganisms is the same as when they are used to classify higher organisms; that is, a genus is a group of related species, a family is a group of related genera, etc. Relationship is judged on the basis of common characteristics.

The two general types of characteristics useful to the taxonomist are structural details (morphology) and physiology. The taxonomic study of microorganisms is especially difficult because their small size requires microscopic examination (and many bacteria have no distinctive morphological characteristics), while study of their physiology generally requires that they be grown in pure culture in the laboratory. The taxonomy of eucaryotic microorganisms is based primarily on morphological differences since their cellular structures are complex and well differentiated, whereas the bacteria are classified on the basis of morphology in some cases and physiology in others. The taxonomic schemes presently used for the procaryotic microorganisms are constantly being revised as more is learned about their biochemistry and physiology and as the electron microscope allows more detailed study of cellular structure.

We do not intend to present an extensive review of microbial taxonomy. The identification of most organisms, particularly of eucaryotes, at the species level requires specialized expertise and much laboratory experience, and the identification of most microbial species should seldom be attempted by those not trained to do so. It is possible, however, to take a broader approach to classification by defining categories of microorganisms based on their metabolic capabilities. Since these capabilities determine the types of environments in which microbial species can flourish, it is important and useful to the environmental engineer and scientist to be familiar with these groups and with some of the more common genera and species included in each group. This information, combined with the ability to recognize the distinguishing morphological characteristics of various types of microorganisms under the microscope, is valuable in many situations.

Periodic microscopic observation of the biomass and notation of the predominating forms and particularly of changes in predominating types are useful tools when correlated with chemical and biological tests for assessing the condition of the treatment plant as well as that of the receiving stream. Such data as the sizes of flocs, types of organisms included in the flocs, types and relative numbers of filamentous organisms, and types and relative numbers of protozoa may often warn of an impending upset in the operation of a treatment plant if base-line data from periods of normal operation are available.

THE BROAD METABOLIC GROUPS

In Chap. 5 we discussed the nutritional requirements of microorganisms and their importance in the selection of microbial species. While the presence or absence of a specific growth factor requirement or a specific utilizable carbon

Table 8-1 Broad metabolic groups

Carbon source	
Inorganic	Autotroph
Organic	Heterotroph
Energy source	
Chemical	Chemotroph
Light	Phototroph
Electron donor	
Inorganic	Lithotroph
Organic	Organotroph

source may determine whether an individual species can proliferate in a specific environment, broader considerations determine whether any members of a genus or family are capable of proliferating under a given set of conditions. It is these criteria that we will consider as general determinants of the ecological distribution of microorganisms.

The three nutritional requirements of microorganisms that are quantitatively most important are the *carbon source*, the *energy source*, and the *electron donor*. These, along with the *electron acceptor*, are determined by, and therefore indicative of, the enzymatic makeup of the cell. In other words, the metabolic capabilities of any organism may be broadly described by a classification based on these nutritional categories (see Table 8-1).

The two broad categories of carbon sources available to living organisms are organic and inorganic. Organisms that require organic compounds as their sole or principal source of carbon for synthesis of cell material are classified as *heterotrophs*, while those that can use carbon dioxide as the sole or principal carbon source are *autotrophs*. Organisms may obtain energy either from light or by oxidation of chemical compounds. Those that utilize light energy are *phototrophs*, while those that oxidize organic or inorganic compounds are *chemotrophs*. The electron donor may be organic or inorganic. Organisms that use an organic compound as a source of electrons are *organotrophs*, while those that use an inorganic electron source are *lithotrophs*. These categories are usually combined; for example, an organism that uses an inorganic electron donor might be a photolithotroph or a chemolithotroph, depending on the energy source used.

As we shall see, metabolic categories are not so clearly separated as we might wish. One problem in their use, which presents no great difficulty, is overlapping of categories. When light is the energy source, the electron donor is clearly distinguished from the energy source. When a chemical energy source is used, however, it normally serves also as the electron donor and often also as the sole or principal carbon source if the organism is a heterotroph.

The second problem in the metabolic classification of microorganisms poses greater difficulties. The versatility of microbial metabolism, which becomes more apparent as our knowledge increases, makes it difficult to assign many microbial species to a single metabolic category. It is preferable, and will probably become common, to speak of an autotrophic mode of life or the capacity for autotrophic growth rather than of an autotrophic species. As more was learned about many organisms previously classified as autotrophs, it became necessary to distinguish between strict, or obligate, autotrophs and facultative autotrophs. Similar modifications were necessary for the definitions of lithotrophs and phototrophs. These modifications resulted from reports that organisms previously considered capable of using only a single mechanism for generating energy or a single class of carbon source or electron donor could, with varying degrees of efficiency, utilize alternative sources of carbon, energy, or electrons. The term *mixotroph* has been used to describe organisms that are able to use alternative energy sources (e.g., chemical and light) with equal facility or even to utilize alternative energy sources simultaneously.

In spite of the problems involved in classifying some microbial species, these categories are still the most useful means of grouping organisms with similar metabolic characteristics. In the descriptions of metabolic groups of microorganisms that follow, we will depart from the recognized taxonomic groupings in many cases in order to bring together organisms that have common metabolic characteristics.

PHOTOTROPHS

All microorganisms that are capable of utilizing light to synthesize ATP by either cyclic or noncyclic photophosphorylation are classified as phototrophs. Among microorganisms these include the algae, the cyanobacteria (blue-green algae), and three families of bacteria: the purple nonsulfur bacteria, family Rhodospirillaceae; the purple sulfur bacteria, family Chromatiaceae; and the green sulfur bacteria, family Chlorobiaceae.

Photolithotrophy and Photoautotrophy

Photolithotrophic growth requires *light* and an *inorganic electron donor* and is usually associated with the ability to use carbon dioxide as the sole source of carbon, i.e., *photoautotrophy*. Many of these organisms require specific growth factors, usually the vitamins B_{12}, biotin, or thiamine, which may be supplied in the very small amounts required, approximately 10^{-6} mg/L, by other organisms capable of synthesizing them. Although almost all photolithotrophs are capable of autotrophic growth, they are also able to assimilate various types of organic compounds simultaneously with the fixation of carbon dioxide and use them for synthesis of cell material or storage products such as poly-β-hydroxybutyrate, glycogen, starch, or lipids. Many species are therefore at least potentially mixotrophic.

Algae The algae are primarily aquatic organisms with a variety of fresh-water and marine species. They are the only photosynthetic eucaryotic microorganisms. Algae are separated into taxonomic groups on the basis of differences in photosynthetic pigments, cell wall composition, storage products (mostly starch and fats), and morphology. Some nonphotosynthetic organisms (leucophytes) that are morphologically identical with photosynthetic species are considered to be algae that have lost their photosynthetic apparatus, and these are included with the appropriate morphological groups among the algae. The ability of such organisms to survive is indicative of their previous ability to grow nonphotosynthetically, and this is a property of many algae. Storage products or exogenous organic matter may be metabolized aerobically in the dark. Algae thus use carbon dioxide and produce oxygen in the light but use oxygen and produce carbon dioxide in the dark.* Water is the exclusive *electron donor* under natural conditions in the light for synthesis of the NADPH$_2$ required to reduce carbon dioxide to the level of carbohydrate for autotrophic growth. A few algal species, e.g., *Scenèdesmus*, are able to use hydrogen gas as an electron donor anaerobically in the dark.

Algae are particularly important as primary producers of organic matter and oxygen. The unicellular forms are a vital part of the aquatic food chain. About 1 lb of fish is produced for each 10 lb of plankton that enters the food chain. The marine phytoplankton are responsible for approximately 90 percent of the photosynthesis on earth, which annually converts 150 billion tons of carbon into organic matter and produces 400 billion tons of oxygen. Excessive growth of algae (algal blooms) can be a severe nuisance, causing fish death by toxin production or suffocation, making water unfit for recreational use or use as a water supply, and causing serious organic pollution when the organisms comprising the bloom die. Certain species are responsible for taste and odor problems even though their numbers remain well below the "bloom level." Algae do not fix atmospheric nitrogen and they require inorganic nitrogen in the form of NO_3^- or ammonia; some species are able to use amino acids and other organic nitrogen compounds as a source of nitrogen. The growth of large numbers of algae therefore occurs only in waters that contain sufficient concentrations of combined nitrogen in addition to phosphorus and other minerals to support rapid growth.

Cyanobacteria The cyanobacteria are still classified as blue-green algae (phylum Cyanophyta) by the botanists, and perhaps rightly so since their photosynthesis is of the plant type; that is, they use water as electron donor for noncyclic photophosphorylation and reduction of NADP and produce oxygen. Their morphology is varied but much less so than that of the eucaryotic algae, and their taxonomy has been described as a "Pandora's

---

*Respiration, using O_2 and producing CO_2, can occur concomitantly with photosynthesis in the light. However, the rate is low compared with the rate of photosynthesis, and the net effect is CO_2 utilization and O_2 production.

box" (Lewin, 1974). Like the algae, they are able either to assimilate simple organic compounds while utilizing carbon dioxide as the major carbon source, or to grow in a completely inorganic medium, i.e., as photoautotrophs. A few species require vitamin B_{12}. Unlike many algae, most blue-green bacteria cannot grow in the dark using organic compounds, although a few species can grow very slowly at the expense of glucose. Viability during hours of darkness is maintained by endogenous respiration of stored glycogen (a storage product typical of bacteria but not of algae).

Cyanobacteria are among the most ubiquitous microorganisms, being found in environments that support few other forms of life. This wide distribution is a consequence of the ability of many species of cyanobacteria to use atmospheric nitrogen as the sole nitrogen source. This means that they require for growth only two gases, carbon dioxide and nitrogen, water as electron donor, and the usual inorganic elements, the most important of which, quantitatively, is phosphorus. Another characteristic contributing to their persistence in a variety of habitats is their ability to survive desiccation (in some known cases for many years) and exposure to extremes of temperature, both high and low. Cyanobacteria are found in all aqueous environments, including hot springs up to 75°C since some species are thermophilic. They are important soil organisms, contributing to soil fertility through their ability to fix nitrogen. Like the algae, they often appear in tremendous numbers as blooms in waters that contain sufficient concentrations of the required minerals. Toxins produced by blooms of cyanobacteria render water unfit for consumption by animals or humans by causing sickness and sometimes death, and they are extremely lethal to fish and other aquatic life. In lower numbers, the growth of some species results in unpleasant tastes and odors. Many cyanobacteria contain gas vacuoles, which enable them to float at the surface of the water or at a depth at which light intensity is optimum for photosynthesis. The gas contained in the vacuoles is probably nitrogen.

The green and purple sulfur bacteria The sulfur bacteria are divided into two families on the basis of the color imparted to the cells by the photosynthetic pigments, primarily the carotenoids. Many species of Chromatiaceae, the purple sulfur bacteria, are actually purple, but others may be dark orange to brown or various shades of pink or red. The Chlorobiaceae, or green sulfur bacteria, include two subgroups—those that actually are green and a few brown species. Both families are separated into genera on the basis of cell morphology. The purple bacteria are in general much larger than the green bacteria. Both families include both cocci and rods, and the cells of one genus of purple bacteria are spirals.

The metabolic characteristics of the two families are similar in many respects. All species except one are obligate phototrophs and obligate anaerobes. Energy is generated by cyclic photophosphorylation, and all species are capable of growth in completely inorganic media, i.e., as *photo-*

lithotrophs and *photoautotrophs*, using hydrogen sulfide as *electron donor* for the reduction of NADP by reverse electron transport and synthesizing all cell material from carbon dioxide. Most species can use hydrogen as an alternative electron donor, although growth is usually much slower than with hydrogen sulfide. These species may still require a small amount of hydrogen sulfide as a source of sulfur for biosynthesis. When hydrogen sulfide is used as electron donor, it is oxidized to elemental sulfur, which accumulates in globules located inside the cell in all species of purple bacteria except one or outside the cell in one species of purple bacteria and all green bacteria. These sulfur globules are visible in the light microscope as shiny particles. Many, but not all, species can utilize elemental sulfur, sulfide, and thiosulfate as electron donors and form sulfur from sulfide or thiosulfate as an intermediary product, which is later oxidized to sulfate. Oxygen is, of course, not produced in photosynthesis and water cannot be utilized as an electron donor.

All species of sulfur bacteria are able to assimilate simple organic compounds such as acetate while deriving the major portion of cell carbon from carbon dioxide, and they are therefore potential mixotrophs. The ability to derive cell carbon from organic compounds is more prevalent among the purple than the green bacteria. The green sulfur bacteria can assimilate organic compounds only if both carbon dioxide and hydrogen sulfide are also available. Some species of purple sulfur bacteria can utilize organic electron donors in place of hydrogen sulfide, thus growing photoorganotrophically, and some can utilize acetate as sole source of carbon, growing photoheterotrophically. In all species except one, energy must be derived from light and cannot be obtained by the oxidation of exogenous organic substrates. One species is capable of very slow growth as a heterotroph in the dark. Storage products, which may be utilized in the dark, are commonly poly-β-hydroxybutyrate, polysaccharides, and polyphosphate in the purple bacteria, and polyphosphate and sometimes polysaccharide in the green bacteria. Most species of both families are capable of using nitrogen gas as sole nitrogen source when other forms of inorganic nitrogen are not available. Many species are also capable of using organic nitrogen.

The habitats of the two families are similar, as might be expected from the similarity of their metabolic capabilities. Growth may occur in any aquatic environment to which light of the required wavelength penetrates if carbon dioxide, nitrogen, and a reduced form of sulfur, or hydrogen, are available, along with required inorganic salts, and if no oxygen is present. These organisms are found in fresh-water and marine habitats, in water or mud, at depths below those occupied by the algae, cyanobacteria, and other aerobic bacteria (see Fig. 8-1). Anaerobic fermentative bacteria at lower depths produce hydrogen sulfide from organic sulfur compounds. In sulfate-containing environments, sulfide is a product of sulfate reduction by organisms that use anaerobic respiration. The fermentative organisms also produce carbon dioxide and hydrogen. The aerobic organisms nearer the surface remove oxygen from solution. The wavelengths of light used by the sulfur bacteria are

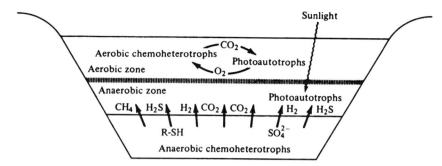

Figure 8-1 Aerobic and anaerobic zones in a deep oxidation pond. In the aerobic zone, aerobic chemoheterotrophic bacteria oxidize organic matter to CO_2 and H_2O using O_2 as terminal electron acceptor. Aerobic photoautotrophs, green algae and cyanobacteria, fix CO_2 into organic matter, using H_2O as electron donor and producing O_2, completing the carbon-oxygen cycle in this zone. In the anaerobic zone, anaerobic chemoheterotrophs ferment organic matter to form a variety of fermentation products. Among these are H_2S and CO_2, which can be used by the anaerobic photoautotrophs, the green or purple sulfur bacteria. CO_2 is fixed to form organic matter, using H_2S as electron donor. H_2 can serve as an alternative electron donor.

different from those used by the cyanobacteria or algae, and the sulfur bacteria are thus able to grow using light that has passed through the surface layer of water or mud occupied by aerobic photosynthetic organisms. The sulfur bacteria are commonly found at a specific depth, often in a fairly thin layer where conditions of light and nutrition are optimum.

Both the purple and green sulfur bacteria include two subdivisions that have somewhat different ecological distributions. Within each family, one subgroup requires low light intensity, low temperatures (10 to 20°C), and low sulfide concentration. The second group prefers higher temperatures, light intensity, and sulfide concentration. The two groups within each family can be distinguished morphologically, since the groups that prefer lower light intensity, sulfide concentration, and temperatures have gas vesicles, which may have survival value in allowing them to live where conditions are optimal. The organisms of the other groups do not form gas vesicles. Light intensity and sulfide concentration also determine the mode of growth of these organisms; this led to difficulty in their classification before it was found that single cells that are motile by means of flagella become immotile and form large slime-embedded aggregates at high light intensity and high sulfide concentration. The two environmental factors interact, and each species develops flagella and becomes motile at specific combinations of light intensity and sulfide concentration. The lower the light intensity, the higher is the sulfide concentration at which cells remain motile. Both green and purple sulfur bacteria are capable of proliferating sufficiently to produce a readily visible colored layer in water, i.e., a bloom. This occurs in environments where sulfate (and sulfate-reducing bacteria) is abundant. Since sulfate seldom is entirely absent from any natural water, these organisms are found in almost any anaerobic environment exposed to light.

Photoorganotrophy and Photoheterotrophy

Photoorganotrophic growth requires light as energy source and an organic compound as electron donor. Since ATP is generated by cyclic photophosphorylation, requiring no external electron input, the need for electrons is related to the nature of the carbon source utilized. To the extent that carbon dioxide is the carbon source, electrons are required for its reduction. Organic compounds generally serve two purposes in these organisms—they donate electrons for the reduction of carbon dioxide and serve as carbon source along with carbon dioxide.

The *purple nonsulfur bacteria*, the Rhodospirillaceae, are the only common microorganisms included in the photoorganotrophic category. These organisms comprise one family with three genera differentiated primarily on the basis of morphology. All have similar metabolic capabilities but differ in some respects among species. This family is an outstanding example of the metabolic versatility that creates difficulties for the bacterial taxonomist. Their normal mode of growth, as determined by the natural habitats in which they occur and by optimum growth conditions in laboratory cultures, is photoorganotrophic and photoheterotrophic. Both organic compounds and carbon dioxide are used as carbon source; the carbon dioxide is reduced by electrons donated by organic compounds. Unlike the sulfur bacteria, however, these organisms are generally facultative phototrophs and facultative anaerobes. Most species can grow well in the light or the dark under aerobic conditions using organic compounds as carbon and energy source, i.e., chemoheterotrophically. Synthesis of the photosynthetic pigments does not occur in the presence of oxygen. The nonsulfur bacteria are inhibited by the presence of hydrogen sulfide except at very low concentrations, but some species are able to grow using hydrogen sulfide as electron donor and carbon dioxide as carbon source at low sulfide concentrations, i.e., as photolithotrophic autotrophs. However, the sulfur formed in this way is not oxidized further. Many species are also capable of using hydrogen as the electron donor for the reduction of carbon dioxide. Like the other photosynthetic bacteria, most purple nonsulfur bacteria are able to fix nitrogen.

The purple nonsulfur bacteria do not proliferate to an extent that could be classified as a bloom, probably because they require carbon dioxide, organic matter, light, and anaerobic conditions for good growth. The sulfur bacteria require light of the same wavelengths, but they have less complex growth requirements and normally can compete successfully with the nonsulfur bacteria. The inhibition of the latter by hydrogen sulfide, which is usually produced in large amounts in any anaerobic environment containing organic matter, contributes to the predominance of the sulfur bacteria. The nonsulfur bacteria are often found growing in association with sulfur bacteria that are present in sufficient numbers to reduce the sulfide concentration to a level that the nonsulfur bacteria can tolerate.

A *green nonsulfur bacterium* that is photoorganotrophic and photo-

Table 8-2 Phototrophic microorganisms

	Photophos- phorylation	Electron Donor	Relation to O_2	Major Carbon source	N_2 fixation	Growth in the dark
Eucaryotes						
Algae	Noncyclic	H_2O	Aerobic	CO_2	No	Yes
Procaryotes						
Cyanobacteria	Noncyclic	H_2O	Aerobic	CO_2	Yes	No
Purple bacteria						
Sulfur	Cyclic	H_2S, H_2	Anaerobic	CO_2	Yes	No
Nonsulfur	Cyclic	Organic	Facultative	Organic	Yes	Yes
Green bacteria	Cyclic	H_2S, H_2	Anaerobic	CO_2	Yes	No

heterotrophic has recently been discovered growing in hot springs in association with cyanobacteria that apparently supply it with the required organic nutrients. This is a filamentous gliding bacterium, *Chloroflexus*. Little is known of its metabolism.

The metabolic characteristics of the photosynthetic microorganisms are summarized in Table 8-2.

CHEMOTROPHS

All organisms that obtain energy by oxidizing chemical compounds or elements are classified as chemotrophs. Depending on the individual species, the energy source may be organic or inorganic, and oxidation may occur under aerobic or anaerobic conditions. ATP may be synthesized by oxidative phosphorylation or by substrate level phosphorylation. Among the microorganisms, the chemotrophs include all the fungi, all the protozoa, and the vast majority of bacteria.

Chemolithotrophy and Chemoautotrophy

Chemolithotrophic growth requires an inorganic compound or element that serves both as energy source for the generation of ATP through oxidative phosphorylation and as electron donor for the reduction of the carbon dioxide that serves as sole or principal carbon source. Energy sources used for chemolithotropic growth include NH_3, NO_2^-, Fe^{2+}, H_2S, S^0, $S_2O_3^{2-}$, and H_2. Almost all chemolithotrophs are obligate aerobes, although a few aerobic species can use nitrate as an alternative terminal electron acceptor. The chemolithotrophs include only about 30 species, all bacteria, but they are an important group of microorganisms because their activities are essential to the cycling of inorganic matter, particularly nitrogen and sulfur.

The nitrifying bacteria The family Nitrobacteraceae includes two groups—one that oxidizes ammonia to nitrite and another that oxidizes nitrite to nitrate (the actual nitrifiers). Within each group, several genera, most of which include only one species, are differentiated on the basis of cell morphology. *Nitrobacter*, which oxidizes nitrite, and *Nitrosomonas*, which oxidizes ammonia, are the most common and well-studied genera. Both are small Gram-negative rods found in soil and fresh water. *Nitrobacter* is also found in salt water. Both organisms grow either as single cells or as masses embedded in slime. Two other genera of nitrite-oxidizing bacteria, *Nitrospina* and *Nitrococcus*, grow only in salt water. Besides *Nitrosomonas*, two other genera that oxidize ammonia, *Nitrosospira* and *Nitrosolobus*, are soil and fresh-water inhabitants, while the fourth genus, *Nitrosococcus*, includes one fresh-water species and one marine species. The nitrifying bacteria as a group are thus widespread in soil and fresh and salt water and may be expected to be present wherever ammonia, carbon dioxide, and oxygen are present along with other required inorganic salts. Almost all nitrifying bacteria have an optimum pH in the alkaline range, usually near 8.0, and grow only slowly at pH values much below neutral. They require no organic growth factors and are capable of growing in completely inorganic media using carbon dioxide as the sole source of carbon.

Chemolithotrophs and organic compounds The nitrifying bacteria were the first chemolithotrophs to be isolated. Winogradsky, who first succeeded in growing them in pure culture in the early 1890s, succeeded only by using a completely inorganic medium. He described these organisms as being totally dependent on the oxidation of an inorganic compound as the sole source of energy and upon carbon dioxide as the sole source of carbon, and he further stated that organic compounds could not be used and were in fact inhibitory. This early description of the nitrifying bacteria as obligate chemolithotrophic autotrophs unable to grow in the presence of organic matter was extended to the sulfur-oxidizing bacteria when they were discovered a few years later and was the basis for classifying many chemolithotrophic bacteria as obligate autotrophs.

Recent studies have shown that two of the generalizations stated by Winogradsky are incorrect. With the exception of one strain of *Nitrobacter* that grows very slowly using acetate as the sole source of carbon and energy, it is true that no organic compound has been shown to support the growth of the nitrifying bacteria in the absence of carbon dioxide and the normal inorganic energy source. However, the testing of possible growth substrates has by no means been exhaustive, and it is possible that some organic compounds are able to support the growth of these organisms. The statements that nitrifying bacteria cannot use organic compounds and that organic compounds inhibit their growth have proved to be incorrect. These organisms have the same relationship to organic compounds as do the obligate photolithotrophs, i.e., the cyanobacteria and the purple and green sulfur bacteria. They are able to assimilate organic compounds only if carbon dioxide and the inorganic energy source are also available. The bulk of the cell carbon is always derived from carbon dioxide, but up to 10 percent of the cell carbon may be derived from such organic compounds as simple sugars, organic acids, and amino acids. Organic carbon is thus not required by any of the nitrifying bacteria, but all are capable of utilizing at least certain organic compounds, when they are available, for the synthesis of some cell constituents. Organic compounds are not generally inhibitory, as they were thought to be for many years. Specific compounds, sometimes in concentrations of only a few milligrams per liter, do inhibit individual species. Amino acids are the most common inhibitory compounds, but in several cases it has been shown that the inhibitory effect of one amino acid can be counteracted by the presence of other amino acids. This type of inhibition is not limited to chemolithotrophs but occurs in chemoorganotrophs as well, e.g., in *Escherichia coli*, and probably can be explained on the basis of a common metabolic control mechanism, which we will discuss in Chap. 12.

The inability of the photolithotrophic and chemolithotrophic autotrophs to grow on organic compounds is a puzzle that has not been solved. Several theories have been proposed, but none has been confirmed by experimental evidence. Among these theories are impermeability to organic compounds, accumulation of toxic metabolic products, and lack of the electron transport components or enzymes required to couple oxidation of organic compounds to energy generation. The fact that some organic compounds used, along with carbon dioxide, by these organisms do not contribute carbon to all constituents of the cell, as shown by radioactive labeling experiments, suggests that some organisms may lack the enzymes required for conversion of carbon sources other than carbon dioxide to all of the compounds that make up the cell. It appears likely that there may be different reasons why different species are unable to use organic compounds as the sole source of carbon and energy. Whatever the reason, it now seems certain that all microorganisms that are capable of autotrophic growth are able to use at least some organic compounds for some synthetic reactions, although many can use organic carbon to only a limited extent and only as a supplement to the autotrophic mode of growth. Thus, there are no obligate autotrophs in the original meaning of the term, that is, no organisms that must obtain all their carbon from carbon dioxide. There are a number of species that *must* use carbon dioxide as a major carbon source and *can* use it as the sole carbon source.

We have discussed this subject at some length because the concept of obligate autotrophy is an important idea, which has recently been much discussed (Smith and Hoare, 1977; Whittenbury and Kelly, 1977). The relationship of chemolithotrophs, particularly the nitrifying bacteria, to organic compounds is an important consideration to the environmental engineer and scientist.

The colorless sulfur bacteria The aerobic bacteria that oxidize sulfur are often called the *colorless sulfur bacteria* to distinguish them from the sulfur-oxidizing photosynthetic bacteria. The chemotrophic sulfur-oxidizing bacteria have a greater variety of metabolic capabilities than do the nitrifying bacteria, but we will discuss them as one group because of their ecologically important role in the sulfur cycle. The colorless sulfur bacteria include four genera, *Thiobacillus, Thiodendron, Beggiatoa,* and *Sulfolobus,* that are capable of using reduced sulfur as an energy source, i.e., of lithotrophic growth, and seven genera that form internal sulfur granules but whose metabolism has not been investigated because they have not been grown in pure culture. The latter are *Thiothrix,* a filamentous nongliding bacterium that grows attached to solid surfaces in sulfur springs and other sulfide-rich habitats; *Thioploca,* a segmented, filamentous gliding bacterium with a common sheath surrounding a variable number of filaments; *Achromatium,* very large single cells that contain crystals of calcium carbonate as well as sulfur granules; *Thiobacterium,* rods embedded in a gelatinous material; *Macromonas,* large motile cylindrical cells that contain both sulfur and calcium carbonate; *Thiovulum,* large motile ovoid cells that grow in a web at the interface of oxygen-containing and sulfide-containing layers in water; and *Thiospira,* motile spiral cells. Since these organisms occur in fresh or salt water containing hydrogen sulfide and deposit sulfur internally, it is assumed that they oxidize sulfide to sulfur.

Beggiatoa is a gliding filamentous organism that can grow as a chemolithotrophic autotroph but is classified as a mixotroph because it grows better when supplied with acetate. It deposits sulfur internally. *Sulfolobus* is an unusual organism, which was discovered only a few years ago in a very acid hot spring. Its temperature range is 55 to 85°C, with the optimum at 70 to 75°C, and its pH range is 0.9 to 5.8, with an optimum of 2 to 3. It is capable of growing as a chemolithotrophic autotroph using elemental sulfur as an energy source, but it can also grow as a chemoorganotrophic heterotroph using organic compounds for carbon and energy. *Thiodendron,* a stalked bacterium, is known to require hydrogen sulfide for growth, but its metabolism has not been studied.

The remaining sulfur bacteria are all species of *Thiobacillus,* which is the most thoroughly studied of the colorless sulfur bacteria. The eight species of *Thiobacillus* exhibit an interesting variety of metabolic patterns. All are aerobes, using oxygen as terminal electron acceptor, but one species, *T. denitrificans,* is capable of using nitrate as terminal electron acceptor and reducing it to nitrogen gas. Four of the eight species are capable of growing as chemolithotrophic autotrophs, require carbon dioxide and sulfur or a reduced sulfur compound as energy source and electron donor, and can assimilate only small amounts of organic carbon. A fifth species, *T. ferrooxidans,* has similar metabolic properties but is able to oxidize either ferrous iron or sulfur compounds. Two of these five species, *T. thiooxidans* and *T. ferrooxidans,* are capable of growing well at very acid pH values, whereas the other three grow best at pH values closer to 7.

The three remaining species can grow in organic media but differ from each other in metabolic capabilities. *T. novellus* may be classified as a facultative autotroph, since it is capable of growing in either completely inorganic or organic medium. The only organic compounds that allow good growth are two amino acids, glutamic and aspartic. Sugars are not used. When organic carbon and energy sources are present, the chemolithotrophic energy source, thiosulfate, is not used. *T. intermedius* is able to use any of several sugars, aspartate, or glutamate as carbon and energy source and grows best with both glucose and glutamate. Unlike *T. novellus*, it is able to oxidize sulfur compounds and organic compounds simultaneously and is therefore capable of three modes of growth—as a chemolithotrophic autotroph, a chemoorganotrophic heterotroph, or a mixotroph. The eighth species is especially interesting and unusual and can be classified as an obligate mixotroph. This organism, *T. perometabolis* (which means crippled metabolism), cannot grow autotrophically and grows very poorly on organic medium. Both organic compounds and reduced sulfur compounds are required for good growth and both are used simultaneously.

Most species of *Thiobacillus* are able to oxidize several sulfur compounds as well as elemental sulfur. In oxidizing either hydrogen sulfide or thiosulfate, $S_2O_3^{2-}$, elemental sulfur is formed and accumulates as globules, which are later oxidized (except by *T. novellus*, which cannot utilize S^0). All *Thiobacillus* species are small rods, and the sulfur is deposited outside the cell. The final product of oxidation is sulfate. This formation of sulfuric acid produces the very low pH values, 1.0 or less, in cultures of *T. thiooxidans* and is responsible for the very corrosive effects caused by the growth of this species. Concrete sewer pipe and concrete installations, such as pumping stations, are rapidly destroyed by the acid produced by the growth of *Thiobacillus* when sewage becomes anaerobic, leading to the production of large quantities of hydrogen sulfide (see Fig. 8-2).

The iron bacteria The iron bacteria are so designated because of one common characteristic that may be incidental to their metabolism. All of these organisms are commonly found in water that contains iron and/or manganese, and the growths are encrusted with oxides of iron and/or manganese. One family, Siderocapsaceae, including four genera, is tentatively classified among the chemolithotrophs on the basis of the deposition of iron or manganese oxides, although little information is available on metabolism. Most or all probably grow on organic compounds. Capsules are generally formed, and it is in or on these capsules that iron oxide is deposited.

Several genera of the sheathed bacteria, *Sphaerotilus*, *Leptothrix*, *Lieskella*, *Crenothrix*, and *Clonothrix*, also are found frequently in water containing iron or manganese and have encrustations of metallic oxides on the sheath or on the capsular material that often surrounds the sheath. *Leptothrix* has been reported to oxidize manganese, though only outside the cell through excretion of a protein. This would not constitute an energy-yielding, or even a physiologically significant, reaction.

It is extremely difficult to show that any of these organisms can use the

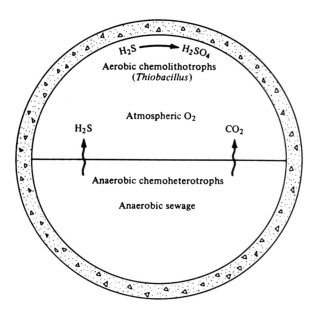

Figure 8-2 Crown corrosion in a concrete sewer. Anaerobic chemoheterotrophs produce H_2S, which is absorbed in condensate on the walls above the water line. Aerobic chemolithotrophs oxidize the reduced sulfur to SO_4^{2-}, forming H_2SO_4, which attacks and erodes the concrete, often to the point of structural failure.

oxidation of ferrous iron as an energy source, since none grows in acid pH environments and ferrous iron is rapidly oxidized nonbiologically at neutral or alkaline pH in the presence of oxygen, which all iron bacteria require.

Gallionella, a stalked bacterium, has been grown in the laboratory in inorganic media containing ferrous sulfide. The twisted stalk, which bears a small bean-shaped cell at the end, is covered with ferric hydroxide, which may account for 90 percent of the total mass (dry weight). *Gallionella* is probably capable of chemolithotrophic growth using ferrous iron as energy source and electron donor and carbon dioxide as carbon source. The only other chemolithotrophic organism that can use ferrous iron as the sole energy source may well be *Thiobacillus ferrooxidans*, which grows well at the low pH values at which ferrous iron is stable to nonbiological oxidation.

The hydrogen bacteria The hydrogen-oxidizing bacteria were previously placed in the genus *Hydrogenomonas*. These bacteria are capable of chemolithotrophic growth using hydrogen as energy source and electron donor and carbon dioxide as sole carbon source. However, all hydrogen-oxidizing bacteria are also capable of growing chemoorganotrophically as heterotrophs, using a wide variety of organic compounds as sources of carbon and energy. Based on other characteristics, these organisms can be placed in several chemoorganotrophic genera, and the ability to grow chemolithotrophically with hydrogen is no longer considered a sufficiently unique characteristic to warrant separate classification. Six hydrogen-oxidizing, facultative chemoautotrophs are classified as species of *Pseudomonas* and two as species of *Alcaligenes*.

The methane-forming bacteria (methanogens) The methane bacteria are a unique group included here because they share some properties with the aerobic chemolithotrophs, although other characteristics are very different. All methane-producing bacteria are included in the family Methanobacteriaceae, which is divided into three genera on the basis of cell morphology. Progress in understanding the metabolism of these organisms has been very slow because few pure cultures have been available until recently and work with them is very difficult because of their extreme sensitivity to oxygen. They are strict anaerobes that occur naturally only in completely anaerobic environments such as the bottom sediments in lakes, ponds, and swamps, the rumen of cattle and sheep, and anaerobic sewage digesters. Methane formation in anaerobic environments rich in organic matter is vigorous, and rapid bubbling often can be observed in swamps and polluted ponds. When ignited by lightning, the methane released may burn for some time. These fires were called "will-o'-the-wisp," but at least one author has suggested that they are now called UFOs (Stadtman, 1967).

The metabolic capability shared by the methanogenic bacteria and other lithotrophic bacteria is the use of hydrogen as energy source and carbon dioxide as electron acceptor. This mode of energy generation fits the definition of chemolithotrophy. The mechanism by which energy is trapped as ATP is not known. The use of carbon dioxide as electron acceptor is analogous to the use of nitrate by the chemolithotrophic *Thiobacillus denitrificans*, but the methane bacteria have not been shown to contain the usual components of an electron transport chain that could be coupled to ATP formation by oxidative phosphorylation using carbon dioxide as terminal electron acceptor for anaerobic respiration. No cytochromes or quinones have been detected in the methane bacteria. Several of the methane bacteria that have been grown in pure culture are able to use carbon dioxide as both sole carbon source and electron acceptor, can grow in completely inorganic media using ammonia as nitrogen source, and are therefore chemolithotrophic autotrophs. Organic compounds such as acetate may stimulate their growth and may be required as carbon source by other species. Formate, HCOOH, is used by most methanogenic bacteria but is probably first decomposed to carbon dioxide and hydrogen. The use of carbon dioxide as electron acceptor results in the formation of methane, and approximately 90 percent of the carbon dioxide used is converted to methane in organisms that use carbon dioxide also as carbon source. Several of these organisms require vitamins, and the rumen organisms require a "rumen factor" of unknown nature.

The methane bacteria play an important role in the degradative leg of the carbon cycle and have unique ecological relationships with other organisms. Some of these relationships are so stable that two- or three-membered cultures may be maintained for long periods of time. This has caused problems in trying to isolate pure cultures of methane bacteria. One such culture was maintained in various laboratories for more than 30 years as *Methanobacterium omelianskii*; it was thought to be a pure culture and was used in many studies of methane formation. In 1967, it was reported that the culture was actually an association of two organisms, neither of which could grow alone on the ethanol-carbonate medium in which it had been maintained for decades (Wolfe, 1971). One organism, a chemoorganotroph, ferments ethanol to acetate and hydrogen, which are utilized by the second organism, *Methanobacterium* M.O.H. The organotroph cannot grow in the absence of the methane-forming organism because it is inhibited by the hydrogen it produces. The *Methanobacterium* cannot utilize ethanol. Many early reports of the utilization of various organic compounds by methane bacteria were undoubtedly based on studies of similar mixed cultures.

Recent studies of the methane bacteria have led to the conclusion that they should be classified as a third form of life, more primitive even than the procaryotes. They differ from bacteria in several significant characteristics: (1) their cell walls contain no muramic acid, (2) their metabolism differs in a number of respects, and (3) their ribosomal RNA contains different base sequences.

The methane-oxidizing bacteria (methylotrophs) The methylotrophs are similar in many ways to other organisms classified as chemolithotrophic autotrophs. The family Methylomonadaceae includes two genera: *Methylomonas* (rods) and *Methylococcus* (cocci). The distinguishing features of these organisms are their absolute requirement for methane or methanol as carbon and energy source and their inability to grow on "organic media." Since methane and methanol are considered to be organic compounds, the organisms are technically organotrophs and heterotrophs. However, they are included among the "specialist procaryotes" (Smith and Hoare, 1977) along with the photolithotrophs and chemolithotrophs that have extremely restricted ability to use organic compounds, e.g., the nitrifying bacteria, *Thiobacillus*, and the cyanobacteria.

The obligate methylotrophs are obligate aerobes, using oxygen as the terminal electron acceptor, and are capable of growth in simple inorganic media with the addition of methanol or methane. No organic growth factors are required. One species can fix atmospheric nitrogen; others use ammonia or nitrate as nitrogen source. These organisms commonly occur in aerobic layers of soil or water above the anaerobic sediments where methane is produced. Like the other specialist organisms discussed previously, the obligate methylotrophs can assimilate a limited amount of some organic compounds and use them for synthesis of cell material, but only if the required C_1 compound, methane or methanol, is also available. Like the others, they are inhibited by certain organic compounds, particularly amino acids, and the inhibition can be relieved by other amino acids. The very striking similarities between the organisms that require carbon dioxide as principal carbon source and those that require methane or methanol have prompted recent suggestions for a redefinition of metabolic terminology (Kelly, 1971; Smith and Hoare, 1977; Whittenbury and Kelly, 1977). It has

Table 8-3 Bacteria capable of growing as chemolithotrophs and/or chemoautotrophs

	Carbon source	Energy source	Electron acceptor	Relation to O_2	Use of organic compounds
Nitrifying bacteria					
Ammonia oxidizers	CO_2	NH_4^+	O_2	Aerobic	Very limited
Nitrite oxidizers	CO_2	NO_2^-	O_2	Aerobic	Very limited
Sulfur bacteria					
Obligate	CO_2	H_2S, S, $S_2O_3^{2-}$	O_2	Aerobic	Very limited
Facultative or mixotrophic	CO_2 and/or organic compounds	H_2S, S, $S_2O_3^{2-}$, and/or organic compounds	O_2	Aerobic	Limited
Iron bacteria	CO_2 or organic compounds	Fe^{2+}	O_2	Aerobic	Probably unlimited in most
Hydrogen bacteria	CO_2	H_2	O_2	Aerobic	Unlimited
Methanogens	CO_2	H_2	CO_2	Anaerobic	Very limited
Methylotrophs	CH_4, CH_3OH	CH_4, CH_3OH	O_2	Aerobic	Very limited

been suggested that the term *heterotroph* be used to describe organisms that use compounds containing two or more carbon–carbon bonds as principal or sole carbon source. A new definition of *autotrophy* would then include the methylotrophs: "Autotrophs are microorganisms which can synthesize all their cellular constituents from one or more C_1 compounds" (Whittenbury and Kelly, 1977). While such a change will probably find its way into common usage only slowly, it seems a very logical way to solve some of the problems encountered in past attempts to classify specialist organisms. Auxotrophic requirements such as vitamins would not affect the classification.

Important metabolic characteristics of the chemolithotrophic microorganisms are summarized in Table 8-3.

Chemoorganotrophy and Chemoheterotrophy

Chemoorganotrophic growth is the most common mode of life among microorganisms. Only the specialist microorganisms we have discussed thus far, i.e., some algae, the cyanobacteria, the photolithotrophic bacteria, and some chemolithotrophic bacteria, are unable to utilize organic energy sources for growth. For all other microorganisms, chemoorganotrophy is the sole available mechanism for obtaining the energy to support growth, or else it is a readily used alternative mechanism. Many algae, the photoorganotrophic bacteria, and some chemolithotrophic bacteria grow well as chemoorganotrophs. For all protozoa, all fungi, and the great majority of bacteria, chemoorganotrophic growth is the only possible means of existence. With the chemoorganotrophs, the distinctions between metabolic categories become blurred, but as we pointed out previously, this presents no great problem if we understand the multiple uses of organic compounds in these organisms.

First, as is true for all chemotrophs, the energy source and the electron donor are identical so we may use these terms interchangeably. The electron donor function is quantitatively much more important in the chemolithotrophs, which must generally reduce carbon dioxide to the level of oxidation of cell material, CH_2O, than in the chemoorganotrophs, which use carbon sources that are seldom highly oxidized relative to cell material. In both types of metabolism, electrons removed from the energy source are used for two purposes *if* the organism has a respiratory metabolism: (1) for electron transport coupled to oxidative phosphorylation, i.e., for the synthesis of ATP, and (2) for reductive steps in the biosynthesis of cellular constituents from the carbon source. If the organism has a fermentative metabolism, electrons that are not used in reductive biosynthetic reactions must be used to reduce organic electron acceptors to produce fermentative end products that are not used by the organism. As discussed previously, it is the oxidative steps in which electrons are removed from the organic energy source that furnish the energy for ATP synthesis via substrate level phosphorylation. Thus chemotrophs, whether using an inorganic or an organic energy source, obtain energy by oxidation of that energy source and use some of the electrons removed in that oxidation in the reductive reactions required to convert the carbon source to cellular constituents. The same compound (or element) then serves as both energy source and electron donor.

Second, organisms that use organic energy sources also use organic carbon sources, so we may use the metabolic categories chemoorganotroph and chemoheterotroph almost interchangeably. As we have seen, some organisms use energy from light or from an inorganic energy source to assimilate organic compounds. While this almost always amounts to a minor fraction of carbon source utilization, we should exercise some caution in equating the two categories. We can safely assume that a chemoorganotroph is also a chemoheterotroph, but in a few cases chemoheterotrophy does not imply chemoorganotrophy. For all protozoa, all fungi, and all but a few bacteria, the two terms may be used interchangeably.

For organisms that use organic compounds as sources of both carbon and energy, there is a great variation in their ability to use the same compound for both purposes. Bacteria such as *Pseudomonas aeruginosa* and *Escherichia coli* are *prototrophs*, which means that they can form all their cell material from a single carbon source. These organisms grow well in a simple medium with one organic compound plus all the required elements as inorganic salts. In such a medium, the single organic compound serves as carbon source, energy source, and electron donor. We will examine the metabolic steps involved in the growth of the organism under these conditions in the following chapters. For now, it is sufficient to say that each organism that uses an organic energy source does so by a series of enzymatic steps that, as we have said before, release energy but also lead eventually to the formation of the metabolic end products typical of that organism. In aerobic metabolism, the energy source is typically oxidized, through a long series of reactions, to carbon dioxide. The metabolic pathway most commonly used for this purpose is the Embden-Meyerhof-Parnas, or glycolytic, pathway, which feeds into the tricarboxylic acid (TCA) cycle in aerobic organisms. Several other pathways for catabolism of the organic energy source are used by various microorganisms, but the important considerations here are that each organism possesses the enzymatic capability for degrading energy sources by one or more "central pathways" (see Chap. 9) and that its ability to utilize any specific organic energy source generally depends on its ability to form additional enzymes to convert that energy source into one of the intermediary metabolites of a central pathway or, in other words, to convert the energy source into a form in which it can enter the mainstream of catabolic reactions. These abilities are genetically determined.

As we shall see in the following chapters, the biosynthetic pathways utilize as starting material, or building blocks, the intermediary metabolites of the central catabolic pathways. Thus, as the energy source is degraded through a series of enzymatic reactions, at various points in the pathway an enzymatic product may be used for either of two purposes: (1) it may continue through the catabolic pathway, yielding maximum energy and being converted finally to the catabolic end product, e.g., carbon dioxide, or (2) it may be acted upon by a different enzyme that catalyzes the initial step in a biosynthetic pathway leading to the formation of a required cellular component, e.g., an amino acid. The prototrophic cell obviously must be able to synthesize all the enzymes required to convert catabolic intermediates to cell material, and in such a cell the same compound serves as both energy source and sole carbon source. As we pointed out in Chap. 5, microorganisms differ greatly in their biosynthetic capabilities. Auxotrophs require organic growth factors, and these requirements may vary from a single vitamin, amino acid, or organic base to a complete assortment of all the vitamins, amino acids, bases, and other compounds needed to form the polymers of which the cell is composed. An organism requires an organic growth factor because it does not have the genetic information for making the enzymes of the biosynthetic pathway that would synthesize the required molecule. If the organism possesses most of the biosynthetic pathways and thus requires only one or a few organic growth factors, the energy source still furnishes almost all the cellular carbon and is therefore the major carbon source. In organisms with very limited synthetic abilities, almost none of the cellular carbon can be derived from the energy source. Such organisms use a single organic energy source and a complex mixture of growth factors as carbon source. Between the two extremes of prototrophy and complete auxotrophy lies a full spectrum of organisms with varying degrees of auxotrophy, so that the energy source in different organisms may supply all, none, or varying amounts of the cellular carbon. The determining factor is not the nature of the energy source but the synthetic capability of the organism.

As we shall see in Chap. 12, the fact that a microorganism is able to use the energy source for the synthesis of cell material does not mean that it will necessarily do so, because if organic growth factors are available in the environment the organism uses these in preference to synthesizing them. Since the organic matter in many natural environments is supplied by decaying plant or animal material, it is probably unlikely that much cellular carbon is derived from the energy source in such environments because most required growth factors would be available. In other circumstances, for example in rivers or lakes contaminated with many types of industrial wastes, prototrophic organisms may enjoy a definite selective advantage because few, if any, organic growth factors are available unless other natural sources of organic matter are also present.

Since most microorganisms are chemoorganotrophs and chemoheterotrophs, we will not attempt to describe all of them. We will limit ourselves to describing briefly some of the secondary metabolic characteristics that distinguish the more important or most commonly occurring subgroups among the chemoheterotrophs. We will focus our attention primarily on the bacteria, because they exhibit much greater metabolic variability than is found among the eucaryotes, and this allows them to flourish in a greater variety of environments. The eucaryotic chemoheterotrophs include the protozoa, the fungi, the slime molds, and the leucophytes, the colorless algae that are incapable of photosynthesis but are still classified as algae on the basis of morphology.

Protozoa The protozoa, with the exception of a few anaerobic parasitic species, are aerobic organisms that use organic energy sources and organic carbon sources. They are classified on the basis of morphology, primarily the means of locomotion, which may be amoeboid movement, swimming by means of flagella, or swimming by means of cilia. Very little information is available on the specific growth factor requirements of individual species. A few protozoa have been grown in pure culture in defined media containing growth factors such as amino acids and vitamins, but most protozoa feed on bacteria and have not been grown except in mixed cultures with bacteria. Other protozoa are parasitic on humans or other animals and obtain nutrients from the cells or body fluids of the host. All these species utilize complex carbon sources and probably have complex growth requirements. Species that are able to use nutrients in solution are called *osmotrophic* or *osmophilic*; these include many of the parasitic protozoa and many of the nonparasitic flagellated species. The amoeboid and ciliated species are generally *phago-trophic*; they ingest bacteria by a process called *phagocytosis*, in which the bacteria are enclosed in an invaginated portion of the cell membrane, which forms at any point on the cell surface in amoeboid species or at the "mouth" in ciliated species. This membranous sac detaches and becomes a food vacuole, which circulates through the cytoplasm as the bacterial cell is digested by enzymes secreted into the vacuole from the cytoplasm. Osmotrophs may take up drops of liquid that contain macromolecules or other nutrients in much the same way.

Protozoa are primarily aquatic organisms, although they are found also in soil. The ability of many species to form cysts, which are resistant to desiccation, allows them to survive under unfavorable conditions. The Sporozoa, which include all obligate parasites, also form a resting stage, the sporozoite, which aids in transmission of the organism from one host to another. A number of human diseases are caused by protozoa, and water contaminated with sewage is often a source of infection with intestinal protozoa, e.g., *Entamoeba histolytica*, which causes amoebic dysentery. The free-living protozoa occur in almost any aerobic environment in which bacteria are present in sufficient numbers to support their growth. They are especially abundant in stagnant or slow-moving fresh or salt water, e.g., ponds, marshy areas, or tidal flats, where organic matter from decaying vegetation supports a large population of bacteria. In such environments, protozoa are an important link in the food chain. Waste treatment facilities are an ideal feeding ground for protozoa, and protozoa perform an important function in clarifying the effluent by ingesting bacteria that remain in suspension after sedimentation. Attached protozoa also aid settling by increasing the weight of sludge particles.

Fungi The fungi are chemoorganotrophs and chemoheterotrophs. All are capable of growing aerobically; many are obligate aerobes, but some species are facultative anaerobes. The fungi are primarily soil organisms. Some of the more primitive forms, often called *lower fungi*, are aquatic organisms. These water molds attack decaying plant or animal material in the aquatic environment and may also attack algae and protozoa. They form motile, flagellated spores very much like protozoa. Some of the water molds are capable of fermentation and therefore can grow in the anaerobic portions of ponds or lakes, contributing to anaerobic decomposition of organic matter in the sediment. Other fungi may be parasitic on or pathogenic to animals or plants. Many of the most economically important plant diseases are caused by fungi. They are less important as human pathogens. Of approximately 100,000 species of fungi, only about 100 are pathogenic to humans and animals, and only a dozen species cause frequent infections. Fungi cause infections of the hair, skin, and nails, e.g., scalp ringworm and athlete's foot, as well as some more serious infections of internal organs such as the lung, often causing symptoms that are wrongly diagnosed as tuberculosis or cancer. Infection probably occurs by a person breathing in spores carried through the air along with dust from the soil in which the organism has grown. Toxins produced by fungi growing on improperly dried crops such as peanuts and grains can have serious effects in humans. The aflatoxins produced by *Asperigillus flavus* can cause fatty degeneration of the liver and liver cancer in persons who eat contaminated food. Very low concentrations are toxic.

The nutritional requirements of most fungi are very simple. Many species can grow with a single organic compound as the source of carbon and energy and with inorganic sources of all other required elements. Ammonia or nitrate can be used as the nitrogen source by most species; none has been shown to fix nitrogen. Frequently, one or a few organic growth factors may be required in addition to the major carbon source. Carbohydrates are generally the preferred carbon and energy sources, but many exotic compounds, e.g., pesticides, are at least partially degraded by fungi. Cellulose-degrading fungi are important in the decomposition of wood and other plant materials, and certain fungi may be the only organisms able to degrade lignin.

Fungi have growth requirements similar to those of many bacteria, and they generally are capable of growing in many of the same types of environments. However, fungi grow much more slowly than bacteria and often are not able to compete successfully under conditions optimal for bacterial growth. Since fungi prefer somewhat lower pH values than do bacteria, slightly acid conditions may allow fungi to predominate in a mixed culture. Most fungi grow as long branching filaments that are divided by crosswalls or septa in most species; the septa are incomplete so that the entire cytoplasmic mass is continuous within the tubular branched mycelium. The yeasts grow as single cells, which reproduce by budding. Dimorphic fungi can grow in either the mycelial or the yeast form, and animal pathogens are often dimorphic, growing in the yeast form in the body. Fungi are classified on the basis of their sexual reproductive cycle, which is not frequently observed in nature. Mushrooms and other large fruiting bodies that bear sexual spores, such as puffballs, have the most readily seen sexual stages. Such spores are formed by the Basidiomycetes. The other class of higher fungi, the Ascomycetes, forms sexual spores within a sac, or ascus. Fungi that cannot be classified because a sexual stage has never been observed are placed in the class Fungi Imperfecti. Most pathogenic fungi are included in this group.

The *procaryotic chemoheterotrophs* are a diverse group of aerobic, facultative, and anaerobic bacteria, which are separated into families and genera on the basis of a variety of characteristics. The current classification scheme (Buchanan and Gibbons, 1974) emphasizes separation on the basis of cellular morphology and the Gram-staining reaction. In the following brief descriptions of the chemoheterotrophic bacteria, we shall adhere to the groupings presently recognized by bacterial taxonomists, but we will attempt to emphasize the metabolic and ecological rather than structural characteristics.

Gliding bacteria The gliding bacteria are divided into two orders, the Myxobacterales, which aggregate to form fruiting bodies under nutrient-limited conditions, and the Cytophagales, which never form fruiting bodies. Most of the Myxobacterales are cannabalistic; that is, they lyse and use as food supply other microbial cells, bacteria, algae, and fungi. A few species are able to digest cellulose. All are strictly aerobic. Genera and species are classified according to the characteristics of the fruiting body, and little is known of their biochemical characteristics except for the cellulolytic species, which grow well in inorganic media with a single organic compound, cellulose or a simple sugar, as carbon and energy source and with ammonia or nitrate as nitrogen source. The individual cells are Gram-negative rods that have no flagella but are able to glide along solid surfaces. The usual habitat of these organisms is soil that contains decaying plant material. Fruiting bodies often may be seen on the bark of trees. A tough slime layer surrounds the cells that form the fruiting body, and the cells often differentiate into microcysts, which are resistant to desiccation and more resistant to heat than the vegetative cells.

The second order of gliding bacteria, the Cytophagales, includes chemo-heterotrophs, chemolithotrophs, and mixotrophs, all classified in a single order on the basis of their gliding motility. We have described the chemo-lithotrophic and mixotrophic species. The chemoorganotrophic species are found, often in large numbers, in soil and in fresh and salt water. Many are able to digest cellulose, agar, or chitin. All species are Gram-negative. Most species are filamentous, and some are enclosed in sheaths. *Leucothrix*, a marine form, is ordinarily found attached to algae or marine animals. Some species of *Cytophaga* have been implicated in diseases of fish. Two genera, *Simonsiella* and *Alysiella*, are fermentative filamentous organisms that have been found only in the oral cavities of humans and animals. Other members of the order are strict aerobes, which use molecular oxygen as the terminal electron acceptor, or facultative anaerobes.

Sheathed bacteria The sheathed bacteria are a group of rod-shaped bacteria that grow in chains of cells enclosed in a tubular transparent sheath composed of proteins, polysaccharides, and lipids. The sheath may be surrounded by a polysaccharide slime layer. Genera are differentiated on the basis of the motility of single cells, the tendency of the sheaths to attach to solid surfaces, and the encrustation of the sheath with oxides of iron or manganese. We have discussed the possible chemolithotrophy of the encrusted organisms. Most of these organisms have not been obtained in pure culture and their metabolism is not well understood. Those that have been grown in the laboratory, *Sphaerotilus*, *Leptothrix*, and *Streptothrix*, are strict aerobes, using oxygen as the terminal electron acceptor, although *Sphaerotilus* grows well at low oxygen tensions. Of these three genera, only *Leptothrix* deposits iron and manganese oxides on the sheath, but there has been considerable dis-agreement as to whether *Sphaerotilus* and *Leptothrix* are really different organisms. Some authors have argued that *Leptothrix* is simply an encrusted form of *Sphaerotilus*.

We have mentioned that filaments of *Sphaerotilus* are commonly found in activated sludge and may be responsible for poor settling (a bulking sludge)

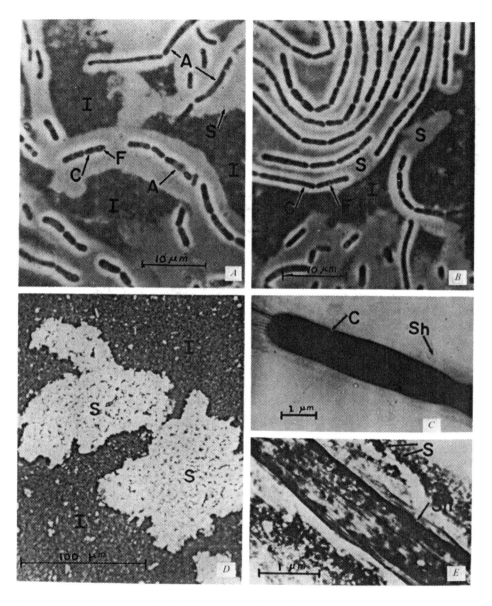

Figure 8-3 The filamentous, sheathed bacterium *Sphaerotilus natans*, which is often implicated in the bulking of activated sludge. (*A*) and (*B*) Phase contrast photomicrographs of India ink wet mounts of *S. natans*. C, cell; S, slime layer; I, India ink; A, abnormal or lysed cells; F, fat (poly-β-hydroxybutyrate). (*C*) Phase contrast photomicrograph at lower magnification showing large masses of slime with embedded cells. S, slime; I, India·ink. (*D*) Electron micrograph of *S. natans*, showing cell, C, and sheath, Sh. (*E*) Electron micrograph of *S. natans* showing the cell closely surrounded by the sheath, Sh, and a large amorphous deposit of dried slime (S) surrounding the filament. (*From Gaudy and Wolfe 1962.*).

when present in large numbers. *Streptothrix*, which differs from *Sphaerotilus* in that free motile cells of *Streptothrix* have not been observed, also occurs in activated sludge and may cause bulking. *Sphaerotilus* and other sheathed bacteria are most easily recognized by examination of wet mounts under oil, using a phase contrast microscope. Empty portions of the sheath are clearly visible. The sheathed bacteria that form holdfasts for attachment to solid surfaces, such as *Sphaerotilus*, can grow in low concentrations of organic matter, since nutrients are continually supplied by water flowing past. These organisms often produce massive growths in slow-running streams that are contaminated with carbohydrate-containing wastes such as paper mill effluents and dairy wastes. The nets of salmon fishermen in the Columbia River have often been fouled with large slimy masses of *Sphaerotilus*, which grows profusely even though the concentration of paper mill waste may be quite low (see Fig. 8-3).

The budding and/or appendaged bacteria This is a large and varied group of organisms classified on the basis of morphology. Included are aerobic, microaerophilic, and facultatively anaerobic chemoheterotrophs that have in common an unusual morphological feature—stalks, hyphae (thin filamentous outgrowths), or other forms of cytoplasmic extrusions or appendages. These cytoplasmic extrusions are called *prosthecae* and the bacteria forming them are *prosthecate*. Some of these bacteria, e.g., *Hyphomicrobium*, reproduce by forming buds at the ends of hyphae. The new cells may remain attached to the mother cell through the hyphae or may break away as free individual cells. The stalked bacteria, typified by *Caulobacter*, attach to solid surfaces by a holdfast at the end of the stalk, which is an extension of the cell itself, having the usual cell wall, cell membrane, and internal twisted membranes forming a core. These organisms are strict aerobes and grow best in very low concentrations of organic matter in fresh or salt water or in soil. Other organisms in this group excrete long noncytoplasmic fibrils, which form stalks. Two such organisms, *Gallionella* and *Thiodendron*, were described among the chemolithotrophs. The budding and appendaged bacteria apparently are widely distributed in soil and water, but their metabolism has been studied only in the few species that have been grown in pure culture.

Spirochetes The spirochetes, order Spirochaetales, are helically coiled, flexible motile organisms, some of which attain lengths of 500 μm. All are chemotrophs. Five genera are recognized and are differentiated on the basis of size, shape, degree of coiling, habitat, and relation to oxygen. Metabolically and ecologically, the spirochetes are a diverse group including aerobes, facultative and obligate anaerobes, free-living, parasitic, and pathogenic species. These organisms have an unusual structure consisting of a cytoplasmic cylinder intertwined with axial fibrils that originate near each end of the cell and extend toward the opposite end. An outer envelope encloses the cylinder and fibrils. Motion may be of several types—rotation, bending, or serpentine locomotion.

Members of the genus *Spirochaeta* are obligate or facultative anaerobes, free living in fresh water and marine environments containing hydrogen sulfide, i.e., in anaerobic environments, and in mud, sewage, and polluted water. This genus includes the longest of the spirochetes. They have complex growth requirements and therefore survive where organic matter is available. The genus *Treponema* includes *T. pallidum*, which causes syphilis in humans, and other pathogenic and parasitic species. Most of the parasitic species are found in the human oral cavity. None of these species is free living, and all are strict anaerobes. The genera *Borrelia* and *Cristispira* are also parasitic or pathogenic. *Leptospira* includes pathogenic species as well as some free-living aerobic species that require long-chain fatty acids as carbon source.

Spiral and curved bacteria The members of the family Spirillaceae differ morphologically from the spirochetes in two respects. The Spirillaceae are rigid rather than flexible, and they have flagella rather than axial filaments. Like the spirochetes, they are helically wound or spiral cells and, like the spirochetes, the group includes free-living, parasitic, and pathogenic species.

The Spirillaceae are divided into two genera: *Spirillum* and *Campylobacter*. *Spirillum* species are free-living, aquatic bacteria found in fresh or salt water that contains organic matter. They are aerobic; some species require low oxygen tension, i.e., are microaerophilic, and most have simple nutritional requirements. Organic acids, alcohols, or amino acids serve as the sole sources of carbon and energy; carbohydrates are generally not utilized. Cells have tufts of flagella at one or both ends. *Campylobacter* species have single flagella at one or both ends and are microaerophilic or anaerobic. Carbohydrates are not utilized as energy source; amino acids or organic acids are used. *Campylobacter fetus* causes abortion in sheep and cattle. Related but unclassified genera are *Bdellovibrio*, a tiny curved motile rod that feeds on other bacteria; *Microcyclus*, aerobic curved rods that form rings and are found in fresh water; *Pelosigma*, S-shaped cells found in mud and fresh and brackish waters and often in aggregates held together by mucoid material; and *Brachyarcus*, bow-shaped cells that may form rings, cloverleaf, or pretzel shapes and have been isolated from several lakes and ponds.

The groups of chemoheterotrophs described thus far are those with morphological characteristics so different from those of the majority of bacteria that their classification on the basis of morphology alone has been considered justified. Many of these organisms have been described from observations made on samples obtained from what is assumed to be their natural habitat and have never been cultured in the laboratory. Some tentative conclusions regarding their metabolic capabilities can be based on a knowledge of the conditions in the environments in which they are found, but much more information based on pure cultures studied in the laboratory is needed before these organisms can be properly classified and their potentials for growth in various environments and for modification of their environments can be understood. The remaining chemoheterotrophic bacteria are, in general, not remarkable for their morphological features, so it has been necessary to determine their metabolic capabilities in order to differentiate them.

Gram-negative aerobic rods and cocci This group includes five families of chemoheterotrophs differentiated on the basis of their distinctive metabolic capabilities. The family Pseudomonadaceae includes four genera: *Pseudomonas*, *Xanthomonas*, *Zoogloea*, and *Gluconobacter*. These are small Gram-negative rods that are motile by means of polar flagella. Their metabolism is strictly respiratory, never fermentative. The genus *Pseudomonas* is noted for versatility in its use of carbon and energy sources. One species is able to use more than 100 different compounds as sole source of carbon and energy, and several species can utilize 50 or more. The ability to grow without organic growth factors is common. Five species of *Pseudomonas* are able to grow as chemolithotrophic autotrophs, using hydrogen as energy source and carbon dioxide as carbon source. Carbon monoxide is also used as energy source by one species. Because of their ability to use a great variety of organic compounds and their lack of requirements for growth factors, *Pseudomonas* species are widely distributed in soil and in fresh and salt water. Some species are pathogenic to plants and two species are pathogenic to humans, but these are probably accidental pathogens that are primarily soil organisms. Some species of *Pseudomonas* are able to use nitrate as terminal electron acceptor and are active denitrifiers.

The genus *Xanthomonas* includes organisms that are plant pathogens and are found only in association with plants. They are strict aerobes that use oxygen as terminal electron acceptor and have complex growth requirements. They produce a yellow pigment and usually a polysaccharide slime.

The genus *Zoogloea* has been considered the primary floc-forming organism in activated sludge. The cells are small, polarly flagellated rods that form visible floc particles in water, in sewage, or in laboratory cultures. Floc formation in the laboratory has been reported to depend on the C/N ratio of the medium. Examination of flocs in the electron microscope has revealed that the aggregates of cells are held together by extracellular fibrils, rather than by a slime layer, as was originally thought to be the case. *Zoogloea* is a strict aerobe that requires oxygen as terminal electron acceptor. If vitamins B_{12} and biotin are supplied, a variety of sugars, amino acids, or organic acids can be used as the sole source of carbon and energy. The fourth genus, *Gluconobacter*, is noted primarily for its incomplete oxidation of sugars and its ability to oxidize ethanol to acetic acid. It is a soil organism and is found on fruits and vegetables and in wine, cider, and wine vinegar. The genus *Acetobacter*, which is metabolically quite similar to *Gluconobacter* but has peritrichous flagella, is more active in the oxidation of ethanol to acetic acid, is more tolerant of acid, and is used commercially in the manufacture of vinegar.

The family Azotobacteraceae includes four genera: *Azotobacter, Azomonas, Beijerinckia,* and *Derxia.* Organisms of this group are the only free-living, aerobic bacteria capable of fixing nitrogen and growing in an environment devoid of combined nitrogen. All are aerobic and require oxygen as terminal electron acceptor, but they grow well and fix nitrogen more efficiently under reduced oxygen tension. This results from the fact that nitrogenase, the enzyme involved in nitrogen fixation, is inhibited by oxygen. Most species are also able to use combined nitrogen in the form of ammonia, nitrate, or amino acids. One important differentiating characteristic of the four genera is habitat. *Azotobacter* is primarily a soil organism, whereas *Azomonas* has three species, one of which is found most frequently in soil and two in water. Both are adapted to neutral or alkaline soils. *Beijerinckia* and *Derxia* are tolerant of acid and are found in acid, tropical soils. All produce large amounts of slime; *Azotobacter* and *Azomonas* produce a loose, flowing slime, while that of *Beijerinckia* and *Derxia* is tough and gummy. All store large amounts of poly-β-hydroxybutyrate. *Azotobacter* forms cysts, which are resistant to desiccation.

The family Rhizobiaceae includes two genera, *Rhizobium* and *Agrobacterium,* which produce nodules on plants. *Rhizobium* produces root nodules on legumes within which it grows symbiotically, fixing nitrogen and supplying the host plant with a utilizable nitrogen source. *Agrobacterium* produces growths on plant stems and does not fix nitrogen.

The family Methylomonadaceae includes two genera: *Methylomonas,* which are rods, and *Methylococcus,* which are cocci. These organisms utilize only one-carbon compounds, methane and methanol, as a source of carbon and energy, and were discussed previously as specialist organisms that should possibly be included in a redefined category of autotrophs. They are strict aerobes and require no organic growth factors.

The family Halobacteriaceae includes two genera: rods, *Halobacterium,* and cocci, *Halococcus.* These organisms are found only in habitats where the sodium chloride concentration is at least 12 percent.

Gram-negative facultatively anaerobic rods This group includes two families: the Enterobacteriaceae, which have peritrichous flagella, if motile, and a negative reaction in the oxidase test, and the Vibrionaceae, which have polar flagella when motile and are oxidase-positive. From the public health standpoint, this group of organisms is perhaps the most important of all the major bacterial groups to the environmental engineer and scientist. Many of these organisms are common, usually harmless inhabitants of the intestinal tracts of humans and animals; others are common inhabitants of fresh or salt water; and a significant number of them are important human pathogens, which are commonly spread by fecal contamination of water. Because of their different roles in nature, the isolation of organisms of this group from water (or food) is of varying significance, which depends on the specific organism isolated. Therefore, differentiation between normal water species and those of fecal origin is essential.

As a group, these organisms differ from other Gram-negative bacteria in that they are able to obtain energy in three ways: (1) aerobically by oxidative phosphorylation coupled to electron transport using oxygen as terminal electron acceptor, (2) anaerobically by fermentation of carbohydrates, and (3) by anaerobic respiration using nitrate as terminal electron acceptor. Nitrate reduction is a property of all the Enterobacteriaceae, except a few species of the plant pathogen *Erwinia*, and of many of the Vibrionaceae. Many of these organisms, including even some pathogenic species, have very simple nutritional requirements and are able to grow in a defined medium that contains a single organic compound as carbon and energy source and all other required elements as inorganic salts. Some species require various organic growth factors, but these requirements are usually not numerous.

The enteric bacteria, as this group is often called, have been more thoroughly studied than most other bacteria, but in spite of this, or perhaps because of it, their classification has been very difficult and is still unsatisfactory. Numerous criteria have been used, such as the ability to ferment lactose; production of gas (hydrogen) as well as acid from carbohydrates; ability to use specific carbon and energy sources, especially citrate; production of hydrogen sulfide; production of indole from tryptophan; ability to hydrolyze urea; and sensitivity to cyanide. One important criterion for identification of the enteric bacteria is determination of their fermentation pattern. All of these organisms ferment glucose through the Embden-Meyerhof-Parnas pathway to pyruvic acid but, as we will see in Chap. 11, the final products made from pyruvate are often characteristic of the organism and are useful in identification.

Two fermentation patterns are found in the enteric bacteria: mixed acid fermentation and butanediol fermentation. In the mixed acid fermentation, lactic, acetic, and succinic acids are formed in large amounts, producing a final pH in glucose broth that is sufficiently low to give an acid reaction with the acid-base indicator methyl red. This indicator changes color in the pH range 4.4 to 6.2, so a reaction that is positive for methyl red requires a pH value of approximately 4.5 or less. The indicators normally used to detect acid production in tests for fermentation require much less acid to produce a positive reaction; for example, bromthymol blue changes color in the pH range 6.0 to 7.6. Thus, organisms that ferment carbohydrates to form acid detectable in the standard fermentation tests would not necessarily cause positive results in the methyl red test. Organisms with a butanediol fermentation of glucose are of this type. While acids are produced, their amounts are much smaller than in the mixed acid fermentation, and a large proportion of the glucose is converted to 2,3-butanediol, a distinctive product of this type of fermentation. Butanediol (butylene glycol) and its precursor acetoin are detected by the Voges-Proskauer test. Thus, determination of the fermentation pattern is a simple matter.

The 12 genera of the family Enterobacteriaceae are separated into five groups or tribes, three of which include only one genus, while the other two include genera that are very closely related. Tribe I, Escherichieae, includes five genera: *Escherichia*, *Edwardsiella*, *Citrobacter*, *Salmonella*, and *Shigella*. The first three are normal inhabitants of human and animal intestines and are only infrequently involved in human infections. *Salmonella* and *Shigella* are pathogenic to humans and are responsible for several of the serious, sometimes fatal illnesses that result from poor sanitation. Historically, typhoid fever, caused by *Salmonella typhi*, was one of the severe epidemic diseases of humans, and it still occurs in epidemic form in countries where sanitation is lacking and immunization is not practiced, and occasionally in countries where sanitation is supposedly good. All species of *Salmonella* are capable of causing enteric fevers similar to but usually less severe than typhoid fever. *Salmonella* species also cause gastroenteritis, commonly called "food poisoning," and various other types of infections. The closely related genus, *Shigella*, includes four species, all of which cause bacillary dysentery in humans. Prevention of infection by either *Salmonella* or *Shigella* is dependent on strict sanitary controls of both water and food, since vaccines are not reliably effective. Humans who recover from infections frequently become carriers and continue to shed the organisms in feces. Epidemics may be started by such human carriers when sanitary controls are relaxed or when an untreated or insufficiently treated water supply is used. All the genera of the tribe Escherichieae have the mixed acid type of fermentation. They are differentiated on the basis of other biochemical tests.

The tribe Klebsielleae includes four genera, all of which are commonly found both in the intestines of humans and animals and in soil and water. Finding these organisms in water is not considered evidence of fecal contamination. *Klebsiella* and *Enterobacter* are closely related species that are differentiated primarily on the basis of motility and pathogenicity. Both genera are capable of fixing nitrogen, but only under anaerobic conditions. *Klebsiella* species cause upper respiratory and other infections including one type of pneumonia, whereas *Enterobacter* (formerly *Aerobacter*) is nonpathogenic. The other two genera, *Hafnia* and *Serratia*, are also generally nonpathogenic. These organisms have the butanediol fermentation pathway.

The other three tribes include only one genus each. *Proteus* (tribe Proteeae) is a common intestinal organism, which is also found in soil. It is noted primarily for its ability to degrade proteins. The genus *Erwinia* (tribe Erwinieae) includes a number of species that are associated with plants and often cause plant diseases such as soft rot, fire blight, or wilt. The genus *Yersinia* (tribe Yersinieae) includes three species, one of which has been responsible for some of the world's most fatal epidemics. This is *Y. pestis* (*Pasteurella pestis*), the organism that causes bubonic plague, known as the

Black Death. In one European epidemic in the fourteenth century, one-fourth of the population died of the disease. While it still is found sporadically in various parts of the world, the organism appears to have become much less virulent. The other two species of *Yersinia* are primarily animal pathogens but may infect humans. *Y. enterocolitica* causes gastrointestinal disease in humans and has been spread through contaminated water supplies. The organisms survive best in cold weather, and most human cases occur under these conditions. Bubonic plague is not a waterborne disease but is spread from rodents to humans by fleas.

The second family of this group of organisms, the Vibrionaceae, includes another of the great killers of mankind, which is spread by fecal contamination of water. The genus *Vibrio* includes the species *V. cholerae*, which causes human cholera. As in other waterborne diseases, prevention requires adequate treatment of sewage and purification of water supplies as well as regulations for food handling. Untreated cases have a fatality rate of approximately 60 percent, and individuals who recover continue to excrete large numbers of the bacteria in feces for as long as 1 year. Some persons may become permanent carriers. No really effective vaccine is available. Other species of *Vibrio* are marine and fresh-water organisms that may cause disease in fish and enteritis in humans who eat seafood. The other genera of this family are: *Aeromonas*, a fresh-water organism, some strains of which are pathogenic for frogs, snakes, or fish and occasionally humans; *Plesiomonas*, an intestinal organism that has caused outbreaks of gastroenteritis in humans; and *Photobacterium* and *Lucibacterium*, luminescent marine bacteria.

Two genera that are frequently isolated from soil and water are not included in the enteric group although they have some of the same characteristics. These are *Chromobacterium*, which forms a purple pigment, and *Flavobacterium*, which forms yellow, orange, or red pigment. Both are Gram-negative rods and include aerobic species and facultatively anaerobic species.

Gram-negative anaerobic rods These include three genera that are not easily cultivated on ordinary laboratory media. They are strict anaerobes and grow best in a medium to which serum or bile and a reducing agent have been added. They are parasitic and sometimes pathogenic to humans and other animals. The genera are *Bacteroides*, *Fusobacterium*, and *Leptotrichia*. *Bacteroides* species are possibly the most numerous of the bacterial inhabitants of the human intestine, but their prolonged survival outside the body is prevented by their sensitivity to oxygen.

A Gram-negative anaerobic chemoheterotroph that is important in the sulfur cycle is *Desulfovibrio*. Some strains are psychrophilic, whereas others are mesophilic. Since these organisms are strict anaerobes, they are found primarily in anaerobic bottom muds in both fresh and salt water. They obtain energy by anaerobic respiration, using sulfate as a terminal electron acceptor and reducing it to hydrogen sulfide. These organisms also are generally capable of fixing nitrogen. No organic growth factors are required. Carbohydrates are not utilized, but organic acids and alcohols are generally used as sources of carbon and energy. Unlike most strict anaerobes, these organisms contain cytochromes.

Gram-negative cocci The family Neisseriaceae includes four genera: *Neisseria*, which is aerobic or facultatively anaerobic, and *Branhamella*, *Moraxella*, and *Acinetobacter*, all strictly aerobic. *Neisseria*, *Branhamella*, and *Moraxella* are pathogenic or parasitic on humans and animals and are not encountered outside the body. *Acinetobacter* and two other genera of Gram-negative cocci, *Paracoccus* and *Lampropedia*, are common inhabitants of soil and water. *Paracoccus* includes two species, both of which are denitrifiers, using either oxygen or nitrate as the terminal electron acceptor. *P. denitrificans* can grow chemoorganotrophically using a wide range of organic compounds as sole sources of carbon and energy and either ammonia or nitrate as nitrogen source. It can also grow as a chemolithotrophic autotroph, using hydrogen as energy source and carbon dioxide as carbon source in completely inorganic media. The second species, *P. halodenitrificans*, requires organic growth factors and high salt concentrations and is found in brines. *Lampropedia* is an obligate aerobe that forms sheets of cells enclosed in an envelope. These float on the surface of the water and have been found in waters that contain organic material, e.g., in swamps.

The Gram-negative anaerobic cocci are *Veillonella*, *Acidaminococcus*, and *Megasphaera*. These are strict anaerobes and are parasitic in humans and animals in the intestinal tract and rumen. They have complex growth requirements and probably can survive in few environments outside the body.

Gram-positive cocci Three families are distinguished on the basis of morphology and relation to oxygen. The members of the family Micrococcaceae are aerobic or facultatively anaerobic. All are able to use oxygen as terminal electron acceptor for respiratory metabolism, and some are able to carry out anaerobic respiration using nitrate as terminal electron acceptor. The characteristic arrangement of cells is a result of division in more than one plane to form irregular clusters or cubical packets. The genera are *Micrococcus*, commonly found in soil and fresh water and utilizing only respiratory metabolism; *Staphylococcus*, parasites or pathogens of humans and animals and capable of respiration or fermentation; and *Planococcus*, marine organisms that are strictly aerobic.

The family Streptococcaceae includes organisms that have the unique characteristic of indifference to oxygen. These organisms have no cytochromes and most have no catalase. Their metabolism is fermentative in the presence or absence of oxygen. All have complex growth requirements. The genus *Streptococcus* includes human and animal pathogens and parasites as well as such commercially important species as *S. lactis* and *S. cremoris*, which are normal contaminants in milk and are used in dairy product manufacturing. *S. faecalis* is one of several species of streptococci commonly found in the human intestine. Fecal streptococci have been used as indicators of fecal contamination of water. The genus *Leuconostoc* also includes species used in dairy manufacture, and members of this genus are found only in milk and milk products or on plants and fermenting plant materials. Two species form slime from sucrose. One of these, *L. mesenteroides*, forms copious amounts of a slime, dextran, which sometimes creates serious problems in sugar refineries. Both *Streptococcus* and *Leuconostoc* typically form pairs and short to long chains. The genera *Pediococcus* and *Aerococcus* form pairs or tetrads (groups of four cells) and occur on fermenting plant materials, whereas the fifth genus, *Gemella*, is parasitic on humans and animals. All the genera of this family except *Gemella* are included in the group commonly called the *lactic acid bacteria*, since their sole or principal fermentation product is lactic acid.

The third family of Gram-positive cocci, the Peptococcaceae, is strictly anaerobic organisms that have been isolated from soil and mud and also from human and animal intestinal and respiratory tracts. These organisms generally have complex growth requirements and are very sensitive to oxygen. Members of the genera *Peptostreptococcus* and *Sarcina* ferment carbohydrates, whereas *Peptococcus* species may ferment proteinaceous materials as well. The genus *Ruminococcus* includes the rumen organisms that ferment cellulose and cellobiose.

The Lactobacillaceae, genus *Lactobacillus*, are Gram-positive rods that are included among the lactic acid bacteria. They are very similar in metabolism and habitat to the Streptococcaceae; they ferment sugars to lactic acid as a sole or principal product.

Spore-forming bacteria These organisms are very important in the mineralization of organic matter in a variety of habitats. The two principal genera, *Bacillus* and *Clostridium*, are Gram-positive rods that differ in their relation to oxygen. *Bacillus* species are strictly aerobic or facultatively anaerobic, whereas *Clostridium* includes strictly anaerobic species, most of which cannot grow in the presence of oxygen. These organisms as a group are metabolically quite versatile, although some individual species are restricted to one or a few organic substrates. Both genera include species that are able to fix nitrogen. Two species of *Bacillus* fix nitrogen, but only anaerobically, whereas several species of *Clostridium* are able to fix nitrogen. Both genera include species that can hydrolyze polysaccharides, proteins, fats, and nucleic acids and metabolize the products of hydrolysis. At least one species of *Clostridium* is able to ferment cellulose. Both genera include species that are thermophilic. One species of *Bacillus*, *B. licheniformis*, is an active denitrifier. Other species are capable of using nitrate as terminal electron acceptor but reduce it no further than nitrite. The principal habitats of these organisms are soil, bottom muds (for the anaerobic or facultative species), and animal intestines. Their ability to form spores that are resistant to heat and desiccation allows them to survive when nutrients are unavailable or during dry periods, and their spores are easily carried from place to place by wind. Contamination of food or of puncture wounds with spores of *Clostridium* can have serious, often fatal consequences. *C. botulinum* growing in food produces toxins, proteins excreted by the cell, which are lethal in minute quantities. Other species, e.g., *C. perfringens*, cause food poisoning but produce less powerful toxins. Infection of a deep wound with *C. tetani* causes tetanus, the symptoms of which are due to toxin production by the organism growing in the infected wound. Several species, most frequently *C. perfringens*, grow rapidly in deep wounds and cause gas gangrene.

Three other genera of bacteria form spores. *Sporolactobacillus* is microaerophilic and ferments sugars to lactic acid. *Desulfotomaculum* includes two species, one of which is thermophilic. These organisms are strict anaerobes and metabolize simple organic acids. Sulfate and other oxidized sulfur compounds are used as terminal electron acceptors and are reduced to hydrogen sulfide. *Sporosarcina* is a strict aerobe that metabolizes simple organic acids and has complex growth requirements. Both *Sporolactobacillus* and *Desulfotomaculum* are rods, whereas *Sporosarcina* is a coccus that forms cubical packets of cells.

Coryneform bacteria This is a rather ill-defined group of generally rod-shaped organisms that produce cells of irregular shape, often characterized as club-shaped, which in some genera fragment into cocci in older cultures. Rods still attached after division are connected to each other at acute angles, the result of so-called "snapping" division. These organisms are a diverse group including some important human and animal pathogens and some common soil and water bacteria. We shall describe only the more common organisms of the group.

The genus *Corynebacterium*, from which the common name of the group is derived, includes such pathogens as *C. diphtheriae*, which causes diphtheria in humans, and a variety of other animal pathogens and parasites. These species are facultative anaerobes. A second group of *Corynebacterium* species are plant pathogens; these are strict aerobes. A third group is composed of the nonpathogenic species commonly found in soil and water. These have not been studied well, but most are probably aerobic.

A closely related genus is *Arthrobacter*. These organisms are strict aerobes and are able to oxidize a wide variety of organic compounds. Their distinguishing characteristic is the succession of morphological forms that appear as the culture ages. Young cells are irregularly shaped rods, usually in short chains forming "Chinese letter" arrangements due to the angles between the cells. As the culture ages, the rods divide into cocci, which often remain attached in the original configuration and produce angular, rigid chains. Often large spherical cells called *cystites* or *arthrospores* are produced. These bacteria are generally found in large numbers in soil and water, probably because of their versatility in using many different organic compounds as carbon and energy source and the simple nutritional requirements of many species. Another related genus is *Cellulomonas*. The outstanding characteristic of these organisms is their ability to metabolize cellulose.

The coryneform bacteria are related to a series of morphological forms that gradually progress to greater lengths and a branching structure similar to the mycelium formed by fungi. In some genera the mycelium fragments into cocci in older cultures. A very large number of genera are included in this group of bacteria. Most are typically found in soil, although some are parasites or pathogens of humans or animals. The most fully developed branching mycelial growth is found in the actinomycetes, order Actinomycetales. These are common in soil and less common in water, but they have been implicated as the cause of taste and odor problems in water supplies obtained from reservoirs. Some form spores borne on aerial hyphae somewhat similar to those found in fungi. These organisms are procaryotes and are considered to be bacteria in spite of their morphological relationship to the fungi.

A COMMENT ON TAXONOMY AND ITS APPLICATIONS

We have attempted to provide in this chapter a brief overview of the rather bewildering variety of morphological forms and biochemical capabilities that make up the world of microorganisms. We have emphasized the bacteria because they exhibit the greatest metabolic variation. Among the bacteria, we have omitted many genera and have given others only a brief mention. We have described in greatest detail those organisms and groups of organisms that we believe to be of the greatest interest or importance to the environmental technologist.

The information included here is intended as a preliminary guide to the groups of organisms that may be expected to inhabit various types of environments. For additional information, specialized texts should be consulted. The authoritative source for identification of bacteria is *Bergey's Manual of Determinative Bacteriology* (Buchanan and Gibbons, 1974), from which much information used in this chapter was taken. A useful source of information on many of the microorganisms found in water supplies is Whipple's *Microscopy of Drinking Water* (1899), which describes procaryotic and eucaryotic microorganisms as well as larger organisms such as rotifers and crustaceans.

A knowledge of the characteristics of individual genera or species of microorganisms will be of varying usefulness to environmental technologists depending on their specific areas of activity. As was stated earlier, the identification of microorganisms in most cases requires specialized training and experience. However, there are many aspects of environmental technology in which general knowledge and careful observation can be of great value. It is not necessary to identify the species of microorganisms present in an activated sludge, for example, to benefit from application of microbiological techniques. Frequent use of the ordinary light microscope or the phase contrast microscope can provide much useful information. (The phase contrast microscope is preferable for examination of unstained cells.) Microscopic observation of a wet mount preparation of "healthy" activated sludge reveals a flocculated mass of bacteria with some filamentous forms entwined in the floc and various types of grazing protozoa and crustaceans. It is not necessary to identify species or to make a quantitative count to determine whether the types and relative numbers of protozoa and higher forms have diminished since the previous observation and whether this correlates with an increase in the amount of suspended solids in the effluent. It is also simple to determine whether an increase in the relative numbers of filamentous forms correlates with a decrease in the settling rate and compactibility of the sludge. With a little practice, one can distinguish between fungal filaments and the sheathed bacteria such as *Sphaerotilus natans* and combine these observations with the knowledge that most fungi prefer a low pH and that *S. natans* grows well with sugars at low DO levels. Microscopic observations and knowledge of the nutritional preferences of the types of microorganisms observed can be correlated with the quantitative analyses, e.g., pH, DO, and S_i, to set limits and guidelines in the design of strategies for successful operation. For example, one may note a small increase in the number of filamentous fungi some days before the sludge begins to bulk, thereby gaining time to set in motion appropriate countermeasures, for example, make a slight pH adjustment or add flocculating or weighting chemicals.

It cannot be overemphasized that the processes inserted into the decay leg of the carbon-oxygen cycle must be subjected to technological control—they do not operate themselves. Microscopic observation is a far more economical and more easily performed analytical aid to operational engineering than are the other biological, chemical, and physical analyses that are performed routinely, and its use in combination with the more quantitative determinations can enhance the possibilities for more successful control of the aqueous environment.

PROBLEMS

8-1 Seven metabolically different types of microorganisms are described below (A–G). Six media also are given, and the conditions of incubation, i.e., presence or absence of light and air, are specified (I–VI). For each medium list the organisms that could grow under the conditions specified, and for each organism give its carbon source, energy source, nitrogen source, and electron donor.

Organisms:

A. Chemoorganotrophic, fermentative metabolism, has no cytochromes or catalase, requires organic nitrogen, has multiple organic growth factor requirements

B. Photolithotrophic, autotrophic, has bacterial type of photosynthesis

C. Photoorganotrophic, facultative phototroph, facultative anaerobe, prototrophic, has bacterial type of photosynthesis

D. Photolithotrophic, autotrophic, has plant type of photosynthesis

E. Chemoorganotrophic, respiratory metabolism—aerobic or anaerobic, prototrophic

F. Chemolithotrophic, prototrophic

G. Chemoorganotrophic, respiratory or fermentative metabolism, multiple growth factor requirements

Ingredients and conditions of incubation:

I. Glucose, 1 g
   Tryptone, 1 g
   Yeast extract, 1 g
   Water, 1 L
   Dark
   Air

II. Glucose, 1 g
   KH_2PO_4, 0.1 g
   $(NH_4)_2SO_4$, 0.1 g
   $NaNO_3$, 0.1 g
   Water, 1 L
   No air
   Light

III. $(NH_4)_2SO_4$, 10 mg
   KH_2PO_4, 0.1 g
   $NaHCO_3$, 1 g
   $NaNO_3$, 0.1 g
   Water, 1 L
   Light
   No air

IV. Glucose, 1 g
   Tryptone, 1 g
   Yeast extract, 0.1 g
   NH_4Cl, 0.1 g
   Water, 1 L
   No air
   Dark

V. $NaHCO_3$, 1 g
   Na_2S, 1 g
   K_2HPO_4, 0.1 g
   NH_4Cl, 0.1 g
   Water, 1 L
   Light
   No air

VI. $(NH_4)_2SO_4$, 1 g
   K_2HPO_4, 0.1 g
   Water, 1 L
   Air
   Light

8-2 Match each of the following microorganisms to one of the descriptions (A–G) in Prob. 1: (*a*) cyanobacteria, (*b*) purple nonsulfur bacteria, (*c*) green sulfur bacteria, (*d*) *Nitrobacter*, (*e*) *Pseudomonas*, (*f*) *Streptococcus*, (*g*) *Thiobacillus thiooxidans*, (*h*) *Clostridium*, (*i*) yeast.

8-3 List the common characteristics of the organisms that would be classified as autotrophs according to the new definition proposed by Whittenbury and Kelly (1977).

8-4 Consult the 8th edition of *Bergey's Manual* and list the characteristics that differentiate among the genera *Escherichia*, *Salmonella*, and *Enterobacter*. Include only those that are the same for all strains or species of a genus.

8-5 Describe in general terms the metabolic characteristics and nutritional requirements of the following groups:

   (*a*) Algae
   (*b*) Cyanobacteria
   (*c*) Protozoa
   (*d*) Fungi

8-6 Describe the process of crown corrosion in a concrete sewer. What types of bacteria contribute to the process and what are their outstanding metabolic characteristics?

8-7 Discuss the relationship among carbon source, energy source, and electron donor in chemoheterotrophs. What determines whether a single compound can serve all three purposes?

8-8 As a class project, consult research journals in the library and locate reports in which the bacteria present in activated sludges, polluted streams, and oxidation ponds have been identified. Then determine the metabolic requirements of these organisms from descriptions in *Bergey's Manual* and try to draw conclusions as to the selective conditions, if any, in each case.

8-9 Give one or more distinguishing characteristics (e.g., shape, Gram-staining reaction, relation to oxygen, metabolic groups, etc.) for each of the following genera:

(a) *Sphaerotilus*   (g) *Arthrobacter*
(b) *Chromatium*   (h) *Zoogloea*
(c) *Azotobacter*   (i) *Shigella*
(d) *Clostridium*   (j) *Staphylococcus*
(e) *Streptococcus*   (k) *Bacillus*
(f) *Pseudomonas*   (l) *Spirillum*

REFERENCES AND SUGGESTED READING

Alexander, M. 1971. Biochemical ecology of microorganisms. *Annu. Rev. Microbiol.* 25:361–392.

Brock, T. D. 1979. The biology of microorganisms. 3d ed. Prentice-Hall, Englewood Cliffs, N.J.

Buchanan, R. E., and N. E. Gibbons (editors). 1974. Bergey's manual of determinative bacteriology, 8th ed. Williams & Wilkins, Baltimore.

Doelle, H. W. 1975. Bacterial metabolism, 2d ed. Academic Press, New York.

Frobisher, M., R. D. Hinsdill, K. T. Crabtree, and C. R. Goodheart. 1974. Fundamentals of microbiology, 9th ed. W. B. Saunders, Philadelphia.

Fry, B. A., and J. L. Peel (editors). 1954. Autotrophic microorganisms. Fourth Symposium of the Society for General Microbiology. Cambridge University Press, Cambridge, England.

Gaudy, E., and R. S. Wolfe. 1962. Composition of an extracellular polysaccharide produced by *Sphaerotilus natans*. *Appl. Microbiol.* 10:200–205.

Gest, H. 1972. Energy conversion and generation of reducing power in bacterial photosynthesis. *Adv. Microb. Physiol.* 7:243–282.

Gest, H., A. San Pietro, and L. P. Vernon. 1963. Bacterial photosynthesis. Antioch Press, Yellow Springs, Ohio.

Gibson, J. 1957. Nutritional aspects of microbial ecology. 7th Symposium of the Society for General Microbiology. Cambridge University Press, Cambridge, England, pp. 22–41.

Holm-Hansen, O. 1968. Ecology, physiology, and biochemistry of blue-green algae. *Annu. Rev. Microbiol.* 22:47–70.

Joklik, W. K., and Hilda P. Willett (editors). 1976. Zinsser microbiology, 16th ed. Appleton-Century-Crofts, New York.

Kelly, D. P. 1971. Autotrophy: Concepts of lithotrophic bacteria and their organic metabolism. *Annu. Rev. Microbiol.* 25:177–210.

Lewin, R. A. 1974. Biochemical taxonomy. *In* Algal physiology and biochemistry, W. D. P. Stewart (editor). University of California Press, Berkeley and Los Angeles.

Painter, H. A. 1970. A review of literature on inorganic nitrogen metabolism in microorganisms. *Water Res.* 4:393–450.

Parson, W. W. 1974. Bacterial photosynthesis. *Annu. Rev. Microbiol.* 28:41–59.

Peck, H. D. Jr. 1968. Energy-coupling mechanisms in chemolithotrophic bacteria. *Ann. Rev. Microbiol.* 22:489–518.

Pfennig, N. 1967. Photosynthetic bacteria. *Annu. Rev. Microbiol.* 21:285–324.

Prescott, G. W. 1968. The algae: A review. Houghton Mifflin, Boston.

Rittenberg, S. C. 1969. The roles of exogenous organic matter in the physiology of chemolithotrophic bacteria. *Adv. Microb. Physiol.* 3:159–196.

Round, F. E. 1965. The biology of the algae. St. Martin's Press, New York.

Smith, A. J., and D. S. Hoare. 1977. Specialist phototrophs, lithotrophs, and methylotrophs: A unity among diversity of procaryotes? *Bacteriol. Rev.* 41:419–448.

Stadtman, T. C. 1967. Methane fermentations. *Annu. Rev. Microbiol.* 21:121–142.

Stanier, R. Y., E. A. Adelberg, and J. Ingraham. 1976. The microbial world. 4th ed. Prentice-Hall, Englewood Cliffs, N.J.

Stewart, W. D. P. 1973. Nitrogen fixation by photosynthetic microorganisms. *Annu. Rev. Microbiol.* 27:283–316.

Stewart, W. D. P. (editor). 1974. Algal physiology and biochemistry. University of California Press, Berkeley.

Whipple, G. C. 1899. Microscopy of drinking water. John Wiley, New York. (Revised by G. M. Fair and M. C. Whipple, 1927.)

Whittenbury, R., and D. P. Kelly. 1977. Autotrophy: A conceptual phoenix. 27th Symposium of the Society for General Microbiology. Cambridge University Press, Cambridge, England, pp. 121–149.

Wolfe, R. S. 1971. Microbial formation of methane. *Adv. Microb. Physiol.* 6:107–146.

Zeikus, J. G. 1977. The biology of methanogenic bacteria. *Bacteriol. Rev.* 41:514–541.

NINE

THE CENTRAL PATHWAYS OF METABOLISM

Chapters 9, 10, and 11 deal with the orderly processing of material through the cell (metabolism), which results in the return of some substrate carbon to carbon dioxide for recycling in the environment and the rearranging and assembling of some substrate carbon into new cell substance. Figure 9-1 will help to bring into focus the major concerns of this chapter and define some general terms that describe the reactions we shall discuss. The dashed oval represents the boundary of a microbial cell, the "littlest reactor."

Metabolism is a term that embraces all the diverse reactions by which a cell processes food materials to obtain energy and the compounds from which new cell components are made. The degradative reaction sequences that yield energy and building blocks for biosynthesis in chemoheterotrophs are collectively termed *catabolism*, while the biosynthetic reactions constitute *anabolism*. Catabolism involves the conversion of a variety of substrates belonging to various classes of compounds into intermediate products that can enter one of a few central catabolic pathways. These central pathways degrade organic compounds to carbon dioxide under aerobic conditions. Anabolism involves the use of various intermediates in these central pathways as starting points for a variety of biosynthetic pathways that lead to the many different compounds required for the manufacture of new cells. Thus, the central pathways of catabolism, which are common to a wide range of organisms, might be considered as the essential core of the cell's metabolism. Metabolic reactions specific for individual substrates used as sources of energy and carbon converge toward the central pathways, while synthetic reactions diverge from them.

In Chap. 3 we examined the structures of many of the types of compounds that comprise living cells. These compounds were of interest in the

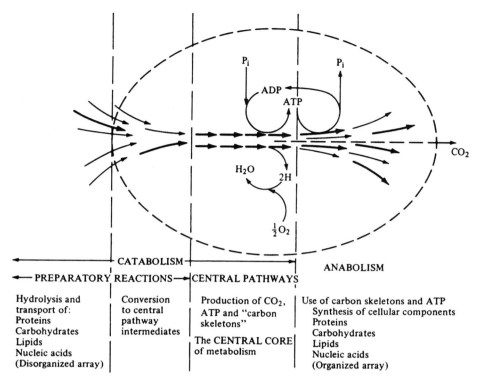

CATABOLISM			ANABOLISM
← PREPARATORY REACTIONS →		CENTRAL PATHWAYS	
Hydrolysis and transport of:	Conversion to central pathway intermediates	Production of CO_2, ATP and "carbon skeletons"	Use of carbon skeletons and ATP Synthesis of cellular components
Proteins			Proteins
Carbohydrates		The CENTRAL CORE	Carbohydrates
Lipids		of metabolism	Lipids
Nucleic acids			Nucleic acids
(Disorganized array)			(Organized array)

Figure 9-1 The orderly flow of carbon into a cell; its rearrangement, oxidation, and utilization for the manufacture of new cell components. The pathways for catabolism of organic compounds converge toward the central pathways, and the anabolic pathways for synthesis of new cell material diverge from the central core.

context of environmental engineering and science because of their position in the carbon-oxygen cycle. They are continually and repeatedly synthesized and degraded by living cells of all types. Photoautotrophs, with a minor contribution from chemoautotrophs, convert carbon dioxide to all the compounds that are required for the production of new cells. The compounds synthesized by these primary producers of organic matter serve as carbon and energy sources for heterotrophic organisms of all types, from bacteria to humans. The macromolecules synthesized by one species to reproduce itself are degraded by another, and the monomers are reassembled according to the genetic pattern of the new cell or are degraded to supply energy and simpler building blocks for the synthesis of different compounds. In degrading the material of other cells, a portion of the organic carbon is oxidized to carbon dioxide and is released for use again by autotrophs to complete the cycle.

Let us assume that the aqueous medium in which the cell depicted in Fig. 9-1 is suspended contains macromolecular organic substrates such as polysaccharides, proteins, or lipids. If substrates are to be metabolized, they must be brought into the cell, but macromolecules cannot enter the cell. Catabolism of large molecules begins outside the cell with their hydrolysis to fragments small enough to be transported into the cell. Transport of substrates requires energy and depends on the presence of specific transport proteins in the cell membrane. As the first step that leads to utilization of a substrate by the cell, transport is a vital link in the catabolic process. The ability of an organism to utilize a specific substrate depends not only on its ability to synthesize the required enzymes for metabolism of the substrate but also on its ability to transport the substrate into the cell. A well-known example of the lack of a transport system is found in the inability of *Escherichia coli* to use citrate as a substrate for growth. As we shall see later in this chapter, citrate is an intermediate in the TCA cycle, and any organism that uses the TCA cycle, as *E. coli* does, has the enzymes required to metabolize citrate. However, *E. coli* has no transport system for citrate and therefore cannot utilize citrate that is supplied as an exogenous substrate. The test for citrate utilization is one of the characteristic reactions used to identify *E. coli.*

After gaining access to the "reactor," the raw material is prepared for entry into a set pattern (pathway) of reactions that convert diverse types of organic substrates to carbon dioxide and to building blocks, carbon compounds of particular types and sizes. The specific preparatory reactions used by the cell depend on the nature of the raw material and the organism. These will be considered in the following chapters. Here we shall consider the major metabolic reaction sequences, which are an important aspect of the unity that characterizes the chemistry of living matter, i.e., the central pathways of catabolism. These provide, from the diverse bits and pieces of raw material entering the cell, the uniform building blocks of carbon "skeletons" and the energy (ATP) needed to combine these building blocks to form the cellular proteins, carbohydrates, and other components.

THE UNITY OF BIOCHEMISTRY

All living cells are composed of the same types of materials: proteins, lipids, carbohydrates, RNA, and DNA. All require the same elements in roughly the same proportions. All utilize the same monomers to synthesize their proteins and nucleic acids, and many different types of cells use the same types of monomers for the synthesis of lipids and carbohydrates. The finding that amazed and delighted early investigators of metabolism was the "unity of biochemistry"; that is, all heterotrophic cells, from bacteria to humans, share many common reactions, both anabolic and catabolic.

Many anabolic pathways are shared by a great variety of species. The pathways for the synthesis of most amino acids are the same in most living cells. Indeed, many of these pathways were first investigated in bacteria and fungi because of the relative ease of working with these organisms, and the same reactions were then sought and found in higher organisms. While many species, from bacteria to humans, are unable to synthesize all the amino acids required for protein synthesis, pathways for synthesis of the amino acids that can be synthesized are generally similar or identical in all species. Syntheses of other small molecules such as sugars, organic acids, and nucleic acid bases are similarly achieved by identical or almost identical reactions in all organisms able to carry out these processes. Thus, in the entire gamut of living organisms, from the simplest—bacteria—to the most complex—higher plants and animals, many of the same compounds are made and many of the same reactions occur. The differences among species or individual organisms in relation to the biosynthesis of cellular components thus lies almost entirely in the ability of the organisms to synthesize required molecules, i.e., not in the reactions of a pathway but in the presence or absence of the pathway. After synthesizing the required monomers or acquiring them preformed as a product of the degradation of another cell, all cells use essentially the same reactions for polymerizing monomers to form the macromolecules of the cell. The anabolic reactions of metabolism thus have many common features in all living organisms.

Catabolism exhibits some equally striking similarities in a broad range of organisms. The central catabolic pathways are identical in all aerobic eucaryotic cells and in many procaryotic cells; that is, *E. coli* or yeast, growing aerobically, uses exactly the same reactions in the degradation of glucose to carbon dioxide and water as does the human muscle cell, using glucose as an energy source; both use the EMP pathway and the TCA cycle. However, there are differences in the catabolic capabilities of microorganisms and higher organisms, which are particularly important in the degradative leg of the carbon-oxygen cycle. Many individual species or genera of bacteria and fungi possess very specialized catabolic pathways. Some have central catabolic pathways not found in higher organisms, e.g., the Entner-Douderoff pathway of the pseudomonads and the phosphoketolase pathway of the heterolactics.

Aerobic microorganisms, as we emphasized in previous chapters, are capable of degrading many compounds that are not metabolized by higher organisms. Hydrocarbons, for example, are metabolized rapidly by various species of bacteria and fungi but only very slowly or not at all by animal tissues. Thus, microorganisms exhibit a greater variety of enzymatic capabilities in preparing compounds for entry into the central pathways while utilizing in most cases the same central pathways as do higher organisms. As we stated previously, most microbiologists believe that there are few synthetic or naturally occurring compounds that are not subject to microbial degradation. It is this metabolic versatility that makes the aerobic microorganisms so useful in biological processes for the treatment of a great variety of wastes of domestic and industrial origin.

A variety of catabolic reactions is found also among fermentative bacteria. Many of these begin the degradation of substrate through the same central pathway used by aerobic eucaryotes (EMP). However, as we discussed in Chap. 7, fermentation requires the use of organic electron acceptors, which the cell must make from its energy source. This leads to a great variety of additional reactions in the formation of fermentation products. Other fermentative bacteria use none of the common catabolic reactions but possess specialized pathways restricted to the use of one or a few types of substrates. The anaerobic microorganisms (both obligate and facultative) perform important functions in the recycling of carbon, particularly in nature where anaerobic environments preclude the aerobic degradation of organic material. The engineered subcycles emphasize aerobic processes because these are necessary for the complete degradation of carbon compounds to carbon dioxide and because some types of compounds can be attacked only aerobically. Hydrocarbons, for example, cannot be fermented. The persistence of underground reservoirs of petroleum is made possible by the fact that they are not exposed to oxygen, which would encourage their metabolism by aerobic microorganisms.

In this and the following two chapters we shall consider some aspects of metabolism that are essential to the understanding of the processes by which organic carbon is recycled in the natural cycles and the engineered subcycles described in Chaps. 1 and 2. In this chapter we will examine the central catabolic pathways used by the great majority of heterotrophic organisms for the complete catabolism of organic compounds to carbon dioxide and water. These are central not only to the carbon and energy metabolism of the cell but also to the functioning of the carbon-oxygen cycle. We will also examine the reactions of the synthetic leg of the carbon-oxygen cycle by which autotrophic organisms convert carbon dioxide to organic matter. In Chap. 10 we will consider some of the reactions required to prepare various classes of substrates for entry into the central pathways in aerobic organisms and the aerobic processes utilized in the technological control of pollution by organic wastes. In Chap. 11 we will examine the anaerobic processes important in the recycling of carbon in anaerobic environments and the anaerobic treatment of organic matter.

The more specialized metabolic pathways used by a minority of species for the degradation of one or a few substrates will not be considered unless they are of major importance in biological waste treatment or in some aspect of the carbon-oxygen cycle in nature. We shall consider the biosynthesis of the major cellular components only from the viewpoint of their origins, i.e., the use of catabolic intermediates as precursors of cellular materials. Readers interested in specific metabolic pathways may see the references listed at the end of this chapter.

Three pathways are central to the use of organic substrates as sources of energy and carbon in most heterotrophic microorganisms. These pathways provide energy to the cell and also provide most of the initial substrates for whatever biosynthetic pathways the cell possesses. These pathways are: (1) the Embden-Meyerhof-Parnas (EMP) pathway, often called *glycolysis*; (2) the hexose monophosphate (HMP) pathway; and (3) the cyclic pathway for terminal oxidation, which has been given several names—the tricarboxylic acid (TCA) cycle, the citric acid cycle, and the Krebs cycle in honor of Sir Hans Krebs, its discoverer. These pathways are interrelated, as are all metabolic pathways. These interrelationships, including specialized catabolic pathways and biosynthetic pathways, are often presented as a "metabolic map," which appears as a maze of reactions more complex than a map of the rapid transit system of New York City. The seemingly bewildering variety of reactions in such a map or in the pages of a biochemistry text reduces, upon closer examination, to a really rather simple group of *type reactions*, that is, reactions that accomplish the same purpose by altering the same groups in the same ways, regardless of the structure of the rest of the molecule. We will examine some of the individual reactions of the three central catabolic pathways in detail, both because of the importance of these pathways and because they afford the opportunity to illustrate many of these repeated reactions. Other important type reactions occur during the preparation of molecules for entry into the central pathways; these will be described in subsequent chapters.

GLYCOLYSIS—THE EMBDEN-MEYERHOF-PARNAS PATHWAY

The glycolytic (EMP) pathway is the major route for catabolism of carbohydrates and related compounds in most microorganisms. In Chap. 7 we described this pathway briefly in relation to its function in the energy metabolism of the cell, and we examined two reactions of the pathway as examples of ATP formation by substrate level phosphorylation [Eqs. (7-7) to (7-10)]. We shall now examine the overall pathway, using glucose as a starting point. Glucose occurs widely in nature as the monomer comprising the major storage carbohydrates, starch in plants and glycogen in animals, as well as structural polysaccharides such as cellulose. Perhaps because of its widespread occurrence, it is used readily by most microorganisms and is the preferred energy source for many.

In the EMP pathway, a molecule of glucose is converted to two molecules of pyruvic acid through the series of reactions shown in Fig. 9-2. In the process, two molecules of ATP are produced by substrate level phosphorylation and two molecules of NAD are reduced. No oxygen is involved in this series of reactions, and the process can be utilized equally well by both aerobic and anaerobic organisms. However, the further metabolism of pyruvic acid differs greatly in aerobic and anaerobic cells. The two molecules of $NADH_2$ must be reoxidized to NAD; that is, the electrons removed from the energy source must be passed from the electron carrier, NAD, to a final electron acceptor. In aerobic cells, pyruvic acid can be completely oxidized to carbon dioxide by entering another of the central pathways, the TCA cycle. In anaerobic cells, pyruvic acid may serve as the final electron acceptor, forming lactic acid as the end product of the pathway, or it may be converted to other

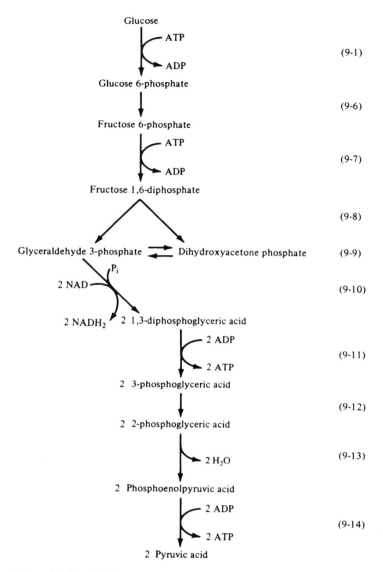

Figure 9-2 The Embden-Meyerhof-Parnas (EMP) pathway for catabolism of glucose. Numbers in parentheses are for the text equations of individual reactions.

electron acceptors, which differ with the species involved. Microorganisms, primarily bacteria, exhibit far greater ingenuity in the metabolism of pyruvate than do higher organisms, as we shall see in Chap. 11. Because of the variety of pathways into which pyruvate may enter, it may be considered the pivotal compound of the central catabolic pathways for carbohydrate degradation.

As in many metabolic sequences, most of the reactions of the EMP pathway may be considered as preparatory reactions leading up to the reactions that fulfill the real purpose of the sequence—the release and capture of energy. It may appear paradoxical that in a pathway designed to synthesize ATP, two molecules of ATP are used. The cell invests some of its precious store of usable energy in the hope of making a profit. One reason for the immediate phosphorylation of the energy source, which is common in catabolic pathways, may be to prevent its loss by diffusion from the cell, since the membrane is almost impermeable to ionized compounds. Whatever the reason, the investment of ATP is a wise one since it is repaid with 100 percent interest almost immediately.

The initial reaction of the pathway [Eq. (9-1)] is one of the reactions in which ATP is expended. Thus free glucose entering the cell is immediately converted to glucose 6-phosphate.

$$
\begin{array}{ccc}
\text{HC}\!=\!\text{O} & & \text{HC}\!=\!\text{O} \\
| & & | \\
\text{HCOH} & \text{ATP} \quad \text{ADP} & \text{HCOH} \\
| & & | \\
\text{HOCH} & \longrightarrow & \text{HOCH} \\
| & & | \\
\text{HCOH} & & \text{HCOH} \\
| & & | \\
\text{HCOH} & & \text{HCOH} \\
| & & | \\
\text{H}_2\text{COH} & & \text{H}_2\text{COPO}_3\text{H}_2 \\
\text{glucose} & & \text{glucose 6-phosphate}
\end{array}
\tag{9-1}
$$

The enzyme *hexokinase* is able to phosphorylate other sugars in addition to glucose. In this reaction, the energy of hydrolysis of the terminal phosphate group of ATP is more than sufficient to supply the energy required for the transfer of the phosphate group to the glucose molecule, forming an ester bond between the alcohol group on carbon-6 and the acidic phosphate (in effect, H_3PO_4, phosphoric acid).

Other mechanisms of formation of glucose 6-phosphate are utilized under certain conditions. If glucose is obtained from an internal store of glycogen or starch, the polymer may be converted directly to glucose 1-phosphate by *phosphorolysis*, as opposed to hydrolysis. In that case, cleavage of the bonds between successive glucose units in the polymer occurs by the addition of phosphoric acid rather than water [Eq. (9-2)].

This reaction is catalyzed by a *phosphorylase*, either glycogen phosphorylase or starch phosphorylase, depending on the substrate. Although this mechanism of phosphorylation of glucose would seem at first glance to represent a savings of energy to the cell, since ATP is not required as a source of the phosphate group, the ultimate cost in energy to the cell is the same. The formation of the polysaccharide requires the expenditure of two high-energy phosphate bonds for the addition of each glucose monomer. Thus, phosphorolysis of the bond in effect conserves half the energy expen-

glycogen

$$(9-2)$$

glucose 1-phosphate   glycogen (less one glucose)

ded in its formation, and the phosphorylation of glucose requires the use of the bond energy of ATP by either mechanism. For entry into the EMP pathway, the phosphate group must be transferred from carbon-1 to carbon-6, and this is accomplished by the enzyme *phosphoglucomutase* [Eq. (9-3)].

glucose 1-phosphate   glucose 6-phosphate

$$(9-3)$$

In some microorganisms, free glucose is not transported into the cell. Phosphorylation of glucose and other sugars can occur as an integral component of the entry process, catalyzed by the *phosphotransferase system*.

$$\text{Phosphoenolpyruvate} + \text{HPr} \rightarrow \text{pyruvate} + \text{phospho-HPr} \qquad (9-4)$$

$$\text{Phospho-HPr} + \text{glucose} \rightarrow \text{HPr} + \text{glucose 6-phosphate} \qquad (9-5)$$

HPr is a small protein (molecular weight 9400) containing a histidine, which is phosphorylated in reaction (9-4). This phosphorylation is unusual in that a metabolic intermediate, phosphoenolpyruvate (PEP), rather than ATP is the phosphate donor. Again, however, the energy cost to the cell is the same as if ATP had been used, since phosphoenolpyruvate would otherwise be used as a phosphate donor in the synthesis of ATP [see Eq. (9-14)].

Reaction (9-5) is catalyzed by an enzyme located in the cell membrane. This enzyme binds free glucose at the outer surface of the membrane and transfers it to the inner surface, where it is phosphorylated and released into the cytoplasm.

After glucose 6-phosphate has been formed, a rearrangement of the molecule is catalyzed by *glucose phosphate isomerase*, which converts the aldohexose to a ketohexose [Eq. (9-6)].

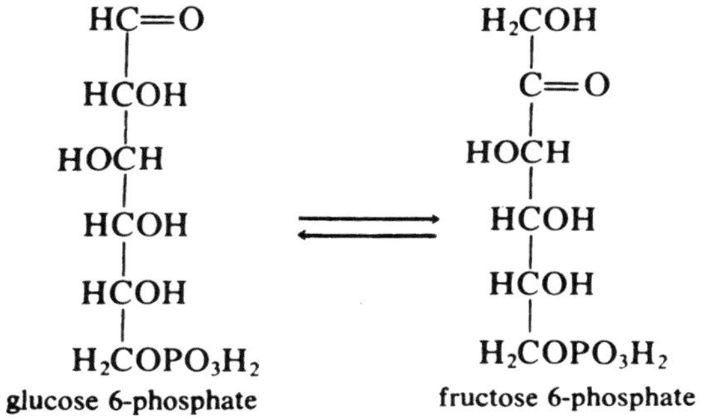

$$
\begin{array}{ccc}
\begin{array}{c}
HC{=}O \\
| \\
HCOH \\
| \\
HOCH \\
| \\
HCOH \\
| \\
HCOH \\
| \\
H_2COPO_3H_2 \\
\text{glucose 6-phosphate}
\end{array}
&
\rightleftharpoons
&
\begin{array}{c}
H_2COH \\
| \\
C{=}O \\
| \\
HOCH \\
| \\
HCOH \\
| \\
HCOH \\
| \\
H_2COPO_3H_2 \\
\text{fructose 6-phosphate}
\end{array}
\end{array}
\tag{9-6}
$$

The resulting molecule, *fructose 6-phosphate*, now has the same configuration at carbon-1 as that originally present at carbon-6, and the same reaction can occur. Thus, the isomerization is a simple internal rearrangement that prepares the molecule for the reaction that follows.

The second investment of phosphate bond energy occurs when ATP donates its terminal phosphate group to carbon-1 of fructose 6-phosphate in a reaction catalyzed by *6-phosphofructokinase* [Eq. (9-7)].

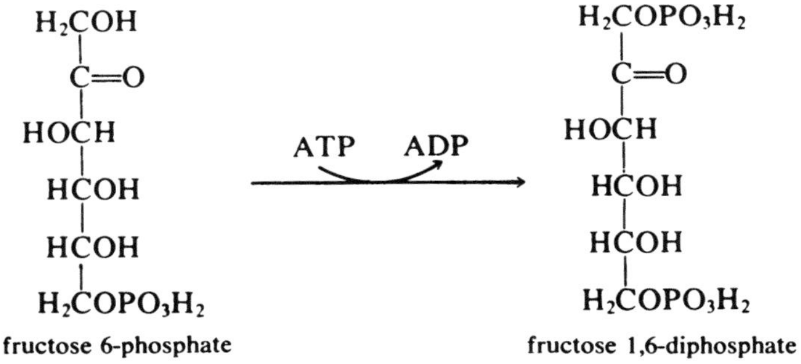

$$
\begin{array}{ccc}
\begin{array}{c}
H_2COH \\
| \\
C{=}O \\
| \\
HOCH \\
| \\
HCOH \\
| \\
HCOH \\
| \\
H_2COPO_3H_2 \\
\text{fructose 6-phosphate}
\end{array}
&
\xrightarrow{\text{ATP} \quad \text{ADP}}
&
\begin{array}{c}
H_2COPO_3H_2 \\
| \\
C{=}O \\
| \\
HOCH \\
| \\
HCOH \\
| \\
HCOH \\
| \\
H_2COPO_3H_2 \\
\text{fructose 1,6-diphosphate}
\end{array}
\end{array}
\tag{9-7}
$$

A comparison of this reaction with Eq. (9-1) shows that the two are identical. In both, ATP is the phosphate donor, resulting in esterification of a primary alcohol.

The next reaction [Eq. (9-8)] is a reversal of the aldol condensation reaction, which is so useful in organic syntheses. The enzyme *fructose diphosphate aldolase* splits the six-carbon molecule into two triose phosphates: an aldehyde, glyceraldehyde 3-phosphate, and a ketone, dihydroxyacetone phosphate.

$$
\begin{array}{c}
\text{H}_2\text{COPO}_3\text{H}_2 \\
| \\
\text{C}=\text{O} \\
| \\
\text{HOCH} \\
| \\
\text{HCOH} \\
| \\
\text{HCOH} \\
| \\
\text{H}_2\text{COPO}_3\text{H}_2 \\
\text{fructose 1,6-diphosphate}
\end{array}
\rightleftharpoons
\begin{array}{c}
\text{H}_2\text{COPO}_3\text{H}_2 \\
| \\
\text{C}=\text{O} \\
| \\
\text{H}_2\text{COH} \\
\text{dihydroxyacetone phosphate}
\end{array}
+
\begin{array}{c}
\text{HC}=\text{O} \\
| \\
\text{HCOH} \\
| \\
\text{H}_2\text{COPO}_3\text{H}_2 \\
\text{glyceraldehyde 3-phosphate}
\end{array}
$$

$$(9\text{-}8)$$

This reaction is reversible and would actually proceed toward condensation if the products were not removed by subsequent reactions.

The two triose phosphates formed by the aldolase reaction are isomers, just as were the two hexose phosphates encountered earlier in the pathway, and they can be interconverted by an isomerase [Eq. (9-9)] in a reaction analogous to Eq. (9-6).

$$
\begin{array}{c}
\text{H}_2\text{COPO}_3\text{H}_2 \\
| \\
\text{C}=\text{O} \\
| \\
\text{H}_2\text{COH} \\
\text{dihydroxyacetone phosphate}
\end{array}
\rightleftharpoons
\begin{array}{c}
\text{H}_2\text{COPO}_3\text{H}_2 \\
| \\
\text{HCOH} \\
| \\
\text{HC}=\text{O} \\
\text{glyceraldehyde 3-phosphate}
\end{array}
\qquad (9\text{-}9)
$$

The enzyme catalyzing this reversible internal rearrangement is *triose phosphate isomerase*. The equilibrium of the isomerase reaction favors conversion of the aldehyde to the ketone, but the reaction is pulled toward the aldehyde because the substrate for the reaction that follows is glyceraldehyde 3-phosphate. Some dihydroxyacetone phosphate may be diverted to the synthesis of lipids, as we shall see later.

We have already described the next two reactions [Eqs. (7-7) and (7-8)]. All the reactions that have occurred to this point prepare the molecule for the single oxidative reaction in the entire EMP pathway. We discussed in Chap. 7 that glycolysis releases only a small proportion of the energy potentially available from the catabolism of glucose. This is due to the fact that only one oxidative step is included in the pathway—that catalyzed by *glyceraldehyde 3-phosphate dehydrogenase* [Eq. (9-10)].

$$
\begin{array}{ccc}
\underset{\text{glyceraldehyde}}{\underset{\text{3-phosphate}}{\begin{array}{c} HC{=}O \\ | \\ HCOH \\ | \\ H_2COPO_3H_2 \end{array}}}
&
\xrightleftharpoons[\quad H_3PO_4 \quad]{\quad NAD \qquad NADH_2 \quad}
&
\underset{\substack{\text{1,3-diphospho-}\\\text{glyceric acid}}}{\begin{array}{c} O \\ \diagup\!\!\diagdown \\ C{-}OPO_3H_2 \\ | \\ HCOH \\ | \\ H_2COPO_3H_2 \end{array}}
\end{array}
\tag{9-10}
$$

In this reaction, the aldehyde is oxidized to the acid, and the energy released by the oxidation is utilized to add inorganic phosphate and form the mixed anhydride of phosphoric acid and 3-phosphoglyceric acid. The two hydrogen atoms removed in the oxidation are accepted by the electron carrier, NAD.

We have noted that an acid anhydride is a highly unstable and reactive compound that can donate its phosphate group to ADP to form ATP. This occurs in the reaction catalyzed by *phosphoglycerate kinase* [Eq. (9-11)]. Reactions (9-10) and (9-11) result in the phosphorylation of ADP by inorganic phosphate using energy supplied by the oxidation of glyceraldehyde 3-phosphate—*substrate level phosphorylation.*

$$
\begin{array}{ccc}
\underset{\substack{\text{1,3-diphospho-}\\\text{glyceric acid}}}{\begin{array}{c} O \\ \diagup\!\!\diagdown \\ C{-}OPO_3H_2 \\ | \\ HCOH \\ | \\ H_2COPO_3H_2 \end{array}}
&
\xrightleftharpoons[\qquad\qquad]{\quad ADP \qquad ATP \quad}
&
\underset{\text{3-phosphoglyceric acid}}{\begin{array}{c} O \\ \diagup\!\!\diagdown \\ C{-}OH \\ | \\ HCOH \\ | \\ H_2COPO_3H_2 \end{array}}
\end{array}
\tag{9-11}
$$

The two molecules of 1,3-diphosphoglyceric acid thus are utilized to synthesize two molecules of ATP, which replace the two molecules utilized earlier in the pathway. A series of rearrangements designed to prepare the molecule for acting again as phosphate donor now ensues. This is necessary because the remaining phosphate group is in an ester linkage that is stable and not active as a donor.

The first preparatory reaction is one in which the phosphate group is moved from carbon-3 to carbon-2 [Eq. (9-12)].

$$
\begin{array}{ccc}
\underset{\text{3-phosphoglyceric acid}}{\begin{array}{c} O \\ \diagup\!\!\diagdown \\ C{-}OH \\ | \\ HCOH \\ | \\ H_2COPO_3H_2 \end{array}}
&
\xrightleftharpoons[\qquad\qquad]{\qquad\qquad}
&
\underset{\text{2-phosphoglyceric acid}}{\begin{array}{c} O \\ \diagup\!\!\diagdown \\ C{-}OH \\ | \\ HCOPO_3H_2 \\ | \\ H_2COH \end{array}}
\end{array}
\tag{9-12}
$$

This reaction is catalyzed by *phosphoglyceromutase* and is analogous to the reaction in which glucose 1-phosphate was converted to glucose 6-phosphate by a mutase [Eq. (9-3)]. The conversion of 3-phosphoglyceric acid to 2-phosphoglyceric acid prepares the molecule for a reaction that converts it from a poor phosphate donor to an active one.

The destabilization of the phosphate group is accomplished by the removal of a molecule of water (dehydration) catalyzed by the enzyme *enolase* [Eq. (9-13)].

$$
\begin{array}{ccc}
\underset{\substack{\text{2-phosphoglyceric acid}}}{
\begin{array}{c}
\overset{O}{\overset{\|}{C}}-OH \\
\mid \\
\overline{H}COPO_3H_2 \\
\mid \\
H_2C\overline{OH}
\end{array}}
&
\underset{}{
\begin{array}{c}
H_2O \\
\rightleftharpoons
\end{array}}
&
\underset{\substack{\text{phosphoenolpyruvic acid} \\ \text{(PEP)}}}{
\begin{array}{c}
\overset{O}{\overset{\|}{C}}-OH \\
\mid \\
COPO_3H_2 \\
\| \\
CH_2
\end{array}}
\end{array}
\qquad (9\text{-}13)
$$

This internal oxidation-reduction reaction, involving carbon atoms 2 and 3, creates a molecule with a very high free energy of hydrolysis that is capable of donating phosphate to ADP. This occurs in reaction (9-14).

In the reaction catalyzed by *pyruvate kinase*, phosphate is transferred from phosphoenolpyruvate to ADP, forming ATP and pyruvic acid [Eq. (9-14)].

$$
\begin{array}{ccc}
\underset{\substack{\text{phosphoenolpyruvic acid}}}{
\begin{array}{c}
\overset{O}{\overset{\|}{C}}-OH \\
\mid \\
COPO_3H_2 \\
\| \\
CH_2
\end{array}}
&
\underset{}{
\begin{array}{c}
ADP \quad ATP \\
\longrightarrow
\end{array}}
&
\underset{\substack{\text{pyruvic acid}}}{
\begin{array}{c}
\overset{O}{\overset{\|}{C}}-OH \\
\mid \\
C=O \\
\mid \\
CH_3
\end{array}}
\end{array}
\qquad (9\text{-}14)
$$

Since two molecules of PEP are formed per molecule of glucose, two molecules of ATP are formed in this step.

We can now summarize the reactions examined thus far. One molecule of glucose has been converted to two molecules of pyruvic acid with the *net* production of two molecules of ATP. The pathway cannot stop here because two molecules of NAD have been reduced to $NADH_2$, and it is necessary that the hydrogen (two hydrogen ions and two electrons) be passed on to other electron carriers or acceptors so that NAD can continue to function as an electron carrier, i.e., as a temporary acceptor of electrons in oxidative reactions. As we stated above, the further reactions of pyruvate and the regeneration of NAD are different in cells that have respiratory and fermentative metabolism. If a cell has a respiratory metabolism, either aerobic or anaerobic, that uses an external, inorganic electron acceptor, it may regenerate its reduced NAD by passing electrons to an electron transport chain (see Chap. 7). We will examine in Chap. 11 the restrictions placed on the cell by the fermentative mode of metabolism. A fermentative cell generally cannot afford to oxidize pyruvate further by reducing NAD, since its major problem is the oxidation of the $NADH_2$ already formed. However, in a cell that uses electron transport as a means of oxidizing $NADH_2$, the further oxidation of pyruvate, with the concomitant reduction of NAD, is advantageous since it releases additional energy. In such cells, pyruvate undergoes an *oxidative decarboxylation* catalyzed by *pyruvate dehydrogenase* [Eq. (9-15)], forming acetyl-CoA, which enters the second of the central catabolic pathways of respiratory catabolism, the TCA cycle.

$$(9\text{-}15)$$

pyruvic acid   acetyl-CoA

THE TRICARBOXYLIC ACID (TCA) CYCLE

The TCA cycle, as its name implies, is a cyclic series of reactions that accomplishes the complete oxidation of the two-carbon acetic acid residue that is fed into the cycle as acetyl-CoA. Carbohydrates, as we have seen in the preceding section, are degraded to carbon dioxide and acetyl-CoA. Organic acids, including the long-chain fatty acids of neutral fats and phospholipids, are also oxidized to acetyl-CoA. Amino acids that serve as energy and/or carbon source are degraded to acetyl-CoA or to other intermediates of the TCA cycle. Hydrocarbons also give rise to acetyl-CoA. Thus, the TCA cycle serves a central and vital function in recycling organic carbon to the atmosphere. In the oxidation of the two-carbon acetate residue to carbon dioxide, eight hydrogen atoms are removed and are passed through the electron transport system, generating ATP by coupled oxidative phosphorylation. By far the greatest proportion of the release and capture of the energy of the glucose molecule that is degraded through the EMP pathway followed by the TCA cycle is thus dependent on the reactions of this terminal cycle. The close association between the TCA cycle and the electron transport system is evident from the fact that both are located in the mitochondrion in eucaryotic cells. The overall cycle is shown in Fig. 9-3.

The two-carbon acetyl residue enters the cycle by condensing with

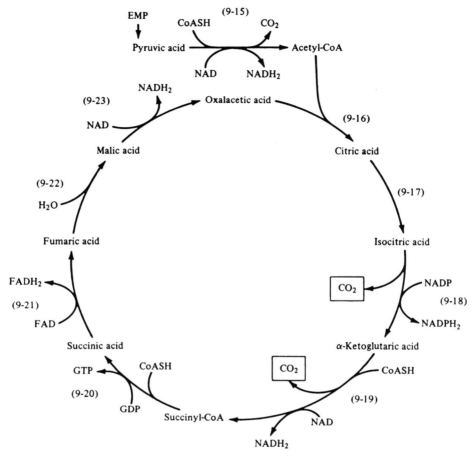

Figure 9-3 The tricarboxylic acid (TCA) cycle for terminal oxidation of the energy source to carbon dioxide in aerobic cells. Numbers on arrows correspond to the text equations for the individual reactions.

oxalacetate to form citric acid, the tricarboxylic acid that gives the cycle its name [Eq. (9-16)].

$$\text{acetyl-CoA} + \text{oxalacetic acid} \xrightarrow[\text{H}_2\text{O}]{\text{CoASH}} \text{citric acid} \tag{9-16}$$

This reaction proceeds readily because of the high free energy of hydrolysis of the thiolester bond

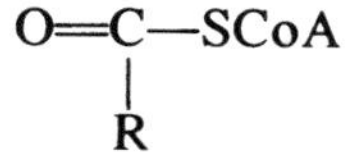

$$O{=}C{-}SCoA$$
$$|$$
$$R$$

which is comparable to that of the terminal phosphate of ATP. The enzyme is *citrate synthase.*

The second step is an internal rearrangement of the molecule catalyzed by the enzyme *aconitase* [Eq. (9-17)], which prepares the molecule for the first oxidative step. (The name of the enzyme is derived from *cis*-aconitic acid, formerly believed to be an intermediate in the reaction.)

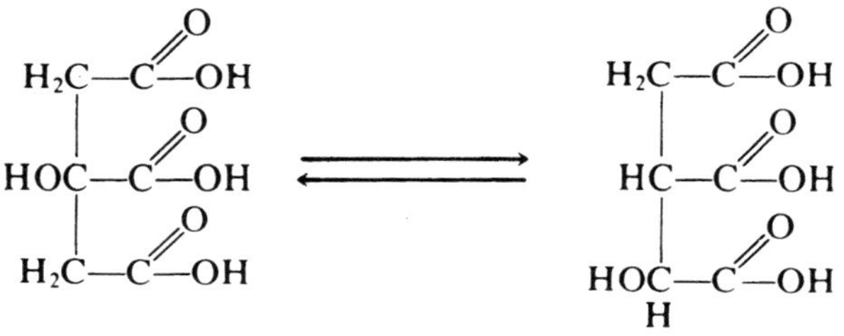

$$\qquad\qquad (9\text{-}17)$$

citric acid isocitric acid

A hydroxyl group must be moved from a β- to an α-carbon to be in position for a reaction similar to one we have seen previously. In the *aconitase* reaction, citrate is converted by rearrangement to *isocitrate.*

In the following step [Eq. (9-18)], the α-hydroxyl group is oxidized to a keto group with the reduction of a molecule of NADP, and the first carbon atom lost in the cycle is removed as carbon dioxide by β-decarboxylation.

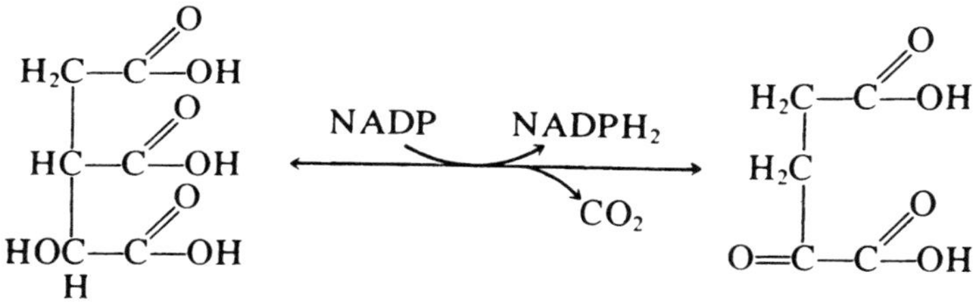

$$\qquad\qquad (9\text{-}18)$$

isocitric acid α-ketoglutaric acid

The reaction is catalyzed by *isocitrate dehydrogenase.*

The resulting α-keto acid, *α-ketoglutarate,* is a structural analogue of pyruvic acid, and it undergoes a reaction identical with that by which acetyl-CoA was formed from pyruvate [compare Eqs. (9-15) and (9-19)].

$$\underset{\alpha\text{-ketoglutaric acid}}{\begin{array}{c} \overset{O}{\underset{\parallel}{C}}\text{-OH} \\ | \\ C\text{=}O \\ | \\ CH_2 \\ | \\ CH_2 \\ | \\ \overset{O}{\underset{\parallel}{C}}\text{-OH} \end{array}} \quad \xrightarrow[\substack{\text{CoASH} \quad CO_2}]{\text{NAD} \quad \text{NADH}_2} \quad \underset{\text{succinyl-CoA}}{\begin{array}{c} \overset{O}{\underset{\parallel}{C}}\text{-SCoA} \\ | \\ CH_2 \\ | \\ CH_2 \\ | \\ \overset{O}{\underset{\parallel}{C}}\text{-OH} \end{array}} \qquad (9\text{-}19)$$

The product of the oxidative decarboxylation in this case is *succinyl-CoA*, and the enzyme is *α-ketoglutarate dehydrogenase*, which uses NAD as electron acceptor. At this point in the cycle, two carbon atoms, equivalent to the entering acetyl residue, have been lost as carbon dioxide, and four hydrogen atoms have been removed. The remainder of the cycle operates to regenerate oxalacetate, which must be available for condensation with another molecule of acetyl-CoA.

Succinyl-CoA, as in the case of acetyl-CoA, contains a high-energy (unstable) thiolester bond, which is capable of generating ATP. This is possible because CoA and phosphate form energetically equivalent bonds with a carboxyl carbon and are therefore interchangeable without a loss or input of energy. As we shall see later, this exchange mechanism is important in several reactions. In the case of succinyl-CoA, phosphate is exchanged for the CoA, forming succinylphosphate, which remains bound to the enzyme *succinyl-CoA synthetase*. The succinate is released and the phosphate is used to phosphorylate GDP to form GTP [Eq. (9-20)].

$$\underset{\text{succinyl-CoA}}{\begin{array}{c} H_2C\text{-}\overset{O}{\underset{\parallel}{C}}\text{-OH} \\ | \\ H_2C\text{-}\overset{O}{\underset{\parallel}{C}}\text{-SCoA} \end{array}} \quad \xrightarrow[\substack{\text{GDP} \quad \text{GTP}}]{H_3PO_4 \quad \text{CoASH}} \quad \underset{\text{succinic acid}}{\begin{array}{c} H_2C\text{-}\overset{O}{\underset{\parallel}{C}}\text{-OH} \\ | \\ H_2C\text{-}\overset{O}{\underset{\parallel}{C}}\text{-OH} \end{array}} \qquad (9\text{-}20)$$

A further exchange then takes place in which the terminal phosphate of GTP is used to form ATP. This is a *substrate level phosphorylation*, which, despite its rather complicated series of transfers, is very similar to that which occurred in the EMP pathway. In both cases an oxidation releases energy that is utilized to form a high-energy bond at the carboxyl carbon. In both cases inorganic phosphate participates in the formation of a mixed anhydride, and this phosphate is donated to a nucleoside diphosphate.

Succinate is oxidized by a flavoprotein enzyme, *succinate dehydrogenase*, which contains FAD as a cofactor [Eq. (9-21)].

$$
\begin{array}{c}
\text{H}_2\text{C}-\overset{\displaystyle O}{\overset{\|}{\text{C}}}-\text{OH} \\
| \\
\text{H}_2\text{C}-\overset{\displaystyle O}{\overset{\|}{\text{C}}}-\text{OH} \\
\text{succinic acid}
\end{array}
\quad
\xrightleftharpoons[]{\text{FAD} \quad \text{FADH}_2}
\quad
\begin{array}{c}
\text{HC}-\overset{\displaystyle O}{\overset{\|}{\text{C}}}-\text{OH} \\
\overset{O}{\underset{\|}{}} \\
\text{HO}-\text{C}-\text{CH} \\
\text{fumaric acid}
\end{array}
\qquad (9\text{-}21)
$$

The product *fumarate* contains a double bond and undergoes another typical reaction found in a variety of pathways. We stated earlier that oxygen in biological products rarely originates from molecular oxygen. The desaturation of a compound by removing hydrogen atoms from adjacent carbon atoms followed by the addition of water is a useful way of introducing an atom of oxygen. The reverse sequence, removal of the elements of water, hydrogen and hydroxide, from adjacent carbon atoms, is also used in certain pathways. The double bond thus formed can be used by fermentative bacteria as an acceptor for two hydrogen atoms in the reoxidation of $NADH_2$.

In the case of fumarate, the elements of water are added to the double bond by the enzyme *fumarase* to form *malate* [Eq. (9-22)].

$$
\begin{array}{c}
\overset{O}{\underset{\|}{}}\;\text{HC}-\overset{\displaystyle O}{\overset{\|}{\text{C}}}-\text{OH} \\
\text{HO}-\text{C}-\text{CH} \\
\text{fumaric acid}
\end{array}
\quad
\xrightleftharpoons[]{\text{H}_2\text{O}}
\quad
\begin{array}{c}
\text{H} \quad \overset{\displaystyle O}{\overset{\|}{}} \\
\text{HOC}-\text{C}-\text{OH} \\
| \qquad \overset{O}{\underset{\|}{}} \\
\text{H}_2\text{C}-\text{C}-\text{OH} \\
\text{malic acid}
\end{array}
\qquad (9\text{-}22)
$$

The substrate needed for the initial condensation reaction, oxalacetate, can then be formed by the removal of two hydrogen atoms by the enzyme *malate dehydrogenase*, using NAD as electron carrier [Eq. (9-23)].

$$
\begin{array}{c}
\text{H} \quad \overset{\displaystyle O}{\overset{\|}{}} \\
\text{HOC}-\text{C}-\text{OH} \\
| \qquad \overset{O}{\underset{\|}{}} \\
\text{H}_2\text{C}-\text{C}-\text{OH} \\
\text{malic acid}
\end{array}
\quad
\xrightleftharpoons[]{\text{NAD} \quad \text{NADH}_2}
\quad
\begin{array}{c}
\overset{\displaystyle O}{\overset{\|}{}} \\
\text{O}=\text{C}-\text{C}-\text{OH} \\
| \qquad \overset{O}{\underset{\|}{}} \\
\text{H}_2\text{C}-\text{C}-\text{OH} \\
\text{oxalacetic acid}
\end{array}
\qquad (9\text{-}23)
$$

If we include the oxidative decarboxylation of pyruvate to form acetyl-CoA, we can summarize the reactions by which the total oxidation of pyruvate occurs as follows: Three molecules of carbon dioxide are released, four molecules of NAD and one of FAD are reduced, making a total of 10 hydrogen atoms placed on electron carriers to be used in the generation of ATP through electron transport phosphorylation, and one ATP is formed by substrate level phosphorylation. The oxalacetate with which the cycle began has been regenerated and the cycle has returned to its starting point.

The two pathways we have described—the EMP pathway and the TCA cycle—are the major routes by which organic compounds of many classes are oxidized completely to carbon dioxide with the concomitant production of ATP. Energy is furnished by the substrate level phosphorylation of the EMP pathway in many fermentative microorganisms; it is produced by oxidative phosphorylation coupled to transport of the electrons removed in the EMP and especially in the TCA cycle in organisms that use respiratory metabolism. These pathways, therefore, are the major energy-yielding routes of catabolism for the majority of chemoorganotrophic microorganisms.

Most of these organisms possess another pathway that has the potential for total oxidation of carbohydrates. This is the *hexose monophosphate* (*HMP*) *pathway*, also called the *pentose phosphate* pathway or the *phosphogluconate* pathway.

THE HEXOSE MONOPHOSPHATE (HMP) PATHWAY

Although the HMP pathway can be used to oxidize glucose completely, it probably is seldom used for this purpose in most organisms. Two important functions that the pathway can perform are essential for the biosynthetic reactions of the cell. One is the provision of reduced NADP, which is specifically required as a cofactor in certain reductive reactions of biosynthesis, e.g., in the synthesis of lipids. The pathway is also a source of pentose phosphate for the synthesis of the nucleotides required for making RNA and DNA. The first few reactions of the pathway (see Fig. 9-4) convert hexoses to the ribose 5-phosphate used in nucleotide synthesis. If pentoses are available as carbon and energy source, they can be converted to hexoses and to other

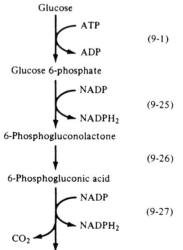

Figure 9-4 The oxidative segment of the hexose mono-phosphate (HMP) pathway. Numbers refer to the text equations for individual reactions.

glycolytic intermediates through other reactions of the HMP pathway (see Fig. 9-5).

The hexose monophosphate pathway is rather complex, involving several condensing and splitting reactions by which sugars with three to seven carbon atoms are interconverted. It is not necessary that the glucose entering the pathway proceed through the entire sequence of reactions, since there are alternative routes utilizing various combinations of the enzymes that can be used for the sole purpose of forming the required biosynthetic precursors and $NADPH_2$. If $NADPH_2$ is required in greater amounts than are the sugars formed as intermediates, the pathway can operate in a cyclic pattern, regenerating glucose 6-phosphate. Since one molecule of carbon dioxide is formed from each molecule of glucose 6-phosphate that enters the cycle, the

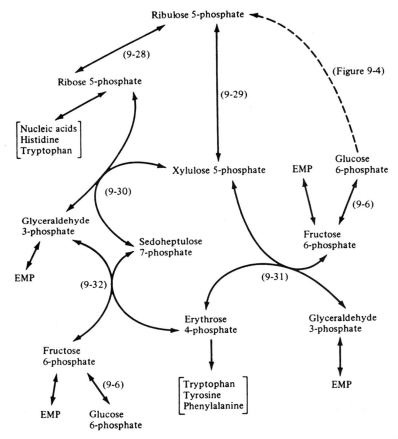

Figure 9-5 The nonoxidative segment of the hexose monophosphate (HMP) pathway. The connections between the HMP and EMP pathways and the reversibility of the reactions allow carbon to flow between the pathways or into the biosynthetic pathways indicated in brackets. Numbers are text equations.

equivalent of one molecule of glucose is completely oxidized to carbon dioxide for each six molecules of glucose that enter. The remaining six molecules of pentose can be used to regenerate five molecules of glucose 6-phosphate. The overall equation for this cyclic mode of operation is:

$$6 \text{ Glucose 6-phosphate} + 12\text{NADP} + 7\text{H}_2\text{O} \rightarrow$$
$$5 \text{ glucose 6-phosphate} + 6\text{CO}_2 + 12\text{NADPH}_2 + \text{H}_3\text{PO}_4 \qquad (9\text{-}24)$$

This is equivalent to the complete oxidation of one molecule of glucose to carbon dioxide, and 12 molecules of reduced NADP are made available for use in reductive biosynthetic reactions. It should be apparent that the glucose 6-phosphate, or the fructose 6-phosphate from which it is formed (see Fig. 9-5), could enter the EMP pathway when NADPH_2 is not needed. Thus, in cells that have both pathways, as is generally the case, the fraction of glucose that is metabolized through all or part of the reactions of the HMP pathway is determined by the momentary needs of the cell for reduced NADP, for ribose 5-phosphate for the synthesis of nucleic acids, or for erythrose 4-phosphate, which is used in the synthesis of the aromatic amino acids, phenylalanine, tyrosine, and tryptophan.

Glucose entering the HMP pathway is phosphorylated in one of the same reactions described previously as the initial reactions in the EMP pathway [Eqs. (9-1) to (9-5)]. The pool of glucose 6-phosphate thus formed can then proceed through either pathway, EMP or HMP, depending on the requirements of the cell.

The Oxidative Reactions

The first reaction specific to the HMP pathway is the oxidation of glucose 6-phosphate by *glucose 6-phosphate dehydrogenase* with the reduction of the cofactor NADP [Eq. (9-25)].

The product 6-*phosphogluconolactone* is hydrolyzed by *lactonase* to produce 6-*phosphogluconic acid* [Eq. (9-26)].

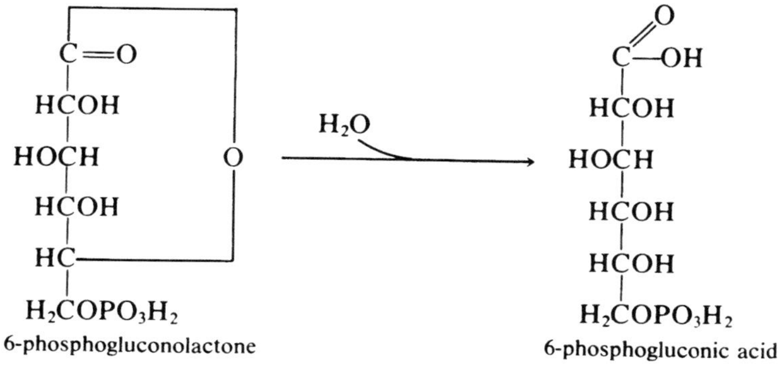

$$
\begin{array}{ccc}
\begin{array}{l}
\text{C}=\text{O} \\
\text{HCOH} \\
\text{HOCH} \qquad \text{O} \\
\text{HCOH} \\
\text{HC}\underline{\quad} \\
\text{H}_2\text{COPO}_3\text{H}_2
\end{array}
&
\xrightarrow{\text{H}_2\text{O}}
&
\begin{array}{l}
\text{C}{\overset{\text{O}}{\diagup}}\text{OH} \\
\text{HCOH} \\
\text{HOCH} \\
\text{HCOH} \\
\text{HCOH} \\
\text{H}_2\text{COPO}_3\text{H}_2
\end{array}
\qquad (9\text{-}26)
\end{array}
$$

6-phosphogluconolactone (left)   6-phosphogluconic acid (right)

The second oxidative step again produces $NADPH_2$ in an oxidative decarboxylation reaction catalyzed by *6-phosphogluconate dehydrogenase* [Eq. (9-27)].

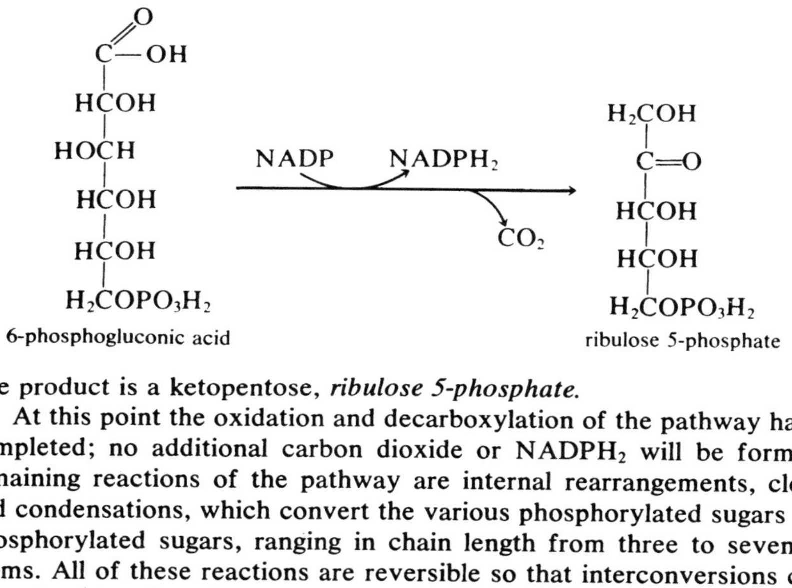

$$
\begin{array}{ccc}
\begin{array}{l}
\text{C}{\overset{\text{O}}{\diagup}}\text{OH} \\
\text{HCOH} \\
\text{HOCH} \\
\text{HCOH} \\
\text{HCOH} \\
\text{H}_2\text{COPO}_3\text{H}_2
\end{array}
&
\xrightarrow[\quad\text{CO}_2\quad]{\text{NADP}\quad\text{NADPH}_2}
&
\begin{array}{l}
\text{H}_2\text{COH} \\
\text{C}=\text{O} \\
\text{HCOH} \\
\text{HCOH} \\
\text{H}_2\text{COPO}_3\text{H}_2
\end{array}
\qquad (9\text{-}27)
\end{array}
$$

6-phosphogluconic acid (left)   ribulose 5-phosphate (right)

The product is a ketopentose, *ribulose 5-phosphate.*

At this point the oxidation and decarboxylation of the pathway have been completed; no additional carbon dioxide or $NADPH_2$ will be formed. The remaining reactions of the pathway are internal rearrangements, cleavages, and condensations, which convert the various phosphorylated sugars to other phosphorylated sugars, ranging in chain length from three to seven carbon atoms. All of these reactions are reversible so that interconversions can take place in either direction, depending on the sugar requirement of the cell. The reaction shown in Eq. (9-27) is not reversible, however, so hexose phosphate can be formed only by an additional series of reactions.

The Interconversions of Sugars

The ribulose 5-phosphate formed in the decarboxylation step can be acted upon by either of two enzymes, both of which catalyze internal rearrangements of the molecule [Eqs. (9-28) and (9-29)]. To produce the ribose 5-phosphate required for nucleic acid synthesis, the keto sugar is isomerized to an aldo sugar by *ribose 5-phosphate isomerase.*

$$
\begin{array}{ccc}
\begin{array}{l}
H_2COH \\
| \\
C{=}O \\
| \\
HCOH \\
| \\
HCOH \\
| \\
H_2COPO_3H_2
\end{array}
&
\rightleftharpoons
&
\begin{array}{l}
HC{=}O \\
| \\
HCOH \\
| \\
HCOH \\
| \\
HCOH \\
| \\
H_2COPO_2H_2
\end{array}
\end{array}
\qquad (9\text{-}28)
$$

ribulose 5-phosphate ribose 5-phosphate

An alternative rearrangement of ribulose 5-phosphate is catalyzed by *ribulose 5-phosphate 3-epimerase*, which exchanges the positions of the hydrogen and hydroxyl substituents on carbon-3 to form another ketopentose, *xylulose 5-phosphate*.

$$
\begin{array}{ccc}
\begin{array}{l}
H_2COH \\
| \\
C{=}O \\
| \\
HCOH \\
| \\
HCOH \\
| \\
H_2COPO_3H_2
\end{array}
&
\rightleftharpoons
&
\begin{array}{l}
H_2COH \\
| \\
C{=}O \\
| \\
HOCH \\
| \\
HCOH \\
| \\
H_2COPO_3H_2
\end{array}
\end{array}
\qquad (9\text{-}29)
$$

ribulose 5-phosphate xylulose 5-phosphate

Xylulose 5-phosphate, like ribulose 5-phosphate, can be involved in either of two reactions [Eqs. (9-30) and (9-31)]. Both are cleavage-condensations in which a two-carbon fragment from xylulose 5-phosphate is transferred to another sugar. Both reactions are catalyzed by the enzyme *transketolase*. The enzyme cleaves xylulose 5-phosphate to form glycolaldehyde, which is bound to the enzyme, and free *glyceraldehyde 3-phosphate*. The enzyme-bound glycolaldehyde moiety is then transferred either to ribose 5-phosphate to form a seven-carbon sugar [Eq. (9-30)] or to erythrose 4-phosphate to form a six-carbon sugar [Eq. (9-31)].

$$
\begin{array}{l}
H_2COH \\
| \\
C{=}O \\
| \\
HOCH \\
| \\
HCOH \\
| \\
H_2COPO_3H_2
\end{array}
+
\begin{array}{l}
HC{=}O \\
| \\
HCOH \\
| \\
HCOH \\
| \\
HCOH \\
| \\
H_2COPO_3H_2
\end{array}
\rightleftharpoons
\begin{array}{l}
H_2COH \\
| \\
C{=}O \\
| \\
HOCH \\
| \\
HCOH \\
| \\
HCOH \\
| \\
HCOH \\
| \\
H_2COPO_3H_2
\end{array}
+
\begin{array}{l}
HC{=}O \\
| \\
HCOH \\
| \\
H_2COPO_3H_2
\end{array}
$$

| xylulose 5-phosphate | ribose 5-phosphate | | sedoheptulose 7-phosphate | glyceraldehyde 3-phosphate |

$$(9\text{-}30)$$

$$
\begin{array}{c}
\text{H}_2\text{COH} \\
| \\
\text{C}=\text{O} \\
| \\
\text{HOCH} \\
| \\
\text{HCOH} \\
| \\
\text{H}_2\text{COPO}_3\text{H}_2
\end{array}
\;+\;
\begin{array}{c}
\text{HC}=\text{O} \\
| \\
\text{HCOH} \\
| \\
\text{HCOH} \\
| \\
\text{H}_2\text{COPO}_3\text{H}_2
\end{array}
\;\rightleftharpoons\;
\begin{array}{c}
\text{H}_2\text{COH} \\
| \\
\text{C}=\text{O} \\
| \\
\text{HOCH} \\
| \\
\text{HCOH} \\
| \\
\text{HCOH} \\
| \\
\text{H}_2\text{COPO}_3\text{H}_2
\end{array}
\;+\;
\begin{array}{c}
\text{HC}=\text{O} \\
| \\
\text{HCOH} \\
| \\
\text{H}_2\text{COPO}_3\text{H}_2
\end{array}
$$

| xylulose 5-phosphate | erythrose 4-phosphate | fructose 6-phosphate | glyceraldehyde 3-phosphate |

$$(9\text{-}31)$$

The erythrose 4-phosphate used as the acceptor of a two-carbon gly-colaldehyde moiety in Eq. (9-31) is formed by a similar cleavage-condensation reaction catalyzed by *transaldolase*. In this reaction, *sedoheptulose 7-phosphate* is cleaved between carbons 3 and 4 and the three-carbon dihy-droxyacetone moiety is transferred to glyceraldehyde 3-phosphate to form *fructose 6-phosphate* [Eq. (9-32)]. This reaction is an aldol condensation, which is essentially the reverse of the cleavage catalyzed by aldolase in the EMP pathway [Eq. (9-8)]. Since both reactions are reversible, both enzymes can catalyze either cleavage or condensation. However, the dihydroxyacetone moiety transferred by transaldolase, like the glycolaldehyde transferred by transketolase, remains bound to the enzyme until it is transferred to an acceptor molecule.

$$
\begin{array}{c}
\text{H}_2\text{COH} \\
| \\
\text{C}=\text{O} \\
| \\
\text{HOCH} \\
| \\
\text{HCOH} \\
| \\
\text{HCOH} \\
| \\
\text{HCOH} \\
| \\
\text{H}_2\text{COPO}_3\text{H}_2
\end{array}
\;+\;
\begin{array}{c}
\text{HC}=\text{O} \\
| \\
\text{HCOH} \\
| \\
\text{H}_2\text{COPO}_3\text{H}_2
\end{array}
\;\rightleftharpoons\;
\begin{array}{c}
\text{H}_2\text{COH} \\
| \\
\text{C}=\text{O} \\
| \\
\text{HOCH} \\
| \\
\text{HCOH} \\
| \\
\text{HCOH} \\
| \\
\text{H}_2\text{COPO}_3\text{H}_2
\end{array}
\;+\;
\begin{array}{c}
\text{HC}=\text{O} \\
| \\
\text{HCOH} \\
| \\
\text{HCOH} \\
| \\
\text{H}_2\text{COPO}_3\text{H}_2
\end{array}
$$

| sedoheptulose 7-phosphate | glyceraldehyde 3-phosphate | fructose 6-phosphate | erythrose 4-phosphate |

$$(9\text{-}32)$$

The two products of the transaldolase reaction are *fructose 6-phosphate* and *erythrose 4-phosphate.*

The metabolic flexibility made possible by the *rearranging enzymes*, the isomerase and epimerase, and the *cleaving-condensing enzymes*, transketolase and transaldolase, should now be apparent. These reactions and their relationships to other pathways are summarized in Fig. 9-5. Two of the products formed by the interconversions of sugars catalyzed by these enzymes are glyceraldehyde 3-phosphate and fructose 6-phosphate, which are also intermediates in the EMP pathway. As mentioned before, ribose 5-phosphate is utilized in the synthesis of all the ribo- and deoxyribonucleotides required for the synthesis of RNA and DNA. It is also used in the synthesis of several important coenzymes (FMN, FAD, coenzyme A, NAD, and NADP) and in the synthesis of the amino acids histidine and tryptophan. Erythrose 4-phosphate is required for synthesis of the aromatic amino acids phenylalanine, tyrosine, and tryptophan. All the reactions involved in the interconversions of the sugars with three to seven carbon atoms are reversible. Therefore, the flow of intermediates in the pathway can be directed toward any one intermediate that is required at any moment. Obviously, if ribose 5-phosphate is used rapidly for the synthesis of nucleic acids, the reaction from ribulose 5-phosphate to ribose 5-phosphate will be "pulled" by removal of the product. Furthermore, ribose 5-phosphate is required for any further reaction involving xylulose 5-phosphate, so that if ribose 5-phosphate is not available, xylulose 5-phosphate will accumulate, if it is formed, and will be converted to ribose 5-phosphate via ribulose 5-phosphate. If none of the intermediates of the pathway is required for synthesis, glucose 6-phosphate can be shunted through the HMP pathway to provide $NADPH_2$, if needed. The glucose carbon can then be returned to recycle through the pathway by the resynthesis of glucose 6-phosphate in a cyclic operation of the pathway, or the glucose carbon can enter the EMP pathway as glyceraldehyde 3-phosphate or fructose 6-phosphate. Operated in the latter manner, the reactions of the pathway serve as a detour or shunt for glucose in its path to carbon dioxide via the EMP pathway and TCA cycle, and the HMP pathway is often called the *HMP shunt*. The sharing of common intermediates with the EMP pathway also allows carbon from the EMP pathway to enter the HMP pathway and to be used for the synthesis of ribose 5-phosphate and erythrose 4-phosphate by reversal of the reactions shown in Fig. 9-5. It is thus possible to provide carbon for the synthetic requirements of the cell by using only the EMP pathway and the reactions shown in Fig. 9-5, i.e., with no loss of carbon dioxide or reduction of NADP. The cell is thus able to utilize whatever portions of this versatile pathway it requires and to move carbon between the EMP and the HMP pathways as needed.

THE ENTNER-DOUDEROFF PATHWAY

The Entner-Douderoff pathway is used by bacteria of the genus *Pseudomonas* and related genera and also by other organisms such as some enterics for the metabolism of certain compounds. It is included here because, although it is not a central catabolic pathway for a wide variety of microorganisms as are those described earlier, it is the central pathway for a very important group of bacteria. As we have noted, *Pseudomonas* species are more versatile than most microorganisms in their ability to metabolize large numbers of different compounds. In addition, the Entner-Douderoff pathway is closely related to the EMP and HMP pathways and shares common reactions and substrates with both pathways, so that we need consider only the reactions unique to the E-D pathway. These are summarized in Fig. 9-6 along with the reactions connecting this pathway with others.

In the Entner-Douderoff pathway, glucose is converted to 6-phosphogluconate as it is in the HMP pathway. However, alternative routes of formation of 6-phosphogluconic acid exist in organisms that utilize the E-D pathway. Some species possess the ability to use all three routes, whereas others utilize only one or two of the possible routes. In any case, it is possible for an organism to utilize the Entner-Douderoff pathway by adding only two enzymes to the complement required for the HMP pathway, the lower EMP, and the TCA cycle.

The two enzymes that are unique to the E-D pathway are those that catalyze the dehydration and cleavage of 6-phosphogluconate [Eqs. (9-33) and

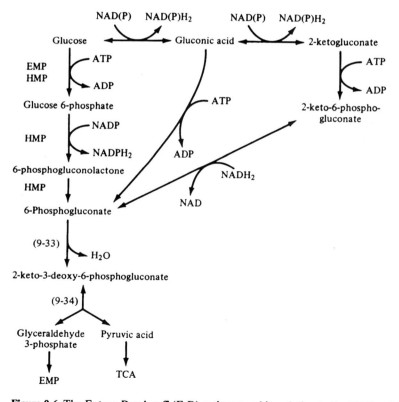

Figure 9-6 The Entner-Douderoff (E-D) pathway and its relation to the EMP and HMP pathways. The three alternate routes for conversion of glucose to 6-phosphogluconate are not present in all organisms that use this pathway.

(9-34)]. In the first of these reactions, catalyzed by *phosphogluconate dehydratase*, a molecule of water is removed in a unique manner; in the reaction product, two hydrogen atoms have been removed from one carbon and the oxygen from another (the usual dehydration reaction involves removal of hydrogen from one carbon and hydroxyl from another).

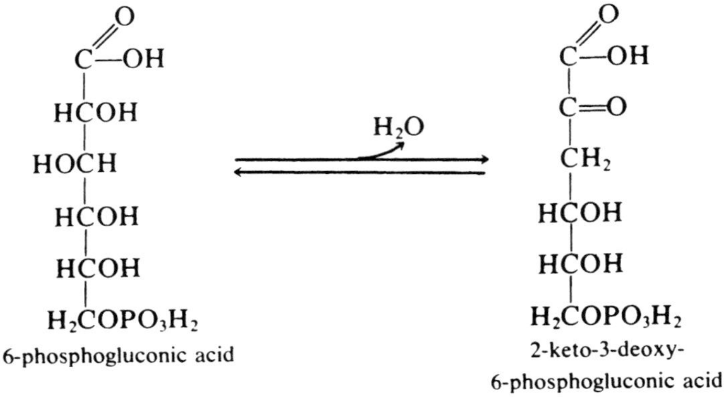

$$
\begin{array}{c}
\text{O} \\
\text{C—OH} \\
\text{HCOH} \\
\text{HOCH} \\
\text{HCOH} \\
\text{HCOH} \\
\text{H}_2\text{COPO}_3\text{H}_2
\end{array}
\quad \xrightleftharpoons{\text{H}_2\text{O}} \quad
\begin{array}{c}
\text{O} \\
\text{C—OH} \\
\text{C=O} \\
\text{CH}_2 \\
\text{HCOH} \\
\text{HCOH} \\
\text{H}_2\text{COPO}_3\text{H}_2
\end{array}
\qquad (9\text{-}33)
$$

6-phosphogluconic acid   2-keto-3-deoxy-6-phosphogluconic acid

The product, *2-keto-3-deoxy-6-phosphogluconate*, is acted upon by an aldolase in a reaction identical with the aldolase reaction in the EMP pathway [compare Eqs. (9-34) and (9-8)].

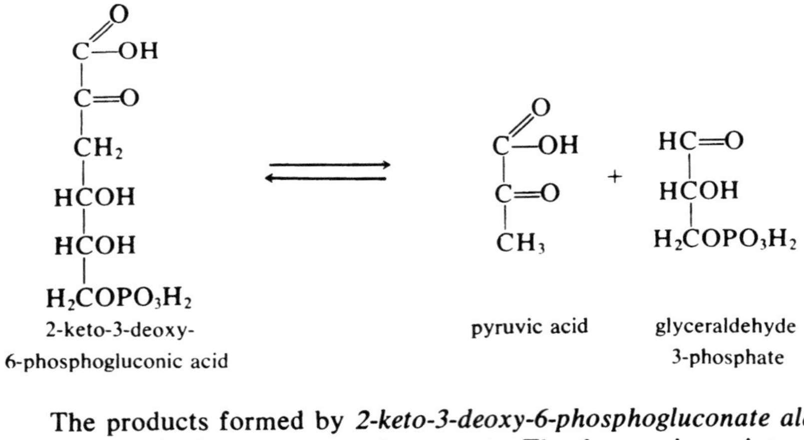

$$
\begin{array}{c}
\text{O} \\
\text{C—OH} \\
\text{C=O} \\
\text{CH}_2 \\
\text{HCOH} \\
\text{HCOH} \\
\text{H}_2\text{COPO}_3\text{H}_2
\end{array}
\quad \xrightleftharpoons{} \quad
\begin{array}{c}
\text{O} \\
\text{C—OH} \\
\text{C=O} \\
\text{CH}_3
\end{array}
\;+\;
\begin{array}{c}
\text{HC=O} \\
\text{HCOH} \\
\text{H}_2\text{COPO}_3\text{H}_2
\end{array}
\qquad (9\text{-}34)
$$

2-keto-3-deoxy-6-phosphogluconic acid   pyruvic acid   glyceraldehyde 3-phosphate

The products formed by *2-keto-3-deoxy-6-phosphogluconate aldolase* are *glyceraldehyde 3-phosphate* and *pyruvate*. The former is an intermediate in both the EMP and HMP pathways. Pyruvate is formed in the EMP pathway and is oxidatively decarboxylated to acetyl-CoA, which enters the TCA cycle. The Entner-Douderoff pathway thus shares intermediates with both pathways, and by utilizing the reactions of the other two pathways, it can channel carbon into biosynthetic reactions or into carbon dioxide. The reactions shown in Fig. 9-6 provide for the complete oxidation of glucose to carbon dioxide and water.

THE CALVIN CYCLE—AUTOTROPHIC CARBON DIOXIDE FIXATION

The pathway by which organic matter is synthesized from carbon dioxide is the same in almost all autotrophic organisms: higher plants, photoautotrophic eucaryotic, and procaryotic microorganisms, and chemoautotrophic bacteria. (In a few higher plants, a different pathway has been discovered recently.) The common pathway is a cyclic one, as first proposed by Melvin Calvin, after whom the pathway is named. It is also called the *reductive pentose cycle* because of its close relationship to the HMP pathway. The Calvin cycle is, in fact, essentially a reversal of parts of the HMP pathway and the EMP pathway and uses many of the same enzymes. We discussed the possibility of utilizing the HMP pathway in a cyclic manner for the complete oxidation of glucose to carbon dioxide. By using many of the same reactions, with the addition of some reactions unique to the Calvin cycle, carbon dioxide can be

Table 9-1 Reactions involved in the net synthesis of one molecule of glucose from six molecules of carbon dioxide via the Calvin cycle in autotrophic organisms

$6CO_2 + 6$ ribulose 1,5-diphosphate $+ 6H_2O \rightarrow 12$ 3-phosphoglycerate	(9-37)
12 3-Phosphoglycerate $+ 12ATP \rightarrow 12$ 1,3-diphosphoglycerate $+ 12ADP$	(9-11)
12 1,3-Diphosphoglycerate $+ 12NADPH_2 \rightarrow$ 12 glyceraldehyde 3-phosphate $+ 12H_3PO_4 + 12NADP$	(9-10)
5 Glyceraldehyde 3-phosphate $\rightarrow 5$ dihydroxyacetone phosphate	(9-9)
3 Glyceraldehyde 3-phosphate $+ 3$ dihydroxyacetone phosphate $\rightarrow$ 3 fructose 1,6-diphosphate	(9-8)
3 Fructose 1,6-diphosphate $+ 3H_2O \rightarrow 3$ fructose 6-phosphate $+ 3H_3PO_4$	(9-38)
Fructose 6-phosphate $\rightarrow$ glucose 6-phosphate	(9-6)
Glucose 6-phosphate $+ H_2O \rightarrow$ glucose $+ H_3PO_4$	(9-39)
2 Fructose 6-phosphate $+ 2$ glyceraldehyde 3-phosphate $\rightarrow$ 2 xylulose 5-phosphate $+ 2$ erythrose 4-phosphate	(9-31)
2 Erythrose 4-phosphate $+ 2$ dihydroxyacetone phosphate $\rightarrow$ 2 sedoheptulose 1,7-diphosphate	(9-41)
2 Sedoheptulose 1,7-diphosphate $+ 2H_2O \rightarrow 2$ sedoheptulose 7-phosphate $+ 2H_3PO_4$	(9-40)
2 Sedoheptulose 7-phosphate $+ 2$ glyceraldehyde 3-phosphate $\rightarrow$ 2 ribose 5-phosphate $+ 2$ xylulose 5-phosphate	(9-30)
2 Ribose 5-phosphate $\rightarrow 2$ ribulose 5-phosphate	(9-28)
4 Xylulose 5-phosphate $\rightarrow 4$ ribulose 5-phosphate	(9-29)
6 Ribulose 5-phosphate $+ 6ATP \rightarrow 6$ ribulose 1,5-diphosphate $+ 6ADP$	(9-36)
SUM: $6CO_2 + 18ATP + 12NADPH_2 + 12H_2O \rightarrow$ glucose $+ 18H_3PO_4 + 18ADP + 12NADP$	(9-35)

fixed to yield glucose. Six turns of the HMP cycle were necessary to achieve the complete oxidation of one molecule of glucose to six molecules of carbon dioxide. Similarly, six turns of the Calvin cycle are necessary to achieve the net synthesis of one molecule of glucose from six molecules of carbon dioxide.

The overall equation for autotrophic carbon dioxide fixation via the Calvin cycle is

$$6CO_2 + 18ATP + 12NADPH_2 + 12H_2O \rightarrow$$
$$glucose + 18H_3PO_4 + 18ADP + 12NADP \quad (9\text{-}35)$$

This equation is the sum of 15 reactions, most of which are rearrangements and interconversions of sugars of various chain lengths, many catalyzed by the same enzymes used for similar reactions in the HMP and EMP pathways. The 15 reactions are shown in Table 9-1. The reactions in the table are identified by the equation numbers used in the text.

The two key reactions of the Calvin cycle are the reaction in which carbon dioxide is incorporated [Eq. (9-37)] and the reaction that forms the molecule to which carbon dioxide is added [Eq. (9-36)].

The substrate to which carbon dioxide is added is *ribulose 1,5-diphosphate*. This is formed from ribulose 5-phosphate, which we saw as an intermediate in the HMP pathway. ATP is required for the phosphorylation, which is catalyzed by the enzyme *phosphoribulokinase* in a reaction very similar to the phosphorylation of fructose 6-phosphate in the EMP pathway [Eq. (9-7)].

$$(9\text{-}36)$$

ribulose 5-phosphate   ribulose 1, 5-diphosphate

The enzyme that fixes carbon dioxide is *ribulose-diphosphate carboxylase*. The addition of carbon dioxide is followed by the cleavage of the molecule to yield two molecules of 3-phosphoglyceric acid, which is also familiar as an intermediate in the EMP pathway.

$$(9\text{-}37)$$

ribulose
1,5-diphosphate   3-phosphoglyceric acid

The 3-phosphoglycerate is then converted to hexose by the reversal of the upper portion of the EMP pathway, and the ribulose 5-phosphate needed for the formation of ribulose 1,5-diphosphate is regenerated by a series of reactions also found in the HMP pathway.

Five of the reactions shown in Table 9-1 deserve comment. Although they accomplish the reversal of EMP or HMP reactions, they are not exact duplicates of those reactions. One [Eq. (9-10)] differs only in the electron carrier used. In the oxidation of glyceraldehyde 3-phosphate in the EMP pathway, the usual electron carrier is NAD, whereas $NADPH_2$ is the electron carrier in the reductive reaction of the Calvin cycle. This difference is consistent with the general rule that NAD usually serves as the electron carrier for catabolic reactions and NADP for anabolic reactions.

In three reactions, phosphate groups are removed by hydrolysis. In the corresponding reaction in which phosphate is added, ATP is required as the phosphate donor. Reversal of such a reaction is not possible because the phosphate bond energy is not conserved in the phosphorylation by ATP; that is, the phosphate groups removed from fructose 1,6-diphosphate [Eq. (9-38)], glucose 6-phosphate [Eq. (9-39)], and sedoheptulose 1,7-diphosphate [Eq. (9-40)] are linked to the sugars by stable ester bonds, not by high-energy bonds capable of donating phosphate to ADP. Enzymes that remove phosphate groups by hydrolysis are called *phosphatases*.

The reaction shown in Eq. (9-41) is catalyzed by an enzyme of the EMP pathway, fructose diphosphate aldolase [see Eq. (9-8)]. Erythrose 4-phosphate replaces the EMP substrate, glyceraldehyde 3-phosphate, but the reactions in Eqs. (9-8) and (9-41) are otherwise identical.

RELATION OF CENTRAL PATHWAYS TO BIOSYNTHESIS

We have examined the most important pathways by which carbon is cycled from the atmosphere to organic compounds and back again to the atmosphere. Perhaps the most striking aspect of these pathways is the economy of enzyme synthesis and the versatility afforded by the ability of the living cell to utilize the same enzymes in several pathways and for more than one purpose, and to shift common intermediates from one route to another as dictated by the needs of the moment. Intermediates may flow from one catabolic pathway to another, and they may be withdrawn from catabolic pathways to serve as building blocks for the biosynthesis of cell material.

Table 9-2 summarizes the major points at which intermediates of the central pathways may be channeled into biosynthesis.

We have discussed the ways in which carbon flow through the HMP and EMP pathways may be varied to provide $NADPH_2$, ribose 5-phosphate, and erythrose 4-phosphate as needed for biosynthetic reactions. The interconnections between these pathways and between them and the Entner-Douderoff pathway, and the reversibility of the reactions that catalyze the interconversions of sugars in the HMP pathway allow intermediates to be

Table 9-2 Flow of intermediates of the central pathways into biosynthesis of cellular constituents

Intermediate	Products
From EMP	
Hexose phosphates	Polysaccharides
Dihydroxyacetone phosphate	Lipids
Phosphoglyceric acid	Serine, glycine, cysteine, purines, chlorophyll, heme
Phosphoenolpyruvate	Tryptophan, tyrosine, phenylalanine, lysine, isoleucine, muramic acid
Pyruvate	Alanine, valine, leucine
From TCA	
Acetyl-CoA	Fatty acids, lysine, leucine, carotenes, sterols
α-Ketoglutarate	Glutamic acid, glutamine, proline, arginine
Succinyl-CoA	Methionine, chlorophyll, heme
Oxalacetic acid	Aspartic acid, asparagine, methionine, threonine, lysine, isoleucine, pyrimidines
From HMP	
Ribose phosphate	Nucleotides, histidine, tryptophan
Erythrose 4-phosphate	Tryptophan, tyrosine, phenylalanine

withdrawn as needed. The withdrawal of intermediates from these pathways does not affect the continued functioning of the pathways. The TCA cycle, however, presents a special problem because of its cyclic nature. When this pathway is used only for the terminal oxidation of substrates that are degraded to acetyl-CoA, it operates in the cyclic manner shown in Fig. 9-3. Oxalacetate is regenerated from succinate by reactions (9-21) to (9-23) and becomes available for condensation with acetyl-CoA to initiate another turn of the cycle. However, there are several situations in which the TCA cycle must be modified to accommodate the biosynthetic demands of the cell. As shown in Table 9-2, a number of cellular components that are required in rather large amounts are synthesized from intermediates of the TCA cycle. The withdrawal of these intermediates from the cycle interrupts the cyclic flow and would prevent the terminal oxidation of the energy source if there were no alternate reactions by which oxalacetate could be fed into the cycle to maintain its operation. Thus, chemoheterotrophs that must utilize TCA cycle intermediates for biosynthetic reactions must be able to synthesize C_4 intermediates to replace those siphoned off. Reactions that replace the catabolic intermediates withdrawn for use in biosynthetic pathways are called *anaphlerotic reactions.*

A different situation occurs in autotrophs. When synthesizing organic matter from carbon dioxide and obtaining energy from light or from an inorganic energy source, it would not be profitable for autotrophs to utilize the catabolic pathways in the same way that the pathways are used by

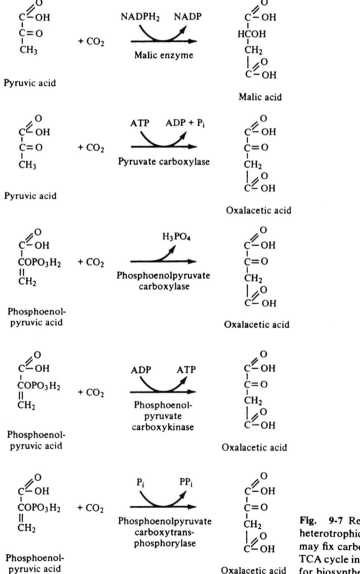

Fig. 9-7 Reactions by which heterotrophic microorganisms may fix carbon dioxide to replace TCA cycle intermediates removed for biosynthetic reactions.

chemoheterotrophs; yet autotrophs must synthesize all their cell material and must use, in general, the same pathways used by heterotrophs. Under these circumstances, only a portion of a pathway may be utilized, i.e., the portion leading to a required precursor. Many autotrophs lack α-ketoglutarate dehydrogenase and therefore must form oxalacetate by reactions other than those of the complete TCA cycle. In these organisms, the reactions leading from citric acid to α-ketoglutarate are used in the forward direction, whereas succinate is formed by a reversal of the reactions that usually lead from succinate to oxalacetate. Both halves of the cycle thus operate independently, and both require an input of oxalacetate. In both chemoheterotrophs and autotrophs, oxalacetate can be synthesized by one or more of the reactions shown in Fig. 9-7. All five reactions have been shown to occur in microorganisms and all involve the addition of carbon dioxide to either pyruvate or phosphoenolpyruvate. Oxalacetate is formed in four of the reactions, while the fifth forms another C_4 intermediate of the TCA cycle, malate. This can be converted to oxalacetate, as in the TCA cycle, by dehydrogenation [Eq. (9-23)] or to succinate by reversal of the TCA cycle reactions (9-22) and (9-21).

A different problem involving the TCA cycle occurs in organisms that metabolize acetate, citrate, or substrates such as aliphatic hydrocarbons or fatty acids that are degraded to acetyl-CoA. If no other carbon source is

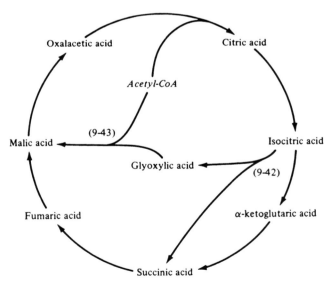

Figure 9-8 The glyoxylate bypass for the synthesis of C_4 compounds in microorganisms that use compounds that are metabolized to acetyl-CoA (acetate, fatty acids, hydrocarbons) as the sole carbon source. The bypass reactions are Eqs. (9-42) and (9-43). The bypass is shown in relation to the TCA cycle (see Fig. 9-3 for details).

available, sugars must be synthesized by reversal of the EMP pathway. However, as was pointed out previously, the oxidative decarboxylation of pyruvate to form acetyl-CoA [Eq. (9-15)] is irreversible. Thus, pyruvate cannot be made from acetate or acetyl-CoA, and the carbon source cannot be used in the synthesis of sugars or of other cellular constituents derived from intermediates of the EMP and HMP pathways. Nor can intermediates be withdrawn from the TCA cycle for use in biosynthesis, because no pyruvate or PEP is available for carboxylation to form a C_4 replacement. Even if pyruvate could be synthesized from acetyl-CoA or a TCA intermediate, it could not be converted to PEP because the conversion of PEP to pyruvate is essentially irreversible. These problems are solved in organisms that can use acetate or fatty acids as sole sources of carbon and energy by two reactions that bypass the oxidative steps of the TCA cycle and form oxalacetate by condensation of two C_2 compounds, acetyl-CoA and glyoxylic acid. These two reactions, which occur only in microorganisms that metabolize two-carbon compounds or their precursors, constitute the *glyoxylate bypass*. The bypass and its relationship to the TCA cycle are shown in Fig. 9-8. The first of the bypass reactions is cleavage of isocitric acid to glyoxylic acid and succinic acid by the enzyme *isocitrate lyase* [Eq. (9-42)].

$$\text{isocitric acid} \longrightarrow \text{succinic acid} + \text{glyoxylic acid} \qquad (9\text{-}42)$$

Both succinate and glyoxylate can be converted to oxalacetate—succinate by the usual reactions of the TCA cycle and glyoxylate by condensation with acetyl-CoA catalyzed by *malate synthase* [Eq. (9-43)].

$$\text{glyoxylic acid} + \text{acetyl-CoA} \xrightarrow[\text{CoASH}]{H_2O} \text{malic acid} \qquad (9\text{-}43)$$

These reactions replace the TCA cycle intermediates that were withdrawn for use in biosynthesis. However, there still remains the problem of feeding part of the substrate carbon into the EMP and HMP pathways for the synthesis of other cellular constituents. This is accomplished by reversal of one of the reactions that convert PEP to oxalacetate (see Fig. 9-7), the PEP carboxykinase reaction.

The reactions of the glyoxylate bypass have been found in a variety of bacteria and fungi and also in higher plants, but they are not present in animals. Together with other reactions previously described, they make possible the synthesis of cellular material from acetyl-CoA, which allows microorganisms to grow at the expense of fatty acids as sole source of carbon and energy. The simultaneous operation of the glyoxylate bypass and the TCA cycle allows acetyl-CoA to be proportioned between the glyoxylate bypass, which provides for the use of acetyl-CoA as carbon source, and the TCA cycle, which provides for its oxidation to carbon dioxide and the generation of ATP for the energy needs of the cell.

In the two following chapters we shall examine the pathways and reactions that feed into and out of the central pathways. We shall also look more closely at the metabolic aspects of the biological degradation of organic wastes.

PROBLEMS

9-1 Discuss the importance of glycolysis in the metabolism of chemoorganotrophic organisms.

9-2 Write the equation for the oxidative reaction of glycolysis. How is the energy released by oxidation used?

9-3 Write the equations for the two reactions in which ATP is formed. Which mechanism of ATP formation is involved here?

9-4 The HMP pathway can be used for the complete oxidation of glucose. The EMP pathway followed by the TCA cycle can be used for the same purpose. Given an electron transport system with a P/O ratio of 2, calculate the net ATP formation for the two routes of oxidation. Assume that all reduced pyridine nucleotide (i.e., $NADH_2$ and $NADPH_2$) can transfer hydrogen ions and electrons directly to the electron transport system.

9-5 Write a pathway by which ribose could be metabolized to carbon dioxide and water using the HMP, EMP, and TCA pathways. How many molecules of ATP (net) would be formed if the organism is an aerobe utilizing oxygen as electron acceptor with a P/O ratio of 3?

9-6 Explain why a two-carbon compound such as acetate cannot be used as the sole source of carbon and energy by an organism that has no glyoxylate cycle.

9-7 Why are reactions by which carbon dioxide can be fixed required in heterotrophic microorganisms?

9-8 Possibly the most important reaction in metabolism is the reaction by which carbon dioxide is converted into a form in which the carbon is available for use by humans, animals, and microorganisms. Write the equation for this reaction.

9-9 In a chemolithotroph that uses carbon dioxide as its sole source of carbon, a complete TCA cycle may not be present, but the organism may be able to make most of the enzymes of the cycle. Explain how this would be advantageous to the organism.

9-10 What is the unique reaction of the Entner-Douderoff pathway? Write the equation for the reaction. What group of organisms utilizes this pathway?

REFERENCES AND SUGGESTED READING

Brock, T. D. 1979. Biology of microorganisms, 3d ed. Prentice-Hall, Englewood Cliffs, N.J.

Doelle, H. W. 1975. Bacterial metabolism, 2d ed. Academic Press, New York.

Gunsalus, I. C., and R. Y. Stanier (editors). 1961. The bacteria. Vol. II: Metabolism. Academic Press, New York.

Hawker, L. E., and A. H. Linton. 1971. Micro-organisms. Function, form and environment. American Elsevier, New York.

Hollmann, S. 1964. Non-glycolytic pathways of metabolism of glucose. Academic Press, New York.

Lehninger, A. L. 1975. Biochemistry, 2d ed. Worth, New York.

Metzler, D. E. 1977. Biochemistry. Academic Press, New York.

Sokatch, J. R. 1969. Bacterial physiology and metabolism. Academic Press, New York.

Stanier, R. Y., E. A. Adelberg, and J. Ingraham. 1976. The microbial world, 4th ed. Prentice-Hall, Englewood Cliffs, N.J.

TEN

AEROBIC METABOLISM

In Chap. 9 we described the central pathways through which carbon flows in the living cell. These pathways provide metabolically usable energy, ATP, and building blocks for biosynthesis. We considered briefly the flow of carbon from the central pathways to the synthesis of cellular constituents, and it is important now to consider the reactions through which carbon flows into the central pathways. We began the central pathways with glucose, which is a commonly occurring and readily used substrate. However, many other forms of organic matter are degraded by microorganisms. For some of these special pathways must be used, but most forms of organic matter can be converted to compounds that are intermediates in one of the central pathways. In this chapter, we shall consider the routes of entry of the major classes of organic matter into the central pathways in aerobic microorganisms and some applications of aerobic metabolism in the biological treatment of wastes.

METABOLISM OF CARBOHYDRATES

The most abundant naturally occurring carbohydrates are polysaccharides. These are linear or branched polymers, often of very high molecular weight, and may contain a single type of monomer (*homopolysaccharides*) or two or more different monomers (*heteropolysaccharides*). The most common monomers in polysaccharides are hexoses and pentoses and their derivatives—uronic acids, amino sugars, and deoxy-sugars. Two major types of polysaccharides are synthesized by living organisms: *storage polysaccharides*, such as glycogen and starch, and *structural polysaccharides*, such as cellulose, the mucopeptide layer of bacterial cell walls, and chitin, which forms the exoskeletons of insects. The capsular polysaccharides synthesized by many microorganisms do not fit in either of these categories. Two disaccharides are relatively abundant in nature: sucrose in fruits and vegetables, and lactose in milk.

In Chap. 9 we described the phosphorylase reaction by which cells utilize their own internal storage polysaccharides by converting them directly to glucose 1-phosphate, using inorganic phosphate and conserving ATP. Microorganisms that use an external source of polysaccharide cannot degrade the polymer by phosphorolysis. Polysaccharides are too large to be taken into the cell and must be degraded extracellularly. The products of degradation are then transported into the cell and used. While some transport systems for phosphorylated compounds exist, most sugars cannot enter the cell if phosphorylated. Therefore, the metabolism of polysaccharides is initiated by the hydrolytic degradation of the polymer catalyzed by enzymes excreted by the cell, i.e., extracellular enzymes. These may be released into the environment or may, in some cases, remain bound to the cell surface. In the latter case, the cell must make contact with the polysaccharide. Since many polysaccharides are insoluble, this is facilitated by growth of the microorganism on the surface of polysaccharide materials, e.g., on cellulose fibers.

Enzymes that hydrolyze polysaccharides are specific for the type of bond to be broken and the monomers involved. We can illustrate this with several examples. Cellulose is a polymer of glucose units with $\beta, 1 \rightarrow 4$ linkages and requires cellulase, a $\beta, 1 \rightarrow 4$ glucanase, for hydrolysis. Starch and glycogen are also polymers of glucose but have $\alpha, 1 \rightarrow 4$ linkages. Both glycogen and one form of starch, amylopectin, are branched, with $\alpha, 1 \rightarrow 6$ linkages at the branch points (see Fig. 3-23). While the linear starch molecule, amylose, and the linear portions of amylopectin and glycogen can be degraded by amylase, complete degradation of the branched molecules requires a "debranching enzyme," an $\alpha, 1 \rightarrow 6$-glucosidase. Lysozyme is specific for the $\beta, 1 \rightarrow 4$ linkage that joins carbon-1 of N-acetylmuramic acid to carbon-4 of N-acetylglucosamine in the mucopeptide layer of bacterial cell walls. Thus, the ability of any microorganism to utilize a polysaccharide depends on (1) its ability to produce the specific enzyme or enzymes required for hydrolysis of the polymer, (2) its ability to transport the products of hydrolysis into the cell, and (3) its ability to convert them to intermediates that can enter the central pathways. Hydrolysis may convert the polysaccharide to monosaccharides, disaccharides, or oligosaccharides. These are small enough to enter the cell and may be taken up and then hydrolyzed or phosphorylyzed by specific intracellular enzymes, as are the naturally occurring disaccharides sucrose and lactose. Starch is degraded to a mixture of glucose and the disaccharide maltose, whereas cellulose is degraded to primarily the disaccharide cellobiose.

Many microorganisms are capable of degrading the commonly occurring storage polysaccharides, glycogen and starch. Many other polysaccharides are probably not widely used, but it is unlikely that any are immune to microbial attack. Cellulose is generally considered to be one of the most difficult polysaccharides for microorganisms to metabolize. Its very high molecular weight, its organization in crystalline structures (micelles) in the cell walls of plants, and its insolubility probably contribute to the difficulty of the attack, but many bacteria fail to utilize cellulose because they are not able to form the necessary enzymes. Some myxobacteria, the cytophagas, and some species of *Clostridium*, as well as some actinomycetes, are the most active cellulose-degrading bacteria. Most decomposition of cellulose is probably due to the action of fungi, however. *Chaetomium globosum*, *Myrothecium verrucaria*, and *Stachybotrys atra* have been reported to be important species in the degradation of cellulose, although more common fungi belonging to the genera *Aspergillus*, *Penicillium*, *Cladosporium*, *Fusarium*, *Alternaria*, and *Trichoderma* are also active. *Trichoderma viride* (recently renamed *T. reesei*) has been used as the source of cellulase in a process designed to convert wastepaper and other waste materials with high cellulose content to glucose (Gaden et al., 1976).

Since the hydrolysis of polysaccharides occurs extracellularly, the products of hydrolysis may be available to organisms other than the ones that produce the hydrolytic enzyme. Complete degradation of a heteropolysaccharide may require the action of more than one microorganism, since several different enzymes may be required to break the different bonds and no one organism may be able to elaborate all of the enzymes required. An example of the "cooperative" degradation of a bacterial polysaccharide has been observed in studies in our laboratory (A. W. Obayashi and A. F. Gaudy, Jr., unpublished results). Other investigators had proposed that the capsular polysaccharides of bacteria would accumulate as inert nonbiodegradable material in extended aeration activated sludge. We isolated and purified capsular polysaccharides from five bacteria, all heteropolysaccharides of different composition. All were usable by mixed microbial populations as sole sources of carbon and energy. These and other studies of the extended aeration process will be discussed later in this chapter. From one heterogeneous population growing on the capsular polysaccharide of *Zoogloea ramigera* (see Fig. 10-34), five different bacteria were isolated. None of the five was able to grow in pure culture with the polysaccharide as the source of carbon and energy, but a mixture of the five organisms grew well and utilized the polysaccharide.

When a polysaccharide has been degraded to its monomers or to small oligosaccharides or disaccharides, the ability of a microorganism to utilize the products depends on its ability to form the transport proteins and the specific metabolic enzymes required. Starch, glycogen, and cellulose, as polymers of glucose, are readily metabolized after hydrolysis. Other hexoses and their derivatives are converted, usually by inducible enzymes (see Chap. 12), to either glucose 6-phosphate or fructose 6-phosphate, most frequently to the latter. Pentoses are generally converted, also by inducible pathways, to ribose

5-phosphate, ribulose 5-phosphate, or xylulose 5-phosphate, intermediates in the HMP pathway. Xylose and arabinose are the most commonly occurring pentoses, since they are found in hemicelluloses and other polysaccharides produced by plants. Arabinose is readily converted to ribulose 5-phosphate, while xylose is converted to xylulose 5-phosphate. The hexuronic acids, e.g., glucuronic, galacturonic, and mannuronic acids, occur as components of microbial capsules and of the acid mucopolysaccharides of animals. They are generally metabolized through the Entner-Douderoff pathway after conversion to 2-keto-3-deoxy-6-phosphogluconate.

METABOLISM OF PROTEINS

Proteins, like polysaccharides, are too large to enter the microbial cell and must be degraded to smaller molecules outside the cell. Many species of microorganisms excrete proteolytic enzymes, *proteinases*, of various types. In general, bacterial proteinases appear to be less specific in their action than are the mammalian enzymes; they do not cleave bonds at only certain sites. The hydrolysis of peptide bonds by many bacterial proteinases apparently occurs at random, and the protein is degraded to individual amino acids and short peptide chains. Proteinases do not hydrolyze small peptides; these are degraded by another group of enzymes, the *peptidases.*

Both small peptides and free amino acids are taken into the cell by specific transport systems. It is interesting that an amino acid may sometimes enter the cell more readily as part of a dipeptide than alone. Most microorganisms probably are able to take up all of the amino acids and many peptides; possible exceptions are some autotrophs that may take up some amino acids but not others. Studies with *Escherichia coli* indicate that the ability to transport peptides into the cell depends on the actual size of the molecule; the smaller the individual amino acid in the peptide, the longer is the peptide that can be transported. On average, the maximum number of amino acids in peptides that can be transported into the cell is five to six. Most peptidases are probably located inside the cell. Thus, when extracellular proteinases attack a protein molecule, it is hydrolyzed progressively to produce both amino acids and peptides. When the size of the peptides decreases to the point at which no further hydrolysis by proteinases is possible, the peptides are transported into the cell if they are small enough. Peptides that are not small enough to be transported may not be available for use by all microorganisms, since most microorganisms probably do not produce extracellular peptidases. Once inside the cell, intracellular peptidases hydrolyze the peptides to amino acids.

If other energy sources are available, most microorganisms use amino acids for only protein synthesis, i.e., as carbon source only. Some anaerobic bacteria prefer amino acids as energy source (see Chap. 11) but in general the "protein-sparing" effect of sugars, noted in early studies of human nutrition, applies equally well to microorganisms. This is demonstrated daily in the routine biochemical tests used for the identification of bacteria. Fermentation broth contains a mixture of amino acids and peptides. A fermentable compound, usually a carbohydrate, is added to determine the ability of the organism to ferment it, as evidenced by the production of acid and/or gas. Bacteria capable of using either the carbohydrate or amino acids as energy source almost invariably utilize the carbohydrate first.

While all chemoheterotrophic microorganisms undoubtedly can use exogenous amino acids for synthesizing proteins, the ability to use amino acids as energy source or to degrade them for the synthesis of other cellular constituents is less widespread. Some bacteria, e.g., *Pseudomonas* species, are able to utilize many single amino acids as the sole source of carbon, nitrogen, and energy, but many other organisms are not able to do so. This may be of little consequence in most natural environments. Proteins that are available to microorganisms in soil, water, or a waste stream would normally be present as components or products of decaying plant, animal, or microbial cells, and other more readily used energy sources usually would be available from the same sources.

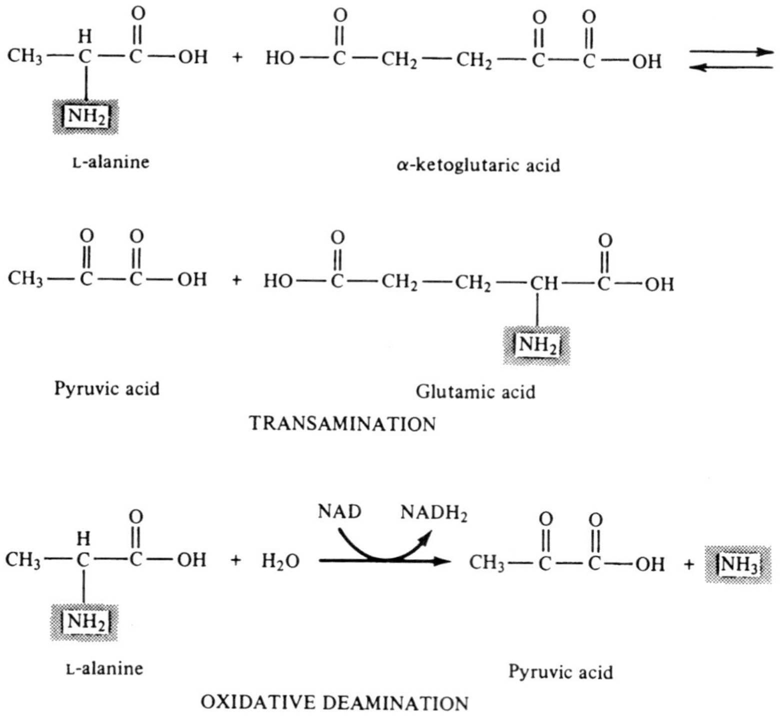

Figure 10-1 Common reactions that remove ammonia from an organic compound.

The pathways of amino acid oxidation in those organisms that can use them as energy source have not been thoroughly studied. In some microorganisms, pathways have been assumed to be the same as those in the human liver because some of the same enzymes have been found. The first step in the degradation of amino acids is generally removal of the amino group. This may occur either by *transamination*, transfer of the amino group to an α-keto acid to form another amino acid, or by *deamination*, release of the amino group as ammonia (see Fig. 10-1). Bacteria that use amino acids as carbon and energy source commonly release considerable amounts of ammonia. The further catabolism of most amino acids probably proceeds, as in mammals, to intermediates of the TCA cycle. Table 10-1 lists the TCA cycle intermediates formed from amino acids in mammals. The degradative pathways leading to these products are rather complex for some amino acids; they can also provide building blocks for the synthesis of other cellular components.

Table 10-1 TCA cycle intermediates formed as products of degradation of amino acids

Amino acid	TCA cycle intermediate
Alanine	Acetyl-CoA
Cysteine	
Glycine	
Isoleucine	
Leucine	
Lysine	
Phenylalanine	
Serine	
Threonine	
Tryptophan	
Tyrosine	
Aspartic acid	Oxalacetate
Asparagine	
Phenylalanine	Fumarate
Tyrosine	
Isoleucine	Succinyl-CoA
Methionine	
Valine	
Arginine	α-Ketoglutarate
Glutamic acid	
Histidine	
Proline	

METABOLISM OF LIPIDS

Lipids, like carbohydrates, have two major roles in the cell; they function as energy storage and as structural components of the cell, the latter primarily in membranes. The most abundant naturally occurring class of lipids is the *triglycerides*, commonly called *neutral fats*. These are esters of glycerol and fatty acids. The fatty acids may be saturated or unsaturated and may have an odd or even number of carbon atoms. Most abundant are the even-numbered chains of 12 to 22 carbons. The triglycerides are the major storage lipids of plants and animals. The second major class of naturally occurring lipids is the *phospholipids* (phosphoglycerides), which are major components of all membranes. These are also esters of glycerol, but mixed esters. Two of the hydroxyl groups of glycerol are esterified to fatty acids, whereas the third is esterified to phosphate. The phosphate is esterified to a small molecule that contains nitrogen, such as ethanolamine, choline, or serine. Many other types of lipids are components or products of plant, animal, and microbial cells. Some of these have special functions, and some are limited to specific types of cells. While all of these molecules are almost certainly subject to degradation by some microbial species, we shall center our attention on the most abundant classes of lipids.

Because of their largely hydrocarbon nature, the triglycerides, as well as other lipids, are excellent energy sources, with the potential of producing a far greater yield of ATP than do carbohydrates or proteins. The initial attack on both triglycerides and phosphoglycerides by microorganisms is extracellular and involves hydrolysis of the ester bonds. *Lipases* remove the fatty acids from the glycerol molecule of the triglycerides. The enzymes that hydrolyze

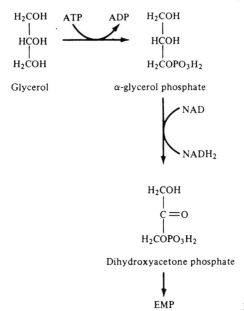

Figure 10-2 Degradative metabolism of glycerol.

the phosphoglycerides are called *phospholipases* and are of two types—those that hydrolyze the fatty acid ester bonds and those that hydrolyze the phosphoglycerol ester bond. The products of these hydrolytic enzymes are then transported into the cell.

Glycerol is metabolized by conversion to dihydroxyacetone phosphate, which is an intermediate in the EMP pathway. The reactions involved are shown in Fig. 10-2. As we shall see in Chap. 11, glycerol is formed as an end product in some fermentative bacteria by a reversal of the same reaction sequence, using a phosphatase to remove the phosphate group. The route from dihydroxyacetone phosphate to α-glycerol phosphate is also used in the synthesis of phospholipids.

The major pathway for the oxidation of fatty acids involves repetition of a sequence of reactions that results in the removal of two carbon atoms as acetyl-CoA with each repetition of the sequence. This process is called *beta-oxidation* because the beta-carbon (second from the carboxyl carbon) is oxidized. The overall reaction sequence is shown in Fig. 10-3. In Chap. 9 we pointed out the usefulness of a series of reactions involving the creation of a double bond to which water or two hydrogen atoms can be added. This mechanism was used in the TCA cycle as a means of introducing the oxygen required to convert succinate to oxalacetate. Beta-oxidation is another example of the use of this mechanism, and we shall see others in Chap. 11.

The first step in the beta-oxidation of a fatty acid is its *activation* by one of several *acyl-CoA synthetases* [Eq. (10-1)]. These enzymes, each capable of activating fatty acids of carbon chain lengths within a specific range, form a thiolester between coenzyme A and the carboxyl group of the fatty acid. The energy for formation of the ester bond is furnished by cleavage of the pyrophosphate from ATP. The reaction is very similar to that involved in the activation of amino acids for protein synthesis (see Chap. 12), involving an enzyme-bound acyl-AMP intermediate.

$$R-CH_2-CH_2-\overset{\displaystyle O}{\underset{}{C}}-OH + ATP$$

fatty acid

$$\downarrow\!\!\!\searrow PPi$$

$$[R-CH_2-CH_2-\overset{\displaystyle O}{\underset{}{C}}-AMP]-\text{enzyme} \qquad (10\text{-}1)$$

enzyme-bound fatty acyladenylate

$$CoASH\searrow\!\!\!\mid$$

$$\downarrow\!\!\!\searrow AMP$$

$$R-CH_2-CH_2-\overset{\displaystyle O}{\underset{}{C}}-SCoA$$

fatty acyl-CoA

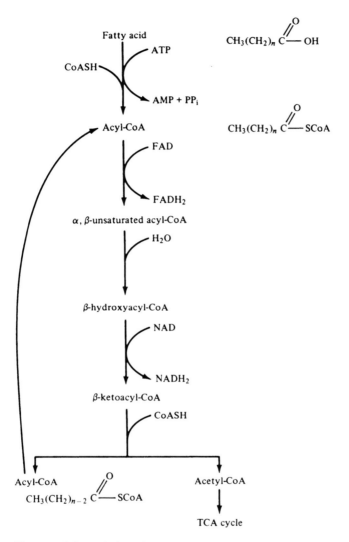

Figure 10-3 Degradation of a fatty acid by β-oxidation.

In organisms with mitochondria, the acyl group is transferred to carnitine. The carnitine-acyl ester is transported into the mitochondrion, where the remainder of the reactions occur after transfer of the acyl group to intramitochondrial CoA.

In the next reaction, hydrogen atoms are removed from the α- and β-carbons, introducing an α,β-unsaturation [Eq. (10-2)]. The enzyme, *acyl-CoA dehydrogenase*, is a flavoprotein enzyme that uses FAD as cofactor.

$$R—CH_2—CH_2—\overset{O}{\underset{}{\overset{\diagup\!\!\!\diagdown}{C}}}—SCoA \quad \xrightarrow{\text{FAD} \quad \text{FADH}_2} \quad R—CH=CH—\overset{O}{\underset{}{\overset{\diagup\!\!\!\diagdown}{C}}}—SCoA$$

acyl-CoA $\qquad\qquad\qquad\qquad\qquad\qquad\qquad$ α,β-unsaturated acyl-CoA

$$(10\text{-}2)$$

The presence of a double bond allows the addition of the elements of water, introducing oxygen into the molecule without the use of molecular oxygen. The enzyme *acyl-CoA hydratase* forms the β-hydroxyacyl-CoA by adding hydroxyl to the β-carbon and hydrogen to the α-carbon [Eq. (10-3)].

$$R—CH=CH—\overset{O}{\underset{}{\overset{\diagup\!\!\!\diagdown}{C}}}—SCoA \quad \xrightarrow{\text{H}_2\text{O}} \quad R—\overset{OH}{\underset{}{\overset{|}{C}H}}—CH_2—\overset{O}{\underset{}{\overset{\diagup\!\!\!\diagdown}{C}}}—SCoA \quad (10\text{-}3)$$

α,β-unsaturated acyl-CoA $\qquad\qquad\qquad\qquad$ β-hydroxyacyl-CoA

Removal of two hydrogen atoms from the same two carbons now creates a β-keto group [Eq. (10-4)]. The enzyme, *3-hydroxyacyl-CoA dehydrogenase*, uses NAD as cofactor. This sequence of three reactions [Eqs. (10-2), (10-3), and (10-4)], by which oxygen is introduced into the molecule, is exactly the same as reactions (9-21), (9-22), and (9-23) in the TCA cycle.

$$R—\overset{OH}{\underset{}{\overset{|}{C}H}}—CH_2—\overset{O}{\underset{}{\overset{\diagup\!\!\!\diagdown}{C}}}—SCoA \quad \xrightarrow{\text{NAD} \quad \text{NADH}_2} \quad R—\overset{O}{\underset{}{\overset{\diagup\!\!\!\diagdown}{C}}}—CH_2—\overset{O}{\underset{}{\overset{\diagup\!\!\!\diagdown}{C}}}—SCoA$$

β-hydroxyacyl-CoA $\qquad\qquad\qquad\qquad\qquad\qquad$ β-ketoacyl-CoA

$$(10\text{-}4)$$

The molecule now has been prepared for cleavage. The enzyme *thiolase*, using coenzyme A instead of water as in hydrolysis or phosphate as in phosphorolysis, catalyzes a thiolytic cleavage [Eq. (10–5)]. The products are *acetyl-CoA*, formed from the original carboxyl and α-carbons, and a CoA ester of a fatty acid shorter than the original by two carbon atoms.

$$R—\overset{O}{\underset{}{\overset{\diagup\!\!\!\diagdown}{C}}}—CH_2—\overset{O}{\underset{}{\overset{\diagup\!\!\!\diagdown}{C}}}—SCoA \quad \xrightarrow{\text{CoASH}} \quad R—\overset{O}{\underset{}{\overset{\diagup\!\!\!\diagdown}{C}}}—SCoA + CH_3—\overset{O}{\underset{}{\overset{\diagup\!\!\!\diagdown}{C}}}—SCoA$$

β-ketoacyl-CoA $\qquad\qquad\qquad\qquad\qquad$ acyl-CoA $\qquad\qquad$ acetyl-CoA

$$(10\text{-}5)$$

The shortened acyl-CoA is now ready to participate in the same series of reactions [Eqs. (10-2) to (10-5)], losing the next two carbon atoms as acetyl-CoA. Repetition of this reaction sequence converts a fatty acid with an even number of carbons totally to acetyl-CoA, which enters the TCA cycle to be oxidized to carbon dioxide and water. Both the FAD and NAD reduced during the reaction sequence transfer hydrogen and/or electrons to the aerobic electron transport system for generation of ATP through oxidative phosphorylation. An additional three molecules of reduced NAD and one of reduced FAD are generated during the oxidation of one molecule of acetyl-CoA through the TCA cycle, and one substrate level phosphorylation occurs.

In eucaryotic microorganisms with a P/O ratio of 3, the net ATP synthesis resulting from the oxidation of one molecule of the commonly occurring 16-carbon fatty acid, palmitic acid, is 129 molecules of ATP. This represents a 40 percent efficiency of conservation of the free energy of oxidation of palmitic acid as ATP.

In the oxidation of fatty acids that contain an uneven number of carbons, a three-carbon residue, propionyl-CoA, remains after removal of the other carbons as acetyl-CoA. This can be carboxylated (using energy furnished by hydrolysis of ATP) to yield methylmalonyl-CoA, which is rearranged to form succinyl-CoA, an intermediate of the TCA cycle [Eq. (10-6)].

propionyl-CoA   methylmalonyl-CoA   succinyl-CoA

$$(10\text{-}6)$$

Unsaturated fatty acids are oxidized through the same series of reactions used for saturated fatty acids but require two additional enzymes. These move the double bond and change its configuration so that it corresponds to the double bond that would be created during the metabolism of a saturated acid by the action of the acyl-CoA dehydrogenase.

METABOLISM OF NUCLEIC ACIDS

The nucleic acids RNA and DNA are hydrolyzed extracellularly to nucleo-tides by enzymes excreted by some bacteria and fungi. Enzymes produced by different organisms may cleave different bonds. A number of these specific enzymes are known to occur intracellularly, but it is not known how many of these may be excreted. Some species of *Bacillus* and *Streptococcus* are particularly active in the extracellular degradation of nucleic acids. Enzymes that degrade DNA are *deoxyribonucleases* (*DNases*) and those that degrade RNA are *ribonucleases* (*RNases*). Both hydrolyze phosphodiester bonds, releasing nucleotides. Before entering the cell, the highly ionized phosphate group must be removed by a phosphatase. The nucleosides are then trans-ported into the cell (see Fig. 10-4).

As is true of proteins, nucleic acids are probably used as an energy source by few organisms if other energy sources, such as carbohydrates or lipids, are available. Growing cells need both ribo- and deoxyribonucleosides for the

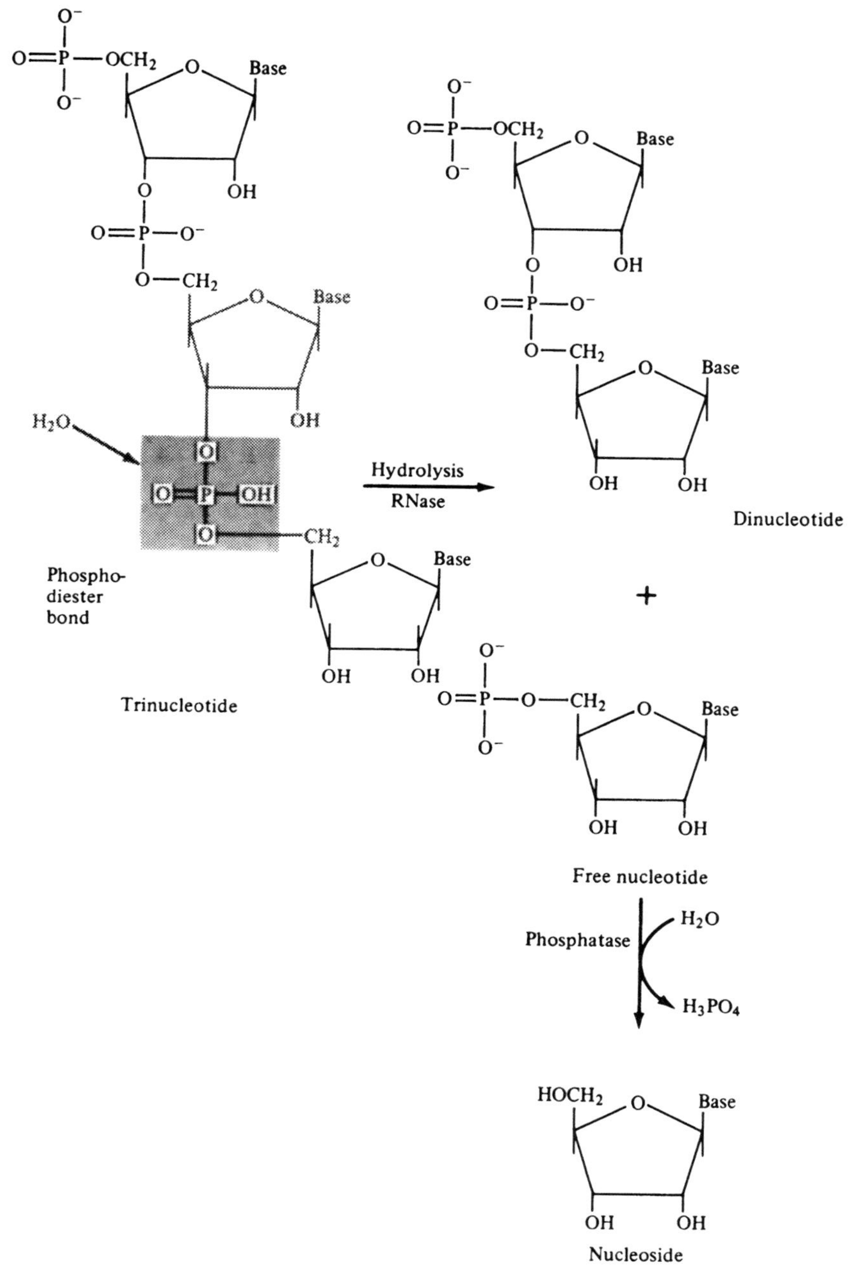

Figure 10-4 The action of ribonuclease in cleavage of the phosphodiester bond, illustrated using a trinucleotide. After the nucleotide is removed, it must be converted to a nucleoside before entering the cell.

synthesis of their own nucleic acids and nucleotide-containing coenzymes, and exogenous supplies of these molecules are used when available; that is, the synthesis of purine and pyrimidine nucleotides is repressed if they are available from an exogenous source (see Chap. 12). Some aerobic micro-organisms are able to utilize one or more of the purines and pyrimidines as sole sources of carbon, nitrogen, and energy. The initial reactions in the catabolism of the nucleosides differ with the organism. In using ribo-nucleosides, the ribose may be removed by hydrolysis or phosphorolysis, forming ribose 1-phosphate. Deamination of the ring often precedes ring cleavage. The purines are degraded to ammonia, carbon dioxide, and gly-oxylic acid. Two molecules of glyoxylate can condense with the loss of carbon dioxide to form tartronic semialdehyde. This can be reduced to glyceric acid, which may be phosphorylated by ATP to form the EMP intermediate 3-phosphoglyceric acid. These reactions have been shown to occur in several species of bacteria and fungi. Pyrimidines are degraded to ammonia and carbon dioxide. An intermediate product of the degradation of uracil is β-alanine, $NH_2CH_2CH_2COOH$, which is required for the synthesis of coenzyme A.

METABOLISM OF HYDROCARBONS

Because of their importance as environmental pollutants and the current interest in their use as substrate in the production of "single-cell protein," we shall consider the metabolism of hydrocarbons at least briefly.

Aliphatic (straight-chain) hydrocarbons are oxidized rather readily by a variety of bacteria and fungi. Molecular oxygen is required for the microbial metabolism of hydrocarbons, and their use is therefore restricted to aerobic environments. Several different modes of attack on aliphatic hydrocarbons have been reported, but the major route of metabolism probably involves oxidation of the terminal methyl group to a primary alcohol; this is the step that requires molecular oxygen. The alcohol is then successively oxidized by dehydrogenases to an aldehyde, and then to an acid (see Fig. 10-5). Once the hydrocarbon has been converted to a fatty acid, it can be metabolized through the beta-oxidation pathway described previously for fatty acids derived from lipids (see Fig. 10-3). Both saturated and unsaturated aliphatic hydrocarbons may be oxidized in this way. Other modes of attack involve oxidation of the subterminal carbon or oxidation at both ends of the molecule.

The aromatic hydrocarbons, including benzene and its derivatives, are metabolized by cleavage of the ring to form a straight-chain acid. Depending on the substituents of the ring, the initial reactions may differ, but molecular oxygen generally is used in one or more reactions. In some cases, substituent groups may be oxidized or removed before the ring is attacked. Ring cleavage commonly occurs between adjacent hydroxyl groups by the addition of oxygen to form a linear dicarboxylic acid. The reactions by which adjacent

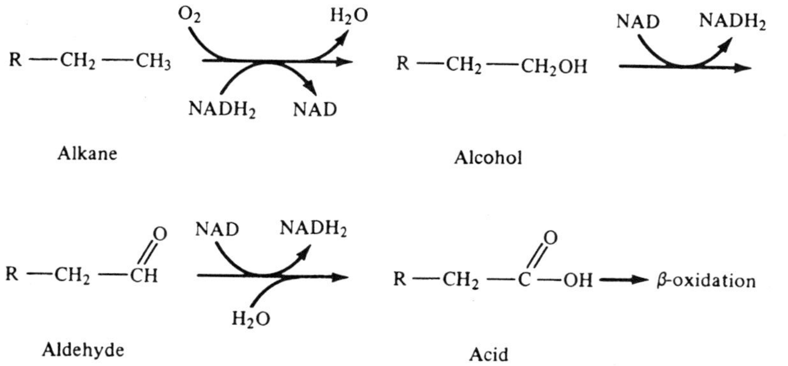

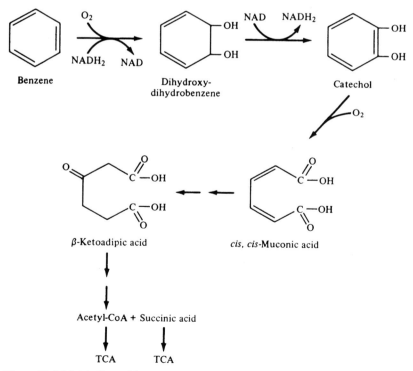

The following reactions appear in the figure:

R—CH₂—CH₃ (Alkane) → with O₂, NADH₂ → NAD, H₂O → R—CH₂—CH₂OH (Alcohol) → with NAD → NADH₂

R—CH₂—CH (Aldehyde) → with NAD → NADH₂, H₂O → R—CH₂—C—OH (Acid) → β-oxidation

Figure 10-5 Metabolism of an aliphatic hydrocarbon by oxidation to a fatty acid.

hydroxyl groups are introduced vary with the compound and the organism. One sequence of reactions that leads to cleavage of the benzene ring is shown in Fig. 10-6. Various pathways of attack on aromatic compounds give rise to the common intermediate β-ketoadipic acid, which is further metabolized to acetyl-CoA and succinate, intermediates in the TCA cycle.

Benzene → with O₂, NADH₂ → NAD → Dihydroxy-dihydrobenzene → with NAD → NADH₂ → Catechol → with O₂ → cis, cis-Muconic acid → β-Ketoadipic acid → Acetyl-CoA + Succinic acid → TCA, TCA

Figure 10-6 Metabolism of benzene.

Many complex hydrocarbons may be metabolized only slowly by micro-organisms. In some cases, complete metabolism cannot be achieved, at least by one organism, and the molecule may be only modified rather than completely degraded. In studies of such reactions using pure cultures, the organism requires a second substrate that can be used as carbon source and possibly to supply additional energy. Such reactions have been called *co-oxidations*.

Much remains to be learned about the metabolism of hydrocarbons. Research has been stimulated in recent years by the problems caused by the persistence of hydrocarbons, particularly those of higher molecular weight and more complex structure, in marine environments following oil spills (Friede et al., 1972). Fortunately or unfortunately, depending on one's viewpoint and the environment in which the degradation occurs, the lower molecular weight fractions obtained from crude oil are readily degraded by microbial action in the presence of water and oxygen. This is a welcome activity when these compounds occur in waste streams or pollute the environment. It has presented a serious problem in storage facilities and fuel lines, however. Jet fuel must be filter-sterilized to prevent the growth of microorganisms in tanks and fuel lines. Fungi are a particular problem since they can survive with very little moisture.

We may summarize our discussion of aerobic metabolism by stating again the belief of microbiologists that all naturally occurring compounds can be metabolized by microorganisms. Some can be metabolized only if oxygen is available, some are metabolized only very slowly, and some only by a few species of microorganisms. Aerobic metabolism is much more efficient and more versatile than anaerobic metabolism, which will be considered in the next chapter. For this reason, aerobic metabolism is the most widely used mechanism for the treatment of organic wastes.

METABOLISM AND GROWTH

Figure 3-1 was a representation of the aerobic decay process. Thus far, we have elucidated in some detail the physiological mechanisms that make these reactions proceed. However, there is much yet to be considered, particularly regarding the cultural conditions and microbial accommodations to them as organic matter is metabolized in natural and engineered environments. In the preceding discussion, it was shown how macromolecular substrates are dissimilated into smaller feedstock carbon chains. These smaller units are channeled through a variety of intermediary biochemical pathways, common to many species, from which the cell obtains energy, ATP, and the carbon chains that are used as starting materials for making nucleic acids, proteins, fats, and carbohydrates, the structural and functional constituents of new cells. A simplified representation of this machinery was shown in Fig. 9-1.

The kinetic description given in Chap. 6 for batch and continuous growth systems was a means of representing the rate of occurrence of the ultimate outcome of the complex system of chemical equations presented up to this point. The results of the operation of this metabolic machinery were expressed in parameters of functional and applied significance: substrate utilization (e.g., wastewater purification), oxygen utilization (BOD exertion), and growth (sludge or biomass production). The kinetic principles for batch and continuous culture systems of microorganisms presented in Chap. 6 were developed to describe the orderly process of balanced growth, i.e., the smooth flow of organic (and inorganic) matter through the metabolic machinery depicted in Fig. 9-1. If there are no snags in the metabolic machinery and if the raw materials are constantly supplied (as in continuous culture) or are initially present in some abundance (as in batch cultures), the reactor of Fig. 9-1 will form its product—a new reactor—and cell replication will proceed in an orderly manner. Units of production are measured in mass units as "growth." The discussion of growth was begun by considering a batch system, and it is appropriate to return to Fig. 6-1 and consider some biochemical events that occur during the cycle of growth.

Balanced Growth

In a batch system in which a small inoculum of fresh, young, acclimated organisms was placed in an aeration vessel containing a concentration of carbon source that was relatively high compared with the initial concentration of cells and excess (but not excessive) concentrations of all other nutrients, the "growth" of the population necessarily involved the synthesis of DNA, RNA, protein, carbohydrate, and lipid from the building blocks obtained from the dissimilatory and energy-yielding intermediary metabolism of the carbon source.

With reference to Fig. 6-1, if we measure growth by dry weight, the biomass grows at an exponential rate very quickly, but if we measure cell numbers, little or no cell division takes place during this early phase; that is, the viable or total count does not rise. The size (volume) of the cells increases and the RNA content of the cells increases. As will be seen in Chap. 12, RNA is needed to make protein, which is the largest single component of the cells. In short, the cells appear to be "tooling up" for rapid multiplication. During this period, the amount of DNA per cell increases. It must approximately double because, upon division, the two daughter cells contain equal amounts of the genetic material. However, the amount of DNA per milligram dry weight of cells decreases during this period because the amounts of other constituents such as RNA are increasing so rapidly. Protein, carbohydrate, and lipid contents also increase. After the first few divisions, the number of cells and the dry weight increase in parallel, in exponential growth, at some constant specific growth rate μ. The percentages of RNA and DNA, protein, carbohydrate, and lipid per cell are constant. This is often referred to as *balanced growth*. This situation does not last for long, because after the inflection point, i.e., the beginning of the phase of decelerating growth, the mass per cell is no longer constant. The biomass concentration X still increases, but the weight per cell and the rate of increase in the number of cells in the system are decreasing. The amount of RNA in each cell decreases as the biomass concentration attains its maximum value; that is, the "stationary" phase is attained as the exogenous carbon source S is exhausted. The system then enters the endogenous or autodigestive phase, which we shall discuss in detail later.

The above description of the biochemical events is somewhat idealized even for a pure culture but, in general, this course of compositional change does correspond to the observed kinetics of changes in X, S, and oxygen utilization when a small inoculum of cells is placed in a suitable medium for laboratory growth studies. We may question, however, whether this is the general case in the real world of environmental microbiology. The experiment described is, after all, a laboratory batch growth experiment, and the growth cycle, as was stated previously, is simply a convenient mode of description for the experimental observations. One application of such an experiment is in understanding the early events during the determination of the biochemical oxygen demand (BOD) of a soluble organic waste. The mechanism and kinetics of exertion of BOD during the BOD test occupy a very important place in environmental microbiology and will receive separate discussion later in this chapter. Another example of this type of system (assuming plug flow) is the receiving stream. The results of BOD testing are often applied in estimating a stream's ability to assimilate organic carbon source without depletion of the dissolved oxygen resource in the stream.

In a continuous culture of microorganisms, the balanced condition, i.e., a steady state in X, S, and biochemical composition, can be obtained also when μ is held constant hydraulically but proportions of DNA, RNA, carbohydrate, protein, and lipid can change according to the value of μ and the nature of the growth-limiting nutrient. The growth-limiting nutrient may be the carbon source, e.g., protein, carbohydrate, or fats, or it may be the available nitrogen or phosphorus or any other constituent needed for growth, i.e., cell replication. The synthesis of various cellular constituents and their relative amounts in the biomass in continuous culture processes are important parameters used in estimating the capacity and healthful status of the treatment process. This is why it is important for us to consider some of the topics discussed in this chapter.

Under balanced growth conditions, we can expect that the specific growth rate μ will be constant. In activated sludge processes, the value of μ is low because cell recycle is used; it is constant because the system is maintained in a steady state due to the constancy of the operational parameters. At any rate, this is the assumption made in deriving and using the various kinetic models.

Under balanced growth conditions μ is constant during an exponential phase in the batch growth cycle. This can be perpetuated indefinitely under steady state conditions in completely mixed continuous culture by hydraulic control of μ; $\mu_n = D$ in once-through reactor systems, and $\mu_n = D[1 + \alpha - \alpha(X_R/X)]$ in cell recycle systems such as activated sludge processes. In the normal case, an activated sludge system is designed so that the carbon source is the growth-limiting nutrient. If it is not the limiting nutrient in the waste, it is made so by the addition of other essential nutrients in which the waste substrate may be deficient, e.g., nitrogen and phosphorus. Since the feed carbon sources are present in low concentrations and the desire is to obtain the lowest possible amount of residual carbon source in the effluent, the values of μ_n (or μ) are necessarily very low (in accordance with principles discussed previously). However, regardless of its low value, μ is held constant (or, in any event, the aim is to hold it so), and activated sludge processes can be thought of as balanced growth systems.* They are very special growth systems, though, because of their low growth rate and the means of obtaining it, i.e., the recycle of massive amounts of cells.

In the remainder of this chapter we shall discuss some aspects of aerobic metabolism that have general fundamental interest in environmental microbiology but have special significance to the subject of biological waste treatment. These include methods of making materials balances on experimental data from treatability studies, microbiological flocculation, some aspects of unbalanced growth (nitrogen requirements, oxidative assimilation, and storage products), endogenous energy requirements, and biochemical oxygen demand—both carbonaceous and nitrogenous.

MATERIAL AND ENERGY BALANCES IN AEROBIC SYSTEMS

In Fig. 6-1, the available carbon and energy source (ΔCOD) was used in general for the synthesis of new cells and for respiration (oxidation) to obtain energy used to synthesize the cells. Consequently, at any time during the experiment, one should be able to account for the mass of the carbon and energy source that has disappeared as the sum of the amount of it that was oxidized and the amount of it that was channeled into the biomass during that time period; one can make a mass balance on the system. Care must be exercised to use some colligative test such as ΔCOD for measuring the organic matter used at any time, because organisms can sometimes elaborate

---

*Much of what is true for activated sludge processes in regard to metabolism is true in general for fixed-bed reactor systems, e.g., packed towers or trickling filters. However, kinetic theory for fixed-bed reactors rests on a much less sound basis than does that for the activated sludge process. It is not possible to say whether balanced growth occurs or whether a steady state can be assumed. Questions as to the degree of homogeneity in the slime film and the degree of activity per unit thickness as well as a changing biomass concentration due to nonsteady sloughing of slime are matters that complicate the development of useful kinetic description and make almost impossible any experimental demonstration of mechanistic models that have been suggested. For this reason, we have omitted discussion of the kinetics of purification or growth in fixed-film reactor systems.

organic metabolic products into the medium, and a test for the specific substrate under question or even for a general class of compound (e.g., the anthrone test if glucose were the substrate) could give a gross overestimate of the amount of organic matter that actually had been removed from solution for respiration and synthesis during the time in question.

Also, the summing of respiration and synthesis to account for the amount of carbon and energy source removed assumes that there are no other removal mechanisms in operation. This is not true for certain compounds. Some organic compounds can be stripped from solution due to aeration and/or agitation. Therefore, care should be taken to check the compounds or the wastewaters to determine their susceptibility to loss of organic content by physical stripping. Even for compounds that are not strippable, it is possible that the microorganisms in the course of metabolism may produce and release into the medium metabolic intermediate products that are strippable (Gaudy, Goswami, Manickam, and Gaudy, 1977). Thus, when one says that the disappearance of the carbon and energy source can be accounted for by respiration and synthesis, there is need for qualification and investigation of each case, and one should also realize that the accounting is an approximation. Quantitative aspects regarding the kinetic description of the stripping process (which usually proceeds at a first-order, decreasing rate, i.e., with kinetics similar to those for reaeration) and factors affecting stripping and metabolism of strippable compounds are somewhat outside our present interest, but the subject has been of continuing investigational concern (see Gaudy, Engelbrecht, and Turner, 1961; Gaudy, Turner, and Pusztaszeri, 1963; and Thibodeaux, 1974). It is enough here to emphasize that some systems contain volatile (strippable) compounds, and one should make a preliminary investigation to determine whether aeration alone, in the absence of microbial metabolism, can bring about significant reduction in the organic content of the wastewater. Also there are some organic compounds in certain wastewaters that may be subject to chemical oxidation by oxygen under the mild chemical oxidizing conditions prevalent in the aeration tank. This, however, does not appear to be a major cause for disappearance of COD. The general assumption is that if the carbon source is respired to carbon dioxide and water, as measured by oxygen uptake, the carbon in carbon dioxide is accounted for as being removed for respiration and is therefore no longer available for incorporation into the biomass. In general, this is true, but some carbon dioxide is used in biosynthesis, e.g., in the synthesis of purines, by heterotrophic organisms. However, this would seldom amount to as much as 10 percent of the carbon dioxide produced.

Thus, accounting for the disappearance of organic matter in a batch growth experiment by respiration and synthesis may be somewhat of an approximation, but it is nonetheless an important and useful concept that is consistent with our understanding of the aerobic decay cycle, and one may use this concept as a basis for materials balances to check on the accuracy of experimental data. On the other hand, if scientists have been convinced of the usefulness of the principle and they have become even more convinced of their infallibility as experimenters, they can use the materials balance principle to calculate values for analyses they have not obtained experimentally. This is one of the ways in which materials balances are used, although it is preferable to use them to check the data that have been obtained rather than to provide missing data. It is recalled that the equation in Fig. 3-1, or any other chemical equation for that matter, is by nature a materials balance or an energy balance. We shall apply this equation, which represents aerobic decay, to any period of time during an experiment such as the one shown in Fig. 6-1. This materials balance principle is the same as that used in Chap. 6, but in that case we were dealing with a constant flux or flow rate of material being processed.

Methods

There are several ways to approach the inventory of materials; we will describe three of them.

1. *Material balance based on weight of carbon source:* If the substrate consists of a known compound, the oxygen uptake and the COD measurements can be calculated to an equivalent weight of substrate; in the example below, hexose is used. The recovery is then the ratio of the sum of the weight of substrate oxidized and the weight of substrate converted to cells (Δ cells) to the ΔCOD, calculated to the weight of substrate. Since the theoretical oxygen demand of hexose is 192/180, oxygen is converted to the weight of substrate by multiplying by 180/192.

$$\text{Recovery} = \frac{(0.94)(O_2 \text{ uptake}) + \Delta \text{ cells}}{(0.94)\Delta \text{COD}} \tag{10-7}$$

2. *Material balance based on carbon:* We may choose to balance the sum of carbon in carbon dioxide and carbon in the cells against the substrate carbon removed from solution. A carbon analyzer would be a handy tool, but several estimates can be used if an analyzer is not available. If one assumes that the carbon content per unit of cell mass is relatively constant, one can estimate the carbon in the cells from the weight of cells produced. In the example below, a carbon content of 51 percent is used. Capture of carbon dioxide for analysis may be difficult depending on the size and type of the experimental reactor used. If the empirical formula for the substrate is known, one can estimate the carbon dioxide production from the oxygen uptake. For hexose, 1 mol of oxygen is required per mole of carbon dioxide produced, and the weight ratio $CO_2/O_2 = 44/32 = 1.375$. Also, one can calculate the carbon content of carbon dioxide, $12/44 = 0.273$, and of the substrate, $72/180 = 0.4$.

$$\text{Recovery} = \frac{(0.273)(1.375)(O_2 \text{ uptake}) + \Delta \text{ cells}(0.51)}{(0.94)(0.4)(\Delta \text{COD})} \tag{10-8}$$

3. *Materials and/or energy balance based on chemical oxygen demand:* Use of the COD test as a measurement of the amount of organic matter has been discussed. The use of oxygen required for oxidation as the standard of expression for the organic matter, rather than some standard organic compound, enabled us to include in our assessment of the amount of organic matter its state of oxidation. Thus, we spoke of measuring organic matter in relation to the fuel energy it contained; the greater the fuel energy, the greater is the amount of oxygen needed to burn it. Thus it can be seen that the measurement of biological oxygen uptake, respiration, is really a measure of the portion of the chemical oxygen demand of the organic substrate that has actually been oxidized by the organisms. One may sum the oxygen uptake for respiration and the COD value of the cells that have been produced, and compare this with the amount of COD that has been removed from solution. The COD of the cells can be estimated if we know the empirical formula of the biomass. The empirical formula for cells used in Eq. (3-10) $(C_5H_7NO_2)$ is often used as an approximation. On the other hand, the COD of the cells may be measured directly by performing a COD analysis on a sample of the biomass (Gaudy, Bhatla, and Gaudy, 1964). One may also obtain the COD of the cells as the difference between the COD of a sample of unfiltered reaction liquor, i.e., mixed liquor, and that of a filtered sample. In fact, if one measures these two COD values periodically as growth proceeds, one has all three pieces of information needed to make the materials balance. The COD removal at any time, ΔCOD, is obtained by running the COD of the filtrate, as before. The ΔCOD of the cells is obtained as outlined above; it can be seen that the ΔCOD that was respired, i.e., the oxygen uptake, is the difference between the mixed liquor COD values at the beginning and end of the time period in question. Thus, in addition to the methods for measuring oxygen uptake given in Chap. 6, the measured ΔCOD of the total reaction liquor, i.e., the mixed liquor, can be used to obtain the biological oxygen uptake. Energy or material balances based on COD can be made in any of the three ways in Eqs. (10-9), (10-10), and (10-11).

$$\text{Recovery} = \frac{O_2 \text{ uptake} + 1.4\,\Delta \text{ cells}}{\Delta COD} \tag{10-9}$$

$$\text{Recovery} = \frac{O_2 \text{ uptake} + \Delta COD \text{ cells}}{\Delta COD} \tag{10-10}$$

$$\text{Recovery} = \frac{(\Delta COD \text{ mixed liquor}) + (\Delta COD \text{ mixed liquor} - \Delta COD)}{\Delta COD} \tag{10-11}$$

Although one can obtain a plot of the growth curve as ΔCOD mixed liquor minus ΔCOD, the growth curve is usually measured by its actual biomass concentration X during an experiment, and measurement of the COD of the mixed liquor along with the COD remaining in the solution is used for estimating the oxygen uptake. Thus, if one measures the COD of the mixed liquor, it is possible to run a fairly rigid experiment without having to determine the reaeration characteristics of the reactor or without using a Warburg apparatus.

Examples of Material Balances Applied to Experimental Data

In Chap. 6 it was mentioned that a Warburg apparatus can be used as a compartmentalized bioreactor. The results shown in Fig. 10-7 were obtained using the Warburg apparatus in this manner. The samples for mixed liquor COD, filtrate COD, and biological solids were obtained by removing a Warburg flask from the apparatus at each sampling time indicated. Oxygen uptake was obtained manometrically. In this figure the time scale used during the substrate removal period, i.e., the first 20 hours, is 10 times longer than that used for the autodigestive phase. Manometer readings were taken for only some of the Warburg flasks, and Fig. 10-8 attests to the high degree of reproducibility of the oxygen uptake values during the substrate removal phase. There is less reproducibility in the autodigestive phase—after the first 20 hours of the experiment. The dotted line represents the oxygen uptake during the first 20 hours transposed to the smaller scale. In general, it can be concluded that during the substrate removal phase, oxygen uptake was

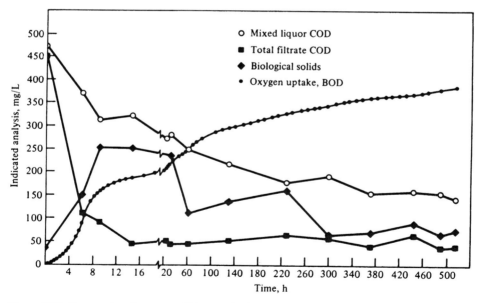

Figure 10-7 Changes in substrate and biomass during exertion of BOD. Data points for CODs and cell mass were obtained by analyzing the contents of Warburg flasks removed sequentially during the experiment. (*Data of A. F. Gaudy, C. L. Goldstein, and P. A. Perkins, unpublished.*)

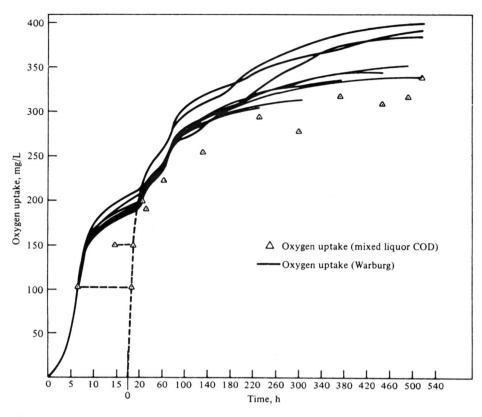

Figure 10-8 Accumulated oxygen uptake (BOD) in replicate Warburg flasks during the experiment shown in Fig. 10-7. The triangles are the oxygen uptakes calculated as the amount of mixed liquor COD removed using the mixed liquor COD curve shown in Fig. 10-7. (*Data of A. F. Gaudy, C. L. Goldstein, and P. A. Perkins, unpublished.*)

essentially the same in all flasks and there was some divergence in the autodigestive phase, although for the most part the divergence was not excessive. The triangles in Fig. 10-8 represent the accumulated oxygen uptake calculated from the mixed liquor COD values. It can be seen that the ΔCOD of the mixed liquor provides a fairly useful estimate of oxygen uptake.

For purposes of making material balances, the manometric oxygen uptake values may be averaged. Table 10-2 shows the percentage recovery calculated from Eqs. (10-7) to (10-11). All of the methods indicate that during the experiment, the sampling and the analyses were conducted with sufficient accuracy to account for the substrate removed as the sum of substrate respired and substrate channeled into cellular material. In general, these balances show that the sum of respiration and synthesis accounted for slightly more material than was removed from solution. This may be due to experimental error, or some of the carbon dioxide produced during respiration may

Table 10-2 Materials and energy balances during the experiment shown in Fig. 10-7

		Substrate recovery, percent*			
Time, h	Weight, Eq. (10-7)	Carbon, Eq. (10-8)	COD, Eq. (10-9)	COD, Eq. (10-10)	COD, Eq. (10-11)
6	62	72	74	96	86
9	109	127	130	100	99
14.5	103	118	121	108	122
24	68	72	73	102	104
29	107	122	125	106	113
60.5	81	86	87	105	109
130	101	109	110	110	116
228	121	130	132	108	110
299	95	97	98	116	122
370	97	100	100	109	112
443	111	115	116	114	117
488.5	99	101	101	114	117
514	104	107	107	110	112
Average	97	104	106	108	111

*The methods used are identified by the text equation numbers.

have been used by the cells. In any event, the recoveries are rather good, and it is seen that the balancing methods are valid for both the growth and autodigestive phases.

This type of study can also be performed in a batch reactor open to the atmosphere, such as that depicted in Fig. 10-9. In this case, oxygen uptake can

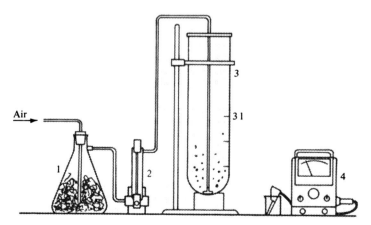

Figure 10-9 Batch reactor setup for treatability study, including measurement of the oxygen uptake. 1, cotton filter; 2, air flow meter; 3, batch reactor, 4-in diameter test tube; 4, DO meter.

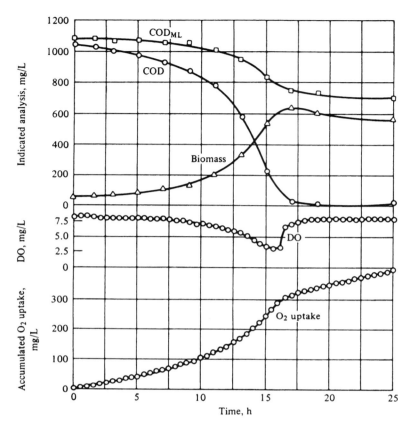

Figure 10-10 Oxygen uptake, COD removal, biomass growth, and mixed liquor COD during a batch experiment conducted in the apparatus shown in Fig. 10-9. (*Data of A. F. Gaudy and T. T. Chang, unpublished.*)
$COD_0 = 1044$ mg/L, $X_0 = 46$ mg/L, $COD_{ML} = 1080$ mg/L, $S_0/X_0 = 23$, $K_2 = 0.1430$ min^{-1} from 0 to 10 h, $K_2 = 0.1320$ min^{-1} from 10 to 25 h.

be computed from a knowledge of the reaeration constant(s) for the reactor, provided one measures the DO profile during the experiment (see the method described in Chap. 6). In Fig. 10-10, data for soluble COD, mixed liquor COD, biomass concentration, and dissolved oxygen in a reactor like that shown in Fig. 10-9 were obtained experimentally, and the oxygen uptake curve was calculated as described in Chap. 6. The soluble COD was removed in this experiment by the seventeenth hour. At that time, the oxygen uptake, calculated from K_2 and the DO profile data, was 313 mg/L. The oxygen uptake calculated as the amount of mixed liquor COD removed was 328 mg/L. Using these experimental results, the COD of the cells can be obtained as (COD of mixed liquor) − (COD of filtrate). The values thus obtained can be checked against the COD values obtained for samples of the biomass. In this experiment, the COD of the biomass was 1.2 mg COD per 1 mg of cells. The recovery at hour 17, obtained as the sum of the oxygen uptake, calculated from the DO sag curve, and the COD of the biomass divided by the amount of filtrate COD removed, was 98 percent. In making such studies, for example, during treatability studies on various wastewaters, one can choose to run either mixed liquor COD or dissolved oxygen profiles to get the oxygen uptake.

In general, the use of the DO meter* to measure the dissolved oxygen represents an easier method than running COD values on the cells, since in order to get an oxygen uptake curve, a fairly large number of CODs must be performed on the mixed liquor. However, the dependability of oxygen uptake data is related to the accuracy of the K_2 values used for the reactor. The accuracy of the numerical computation is also dependent on obtaining a DO profile that has a significant sag. The sag, however, cannot be allowed·to reach a DO value of 0 during the experiment or else the experimental results may be affected by the DO concentration. Thus, the DO sag method has some pitfalls and requires considerable care and effort. Many environmental factors can change the reaeration constant. Prominent among these are the concentration of suspended solids present and the concentration of inorganic salts. In Fig. 10-11, the effect of the concentration of mineral salts in the medium on the

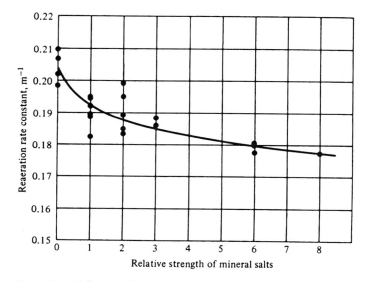

Figure 10-11 Effect of mineral salts on reaeration rate. Air flow rate = 3000 cm³/min; no biological solids in the system: all K_2 values adjusted to 25°C. (*Data of A. F. Gaudy and T. T. Chang, unpublished.*) Relative strength: $1 = (NH_4)_2SO_4$, 500 mg/L; $MgSO_4 \cdot 7H_2O$, 100 mg/L; $FeCl_3 \cdot 6H_2O$, 0.5 mg/L; $MnSO_4 \cdot H_2O$, 10 mg/L; $CaCl_2 \cdot 2H_2O$, 7.5 mg/L; 1.0 M phosphate buffer (KH_2PO_4, 52.7 g/L; K_2HPO_4, 107 g/L), 100 ml/L.

*A DO meter is a device that measures dissolved oxygen. Several types of manufactured DO meters are available.

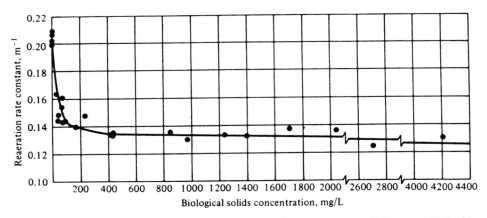

Figure 10-12 Effect of biomass concentration on reaeration rate constant. Cells were killed with Clorox at 2.4 ml/L per 100 mg/L of biomass. All K_2 values adjusted to 25°C. (*Data of A. F. Gaudy and T. T. Chang, unpublished.*)

value of K_2 is shown. The number 1 on the x axis represents the normal strength of the mineral salts medium that was used in the experiments shown in Fig. 10-10; 2 indicates double strength; etc. The effect of increasing the concentration of the biomass on K_2 is shown in Fig. 10-12. A small concentration of Clorox, sufficient to kill the cells, was added so that there was no oxygen uptake during the experiment. Curves such as these can be used to adjust the K_2 values as a batch experiment progresses.

We can see from these examples that while oxygen uptake measurements during the course of an experiment can be obtained, there is really no easy way to do it; it is usually more difficult to obtain oxygen uptake measurements than to obtain the growth curve or the COD removal curve. The manometric method requires special equipment; the sag method requires investigational attention in regard to K_2; and the accumulated decrease in mixed liquor COD requires running many COD determinations in order to obtain enough plotting points to define the oxygen uptake curve. Although it is not a recommended procedure, the oxygen uptake can be determined, by using the aerobic balance principle, as the resultant missing data: Accumulated oxygen uptake $= \Delta COD - 1.4(\Delta X)$. That is, the accumulated oxygen uptake at any time during the experiment can be calculated as the COD removed minus the COD of the cells produced. In lieu of actually determining the COD of the cells, one may use the value 1.4 calculated from the theoretical COD of microbial cells using the empirical formula for the microbial cells [Eq. (3-10)]. This procedure is, of course, not so recommended as the use of the principle of aerobic materials balances for checking experimental results, but it provides an approximation of the oxygen uptake or biochemical oxygen demand exerted during the substrate removal and autodigestive phases. Later we shall see that there is yet another analytical procedure, run under more dilute conditions, to obtain the value of the oxygen uptake, i.e., the BOD test.

MICROBIAL FLOCCULATION

The heterogeneous microbial populations grown in biological reactors such as activated sludge tanks or fixed or rotating contact reactors are commonly separated from the liquor by quiescent settling. This requires that the microorganisms exist not as singly dispersed cells in the liquid but in clumps, aggregates, or flocculated particles with a density sufficiently greater than that of the bulk liquid so that they subside either as individual floc particles or as a blanket or matrix of increasing bulk density due to aggregation of the floc as the total mass settles. On fixed or rotating biological contactors, the clumps of cells that are subject to settling are usually bits of biological slimes that were formerly attached to the reactor surfaces and have broken off due to a combination of their own weight and the hydraulic forces tending to detach and flush them out of the reactor. In activated sludge processes, the clumps or flocs form largely because individual cells have come into contact and coalesced into large aggregates containing billions of cells of different species of bacteria, fungi, and protozoa. If we can think of the microbial cell as the "littlest bioreactor," the floc particle is the "littlest ecological system" or the next to the littlest bioreactor.

Under usual operational conditions, flocculation of cells inevitably occurs in an activated sludge process; unfortunately, it is the unreliability of the inevitable that causes concern. It is clear that the economics of accomplishing the aims of the public mandate to provide adequate wastewater treatment (e.g., Pub. L. 92–500) is completely dependent on a reliable means of separating the cells from the liquid, and it should be of great concern to environmental technologists that so little is known about the mechanism(s) that cause flocculation and permit quiescent settling to be used. Flocculation is seldom observed during batch growth studies when pure cultures are used. Flocculation is sometimes, but not often, observed during batch growth studies with heterogeneous microbial seeds when the seed is obtained from a well-flocculated activated sludge. In general, however, when microbial cells are growing rather rapidly, they are dispersed. Flocculation tends to occur in heterogeneous populations when the ecosystem matures and there are limited growth resources. It can also occur in a pure culture system, and when it does, it usually does not happen until the growth resource, i.e., the carbon and energy source, is severely limited or exhausted. The relative old age or maturity of the system seems to be the common condition for autoaggregation whether in heterogeneous or pure culture populations. For one reason or for many, heterogeneous cultures found in nature tend to aggregate into a biological community.

In activated sludges, the natural community is the floc particle. We use the flocculated condition to advantage even though we cannot say why or how it really comes about. We have seen that if the biochemical ability of microorganisms were the only consideration, we could use much higher specific growth rates in wastewater treatment reactors (see Chap. 6). There are various reasons why very low specific growth rates are used, and prominent among them is the fact that settleable floc particles generally do not develop unless values of μ are very low. Another prominent and related reason for the use of low values of specific growth rate is that the observed yield, i.e., the excess sludge, which represents a waste disposal problem in the treatment process, is minimized by growth of the cells at low values of μ. As we have already seen, low values of μ also provide low values of S_e. Thus, from the standpoint of flocculation and settleability as well as removal of soluble substrate, the slow growth characteristic of the mature ecosystem is advantageous. The floc particle, consisting of bacteria and fungi as primary feeders with grazing protozoa, crustacea, and perhaps other predators, is an ecosystem with some semblance of balance and completeness and is considered to be a healthy natural one, whether it occurs in a biotreatment reactor or in a healthy receiving stream.

A general knowledge of the conditions leading to flocculation and those needed to maintain it, i.e., low specific growth rates or low flux of primary carbon and energy source, is useful but it does not give us a scientific basis for understanding the mechanisms by which flocculation comes about and is maintained. This information is needed to provide design and operational parameters and to suggest remedial measures to ensure floc building.

Early attempts at understanding the process of flocculation consisted of making exhaustive microscopic observations of the particles, followed by isolation and separate study of the predominating species. Studies of this type led Butterfield and his coworkers (1931) to conclude that flocs were formed because certain organisms—in fact, one in particular—capable of producing a copious matrix of capsular slime predominated in activated sludge. One organism, the pseudomonad *Zoogloea ramigera*, proliferated in such a gelatinous matrix; as the cells grew, the slime matrix grew also, presumably because the cohesive forces between polymer molecules were greater than the hydraulic mixing forces tending to shear the floc. Other organisms that ordinarily grew in dispersion as well as bits of nonmicrobial organic matter became entrapped in the zoogloeal mass. This gelatinous mass also provided a grazing area for predators and a surface for attachment of stalked forms, thus completing the microcosm. All flocs examined appeared to contain the zoogloeal masses, and the conclusion was that flocculation occurred because of the presence of this particular species. It became accepted that this microorganism was the cause of flocculation. McKinney and Horwood (1952) demonstrated that flocculation was not dependent on the presence of *Z. ramigera* by showing that various species found in sewage also were capable of clumping or forming floc. In general, floc formation occurred when little carbon and energy source remained in the system. Thus, flocculation was

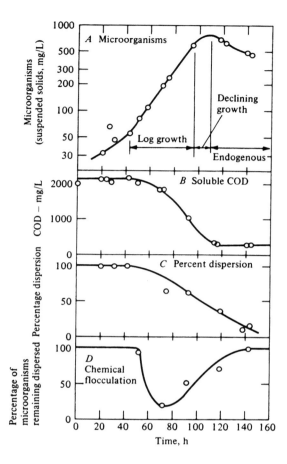

Figure 10-13 Variations of (*a*) concentration of biomass, (*b*) soluble COD, (*c*) biological flocculatability, and (*d*) chemical flocculatability with time in batch studies. (*Adapted from Tenney and Stumm, 1965.*)

envisioned as a more general phenomenon, controlled more by environmental conditions than by the presence of a particular species of microorganism. The development of an aggregated biomass in relation to growth and depletion of the exogenous carbon source is demonstrated in Fig. 10-13.

While the overriding general condition for enhancing flocculation in pure cultures and heterogeneous populations is apparently the age or maturation of the population, this information by itself tells us little about what causes the cells to aggregate. The concepts of colloidal chemistry provide a possible explanation. Bacterial cells are of colloidal size, and they usually carry a negative charge except at very acid pH values; that is, they migrate toward the positive pole (anode) in an electric field. Like any negatively charged colloid, upon charge neutralization they may precipitate on a solid surface or on another cell, thus forming aggregates or flocs. If the difference in electric potential between the surrounding medium and the migrating bacterial colloid (the zeta-potential) is lowered to approximately 15 mV, cells may coalesce;

i.e., the surface charge does not need to be reduced to 0 for flocculation to occur. On the other hand, cells can exhibit a 0 zeta-potential and form stable (nonaggregated) suspensions; hence, one cannot attribute flocculation to simply charge neutralization.

McKinney felt that cellular energy was required to maintain the cell in a stable suspension. Crabtree et al. (1966) concluded that the synthesis of the microbial storage product poly-β-hydroxybutyric acid was the cause of microbial flocculation, but others have isolated floc-forming species from activated sludge that do not produce poly-β-hydroxybutyrate (Angelbech and Kirsch, 1969). Although some bacteria that do not produce any extracellular capsular slimes have been observed to flocculate, the presence of some cohesive or binding material has been observed so often in flocculating populations that it should be accorded a prominent role if not considered the sole mechanism by which floc is made and held together. Some pseudomonad strains have been shown to be held in a flocculated state by intertwined fibrils (flagella-like protrusions), which can be dissolved and the cells dispersed upon contact with the enzyme cellulase (Deinema and Zevenhuizen, 1971). Tago and Aida (1977) have recently shown that a floc-forming bacterium isolated from activated sludge is held in aggregation by a mucopolysaccharide. Deflocculation can be brought about only by the action of a hydrolyzing enzyme, produced by the organism, that is specific for the mucopolysaccharide. Figure 10-14 shows the cells embedded in a film mesh. Deflocculated cells are devoid of this mesh.

The fact that the addition of various commercial polymers (polyelectrolytes) as flocculation aids often results in aggregation also suggests that autoflocculation may be generated by naturally produced polyelectrolytes, e.g., polysaccharides. Thus, while it has been shown that Z. ramigera is not the sole cause of floc formation, current thinking on the subject accords a major role to biologically produced polymers in the form of capsules, fiber-like strands, or film meshes. More generally, there is a growing body of information indicating that in natural, i.e., mixed, populations, as compared with pure cultures of single species, many types of cells produce a matted mass of polysaccharide fibers extending outward from the membrane. This glycocalyx provides the effective surface of the cell and permits it to attach to specific surfaces and/or to bind to the glycocalyx of other cells. Since the glycocalyx is generally found in mixed rather than pure cultures, the assumption is that it is produced at the expenditure of carbon and energy as a protective mechanism and that its presence offers the cell a competitive advantage in a natural environment (see Costerton et al., 1978). The common occurrence of a polysaccharide glycocalyx has only recently been postulated, and much more study of this phenomenon is needed. It is important to emphasize that these and other recent findings demonstrate that it is not sufficient to study microbial species in isolated pure culture. Such studies are important and necessary, but they must be supplemented by studies of mixtures of populations, since pure culture studies cannot detect

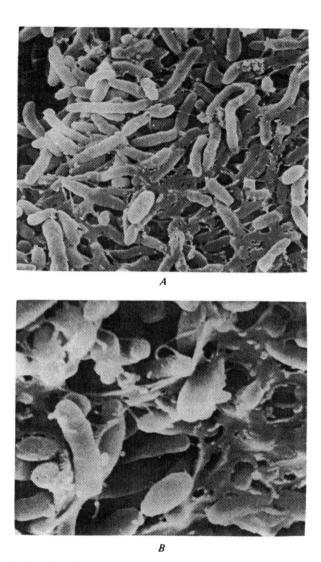

A

B

Figure 10-14 Scanning electron micrograph of a floc of bacteria isolated from activated sludge. Cells are embedded in a film mesh; *A,* 5000*X*; *B,* 12,5000*X.* (*Courtesy of Dr. Y. Tago, University of Tokyo.*)

modifications of metabolic behavior and capabilities of the species in communal situations, such as the flocculated community represented by an activated sludge particle.

The filamentous bacteria and fungi can play a useful role in the construction of the floc particle. An overabundance of such organisms can, however, result in the development of a very loose and large floc particle that is of such bulk and "fluffiness" that it will not subside. Thus, upon quiescent standing, the suspension will not separate into sludge and supernatant layers. This condition, known as a *bulking sludge*, causes biomass to escape in the overflow or effluent from a clarifier. Bulking sludge can be caused by the presence of excessive amounts of extracellular polysaccharide, which prevent the cells from forming a tight floc, as well as by an overabundance of filamentous forms. It could also be brought about by biological attack on film meshes and capsular slimes that bind cells together. There are various possible reasons for the loss of flocculating and settling characteristics; however, excessive numbers of filamentous organisms is probably the most frequent cause.

Early studies in our own laboratories, using commercial polyelectrolytes as flocculating agents for dispersed heterogeneous microbial populations, indicated that rapid flocculation can be made to occur by the addition of such polyelectrolytes. Invariably, the flocculated cells removed the carbon and energy source at a slower rate than did the dispersed suspensions. It could not be determined whether the slower removal rate was due to any metabolic inhibition caused by the polymers or to the lower degree of accessibility of carbon and energy sources to the tightly flocculated biomass (T. Yu and A. F. Gaudy Jr., unpublished results). In other experiments, the addition of each of five extracellular capsular slimes, harvested from bacteria that produced copious capsules, to stable suspensions of heterogeneous microbial populations shortly after the period of exponential growth did not result in any flocculation of the cells (J. Kelley and A. F. Gaudy Jr., unpublished results). Thus, on the basis of admittedly preliminary experimentation, one gains the impression that if biologically produced extracellular polymers are significant causes of flocculation, they are probably rather selective in their action in regard to the type of cells and stage of growth. It now seems certain that no one polymer-producing species is responsible for flocculation, but that many species may play a role in the process. At any given time, one or two species in the system may be largely responsible for floc formation, but these may change as environmental conditions in the process change. The degree of autoflocculation will change from time to time, and the usefulness of a specific remedial measure may change with a shift in the predominance of species.

It is unfortunate that so little is known about the mechanism of autoflocculation. In all probability there is no one overriding means by which it is brought about in heterogeneous populations. The microcosm that the floc particle represents is so complicated that study is indeed difficult. However, greater understanding will surely be required if quiescent settling is to remain the operation of choice for separation. Some of the pressures to study this phenomenon, but certainly none of the need, will be relieved as chemically assisted bioflocculation is used more commonly. The addition of ferric salts and/or alum, with or without polyelectrolyte coagulant aids, during times of inadequate autoflocculation is a useful remedial measure that is now coming into greater use. The formation of chemical flocs such as are produced by the addition of ferric salts and alum is also a useful measure to combat periodic bulking. Often both nonflocculating and bulking systems can be corrected gradually by adjusting the pH and gaining closer control of fluctuations in the concentration and type of incoming substrate S_i. Chemical assistance in the separation during the period required to effect the needed shifts in microbial species may in many cases be the only way a plant can maintain functional operation during the correction period.

Autoflocculation of the biomass forms a bridge between the unit processes of microbial growth in the bioreactor and gravity sedimentation in the clarifier. Although not formally recognized as such, microbial flocculation is in itself a unit process that usually accompanies slow growth in the bioreactor and permits the clarifier to perform its function as a subsidence basin. Not enough is known about the mechanisms of autoflocculation and the means of controlling them to allow us to devise a set of design and operational principles as is done for other unit processes. Microbial flocculation can be controlled to some degree by engineering expedients consisting of add-ons, such as dosing with chemical coagulants and commercial polymers. The use of these aids to flocculation represents a significant increase in costs, and continued study of autoflocculating mechanisms could help eliminate the need for additives.

OXIDATIVE ASSIMILATION—EXTREME UNBALANCED "GROWTH"

As we pointed out, when microbial growth is discussed it is generally assumed that all of the nutrients required for cell replication are present. However, it is possible to remove substrate when this is not the case. The term *oxidative assimilation* pertains to the uptake of organic matter from solution and its use for the synthesis of some cellular constituents under conditions in which cell replication is not possible or is severely retarded. Under such conditions, growth, as measured by an increase in the biomass concentration, is significant and, as we shall see, may proceed to the same extent and at approximately the same rate as growth in a system that is not severely limited in replication. Most experiments conducted on oxidative assimilation have involved the use of so-called resting cells. This simply means that the nitrogen source has been withheld or an inhibitor of protein and/or nucleic acid synthesis has been added. The term *nonproliferating systems* is equally appropriate. Thus, the only major carbonaceous synthesis that is possible in systems in which protein and nucleic acid synthesis is blocked is that of carbohydrates or lipids.

Oxidative assimilation was discovered when experimenters working with nonproliferating pure cultures in defined media could not account for the utilization of the added organic compound by carbon dioxide production alone. That is, the added organic compound disappeared from solution but was not totally oxidized. Some synthesis of cellular materials (oxidative assimilation) was indicated even though growth, i.e., replication, was impossible. In those early studies, removal of organic matter was not measured by a general test for total organic matter such as the COD test but by a specific test for the compound added. It was, of course, possible that cells could change the original compound so that it would not be detected without removing a very large portion of the organic matter it contained. However, sufficient evidence has been amassed to show conclusively that organic matter can be oxidatively assimilated into cell components. This is tantamount to

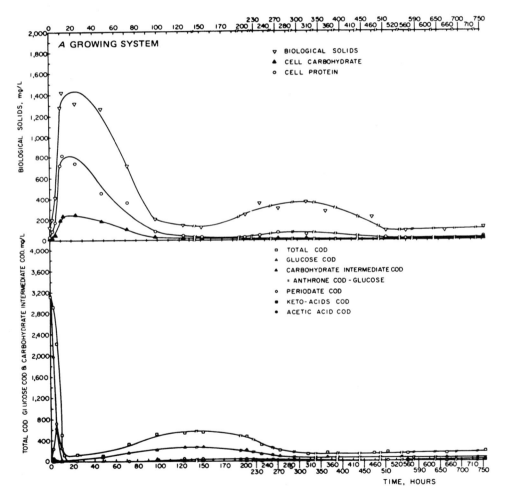

Figure 10-15 Comparison of biomass accumulation and endogenous phases under (A) growing and (B) nonproliferating conditions for initial $S_0/X_0 = 32$. (*From J. J. Su and A. F. Gaudy, unpublished data.*)

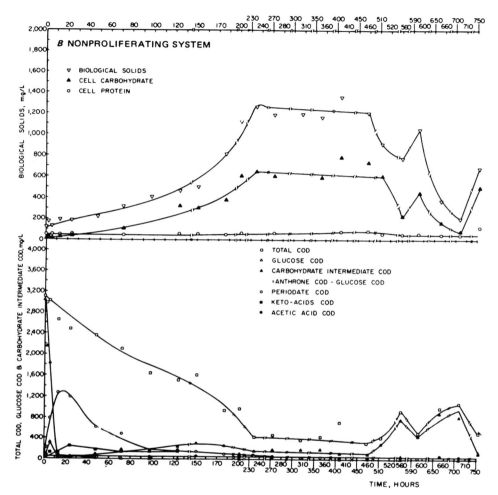

Figure 10-15 (Contd).

saying that it may be possible to purify wastewaters with no cell replication taking place during the purification process.

With this interest in mind we may ask a few questions. How general is the phenomenon? Does it occur in heterogeneous microbial populations or in only a few species of microorganisms and with a few substrates? Is there any relationship between the concentration of biomass present and the oxidative assimilation of the substrate? It would seem safe to surmise that the more resting cells one started with, the faster the rate of substrate uptake would be into nonnitrogenous materials. One might also wonder whether, if the S_i/X ratio were high, most of the substrate might be removed for nonnitrogenous syntheses even with a relatively low concentration of cells; that is, what are the extent and limits of capacity for the expression of this biochemical phenomenon? Furthermore, other than the obvious applied interest previously stated, can this phenomenon give us further insight into understanding the biological treatment processes?

Answers to some of these questions are found in parts B (nonproliferating systems) of Figs. 10-15, 10-16, and 10-17, which show the course of biomass accumulation and substrate removal as well as information on biomass composition. Glucose was used as substrate in the·experiment shown in Fig. 10-15; glycerol was used in those in Figs. 10-16 and 10-17. In all of the nonproliferating systems, the substrate was removed and significant portions were assimilated into the biomass. Essentially none of the substrate was channeled into the net synthesis of cellular protein, but much was channeled into the synthesis of cellular carbohydrates. The rates of biomass accumulation and substrate removal were much affected by the ratio of biomass X_0 to substrate S_0. It should be clear from these experiments, especially from the one shown in Fig. 10-15, that oxidative assimilation is not observed only at high biomass concentrations. The oxidative assimilation capacity of microorganisms is extraordinarily high: for example, in Fig. 10-15, slightly more than 100 mg/L of cells oxidatively assimilated 3200 mg/L of glucose COD. That is, for

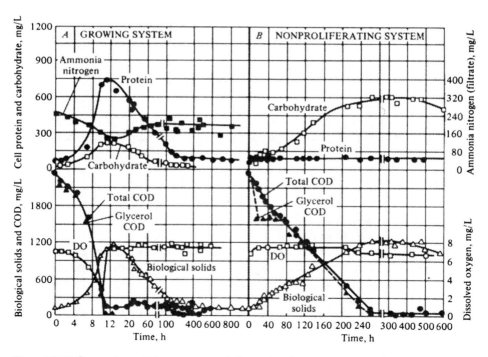

Figure 10-16 Comparison of biomass accumulation and endogenous phases under (A) growing and (B) nonproliferating conditions for initial $S_0/X_0 = 21$. (*From Thabaraj and Gaudy, 1971.*)

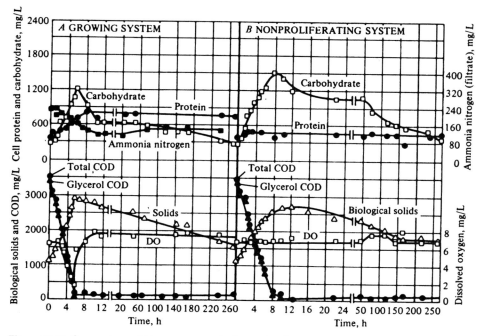

Figure 10-17 Comparison of biomass accumulation and endogenous phases under (*A*) growing and (*B*) nonproliferating conditions for initial $S_0/X_0 = 3$. (*From Thabaraj and Gaudy, 1971.*)

cells capable of producing a nonnitrogenous store (oxidative assimilation) from a particular substrate, it is not the amount of assimilation that is limited by biomass concentration but its rate. It is important to note in this long-term experiment that after the exogenous COD was removed, the carbohydrate that had been synthesized by the cells was available as a substrate. In the "endogenous" phase, the biomass concentration and its carbohydrate content decreased in cycles of increase and decrease accompanied by complementary cycles in soluble COD, as material inside the cells was released and reused in subsequent assimilation. The term *autodigestion* might be better used than *endogenous respiration*, since the downward cycling of solids was accompanied by the release and uptake of soluble exogenous COD. After the time period shown in the figure, a small amount of ammonium ion was introduced and shortly thereafter growth was initiated, using the residual COD and reducing it to a very low value.

As to the growing systems shown in these figures (parts *A*), we shall look at them in descending order of S_0/X_0. In Fig. 10-15*A*, we see that even at the very high S_0/X_0 ratio of 32/1, the organic matter was removed from solution very rapidly in approximately 12 h, compared with several hundred hours in part *B*. Protein and carbohydrate content rose together, along with the biomass concentration. Clearly, the substrate was being removed because the cells were proliferating, in contrast to the events in part *B* where the biomass

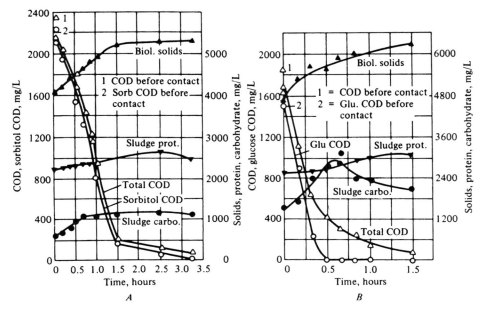

Figure 10-18 (A) Metabolism of sorbitol under growing conditions for initial $S_0/X_0 = 0.6$. (B) Metabolism of glucose under growing conditions for initial $S_0/X_0 = 0.4$. (*From Krishnan and Gaudy, 1968.*)

growth was due to nonnitrogenous syntheses. There are indeed tremendous and obvious differences in the mechanism and rate of substrate removal and sludge growth in the growing and nonproliferating systems, even though the end result in both is sludge accumulation and substrate removal. As we move down the scale to the 21/1 ratio (see Fig. 10-16A), a response much like that at the 32/1 ratio is observed. One can conclude that there are essentially the same differences in mechanism and kinetics in parts A and B for both experiments during the substrate removal phase. However, at the 3/1 ratio (Fig. 10-17), the behavior of both systems obviously changes.

At a 3/1 ratio of S_0/X_0, protein synthesis in the growth system is outpaced by carbohydrate synthesis. In fact, protein continues to be synthesized after substrate has been removed, and the synthesis of this nitrogenous compound, which is indicative of replicative growth, goes on at the expense of the stored carbohydrate. Thus, as we move toward lower S_0/X_0 ratios in the growth system, the mechanism that is most prevalent during the substrate removal phase involves the synthesis of nonnitrogenous cellular materials; that is, it involves oxidative assimilation. The system with nitrogen initially behaves much like the system without nitrogen. Furthermore, at even lower S_0/X_0 ratios, both the kinetics and mechanism of substrate removal and sludge production become the same for both types of system. They both remove substrate at approximately the same rate by oxidative assimilation rather than by protein synthesis. When the S_0/X_0 ratio is less than 1.0, substrate uptake can be extremely rapid, and during the time required to remove the substrate from solution, most of the synthesis involves nonnitrogenous compounds. For example, in Fig. 10-18A, for which the S_0/X_0 ratio was well below 1.0, more than 90 percent of the carbon source and the COD due to it were removed in 1.5 h, and more carbohydrate than protein synthesis occurred. The same occurs in the glucose-fed system shown in part B of the figure.

The Biosorption Process

When one considers that in an activated sludge process biomass concentrations of 2000 to 2500 mg/L have been traditionally used, and the substrate concentrations in many cases are in the range of 200 to 300 mg/L ($S_i/X = 0.1$), it can be assumed that substrate can be oxidatively assimilated very rapidly. If one were to increase X to the range of 5000 to 7000 mg/L, the removal of soluble substrate could take place in several minutes. There should be little wonder then why, in the modification of the activated sludge process often referred to as the *contact stabilization* or *biosorption* process, shown in the flow sheet of Fig. 10-19, a mean hydraulic retention time $\bar{t}$ of approximately 30 to 45 min is sufficient for substrate uptake or purification of the waste. This process scheme was devised three decades ago, when information on the oxidative assimilation of soluble organic compounds by heterogeneous populations such as that shown in Figs. 10-15 to 10-18 was not available. Also, at the time this flow sheet was envisioned, the wastewater used in the experimental operational runs contained considerable amounts of suspended organic solids. The removal of the substrate, measured on the basis of its biochemical

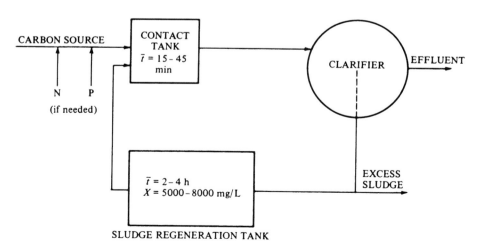

Figure 10-19 Flow diagram for contact stabilization or biosorption activated sludge process.

oxygen demand, was very rapid, and there was reason to hypothesize, based on observations, that the rapid decrease in BOD was due to the adsorption of the oxygen-demanding organic matter onto the flocculated biomass particles. However, when the process was used with some success on soluble industrial wastes that did not contain adsorbable molecules, there was reason to doubt that adsorption was a major mechanism of purification for this process. These doubts were reflected in the tendency to replace the word *adsorption* by the word *sorption*. Nonetheless, the idea persisted that the mechanism of purification in the biosorption process was adsorption, and it was thought that the real purpose or function of the second aeration tank (Fig. 10-19) was for biooxidation, i.e., "burning off" of the "sorbed" material, to regenerate the ability of the sludge to transfer soluble molecules onto or into the biomass. Thus, the sludge was regenerated so that it could be used for another pass through the contact tank. However, the mechanism of substrate removal for soluble organics is more properly described as their oxidative assimilation into nonnitrogenous storage products. And, as demonstrated in Figs. 10-17 and 10-18, for example, when S_0/X_0 is low, these storage products are synthesized and then used for the production of cellular protein characteristic of cellular growth. It should be noted that neither sorbitol nor glycerol reacts in the anthrone test for carbohydrate, which was used to measure the carbohydrate content of the cells. Therefore, the observed increase in the amount of cellular carbohydrate could not have been due to sorption, or simple uptake, of these substrates. Rather, they were metabolized and converted to material that was carbohydrate in nature.

One may reason with validity that even in the absence of the conversion of some of these carbohydrate stores to protein, endogenous respiration (e.g., in the second aeration tank of Fig. 10-19) would burn off (i.e., respire) some of the intracellular storage products. However, it has been shown rather conclusively that the capacity of the biomass to take up more substrate upon recycling is rapidly and progressively diminished unless a significant portion of the stores is converted to protein in the endogenous or regeneration phase (Komolrit et al., 1967). Unfortunately, this regeneration or rejuvenation of oxidative assimilation capacity is seriously hampered in the standard contact stabilization process because the nitrogen source, which is usually ammonium ion, has exited in the effluent and is thus no longer available to participate in the conversion of storage products to protein when a portion of the clarifier underflow is channeled to the regeneration tank. Clearly, the short contact time and high solids content put the contact stabilization process at a distinct biochemical disadvantage. This probably has had much to do with its limited use. One can imagine the chagrin of the industrial manager who is producing a nitrogen-deficient waste and finds not only that much of the ammonium ion he has paid to add to treat the waste has not provided reliably good treatment but also that its addition has made him a nitrogen polluter because the treatment process was designed, built, and operated on the basis of an erroneous concept of the biochemical mechanism of the process. We do not mean to say that adsorption cannot play a role in the purification process, whether it is a high-solids system or not. After all, some wastes—particularly municipal wastes—may contain as much or more colloidal organic material as soluble compounds. Adsorption and entrapment of particulates and highly charged molecules in the biomass matrix undoubtedly play a role in these situations. But the process cannot be perpetuated without protein synthesis somewhere in the flow scheme; i.e., one cannot continue to purify waste without producing new cells. If some of the adsorbed colloidal material contains a source of organic nitrogen, for example, proteins, these may be utilized in the regeneration tank to the extent that they are present in the underflow sludge. Thus, for municipal wastes the process may provide satisfactory results, although not for the reasons formerly believed.

Treatment of Wastes by Continuous Oxidative Assimilation

While an understanding of oxidative assimilation and the effect of the ratio S_i/X on its role in substrate removal aids us in the mechanistic analysis of treatment processes, it also poses a challenge for us to use the knowledge to improve the efficiency of treatment, and it suggests an interesting possibility. If, at high S_i/X ratios, wastewater purification takes place first by uptake and conversion to an internal carbon store followed by the use of this store for production of nitrogenous compounds needed for replicative growth, is it possible to separate the processes physically? This is essentially what the system tried to do in the contact stabilization process, but it cannot accomplish this because of the lack of a nitrogen source where it is most needed. One possible solution would be to add the nitrogen to the regeneration tank rather than to the contact tank. The retention times, i.e., tank volumes, would not necessarily be those used in the biosorption process. The $\bar{t}$ for the first tank would be governed by the time requirements for oxidative assimilation of the particular waste in question, the waste COD concentration S_i, and the amount of biomass X one chose to use. The $\bar{t}$ for the second aeration tank would depend on the time required to convert a sufficient amount of the stored carbon to protein to regenerate the rapid substrate uptake capacity of the biomass.

Could such a process work on a continuous flow basis? Which hydraulic regime, plug or completely mixed flow, would be usable? Laboratory pilot plant investigations have provided some answers and insights into these questions and modes of operation, and they suggest some design considerations. The flow diagram and pilot plant shown in Fig. 10-20A and B have been used to treat continuously synthetic waste with such carbon sources as glucose and acetic acid and an industrial waste from a sugar refinery (Gaudy, Goel, and Gaudy, 1968, 1969; Gaudy, Goel, and Freedman, 1969). In these experiments, the biomass regeneration operation was carried out in a batch or plug flow manner. In the pilot plant diagram (Fig. 10-20B), protein synthesis is carried out in one reactor and a second is used as a biomass hopper or dosing

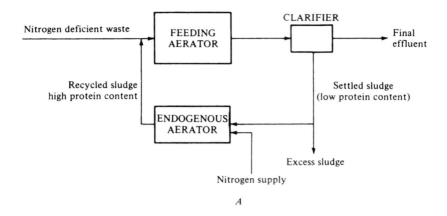

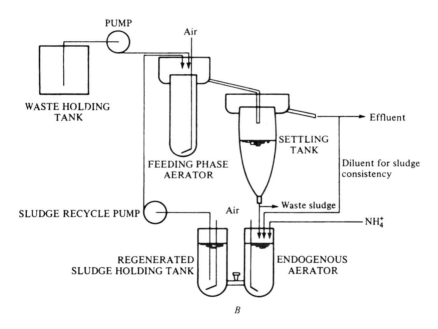

Figure 10-20 (*A*) Process modification for the treatment of nitrogen-deficient wastes by activated sludge. (*B*) Laboratory scale pilot plant. (*From Gaudy, et al., 1968.*)

tank from which the regenerated biomass is pumped continuously into the contact tank. A sample of the system performance obtained is shown in Fig. 10-21. In this case, the carbon source was acetate at $S_i = 1000$ mg/L. It was possible to run this system in a pseudo-steady condition and obtain rather good purification. This capacity for unbalanced synthesis may permit some savings in chemical costs when handling wastes that need addition of

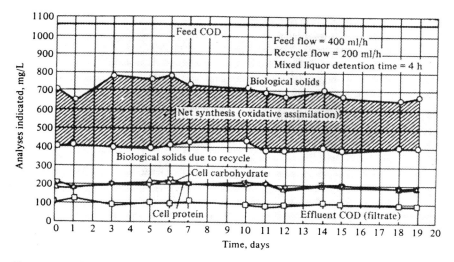

Figure 10-21 Continuous oxidative assimilation of acetic acid in the feeding aerator with biological solids previously subjected to endogenous respiration in the presence of ammonia-nitrogen (COD/N = 70/1). (*From Gaudy, et al., 1968.*)

nutrients before treatment; for example, a great variety of industrial wastes are deficient in nitrogen and phosphorus. The COD/N ratio of 70/1 used in the system shown in Fig. 10-21 is rather high compared with values usually considered to be ideal for "healthful" nitrogen content of the biomass (i.e., COD/N = 25/1 to 20/1), but the process achieved more than 90 percent removal of the substrate COD.

The value of these results lies not so much in describing a new treatment process but in pointing out the importance of understanding the biochemical mechanisms that produce the observed kinetics. Much developmental work is required before the total separation of oxidative assimilation and replicative synthesis can be put into practice in the field. Also, individual wastewater characteristics and the conditions under which the wastes are produced will determine whether it is desirable to make such a separation in any specific case. The point to be emphasized is that when a treatment process is designed, certain physical constraints are placed on the system; e.g., $\bar{t}$ is fixed in relation to F. The operational strategies used when the plant is functioning create environmental conditions that affect the microorganisms responsible for purification. We have just seen that the general design and conditions of operation are such that contact stabilization plants are almost certain to make inefficient use of the nitrogen source. Changes in the design and operation of this process should improve performance, including changing $\bar{t}$ for the contact and regeneration reactors and adding at least some of the nitrogen supplementation (for wastes that require it) to the sludge in the regeneration tank rather than adding it all to the mixed liquor in the contact tank.

Growth under Nitrogen Limitation

Neither the continuous oxidative assimilation process nor the contact stabilization process was considered in Chap. 6 because the kinetic model equations discussed there refer to balanced replicative growth (carbon-limited), even though the growth is measured in mass rather than numbers of cells. Various types of kinetic models might be written to describe these special cases, but demonstration of their validity would require much experimental investigation. Also, it is yet to be determined whether the current kinetic models for balanced growth would more accurately describe events if they included terms for oxidative assimilation, especially in view of the fact that most processes for the biological treatment of organic wastes operate at rather low S_i/X ratios. For municipal wastes, the growth-limiting factor is invariably the carbon source. When treating industrial wastes, it is necessary, if one wishes to use the same kinetic model, to add sufficient nitrogen and phosphorus so that they, rather than carbon, are present in nonlimiting concentrations. One can estimate the amounts of nitrogen and phosphorus needed in relation to carbon on the basis of either the average chemical composition of microorganisms or experimentation; that is, for a certain amount of carbon, how much nitrogen and phosphorus are needed, experimentally, in order to give good purification? If the BOD test is used as a measure of the amount of carbon source, a BOD/N ratio of approximately 20/1 and a BOD/P ratio of 100/1 are usually recommended, although these values have been the subject of some controversy in past years. One may wonder how a system might behave under conditions in which carbon source was not the limiting nutrient. In the discussion that follows, the effect of the nitrogen source on the growth of the heterogeneous biomass is considered.

We have seen that by using cell recycle and adding the nitrogen source only to the biomass rather than to the inflowing feed, we have made use of a totally unbalanced growth situation and were able to operate under steady state conditions in the contact aerator even though growth, as measured by cell replication, does not occur until the biomass has been separated from the reaction liquor. In Chap. 6, when dealing with the Monod relationship, it was stated that the term S really should stand for the concentration of any growth-limiting nutrient, although we invariably added the other major nutrients, nitrogen and phosphorus, in sufficient quantities to make carbon the limiting nutrient. With regard to nitrogen supplementation, for heterogeneous populations a ratio of substrate (COD) concentration to nitrogen concentration (with nitrogen added as ammonium ion) of 25/1 to 20/1 is needed to ensure that carbon is the nutrient that limits μ in balanced growth. This has been shown in batch growth studies similar to those previously described for carbon source but with varying concentrations of nitrogen source (Goel and Gaudy, 1969a). As one would expect, the μ_{max} values one obtains are essentially the same as when carbon is the limiting nutrient, since one approaches the same nonlimiting case in either study. The K_s values for ammonia-nitrogen are much smaller than those for carbon because less nitrogen is needed to produce a cell. Since the same ratio is obtained in such growth studies and in pilot plant studies to determine the best ratios of carbon source to nitrogen source for efficient wastewater treatment, we can have considerable faith in the usefulness of this numerical factor. At this ratio or lower ones, we can be fairly confident of our assumption that carbon is the limiting nutrient. For most municipal wastes, the actual BOD/N ratio is much less than 20/1; there is more nitrogen source than there is carbon source to balance it. However, for most industrial wastes, the ratio is considerably greater than 25/1, and nitrogen is traditionally added to the incoming waste in order to satisfy this ratio.

At this point we may ask a question in regard to balanced growth: Is all of the nitrogen we add to balance the carbon source taken up by the cells, or by maintaining the 25/1 to 20/1 ratio, do we actually add some nitrogen in excess of that which can be utilized merely to ensure that carbon is the limiting nutrient? Does some of the nitrogen leak into the effluent? Perhaps balanced growth is not really needed, since we have already seen that we can remove substrate without the synthesis of nitrogenous components. Do we really need to provide enough nitrogen so that all of the cells can replicate or just those we wish to recycle to the activated sludge tank? The results shown in Fig. 10-21 indicate that we certainly do not require so much nitrogen, since the COD/N ratio was 70/1. However, this was a special case in which we engineered the system for sequential assimilation and replication. What would happen if, for example, a once-through growth system similar to that which produced the results shown in Fig. 6-26 with excess nitrogen source (COD/N = 10/1) were operated under conditions where nitrogen rather than carbon was the limiting nutrient? The results of such an experiment shown in Fig. 10-22, in which nitrogen was added with the feed COD at a ratio of 70/1, indicate that at the dilution rate used $(1/12\,h^{-1})$ rather good COD removal could be achieved (86 percent). Also, it is apparent that a pseudo-steady state was achieved in X, S_e, and the carbohydrate and protein concentrations in the biomass. To appreciate the difference between the steady states developed in the two systems, we must consider the biochemical compositions of the biomasses in the two reactors. In the system shown in Fig. 10-22, the average protein and carbohydrate contents were 12 and 46 percent, whereas under the balanced growth conditions of Fig. 6-26 they were, on the average, 50 and 29 percent, respectively (these were not shown on the figure). Much of the higher carbohydrate content (shown in Fig. 10-22) was concluded to be in extracellular polysaccharide rather than in intracellular carbohydrate stores, because many of the cells in the biomass were heavily encapsulated. This was not the case for the cells of Fig. 6-26 or 10-21. The subsequent fate of the capsular material was not investigated in this particular study, but the biochemical fate of this type of polysaccharide will be discussed later.

The point to be made here is that the soluble substrate was removed biochemically from solution, and part of it was converted to suspended solids

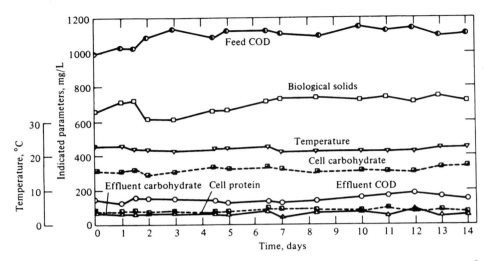

Figure 10-22 Steady state parameters in a once-through reactor during operation at $D = 0.083$ h^{-1} and COD/N = 70/1. (*From Goel and Gaudy, 1969b.*)

as a constituent of the biomass. However, when the system was run at higher dilution rates, increasing amounts of substrate leaked from the reactor. This would be expected as normal dilute-out behavior (see Figs. 6-25, 6-27, and 6-28), but the dilute-out was more severe than that found in subsequent studies at a COD/N ratio of 40/1, and this latter dilute-out was more severe than that at a COD/N ratio of 25/1. There was little difference between the dilute-out behaviors of systems run at COD/N ratios of 25/1 and 10/1. Thus, the growth rate ($\mu = D = 1/\bar{t}$ for once-through systems) has much to do with the effective utilization of the substrate whether carbon or nitrogen is the growth-limiting nutrient. One gets the impression that by running systems at low growth rates, the COD/N ratios can be significantly lower than 25/1 to 20/1. In general, this is a fairly good assumption. Depending on engineering manipulation of μ [see Eqs. (6-72) and (6-75)], it may be possible to slow μ to such an extent that nearly all the nitrogen in the system can be recycled, or reused (see the discussion of total oxidation later in this chapter). The interrelatedness of the carbon and nitrogen content of the feed as well as the effect of the growth rate on the removal of soluble substrate can be seen in Fig. 10-23. The point plotted at the lower left-hand corner is the average effluent COD concentration for the results shown in Fig. 10-22. The other plotting points were obtained from similar experiments. At the 70/1 ratio, the rapid deterioration of effluent quality as μ was increased is apparent, and the deterioration was attenuated at 40/1 and 25/1. Thus, slowing the growth rate exerts a sparing action on the amount of nitrogen supplementation required.

Another aspect of practical metabolic significance is the effectiveness of utilization of the added nitrogen. The need to rid effluents of excess nitrogen

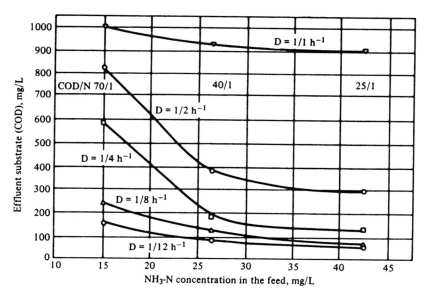

Figure 10-23 Relation between effluent COD and ammonia-nitrogen content in the feed at various dilution rates and COD/N feed ratios of 70/1, 40/1, and 25/1. (*From Goel and Gaudy, 1969b.*)

in reduced form, e.g., NH_4^+, makes it important to consider the efficiency of biological processes in taking up nitrogen, and the cost of adding nitrogen to wastes that are deficient in this nutrient is an additional economic consideration. Figure 10-24 shows the effluent quality with respect to ammonia-nitrogen at the various specific growth rates and COD/N ratios used in the study. Comparing this figure with Fig. 10-23, it is seen that for any particular specific growth rate, whereas the effluent quality with respect to removal of the carbon source increases as COD/N decreases, the quality of effluent with respect to ammonia–nitrogen, and thus the efficiency of nitrogen uptake, decreases at lower COD/N ratios. Also, at the 25/1 ratio, effluent quality with respect to COD removal increases with decreasing μ, whereas the effluent quality with respect to removal of ammonia–nitrogen decreases as μ decreases. In designing the nutrient mix and selecting growth conditions for treatment, forward-looking functional design criteria should take into account and balance the removal efficiency of the carbon source against the efficiency of utilization of the supplemental nitrogen source. For systems with excess nitrogen source one can allow nitrification to proceed by providing for very low specific growth rates, thus exerting the biochemical oxygen demand of the ammonia in the aeration tank. According to Eq. (6-75), the growth rate may be lowered by manipulating α, X_R, and D (i.e., $1/\bar{t}$). It is not always possible to lower the specific growth rate sufficiently by manipulating α and X_R alone, however; an increase in $\bar{t}$ may be required to assure nitrification for a given waste flow F. This requires a larger reaction tank, which is a

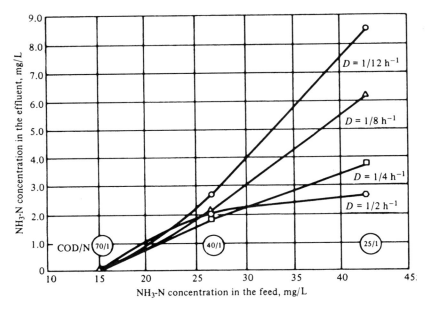

Figure 10-24 Effect of nitrogen level on leakage of nitrogen into the effluent at various dilution rates. (*From Goel and Gaudy, 1969b.*)

significant economic factor and one that should be considered in the design stage when the reactor volume is determined. For a waste originally deficient in nitrogen, it is entirely inappropriate to add excess ammonia–nitrogen and then increase the volume of the reactor to oxidize the excess. It seems far better to explore the possibility of altering the C/N ratio to achieve the best COD removal possible without leaking excessive amounts of the supplemental ammonia–nitrogen.

STORAGE PRODUCTS

In the process of oxidative assimilation, other intracellular storage products can be formed in addition to or instead of carbohydrates. For example, organisms can store lipids and the lipid-like polymer, poly-β-hydroxybutyrate. Stores formed from noncarbonaceous compounds in the medium also can be accumulated, for example, the inorganic polyphosphate volutin, which is of practical interest because of the need to remove excess phosphate from wastewater streams. Carbohydrates are stored in macromolecules, usually in glycogen-like polysaccharides. The term *glycogen-like* is used because these compounds exhibit the $\alpha,1 \rightarrow 4$ glycosyl linkage and the $\alpha,1 \rightarrow 6$ branching links typical of glycogen, but the size of the molecules and the degree of branching vary for different organisms. Both poly-β-hydroxybutyrate and polyphosphate are also macromolecules, while the lipid stores cannot be so classified. The ability to form storage products is typical of many organisms, both procaryotic and eucaryotic, which is undoubtedly why it is so readily demonstrated in heterogeneous populations on a wide variety of substrates. Also storage is not unique to heterotrophs; it is found in autotrophs as well.

The synthesis of glycogen occurs readily with any compound that can be converted to glucose 6-phosphate or glucose 1-phosphate. Figure 10-25 shows the general pathway of synthesis and breakdown of glycogen, which is probably the most widely used storage product among bacteria. In general, glycogen is synthesized and used by procaryotic and eucaryotic organisms in the same manner, but with uridine triphosphate (UTP) rather than ATP providing the energy for synthesis in the latter group of organisms. The energy is partially retained in the form of phosphorylated glucose (glucose 1-phosphate) when the organism draws upon the glycogen store for energy or for carbon skeletons for the synthesis of other compounds (see Chap. 9). Both fatty acids, used in the synthesis of various types of lipids, and β-hydroxybutyric acid, used in the synthesis of the storage polymer poly-β-hydroxybutyrate (PHB), are synthesized from acetyl-CoA.

While carbohydrate stores may be synthesized and drawn upon during growth, it does not appear that the intracellular store of PHB is drawn upon until all or nearly all exogenous carbon has been removed from the medium. Thus, in cells that may be able to synthesize both glycogen and poly-β-hydroxybutyrate, the PHB, like fats in mammals, may be looked upon as a longer term storage than glycogen. When the PHB store is catabolized, the

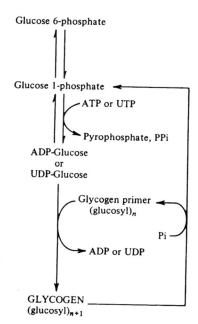

Glucose 6-phosphate

Glucose 1-phosphate

ATP or UTP

Pyrophosphate, PPi

ADP-Glucose
or
UDP-Glucose

Glycogen primer
(glucosyl)$_n$

Pi

ADP or UDP

GLYCOGEN
(glucosyl)$_{n+1}$

Figure 10-25 Synthesis and degradation of glycogen. (*Adapted from Dawes and Senior, 1973.*)

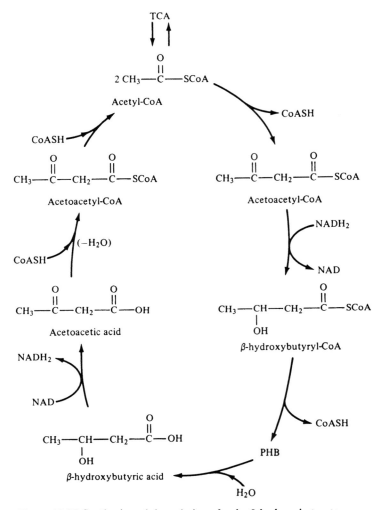

Figure 10-26 Synthesis and degradation of poly-β-hydroxybutyrate.

initial hydrolysis product is β-hydroxybutyric acid. This is oxidized, using NAD as the electron carrier, to acetoacetate, which can be converted to two molecules of acetyl-CoA and enter the TCA cycle. Figure 10-26 shows the synthesis and degradation of poly-β-hydroxybutyrate.

Polyphosphate is another storage product found in both procaryotes and eucaryotes. Purifying wastewaters of excess phosphorus, whether by physical or biological means, is of vital concern in the regulation of the life-support system. The fact that calcium phosphate is insoluble is the basis for treatment of wastewaters with lime, which provides perhaps the surest way of removing phosphorus. The fact that much of the earth's phosphorus exists as calcium phosphate accounts for the generally low phosphorus concentration in the aqueous environment. Organisms that can synthesize a readily available phosphorus storage product may thus have some competitive advantage. A rather wide variety of organisms can store large quantities of phosphorus, and they play a useful role in recycling this vital element, which does not enjoy the complete recycle mechanism available to carbon, hydrogen, oxygen, and sulfur (see Chap. 1). The ability to take up phosphorus in excess.of that normally found in the nucleic acid, nucleotide, and phospholipid complement of the cells also offers a possibility for biological treatment and removal of phosphorus, although efforts to harness and control this phenomenon in biological treatment processes have not been generally successful.

Polyphosphates occur in intracellular volutin granules, which stain metachromatically. For example, the basic dye toluidine blue turns red-violet in combination with polyphosphates. Polyphosphates may occur as a linear chain varying from only a few units to thousands of units long or as cyclic condensed phosphates (see Fig. 10-27A and B). Not all polyphosphates are simple phosphate anhydride polymers; in some organisms (e.g., the green alga *Chlorella*), imidophosphate bonding as well as phosphate anhydride bonds probably exist in the polymer. Also, volutin granules can contain RNA, protein, lipid, and magnesium in addition to polyphosphate.

Study has shown that rapid growth does not promote polyphosphate accumulation; to the contrary, an increase in polyphosphate content accompanies a decline in the growth rate due to exhaustion of some essential

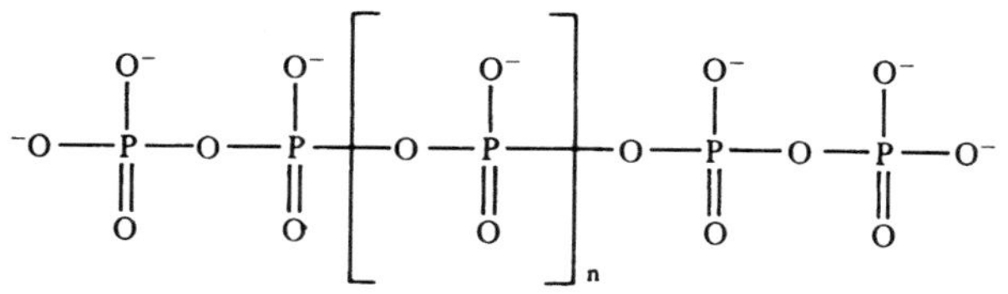

Linear polyphosphate

A

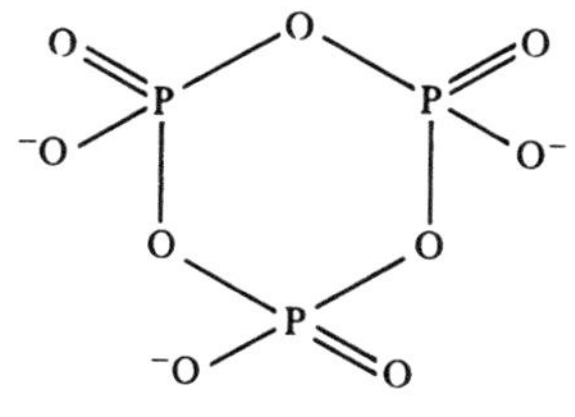

Trimeta phosphate

B

Figure 10-27 Structural configurations of some polyphosphates.

metabolite. Sulfate starvation is particularly effective in enhancing polyphosphate accumulation in media containing a high phosphate concentration. In media containing fairly low phosphate concentrations, any nutritional deficiency that prevents nucleic acid synthesis can lead to volutin accumulation. Holding cells for a time under phosphate starvation followed by exposure to phosphate can also lead to very great accumulation of polyphosphate, e.g., the so-called polyphosphate over-plus, which may account for much of the "luxury uptake" observed at times in the activated sludge process.

Polyphosphate can serve as a source of phosphorus for the synthesis of phospholipids and nucleic acids. Indeed, this appears to be the major role of the polyphosphate store, even though there is a high-energy phosphate transfer potential in the linear anhydride bonds. It appears that the breakdown of polyphosphate is in most cases hydrolytic; that is, it leads to phosphoric acid (Pi). The hydrolysis does not conserve the polyphosphate group transfer potential as it would, for example, if the polyphosphate could be active in the phosphorylation of an organic metabolic intermediate or of ADP.

Polyphosphates can be synthesized by the transfer of Pi from ATP by the enzyme polyphosphate kinase [Eq. (10-12)].

$$(Pi)_n + ATP \rightleftarrows (Pi)_{n+1} + ADP \qquad (10\text{-}12)$$

Also, phosphate in 1,3-diphosphoglycerate may be transferred to polyphosphate in some organisms [Eq. (10-13)].

$$
\begin{array}{c}
\overset{O}{\overset{\|}{C}}\text{--}OPO_3H_2 \\
|\ \\
HCOH \\
|\ \\
H_2COPO_3H_2 \\
\text{1,3-diphospho-} \\
\text{glyceric acid}
\end{array}
\quad + \ (Pi)_n \quad \longrightarrow \quad
\begin{array}{c}
\overset{O}{\overset{\|}{C}}\text{--}OH \\
|\ \\
HCOH \\
|\ \\
H_2COPO_3H_2 \\
\text{3-phospho-} \\
\text{glyceric acid}
\end{array}
\quad + \ (Pi)_{n+1}
\qquad (10\text{-}13)
$$

The degradation of the polymer could be effected by a reversal of reaction (10-12), but it is more generally accomplished by hydrolysis of the polyphosphate by the action of polyphosphatase [Eq. (10-14)].

$$(Pi)_n + H_2O \rightarrow (Pi)_{n-1} + H_3PO_4 \qquad (10\text{-}14)$$

In some microorganisms the polymer may be broken down by transfer of a phosphate to AMP, forming ADP, and phosphate can be transferred from polyphosphate to carbon-6 of glucose. Both of these reactions conserve the energy used in polyphosphate synthesis, but they do not appear to have widespread occurrence.

Based on work with *Aerobacter (Enterobacter) aerogenes*, reactions (10-12) and (10-14) appear to provide a cyclic scheme for synthesis and degradation (see Fig. 10-28). When cells are growing rapidly, nucleic acid synthesis

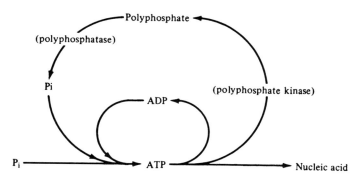

Figure 10-28 Polyphosphate cycle in *Aerobacter aerogenes*. (*Adapted from Dawes and Senior, 1973.*)

competes for the available ATP and polyphosphate is not made; the inorganic phosphate fed to the cycle is channeled to form nucleic acid. Also, any polyphosphate previously stored may be drawn upon and there is no significant accumulation of polyphosphate. However, upon cessation of growth or inhibition of nucleic acid synthesis, this competition for ATP is no longer exerted and polyphosphate synthesis occurs. One would expect that conditions for oxidative assimilation, i.e., prevention of nucleic acid synthesis, should encourage polyphosphate synthesis in organisms capable of this synthesis. If the cells had been previously subjected to phosphate starvation, which fosters elevated levels of polyphosphate kinase, rapid synthesis of large amounts of polyphosphates might ensue when the cells are in contact with a medium that is high in inorganic phosphate. Thus large amounts of phosphorus might be removed and stored along with the carbon source.

Although volutin production has been demonstrated in many species, its occurrence is much less widespread than is storage of organic carbon. Thus luxury uptake of phosphorus as a treatment process is dependent on the maintenance in the biomass of sufficient numbers of cells that can produce volutin. Shifts in species predominance, in predator ratios, and in other conditions could cause the ability to fluctuate or to appear and disappear. Such ecological events may account for some inconsistencies in the results of work with luxury uptake by activated sludge.

The most certain method for the removal of phosphorus is by chemical precipitation, and in fact chemical removal can occur in activated sludge processes, along with biological removal, depending on the concentration of calcium, aluminum, or iron that is in the carriage water or added to the reactor. However, the biological removal of phosphorus deserves even more study than it has already received. The combining of chemical and biological treatment for phosphorus removal has certain advantages, but more study is needed to assure that the chemical additives do not hamper the desirable biological responses to changes in environmental conditions. The development of a chemical sludge carries with it the usual overriding disadvantage of chemical treatment, the formation of voluminous sludges that usually are dewatered with difficulty.

ENDOGENOUS RESPIRATION AND AUTODIGESTION

In the growing systems of Figs. 10-15, 10-16, and 10-17, the period of growth and substrate removal was quite short compared with the endogenous period. While there is decided interest in growth, the endogenous phase of metabolism is of equal importance, since the growth period accounts only for removal of the carbon and energy source from the liquid; it amounts to only 40 to 60 percent of the overall aerobic decay of the organic matter. It is seen from the figures that as the endogenous phase progresses, an increasing portion of the carbon source that was incorporated into new cells during the growth phase is metabolized (endogenated) to carbon dioxide and water. It is also quite apparent that aerobic decay of the organic matter that was synthesized takes place much more slowly than did its synthesis. These are batch systems; in Chap. 6 it was allowed that in a continuous culture system, growth and endogenous metabolism (decay) could occur concurrently as well as in tandem. In a batch system, the endogenous phase is an operational description of the condition in which measurable soluble substrate is absent, whereas it must be considered as a conceptual principle in a continuous culture system.

The Maintenance Constant

To include the endogenous decay of the biomass in the kinetic description of the overall metabolic process, the decay constant was introduced and this "constant" was related to the theory of maintenance energy by interchangeably referring to k_d as the maintenance constant or the endogenous constant.

Maintenance, endogenous metabolism, and *starvation* are terms that help to describe the last stages of the return of organic to inorganic matter. One can relate these terms by envisioning an overweight individual, i.e., one who has stored a carbon and energy source such as a layer of fat. If only enough carbon is fed to maintain the body weight, no weight is gained or lost. If less is fed, some of the fat is "endogenously" oxidized to carbon dioxide and water, and if extremely little carbon is fed, the fat stores are depleted and the specimen may eventually die of "starvation." If a lean individual with no excess carbon store is fed less than a maintenance diet, the required energy for maintenance must be provided from endogenous sources, and functional, but not absolutely necessary, complements of body substances are depleted. The ill effects of starvation may thus begin sooner in the individual with little storage reserve. It is possible that the portion of carbon that must be oxidized for maintenance purposes may decrease as starvation proceeds, but maintenance plots for microorganisms in continuous culture reactors, such as those in Fig. 6-38, usually exhibit a constant slope when tested over a fairly large range of specific growth rates. Thus, there is some indication that k_d can be represented as a constant for continuous culture systems under stated operational or environmental conditions.

The description of endogenous metabolism in microbial populations is not so simple and straightforward as stated in this analogy. In any population, even one consisting of a single species, there are always individuals that are less able than others to withstand starvation conditions. These die first and become a source of carbon for the remaining live cells, which not only may be able to survive but also may replicate. This phenomenon is sometimes referred to as *cryptic growth*. Some cells undergo autolysis very shortly after exhaustion of the exogenous carbon source. On the other hand, some cells approach a resting stage, in which the metabolic processes are very much curtailed, and this may aid that particular species or strain to survive the starvation conditions. In heterogeneous populations, some species may exist in starvation conditions by actively attacking other species, i.e., predation may occur. In addition, there is the natural prey–predator food chain, e.g., bacterial–protozoan, resulting in utilization of previously synthesized microorganisms in the biomass. Thus, many metabolic and ecological phenomena that are in no way related to an organism's use of its own carbon compounds can occur in the so-called endogenous phase, as defined in a batch system. In heterogeneous populations, this phase may more appropriately be termed *autodigestion*. Furthermore, an endogenous or autodigestive phase in a batch system should be spoken of as a separate phase of the growth cycle only when the system no longer contains exogenous carbon source. Autodigestion in a heterogeneous population in a continuous flow system is a natural phenomenon that occurs concurrently with growth, and when the growth rate is slow enough its effect is measured in significantly lower net synthesis, i.e., lower observed yield of biomass. That is, more of the incoming substrate S_i is respired to carbon dioxide and water. Until the excess sludge from wastewater treatment plants becomes a useful resource rather than a waste product, the aim of both design and operation is to foster as much of this "biological incineration" as possible, i.e., to enhance the aerobic decay leg of the carbon-oxygen cycle. Thus, the control of autodigestion has as much practical significance as does the control of microbial growth and substrate removal.

There is considerable variation in the values of k_d reported in the literature (see Tables 10-3 and 10-4). Values for pure cultures are higher than those for heterogeneous microbial populations in spite of the fact that with pure cultures, predation cannot lower the observed cell yield. Part of the difference may lie in the analytical methodologies used or in the ash contents of the biomass. In both heterogeneous populations and pure cultures, values of k_d are higher at higher temperatures.

It should be emphasized that the maintenance concept is just that—a concept! It is reasonable to expect such a phenomenon to exist, but there is little proof for the mechanism beyond the fact that one obtains numerical values from plots of Eqs. (6-50) and (6-54) that allow more accurate prediction of the observed yield Y_o than is obtained when k_d is not included in the calculations.

When the maintenance effect is stated in the form of Eq. (6-48), the familiar rectangular hyperbola brings to mind questions similar to some of those asked about the Monod equation (Gaudy, Obayashi, and Gaudy, 1971;

Table 10-3 Values of decay constants for heterogeneous microbial populations

Substrate	Analysis	k_d, day^{-1}	Y_t	Temp, °C	Reference
Municipal sewage	BOD$_5$	0.025	0.70	13.2	Keyes and Asano (1975)
Municipal sewage	BOD$_5$	0.055	0.50	19–22	Heukelekian et al. (1951)
Municipal sewage	BOD$_5$	0.048	0.67	20–21	Middlebrooks and Garland (1968)
Municipal sewage	BOD$_5$	0.098	0.37	—	Fisher and Baker (1976)
Municipal sewage	COD	0.07	0.67	17–20	Benedek and Horvath (1967)
Synthetic waste	BOD$_5$	0.18	0.65	—	McCarty and Brodersen (1962)
Skim milk	BOD$_5$	0.045	0.48	—	Gram (1956)
Synthetic waste	COD	0.14	0.59	22–24	Srinivasaraghavan and Gaudy (1975)
Pepsi Cola waste	COD	0.14	0.76	24–30	Bonotan-Dura and Yang (1976)
Synthetic waste	COD	0.48	0.34	57.5–58.5	Surucu et al. (1975)

Table 10-4 Values of decay constants for three species of microorganisms

Organism	Substrate	k_d, day^{-1}	Y_t	Temp, °C	Reference
A. aerogenes	Glycerol	1.00*	0.55	—	Pirt (1965)
A. aerogenes	Glycerol	1.64	0.55	40	Postgate and Hunter (1962)
A. aerogenes	Glucose	1.46	0.47	25	Topiwala and Sinclair (1971)
A. aerogenes	Glucose	1.92	0.46	30	Topiwala and Sinclair (1971)
A. aerogenes	Glucose	2.40	0.45	35	Topiwala and Sinclair (1971)
A. aerogenes	Glucose	2.88	0.49	40	Topiwala and Sinclair (1971)
A. cloacae	Glucose	0.99*	0.44	37	Pirt (1965)
C. utilis	Glycerol	0.24	0.55	35	Sinclair and Ryder (1975)

*The k_d value was calculated from the "specific maintenance" using $m = k_d/Y_t$.

Gaudy, Yang, Bustamante, and Gaudy, 1973). For example, if one changes the specific growth rate μ_n, the observed cell yield Y_o should change. Judging from recent experiments in our own laboratories, there may be an inertial effect or kinetic lag, similar to the one in μ_n when S changes, before Y_o responds to a change in μ_n (Gaudy and Srinivasaraghavan, 1974a). It is interesting that in continuous growth studies with heterogeneous microbial populations, when the growth rate for a slowly growing system with relatively low Y_o was increased, the value of Y_o did not increase but remained at the same numerical value as for the lower growth rate. This was true whether the growth rate was increased in the continuous flow reactor or cells were transferred to a batch reactor and grown at or near μ_{max}, which would be expected to give yields close to the maximum Y_t. After approximately a week in the continuous flow reactor at the faster growth rate, the observed cell yield increased. Thus, one could say that there was a considerable inertia or lag with respect to the change in cell yield with the change in μ_n. It is possible that during this time there was a selection of cells with a higher true cell yield Y_t. Subsequent experiments with a pure culture system have provided some indication that a low specific growth rate in continuous culture systems may exert a pressure for selection of a mutant strain with a lower true yield and lower μ_{max} than the wild-type organism (C. W. Jones, A. F. Gaudy Jr., and E. T. Gaudy, unpublished data). Thus, the lowering of Y_o under slow growth conditions may involve genetic and therefore ecological alterations in the population as well as, or instead of, temporary changes in response to environmental conditions. Much more information is needed regarding cell maintenance if we are to understand its effect on growth and decay in the biosphere and in engineered processes. It often occurs in the practice of engineering sciences that concepts are usefully applied long before they are totally understood. New understanding serves sometimes to temper and sometimes to extend the use of a concept. In either case, the further knowledge enhances engineering judgment, which is so important in applied use of the concept.

According to the recommended kinetic model in Eqs. (6-45), (6-75), and (6-85), one could slow the specific growth rate μ to the value of k_d, thus making μ_n approach 0, and there would be no net synthesis. There would be no excess or waste sludge X_w, and the system (at least according to the equations) would achieve total oxidation of the originally fed carbon source S_i or ΔCOD. We may ask: What are the conceptual and experimental ramifications of this manipulation of the equations? Also, we might ask: Does the k_d value inserted into the growth model also hold for the case in which the biomass is simply aerated and not fed any soluble substrate? Or, would the same k_d value apply if the biomass were fed only more biomass, as is the case in the unit process of aerobic digestion of excess sludge?

The extended aeration activated sludge process and the process of aerobic digestion of excess sludge are the two major practical uses of the autodigestive phase in wastewater treatment schemes. How far toward completion of autodigestion can the system proceed? What role does autodigestion play in assessing the biochemical oxygen demand (BOD) of a sample of wastewater? These aspects are considered below.

Total Oxidation—Concurrent Growth and Autodigestion

If one assumes steady state conditions as is done in the kinetic models, μ can be balanced against k_d, and the resulting equations describe a state of total oxidation. Nearly three decades ago, long before the kinetic model for activated sludge recommended here was developed, Porges and his coworkers (1953) concluded that such a balanced condition might be attainable and that a system could be operated by returning all sludge to the aeration tank; that is, a system could be operated without wasting any sludge. It was surmised that longer detention times should be used than for the regular activated sludge processes, and this particular modification became known as "extended aeration" activated sludge. In Fig. 10-29, the heavy lines show the elements and flow scheme of an extended aeration plant. The conventional plant includes all the elements shown.

The idea of total oxidation of S_i (ΔCOD) was controversial, and it encouraged other researchers to study the proposed process. Out of this research, three general conceptual conclusions were reached. The first was that microbial cells synthesized some components that were not available substrates, e.g., extracellular polysaccharides (Symons and McKinney, 1958). This material would build up in the system, thus comprising an ever-increas-

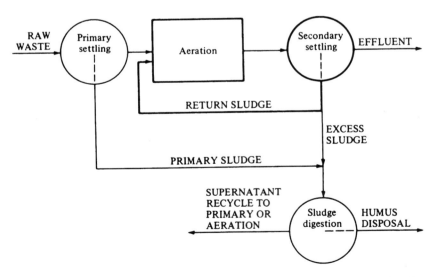

Figure 10-29 Comparison of extended aeration (heavy lines) and conventional activated sludge processes. (*From Gaudy, Ramanathan, Yang, and DeGeare, 1970.*)

ing, metabolically inactive fraction and leading to eventual failure of the process. The second conclusion was that the process would not reach a steady state (wherein $\mu = k_d$); that is, X would continually increase. If no sludge was wasted, suspended solids eventually would leave the system (Busch and Myrick, 1960). The third general conclusion was that in order to maintain the viability of the process, 10 to 20 percent of the sludge should be wasted (e.g., see Kountz and Forney, 1959). This amount became accepted as a residual, inactive fraction. During this time, because of the tremendous appeal of the process especially for small installations, the idea was "packaged" by various equipment manufacturers and many prefabricated plants were put on-line. Also, many built-in-place extended aeration treatment plants were installed. Usually, there was a provision for a small amount of sludge wasting in the form of draining and drying beds. Also, it became general practice to omit the primary treatment, i.e., sedimentation of settleable solids. This was necessary because there was no provision for the disposal of excess organic matter other than the biological incineration of the oxidizable portion of the cells. Thus the process, as installed in the field, was required not only to digest the cells that had been synthesized by growth on the waste, but also to digest essentially all organic matter and remove from solution the inorganic matter coming through the sewer. The digestion of such material as cellulose, lignin, and coffee grounds is, of course, entirely different from aerobic autodigestion of the components of microbial cells. Thus, the field installations were a very severe test of the process but not a good test of the concept. When sludge was found to build up in some of these installations, it was considered a verification of the conceptual conclusions of the researchers. On the other hand, at some extended aeration plants there was no excess buildup of biomass and no excessive carry-over of solids in the clarifier effluent. While the process remained the center of considerable controversy, the number of installations increased and is still increasing. The concept is also used for the treatment of some industrial wastes in reaction basins with rather long detention times of 10 to 20 days—i.e., aerobic lagoons.

Not all experimental observations on the extended aeration process led to negative conclusions in regard to the acceptability of the concept of total oxidation. Some experimenters reported a decrease in biomass concentration in a laboratory pilot plant in which all the settleable solids were returned to the aerator. Since the decrease in aeration solids had occurred during a period when sludge settleability was good, the decrease in solids concentration could not be attributed to their inadvertent escape in the clarifier overflow. It was reasoned that some cells in the sludge could have undergone lysis, and the released organic material could have been taken up by the remaining living cells (Washington, Hetling, and Rao, 1965). From these experimental observations one could not draw definite conclusions about the validity of the total oxidation concept. But the results did indicate that the biomass concentration in an extended aeration process would not necessarily continually rise as had been concluded from earlier results. Other results also indicated the possibility of attainment of total oxidation. For example, in the batch growth system shown in Fig. 10-16A, the biomass concentration essentially returned to the value of X_0. This is a rather clear demonstration of net total oxidation of biomass synthesized in the substrate removal phase. In other work in our laboratories, we have noted this phenomenon on a number of occasions when we extended the length of batch experiments. Biomass concentration did not always return to the value of X_0, but it happened with sufficient frequency to indicate that the total oxidation concept was possibly not so microbiologically unsound as had been previously suspected.

The negative conclusions in regard to the concept of total oxidation arose due to some good engineering studies indicating that there were problems with the process when all suspended solids were returned to the aeration tank. The general operational experience was that, more often than not, the biomass concentration built up to such high concentrations that the sludge would not settle in the clarifier. The finding that aerobic digestion did not always exceed or balance biomass accumulation due to growth was cause for engineering caution in regard to the feasibility of the process, but this was extrapolated to conceptual conclusions about the validity of the concept. This was unfortunate, but it is not uncommon when applied research precedes basic research information. The cliche "necessity is the mother of invention" is a time-honored truism that describes many engineering endeavors and has provided the justification for many basic research investigations as follow-up to experimental observations. Since there were alternatives to a process embodying the concept of total oxidation, there was no necessity to examine it further, and failures of the process that may have been a matter of the rate or frequency of occurrence of a phenomenon were generally accepted as proof that the phenomenon does not occur at all. The point to be emphasized is that if a problem is a matter of rate or frequency, engineering can perhaps do something about it, but if a particular result is physically, biologically, or chemically impossible, no amount of innovative engineering can make the process work—at any rate! When weighing conflicting reports from field installations and results that indicated the concept might have some validity, the only reasonable course of action would seem to be further investigation.*

In the 1960s, it was decided to direct some of the investigational efforts in our own laboratories to a study of the extended aeration process. The aim was first to examine the validity of the concept of total oxidation by seeking answers to several questions. Could a soluble substrate be removed biologically due to growth of the biomass, and could that biomass be autodigested to carbon dioxide and water? If one retained all the cells in the system, would the biomass continue to increase, and would a biologically inert fraction

---

*It is not usual practice to include such description of the reason for seeking information, but our personal involvement in research into the microbial aspects of environmental control coupled with the fact that control of the life-support system is emerging as a new professional discipline and is in need of much investigational activity made it desirable to illustrate the lines of reasoning and rationale that go into the decision-making process before embarking on an investigation.

accumulate until the substrate removal capability was lost or seriously impaired? If total oxidation was impossible, how long could the process operate before it loses its substrate removal capacity? The latter was an important engineering consideration because the process might be usable for a sufficiently long period to warrant its installation even if planned periodic shutdowns for removal of inert sludge were necessary. The removal of an inert sludge would not pose so difficult a disposal problem as would the removal of a putrescible sludge. Most of the questions listed are engineering questions. The basic research questions are related to the engineering questions, but these demand investigation into *why* a particular result occurs, and these were bridges to be crossed after observing the results. In short, the study was designed, like most research in this field, to accomplish two aims: applied and basic. There are those who feel these aims are incompatible (for a discussion of this, see Gaudy, 1971). Following such investigations, there may be follow-up developmental research designed to answer the question of *how* to bring the results to practical fruition.

A laboratory pilot plant was operated and, in order to be certain that no suspended biological solids exited the system, all effluent was centrifuged and the solids were returned to the aeration chamber. The experimental pilot plant is shown in Fig. 10-30. For experimental convenience, separate aeration and clarification tanks were not used, but aeration and settling compartments were separated by a movable baffle as is common practice in wastewater treatability studies. The return sludge was drawn back into the aeration chamber by circulating currents in the aeration chamber. In field installations, this is not a good way to return sludge from the clarifier to the aeration chamber because one has little or no control over the mass rate of return, but in

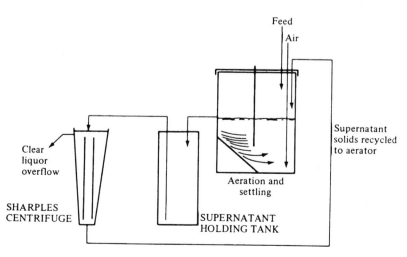

Figure 10-30 Continuous flow extended aeration pilot plant. (*From Gaudy, Ramanathan, Yang, and DeGeare, 1970.*)

laboratory treatability studies the method suffices because one can control the mass rate of biomass wastage (thus μ_n) by momentarily removing the baffle after stopping the effluent line and wasting mixed liquor. In a study of the extended aeration process, no sludge is wasted. The supernatant holding tank served as a "backstop" for any solids in the clarifier supernatant; the centrifuge was used to capture these solids for return to the aeration tank. The aim was to run the process under close operational scrutiny on a soluble feed (glucose and minimal salts with NH_4^+ as a nitrogen source) and observe its performance in regard to the questions listed above. The aim was to run the process until it became metabolically inactive, if this occurred.

It became apparent during the first year of the experiment that the biomass concentration was not at a steady state. There were periods when it remained fairly steady, periods when suspended solids accumulated, and periods when the biomass concentration decreased rather sharply. One such period is seen in Fig. 10-31; between days 285 and 307, the biomass concentration dropped from 8000 to 2000 mg/L. Even during this period, the soluble substrate S_e in the effluent was fairly low, and in general S_e did attain a pseudo-steady state. No biomass (except approximately 0.2 percent removed for analysis) escaped the system, and a reasonable surmise as to the cause of the observed results was that the periods of solids decline were times when the ecological condition of the population was such that many species became substrate for other microorganisms; i.e., autodigestion exceeded growth. There were times when the biomass concentration rose to very high levels, and high X_e values appeared in the effluent. For example, the period shown in Fig. 10-32 during the third year of operation is one in which considerable

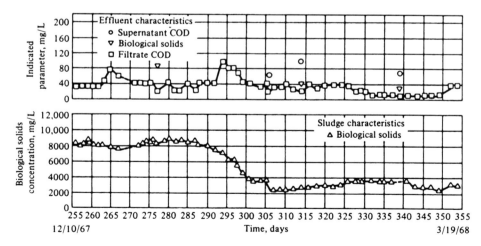

Figure 10-31 Performance of a total cell recycle activated sludge pilot plant during a period of accelerated autodigestion. $S_i = 500$ mg/L of glucose. Note decrease in biomass concentration with no increase in soluble COD in the reactor. (*From Gaudy, Ramanathan, Yang, and DeGeare, 1970.*)

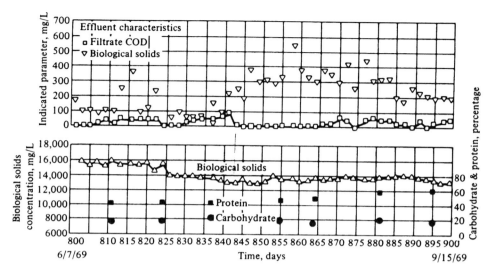

Figure 10-32 Performance of a total cell recycle activated sludge pilot plant during a period of high biomass solids concentration in the effluent. $S_i = 500$ mg/L of glucose. Note the high biomass concentration in the aeration tank. (*From Gaudy, Yang, and Obayashi, 1971.*)

biomass would have escaped to the effluent had it not been captured in the holding tank, centrifuged, and put back into the aeration chamber. After 1000 days of operation, the substrate removal ability of this sludge remained rather good ($S_i = 530$ mg/L of glucose COD and $S_e = 50$ mg/L). It is also significant that the protein and carbohydrate contents of the biomass were consistent (60 and 20 percent, respectively) with those of a normal healthy biomass. There was no buildup of extracellular polysaccharide. The speed with which this very "old" sludge was able to purify the feed is demonstrated in Fig. 10-33. These data were obtained by shutting down the feed pump and adding stock feed as a batch slug dose of 500 mg/L of glucose. Subsequent sampling and analysis for suspended solids and filtrate COD generated the results shown.

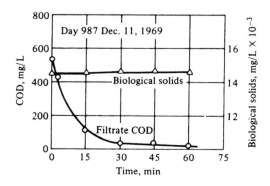

Figure 10-33 COD removal capability of total recycle activated sludge after 987 days of operation. (*From Gaudy, Yang, and Obayashi, 1971.*)

The 987-day-old biomass removed soluble organic matter very rapidly (within a contact time characteristic of the high solids–biosorption process).

The results of these long-term experiments necessitate a conclusion that the concept of the total oxidation of biomass synthesized during growth on an exogenous carbon source is indeed valid, albeit difficult to effectuate by current engineering practice. It should be noted that the effluent between days 900 and 1000 was by no means within acceptable limits for suspended solids; that is, some add-on such as sand filtration would be needed for those times when the effluent was turbid. Chemical flocculants or flocculant aids might be added to help incorporate the dispersed suspended solids into the floc.

The finding that the carbohydrate content of the biomass did not build up is rather interesting. Did it mean that during the 3-year period not much extracellular polysaccharide was produced, or that it was produced and subsequently used as a carbon source, i.e., aerobically digested? Since it is generally found that cells that produce extracellular polysaccharide slimes do not use them as a source of carbon and energy, it is apparent that if they are to be metabolized, other cells in the biomass must degrade them. When each of the extracellular slimes from five capsule-producing organisms was fed as a sole source of carbon and energy to a small inoculum of sewage, each was used as a substrate for growth after a period of acclimation of 2 to 10 days. Figure 10-34 shows the course of growth and substrate removal in one such experiment using the extracellular polysaccharide of *Z. ramigera*. Extracel-

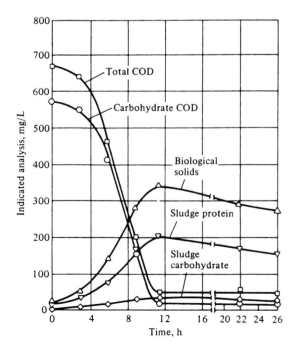

Figure 10-34 Growth of a heterogeneous microbial population on the extracellular polysaccharide of *Zoogloea ramigera*. (*From Obayashi and Gaudy, 1973.*)

lular capsular material is not biologically inert, and it is relatively easy to develop populations of sewage origin that can readily metabolize this material.

The pilot plant studies as well as the studies on polysaccharide metabolism provide convincing evidence that the concept of total oxidation is not unsound, and they encourage further fundamental research on the degradation and metabolism of other cellular components as well as further engineering studies on ways to control and enhance their metabolism. How does one control the composition of the population so that it includes species that can metabolize the living or dead organic matter of other cells? This vital question is presently unanswered and may be unanswerable for many years. Indeed, answers to such questions lie at the heart of our understanding of the decay leg of the carbon-oxygen cycle. One way in which aerobic decay of cellular components is enhanced is to make these materials the sole or major source of carbon and energy in a biological reactor. This is done in the extended aeration process by recycling all the sludge from the clarifier, thus letting μ_n approach 0. If the reactor is being fed a waste with an S_i of 300 mg/L, i.e., equivalent to a moderately strong municipal sewage, and the mixed liquor solids concentration in the aerator is, for example, 6000 mg/L, the major source of nourishment for the living cells in the biomass is the biomass itself, not the incoming carbon source, and enzymes for attacking complicated macromolecules are produced more easily than when there is an abundance of more readily metabolized carbon sources (see Chap. 12).

Control of solids level in extended aeration—the "hydrolytic assist" Although the recycle of all biomass encourages its use as substrate, the extended aeration process does not often attain a steady state solids level. It is possible that k_d changes from time to time; it may be low when the biomass is accumulating and high during a period of solids decline. The subsidence of the biomass floc is dependent on its concentration, and during times when the concentration X builds to high levels, subsidence may be so hampered that biomass is lost in the effluent. It is not known what ecological events trigger the onset of a period of accelerated autodigestion that could relieve the high biomass concentration. However, one of the first events must involve the attack by microbes of the macromolecular structural components of other cells. Hydrolysis of organic molecules by chemical agents is a relatively mild chemical reaction (compared, for example, with oxidation), but biochemically it is rather difficult to perform. Cells that possess the genetic capability for making the required enzymes may not always be present in the population. Thus, there may be little possibility of predicting or regulating the natural autodigestive process. However, one can envision synthetic or engineering control of the biomass concentration by combining elements of chemical and biological attack on the biomass in excess of the concentration one wishes to maintain in the system. For example, one may break open some cells by ultrasonication of a portion of the returned sludge. This would release some biological components to become substrates for the nonsonicated, intact portion of the sludge. Such treatment will not degrade most macromolecules, but chemical hydrolysis of a portion of the return sludge might effect more complete degradation.

Early attempts showed that acid hydrolysis (pH approximately 1.0, 5 h of autoclaving at 121°C, 15 psig) dissolved most of the cell constituents (Gaudy, Yang, and Obayashi, 1971). Filaments disappeared and only small amounts of organic matter measured as COD volatilized during the autoclaving process. Upon neutralization, the hydrolysate was an excellent carbon and energy source for heterogeneous microbial populations. Other conditions of hydrolysis might also suffice, depending on the nature of the substrates and sludges and the degree of hydrolysis required to control the biological solids concentration, e.g., alkaline hydrolysis and both milder and more severe conditions of hydrolysis than those mentioned above.

A chemically assisted extended aeration process such as that shown in Fig. 10-35 could permit some engineering control of the autodigestive process and of the concentration of biomass X in the reactor. One could choose to run the hydrolysis unit on a semicontinuous basis, thus attempting to keep k_d more or less constant and letting the system approach a pseudo-steady state in X. On the other hand, the hydrolytic assist could be used periodically, such as only when X is too high to permit good subsidence in the clarifier. Also, the hydrolysis procedure could be used to encourage changes in the predominating species. For example, if filamentous organisms began to predominate, some return sludge could be hydrolyzed and this rich substrate fed back to the bioreactor in the expectation that the individual flocculatable cells would outgrow the filaments. During the refeeding period, one might also create operational or environmental conditions that would be less favorable to filaments. For example, one could make certain that the pH was well above

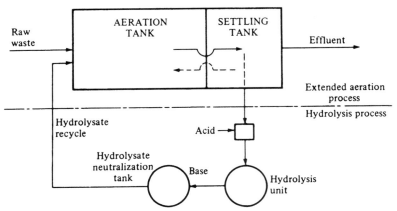

Figure 10-35 Extended aeration activated sludge process incorporating chemical hydrolysis for control of sludge concentration. (*From Yang and Gaudy, 1974a.*)

neutrality so as not to encourage rapid proliferation of fungi. The hydrolysis procedure might have to be administered more than once, but operators would at least have at their disposal some means of attacking the problem other than the repeated addition of coagulating chemicals in the hope that the problem would correct itself eventually.

Extended aeration pilot plants using this mode of control have been operated with much success in controlling both biochemical substrate removal efficiency and autoflocculative separation of the cells. An operational mode was used in which small portions of the biomass were hydrolyzed weekly and were gradually fed to the bioreactor along with the return sludge (Gaudy, Yang, and Obayashi, 1971; Yang and Gaudy, 1974a, 1974b). This type of hydrolytic assist has also been helpful in correcting bulking due to filamentous organisms. Apparently the cell walls and sheaths of fungi and filamentous bacteria are rather resistant to biological attack; ridding the biomass of excess filaments once they have become established is greatly aided by a hydrolytic assist. Such engineering procedures can be of help until biological engineering proceeds to such an advanced state of knowledge to permit operational control of species predominance in the bioreactor. The latter may be many years away, if indeed it is ever attainable.

Wastewaters often contain large amounts of dissolved and colloidal inorganic matter, and a buildup of ash content of the biomass could be a source of concern. By its very nature, the hydrolytic assist adds considerable amounts of inorganic salts; however, pilot plant studies undertaken for over a year and a half using a waste with an extremely high ash content, approximately 70 percent, have amply demonstrated that while the ash content of the biomass can be expected to be high, it will not continually increase nor will the high ash content hamper the efficiency of the treatment process (Gaudy, Manickam, Saidi, and Reddy, 1976). Feedstock in these pilot plant studies consisted of trickling filter sludge taken from the bottom of the secondary settling tank at a municipal treatment plant and subjected to partial hydrolysis. Thus the "wastewater" that was fed represented a highly complex mixture of organic and inorganic components. A sample of the results at a rather high inflowing substrate concentration (S_i = 1000 mg/L, measured by COD) is shown in Table 10-5. In addition to noting the rather good performance with regard to removal of soluble and suspended materials in the waste, it is interesting to compare the feed and effluent nitrogen levels. If one assumes that all the organic nitrogen as well as all the ammonia–nitrogen was available to the cells as a nitrogen source, the BOD/N ratio is 50/1, which is well above the 20/1 ratio normally used for nitrogen-deficient waste. Yet nearly all of the influent nitrogen appeared in the effluent as nitrate nitrogen. One can see that the return of all cells closes the system with respect to nitrogen; it is reusable in the process and hence there is on average no net usage of nitrogen, i.e., no incorporation into excess sludge. The effluent concentration is equal to the input concentration. Furthermore, the nitrogen in the effluent exists in a highly oxidized form. The slow growth rate (longer

Table 10-5 Performance of a hydrolytically assisted extended aeration activated sludge fed a complex organic waste

	Feed, mg/L	Effluent, mg/L	Percent efficiency
Total COD	1096	56	94.9
Filtrate COD	—	42	96.2
Percent soluble	—	75.0	
BOD_5	551	17	96.9
BOD_5 soluble	—	7	
Percent soluble	—	41	
BOD_5/COD	0.5	0.3	
Suspended solids	777	15	98.1
Soluble solids	1367	1759	
Organic-nitrogen	4.3	0	
Ammonia–nitrogen	6.6	0	
Nitrate–nitrogen	0	9.9	

Data from Gaudy, Manickam, Saidi, and Reddy, 1976.

retention of cells) encourages nitrification. In the case of treatment of a waste that did not contain a usable source of nitrogen (e.g., some industrial wastes), the nitrogen supplementation could be minimized by use of the extended aeration process and perhaps even eliminated after initial addition for development of the sludge, provided one can control the autodigestive process (e.g., using a hydrolytic assist) to increase the ability to recycle the nitrogen component. Table 10-5 gives average values over a period of nearly 3 months during the second year of pilot plant operation at the S_i value shown. Even though there was no net utilization of the inflowing nitrogen during this time, the average protein content of the sludge, based on volatile suspended solids, was 35 percent; it was not uncommonly low. Continued work in our own laboratories on this and other possible chemical and physical assists to purification and sludge disposal provide indications that such approaches offer considerable promise.

Other workers have investigated the "hydrolytic assist." Hydrolytically assisted extended aeration pilot plants have been operated successfully on soft drink bottling plant wastes (Yang and Chen, 1977) and on bleached kraft pulp wastes (Lee et al., 1976) as well as to enhance aerobic digestion processes (Singh and Patterson, 1974). Also, in our own laboratories the hydrolytically assisted extended aeration process has been used to treat waste that contains cyanide. In regard to the treatment of waste containing toxic chemicals, care must be taken to ascertain whether the biomass degrades or accumulates the toxic material. In the latter case, hydrolysis of a portion of the sludge could conceivably release the toxicant. In the studies cited above, there was evidence that the cyanide ion was degraded biologic-ally (Y. J. Feng and A. F. Gaudy, unpublished results).

In summary, total oxidation is possible. The major engineering problem is the retention of the biomass until accelerated autodigestion occurs. However, there is no way to predict the periodicity and extent of such a period of autodigestion. Some form of chemical hydrolysis of the macromolecules that are biologically hydrolyzed with difficulty provides a useful means of controlling the sludge concentration. One way that shows promise is either acid or alkaline hydrolysis of some of the return sludge. The mode and severity of the conditions of hydrolysis are best determined for each local condition during treatability studies on the wastewater in question. Sludge disposal will become an increasing problem and combined chemical-biological schemes such as the one described above may prove to be economical when compared with the cost of incineration and other more conventional means of sludge treatment and disposal. It is wise to keep one's options open in this regard, and it is the responsibility of fundamental biological engineering investigators to provide options and to determine their scientific validity.

BIOCHEMICAL OXYGEN DEMAND (BOD)—THE CONCEPT AND THE TEST

Basic concepts regarding BOD were given in Chap. 3, but any discussion of the microbiological aspects of control of the water resource would be sorely lacking if it did not deal with the much-used concept of biochemical oxygen demand and the BOD test in more quantitative detail. It was stated before that BOD is simply the oxygen taken up for aerobic respiration during a batch experiment similar to those shown, for example, in Fig. 10-7 or 10-8. During a BOD test, one measures the amount of oxygen used in a prescribed time, usually at the end of 5 days of incubation, and uses the value as a measure of the strength of the waste. If an untreated (raw) waste is examined, the sample usually must be diluted. Therefore the results obtained are multiplied by the appropriate dilution factor. Dilution is necessary because, instead of using a Warburg apparatus or other manometric device or a reactor of known reaeration characteristics, the oxygen uptake is measured as the decrease in dissolved oxygen concentration in a completely filled and stoppered reaction flask, i.e., a BOD bottle, of approximately 300 ml volume. Since one can dissolve only 8 to 9 mg/L of oxygen in water, the drop in DO during the 5-day incubation period must be small; because most wastes exert more oxygen demand than this, dilution of the sample is usually necessary. The use of such a closed system and the development of the dilution technique were originally aimed at simulating the assimilation conditions in the receiving stream. That is, the primary interest was not in assessing the strength of a wastewater but in gaining some insight into how much oxygen would be used in the dilute condition one would expect in the receiving stream, i.e., when the waste was diluted with river water. Thus, when the test was developed, the aim was not to measure S_i for the design of wastewater treatment plants, but to predict the effects of a wastewater on a receiving stream.

It is not our purpose to review the long history of development of the BOD test. It should suffice to say that many workers have contributed to its development. Indeed, development of the field of environmental engineering science has been dependent on the development of the BOD concept and test. The names Phelps, Streeter, and Butterfield (particularly Phelps) are prominent in the development of the BOD concept. The primary interest of these early workers was to determine the DO profile in the receiving stream in response to a known dilution of wastewater. The interest was in determining the capacity of the receiving stream for assimilation of organic wastes. This capacity depends, in the main, on two overriding and opposing factors: the rate of reaeration in the receiving stream and the rate of oxygen utilization due to microbial metabolism of the organic matter in the stream.

Kinetics of BOD Exertion

In Chap. 6 we used a DO profile and the aeration characteristics of the reactor to estimate the oxygen uptake curve. The problem now is simply to reverse the process and estimate the DO profile, however, not in a reactor with relatively definite reaeration characteristics but in the natural reactor. Aeration, i.e., absorption of oxygen into the water, follows a first-order, decreasing-rate kinetic curve, as discussed in Chap. 6. The mechanism and the kinetics are essentially the same (values of K_2 will differ) whether the reactor is an aeration tank or a receiving stream. On the other hand, as can be seen in Figs. 10-7, 10-8, and 10-10, the oxygen uptake or BOD curve looks very much like the biomass growth curve during the substrate removal phase and thereafter continues on a path that can be only loosely defined; the extent of oxygen uptake in the endogenous or autodigestive phase is dependent on the population present. In any event, it is obvious that there is no one kinetic order or particular mathematical form that can describe the entire curve. Therefore, while we can speak of a specific rate of reaeration over the entire period, it is impossible to speak of a specific rate of oxygen uptake. Obviously more than one kinetic event occurs during the exertion of this biochemical oxygen demand, e.g., there is growth and autodigestion, and no one type of kinetic equation describes the curve. However, this was not the general type of oxygen uptake curve observed by early workers such as Phelps. Accumulated oxygen uptake appeared to exhibit a first-order, decreasing-rate kinetic character such as shown in Fig. 10-36. Equations (10-16) and (10-19) (see legend for Fig. 10-36) became standards of kinetic expression for the accumulated oxygen uptake or BOD exerted, y, and the BOD remaining to be exerted, L_t. Thus, both reaeration and oxygen uptake were considered to occur with the same kinetic order and one could calculate the changes in DO deficit or concentration of DO in the receiving stream in time or distance downstream from the point of entry of the waste into the stream in accordance with the following differential equation:

$$\frac{dD}{dt} \quad = \quad K_1L \quad - \quad K_2D \tag{10-20}$$

| rate of change in DO deficit | O_2 uptake (BOD exertion) | O_2 replenishment (reaeration) |

Integration of Eq. (10-20) yields the sag equation of Streeter and Phelps (1925).

We may ask why there were differences in the observed oxygen uptake then and now. We can get a partial insight by examining the oxygen uptake curve shown in Figs. 10-7 and 10-8. Clearly, during the period of substrate removal the curve is S-shaped. This type of curve can be observed time and time again under conditions of balanced nutrition. However, while a large

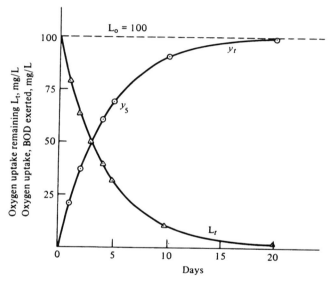

Figure 10-36 Oxygen uptake, BOD, curve according to the first-order, decreasing-rate equation of Phelps (1944).

$$y_t = L_0(1 - e^{-K_1t}) \tag{10-15}$$
$$= L_0(1 - 10^{k_1t} \tag{10-16}$$
$$L_t = L_0 - y_t \tag{10-17}$$
$$= L_0e^{-K_1t} \tag{10-18}$$
$$= L_010^{-k_1t} \tag{10-19}$$

where L_0 = ultimate amount of oxygen uptake to be expressed; $L_0 = 100$ mg/L in example above
L_t = amount of BOD remaining to be expressed at any time t
y_t = amount of BOD expressed or exerted at time t
K_1 = velocity constant for logarithms to base e, $K_1 = 2.3\, k_1$. $K_1 = 0.23$ in example.
k_1 = velocity constant for logarithms to base 10.
y_s = BOD_5, i.e., the standard BOD test value that would be registered for the above oxygen uptake

portion of the oxygen is taken up during this period, this time is relatively short in comparison with the total incubation time. In Fig. 10-8, if we move the early portion of the curve to the right and plot the whole curve on the same scale (see the dotted line), the S-shape is less apparent. This portion of the curve might even be fitted to Eq. (10-16) with the obvious exception that it would have to have a much higher K_1 value than was used in the example of Fig. 10-36. If one had not taken many samplings of oxygen uptake during this period, the decided S-shape of the curve may not have been noticed at all. Indeed, such might also be the case for the pause and acceleration of the rate that takes place between days 1 and 3 in Figs. 10-7 and 10-8; that is, first one might smooth the curve into one that looks something like the first-order, decreasing-rate curve of Eq. (10-16). Second, if not enough samples were taken, one might not even observe these divergences from Eq. (10-16).

A few important questions may be asked at this point. Does it make any practical difference whether one uses the actual curve or approximates it by Eq. (10-16)? Is there any fundamental problem that could arise by using this equation? The answer to the first question is yes; it does make a difference, because when one compares the predicted DO profile (the sag curve) calculated using Eq. (10-16) with the profile obtained using the numerical integration procedure employing the actual BOD curve, the low point of the sag curve, the critical DO, is lower using the latter procedure (Isaacs and Gaudy, 1967; Gaudy, 1972). This comes about largely because in the early stages the rate of oxygen uptake is increasing at an increasing rate rather than a decreasing rate, i.e., the BOD curve is convex upward. The allowable minimum dissolved oxygen concentration in a receiving stream is generally 4 mg/L DO or approximately 50 percent of the DO saturation concentration. Thus, it is important not to use a means of estimating that underestimates the severity of the condition. The answer to the second question is also yes; there are indeed fundamental problems with the use of the BOD technique and its approximation as a first-order, decreasing rate for the purpose of formulating an equation for oxygen uptake. A major problem is that the first-order approximation of BOD has become so embedded in environmental engineering principles that it is used as kinetic principle rather than the approximation it is, and it has erroneously become a starting point or foundation upon which to build new kinetic principles in the field.

We have seen in Chap. 6 that the rate of growth, as the growth curve proceeds, increases up to an inflection point and thereafter decreases. During the increasing portion it may enter an exponential phase (constant μ), and the specific rate of growth has been shown to be dependent on the concentration of the carbon source. Also, we know that the oxygen uptake curve roughly parallels the shape of the growth curve. Thus, the specific rate of oxygen uptake should also be dependent on the concentration of the carbon source. Experimentation has shown this to be true (see Fig. 10-37). When one plots the numerical value of the slope of the semilogarithmic plots of oxygen uptake against the values of S_0, the data appear to fit the rectangular

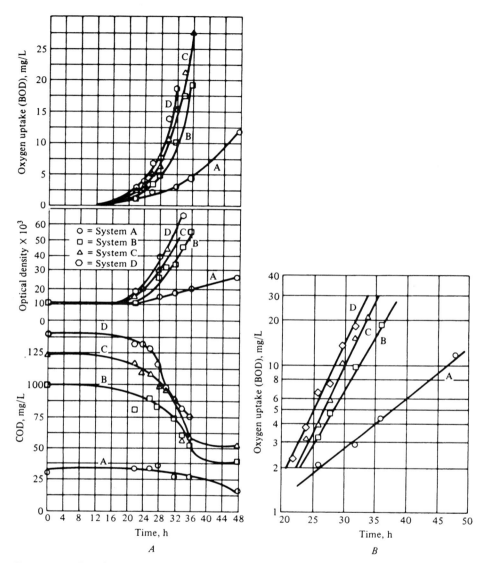

Figure 10-37 Exertion of BOD due to various concentrations of a glucose/glutamic acid substrate in BOD bottles supersaturated with oxygen. (*A*) Oxygen uptake, growth, and COD removal curves. (*B*) Semilogarithmic plot of oxygen uptake. (*Adapted from Jennelle and Gaudy, 1970.*)

hyperbola of the Monod equation rather well (Jennelle and Gaudy, 1970). Thus, during the early portions of the oxygen uptake curve, the rate constant is dependent on S_0. This is simply another way of saying that the specific rate of BOD exertion depends on the dilution factor of the waste sample used in the BOD bottle. One would therefore obtain different kinetic constants, depending on the dilution used. If, for practical reasons, one used a dilution factor sufficient to get an oxygen uptake curve without resorting to supersaturating the water in the BOD bottle (which is the usual case), a very high dilution would be required—perhaps one much higher than the dilution expected in the receiving stream. In this case, one would be underestimating not only because of the use of a first-order, decreasing-rate approximation but also because the numerical value of the velocity constant used would be lower than would have been observed at a lower dilution of the waste. During the early portion of the BOD curve, which is so important to estimating the critical DO or low point on the DO sag, the first-order approximation violates one of the empirical principles established by many experimental observations, i.e., the relationship between specific rate of growth and/or oxygen uptake and the substrate concentration.

There is really no need to use the first-order approximation when predicting the DO profile in a body of water. One can obtain a number of oxygen uptake curves using the expected dilution of wastewater in receiving water and an open-reactor technique (see Chap. 6). The curve can then be integrated numerically with the estimated stream reaeration constants, K_2 values (Peil and Gaudy, 1975; Gaudy, 1975).

The plateau in the BOD curve It is apparent in Fig. 10-7 that for most of the experimental (incubation) period, the oxygen uptake takes place in the endogenous or autodigestive phase. In this phase the portion of the original exogenous substrate that has not been oxidized to carbon dioxide and water exists in the biomass, and the oxygen uptake yet to be exerted will be exerted at the expense of the biomass, which is now the only organic food source in the system.

In the experiment shown in Fig. 10-7 and in many others as well, the beginning of this phase seems to be marked by a slowing and restarting of oxygen uptake, i.e., a "plateau" has developed in the oxygen uptake curve shortly after the time of exhaustion of the original exogenous carbon source. A. W. Busch (1958), who made the pioneering observation of such plateaus during exertion of BOD, concluded that the plateau always signified the end of the substrate removal phase, and that oxygen uptake after the plateau was caused by microbial predators, e.g., protozoa feeding on the bacterial biomass that had grown during the first phase. Both the observation and the conclusion precipitated considerable controversy among investigators. The observation could, of course, be checked easily by other investigators, but the conclusion was contrary to previous concepts. Also, there was no reason to expect that the two phases of substrate utilization, exogenous and autodigestive, would always be so markedly separated that there would be a distinct pause in oxygen uptake between them. The former concept of BOD exertion was due largely to the work of Butterfield, and it held that the exogenous substrate was not removed so rapidly as is shown in Fig. 10-7, i.e., during the first 1 or 2 days, but that it occurred throughout the oxygen uptake period, i.e., throughout the experiment. According to this concept, the filtrate COD curve shown in Fig. 10-7 would be much more shallow; soluble COD would be removed much more slowly. The role of predators, e.g., protozoa, was to continually lower the bacterial numbers so they would continue to grow, thereby removing more and more of the exogenous substrate. Thus, the actual respiratory contribution made by the predators was considered to be negligible. This concept still persists, although the newer concept is more obviously in line with the natural food chain in the decay leg of the carbon-oxygen cycle as well as with recent experimental observations. In any event, science goes forward through research, and the drive to research is fueled by controversy rather than agreement.

Thus, the occurrence of the plateau was challenged by some workers who soon found that the experimental observations of a plateau were repeatable, although not with the quantitative precision predicted by Busch and his coworkers, who felt that since the plateau marked the end of the substrate removal phase, the standard incubation time in the BOD test could be shortened from 5 days to perhaps 1 or 2 days. One of the findings of other workers was that a plateau in oxygen uptake could develop even if predators were not present. This finding of a plateau even in a pure culture study indicated that there were reasons other than predation for a second stage of oxygen uptake (Wilson and Harrison, 1960; Bhatla and Gaudy, 1965a). There was also the question of why there should be a pause rather than a smooth transition between what appeared to be two distinctly different respiratory activities. Indeed, it can be shown that the plateau is not always observed, but it is present with sufficient frequency to cause one to seek explanations for it.

One can envision various possible explanations for the occurrence of a plateau and then systematically try to prove or disprove these hypotheses by experimentation. There are various ways in which plateaus in oxygen uptake could conceivably be generated. If the exogenous substrate has been removed at the time of the plateau, one must look for a reason for the delayed onset of endogenous or autodigestive respiration. If a plateau is observed in a pure culture system, it is possible that some species need an acclimation period before initiating respiration of internal or endogenous stores. If a heterogeneous population is used, one might ask: Is there a period of adaptation before predators such as protozoa or other bacteria can begin to prey upon other cells? Also, one may ask whether it is entirely necessary that the plateau marks the end of the substrate removal phase. For example, holding to the ideas of Butterfield, can a plateau occur during the substrate removal period? The latter situation might exist in a system in which there is more than one type of exogenous carbon source if those carbon sources or groups of carbon sources were metabolized sequentially rather than concurrently (see Chap. 12). If a single carbon source were being metabolized, cells might be able to elaborate metabolic products into the medium that would be metabolized only after an acclimation period, and the acclimation period could be reflected in a pause in oxygen uptake.

In brief, all the above questions were asked and investigated experimentally, and all the possible reasons listed have been shown experimentally to be involved in the generation of plateaus in oxygen uptake, each acting under appropriate conditions (Gaudy, Bhatla, and Abu-Niaaj, 1963; Gaudy, Bhatla, Follett, and Abu-Niaaj, 1965; Bhatla and Gaudy, 1965a, 1965b, 1966). However, when a plateau is observed in oxygen uptake, the events shown in Fig. 10-38A usually occur (Bhatla and Gaudy, 1965b). When the plateau does not occur, the situation is usually that shown in part B. The difference lies in the "ecological kinetics" of the food chain decay process. Protozoa are thought to be somewhat fastidious in regard to the bacteria they use as food material. If any particular species of predator is to increase in numbers, it is possible that a large amount of its particular food supply must be present. Thus, the predator respiration curve may depend not only on the number of protozoa present in the original seed but also on the diversity of species

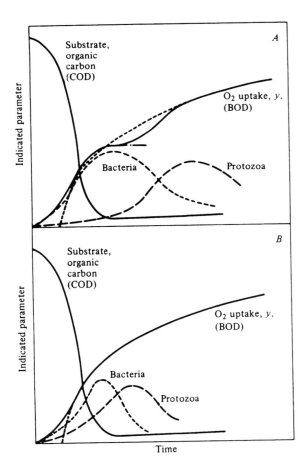

Figure 10-38 Relationship among oxygen uptake, organic carbon removal, and microbial growth. In A, growth of the protozoa lags bacterial growth sufficiently to cause development of a plateau in oxygen uptake, whereas in B, predator growth does not lag sufficiently to allow development of a plateau. (*From Gaudy, 1972.*)

present, in order that the possibilities of obtaining a useful match in the predator–prey relationship are increased. In the bacterial growth portion of the respiration curve, not one but many species are growing. The distribution of species can be expected to change during the early S-shaped portion of the curve. The two phases of metabolism may meld together and no plateau would be observed. On the other hand, the two phases may be separated in varying degrees to yield plateaus of various lengths. If the right match between predator and prey is not made, there may be no second phase of oxygen uptake at all. Furthermore, there is certainly no reason to think that the bacteria–protozoa relationship is the only one that can occur. The next group of predators in the food chain, e.g., crustacea, may cause a third phase, and so on. Also, some of the predators of one bacterial species may be another bacterial species, or some cells may autolyze to release exogenous substrate, thus allowing the growth of some bacteria. The latter types of respiration curves would cause smaller phases of increase in the oxygen uptake because the total energy in the closed system is being depleted constantly by respiration (see Fig. 3-1). Thus, if one smoothed over the plateaus after the exogenous carbon source in the sample has been removed, the curve would approximate first-order, decreasing-rate kinetics. However, attempting to force the entire oxygen uptake curve into this mold can lead to erroneous prediction of the DO profile in receiving streams. The use of BOD to help determine the assimilation capacity of a receiving stream is probably the best use for a BOD curve, and there is no need to use formal integration of a rough mathematical approximation of it when numerical integration of the best estimate of the entire curve is so easily accomplished in this age of calculators and computers.

Use of the BOD test in design and operation Prediction of the effect on the oxygen resources of the receiving stream is not the only use for the BOD concept. For years the concept of BOD has been embodied in the BOD test, which does not measure or trace the BOD curve but simply determines a single point along the curve—the accumulated oxygen uptake after a 5-day incubation period. It should be clear from the foregoing discussion that after 5 days, the system is in the autodigestive phase, i.e., the decreasing-rate portion of the curve. The rate and the mode of kinetics followed in arriving at this point are not known. However, overlooking this problem, it can be said that the standard 5-day BOD test (BOD_5) provides a way to estimate the pollution potential of the waste sample in terms of its demand on the oxygen resources of a body of water. As a relative index, it is useful provided we are aware of the facts outlined above. The 5-day BOD test is used widely in design as a measure of S_i flowing into biological treatment reactors. Here it is a very rough approximation because BOD is a measure, not of S_i, but only of the respiration due to metabolism of S_i. The respiration it measures in 5 days is a composite of that during the substrate removal phase and a portion of the autodigestive phase. The degree of autodigestion, as we have seen, is far less reliably reproduced or predicted than are the occurrences in the substrate removal phase. The ΔCOD of the waste appears to be a far more useful measure of S_i (see Chap. 3) for purposes of designing wastewater treatment plants. When considering the effluent of an operational treatment plant, however, it is the effect the effluent might have on the stream that is important. Running the standard 5-day BOD test on the plant effluent can give a useful indication of the potential hazard of excessive drawdown of oxygen in the stream (which existed 5 days ago). This is not very much help in operating the treatment plant, but it can provide useful information in the regulatory process when effluent BOD_5 values are compared with stream monitoring data, or effluent BOD values may help pinpoint the causes of reported adverse events downstream from the point of entry of the plant effluent.

The BOD test is a bioassay run under very dilute conditions, and the values one obtains depend not only on the types of kinetics exerted during the incubation period but also on the seeding material. Some seeding material from the waste treatment plant itself might be used, but one might well ask whether a low value for BOD would mean that the organic substrate had been removed in the plant or simply that the seed cells could not metabolize it. There is, after all, more than one way to register low effluent BOD values. Consider that some wastes, municipal wastes included, contain small concentrations of toxic materials. The effectiveness of a toxicant is dependent on the concentration of cells in the system. In biological treatment systems, rather high concentrations of cells are used, and the small concentration of toxicant present in the waste may not hurt the system to any great extent. But the presence of these small amounts of toxicants may greatly inhibit respiration under BOD test conditions in which a very small inoculum of cells must be used. Thus, BOD_5 is not necessarily a good assay even for keeping an historical record of plant performance. The fact that must be realized is that there is no single test of effluent quality that suffices for all the needs of designers, operators, and regulatory agencies. The BOD test is obviously of no immediate aid to the operator of a biological treatment plant. Values of COD_i, COD_e, and ΔCOD may be used as operational guides, and the BOD concept, using an open-jar technique, can be used to predict the effects of effluent on receiving streams. These methods are in general much more useful for these purposes than is the BOD_5 (Gaudy and Gaudy, 1971; O'Herron, 1974).

BOD is not the all-purpose measurement it is sometimes considered to be. The search for something better is a continuing effort, as is the search for better understanding of the process by which BOD is exerted. Throughout its long history, a great deal of the research on BOD has been directed toward finding the best ways to standardize the test so that results of various analysts can be compared. Standardization, of course, has its advantages, but without continued work on the kinetics and mechanisms of growth and respiration, standardization could simply indicate that we are making a uniform mistake. One of the current mistakes is that we use the BOD test for too many purposes. If it were not for the fact that there is always some residual COD or TOC in effluents from a biological treatment plant, even when a known, easily metabolizable compound is fed, one could recommend the ΔCOD procedure unequivocally as a measure of S_i in the design of treatment plants.

BOD and Residual COD

Some residual COD (COD_e) is always observed in batch and continuous flow growth reactors. It is expected when the original sample or inflow to the plant contains both metabolizable and nonmetabolizable organic matter. The ΔCOD is a useful measure of S_i in this case; however, when a readily metabolizable compound (e.g., glucose, for which one expects the inflowing COD, COD_i, to equal ΔCOD) is used as carbon and energy source, there is always some remaining organic matter as measured by COD even when it can be shown that there is none of the original substrate present. For example, the residual soluble COD in Fig. 10-7 is approximately 50 mg/L. The carbon and energy source in this case was glucose, and although analyses for glucose were not made in this particular experiment, they have been made in many similar studies and only very small concentrations of glucose are found—usually 5 to 10 percent of the residual COD.

When one runs a BOD test using seed taken from the growth reactor or from sewage, the values observed for this soluble material are very low, in the range of 10 to 20 percent of the COD; thus, if one judged the performance of the growth reactor on the basis of the 5-day BOD test rather than the effluent chemical oxygen demand COD_e, its performance would be judged to be much better. The fact that the organic matter detected as COD_e was not metabolized at the plant and exhibited a low biochemical oxygen demand does not necessarily mean that the organic matter in the effluent will not be metabolized in the receiving stream, however. Two types of observations show that the low BOD "of the moment" does not necessarily indicate that the residual BOD is biologically inert. If one collects such an effluent and subjects the soluble residual COD to long-term aeration in the presence of a small seed from a receiving stream or soil or sewage from the plant that produced the effluent, there is often slow oxygen uptake—much slower than the oxygen uptake due to the COD of the original waste sample. Also, if one allows the soluble residual COD to build up in a batch-fed, closed reactor, at times it will increase and then undergo a decrease, indicating that species have developed that eventually metabolize some portions of the residual soluble organic matter (T. R. Blachly and A. F. Gaudy, unpublished results). However, these materials are generally metabolized slowly. Thus, the aeration capacity of most receiving streams is sufficient to accommodate the "residual" soluble COD from a metabolically efficient biological treatment plant even though this material may represent a significantly higher oxygen demand in the receiving stream than would be indicated by the amount of soluble BOD in the effluent as judged by the standard BOD test.

Much more work is needed to characterize the nature of compounds constituting COD_e and their effect on receiving streams. Such knowledge is vital to eventual successful attainment of technological control of the water resource. While a low soluble BOD_e is no guarantee of low ultimate oxygen demand in the receiving stream, there would not appear to be any immediate cause for concern over depletion of the oxygen resources in receiving bodies due to this slowly metabolized residual COD. Even though technological–political decisions may be made not to use it, receiving streams do possess assimilation capacity and a good use of these slow-rate aerobic reactors is for the oxidation of slowly metabolized organic matter such as the relatively low residual COD from effectively operated biological treatment plants. Whether this approach will be possible depends on gaining further knowledge of the kinetic course of the metabolism and eventual disposition of these residuals. It should be emphasized, however, that depending on the immediate use of the effluent, there may not be time for aerobic decay of these organic residuals in the receiving stream, in which case tertiary treatment at the plant site is necessary. The soluble residual COD generally can be very much reduced by holding the effluent in contact with a high concentration of biomass (Gaudy and Srinivasaraghavan, 1974b; Saleh and Gaudy, 1978). This observation provided justification for assuming that the soluble substrate in the recycle line of an activated sludge plant could, for the most part, be neglected (see Chap. 6). It is not clear how much of this removal was due to sorption of polar molecules on floc surfaces and how much to biological, i.e., metabolic, uptake, but if one contacted the same effluent with activated carbon, somewhat less soluble COD_e was usually removed, suggesting that the result was partially due to metabolic uptake.

The nature of COD_e is uncertain and somewhat difficult to determine because the compounds of which it is composed are present in such small amounts. Analyses made during the experiments shown in Fig. 10-16A throw some light on the subject (Thabaraj and Gaudy, 1971). Note that the COD and the glycerol concentration remaining in solution during the substrate removal period are the same. Thus, there is little or no evidence for the buildup of metabolic intermediate products during the purification phase. After the eleventh hour, no glycerol was found in the medium. During the autodigestion period, the COD, after rising to 200 mg/L, declined in an oscillating pattern. Attempts made to determine the nature of the soluble COD included analyses for various volatile acids, keto acids, and ribose. These types of compounds accounted for only a small fraction of the residual COD. Only traces of acetic acid and keto acids were found, and the ribose concentration varied from 16 to 2 mg/L, with the high concentrations occurring early in the autodigestive phase. The presence of ribose suggests that a portion of the filtrate COD arose from degradation of nucleic acids during the autodigestion period.

In addition to the soluble COD, treatment plant effluents contain suspended organic solids because the secondary clarifier does not yield 100 percent efficiency. Usually the suspended solid's concentration in the overflow amounts to 20 to 30 mg/L. When BODs and CODs are run on effluents, usually the total value, i.e., soluble plus suspended, is obtained. In addition to the obviously greater amount of work involved in determining values for soluble and total BOD and COD separately, the reasoning is that the receiving stream makes no distinction between soluble and suspended oxygen demand. Therefore, prediction of the effect of the effluent on the receiving stream or measurement of the efficiency of plant performance is better based on overall values; that is, the efficiencies of the bioreactor and the settling tank are lumped together. There is some justification for this approach, but it obviously does not help professionals to improve the design and the operation of the individual treatment units involved. There are situations in which 20 to 30 mg/L of suspended solids in the effluent cannot be allowed. In such cases, tertiary treatment or perhaps addition of coagulants to aid in separation of suspended solids can be used. In most cases this low level of suspended solids can be released to a receiving stream.

One may ask whether these solids exert an oxygen demand similar to or different from that of the soluble organic matter in the effluent. It is recalled that the models for design and operation are set up to predict performance in the growth reactor, and they involve only the soluble substrate. For convenience (not because we really know), suspended organic solids in a reactor influent can be thought of as becoming trapped in the biofloc particles, where they are perhaps partially metabolized in the bioreactor but are most probably brought down with the sludge, and their later metabolism is tallied as part of the endogenous or aerobic decay of the sludge. The degree to which any suspended organic solids in the influent to the bioreactor are metabolized may well depend on the degree of starvation in the system, i.e., the net specific growth rate. Suspended solids in the effluent, regardless of their source, are those that have escaped entrapment in the treatment system. If we cannot say with surety how suspended solids in the influent are used in the treatment plant, we are even more uncertain about the metabolic fate of effluent suspended solids in the receiving stream. It would be valuable information in environmental monitoring to determine both the soluble and suspended biological and chemical oxygen demands of plant effluents. This would help answer questions about the similarity or difference in the kinetics of oxygen uptake between the two and the relative contribution of each to oxygen demand in the receiving stream.

The suspended solids usually found in effluents are doubtless a mix of living and nonliving organic matter. Some of the nonliving organic matter may be material that was in the influent to the bioreactor, or it may be cell debris arising from growth and decay in the system. Often the level of suspended solids in an effluent is found to be the same, i.e., 10 to 30 mg/L, whether the feed COD is totally soluble or contains some suspended organic matter, suggesting that the main source of effluent suspended solids is the biomass itself. However, this, like much else in this field, requires closer investigation. It is important to emphasize that the composition of the suspended solids can vary with their concentration in the effluent. We have been considering the makeup of the 10 to 30 mg/L suspended solids found in a "good" effluent, i.e., one from a plant that is working fairly well. However at times, due to bulking or rising sludge, the effluent solids can be expected to have the character of the return sludge if the process is activated sludge, or the character of the attached slime if the reactor is a fixed- or rotating-bed type.

The contribution of effluent suspended solids to the biochemical oxygen demand of an effluent is usually approximated in design calculations for a treatment plant. The general approach is to calculate the stated allowable effluent suspended solids concentration, which has been established by the appropriate regulatory agency, to an equivalent oxygen demand. If, for example, 30 mg/L of suspended solids is allowed, one usually begins by assuming that a certain portion of these solids is metabolizable. For example, one might assume that 80 percent of the suspended solids represents usable substrate. One may further assume that the suspended solids are biological cells. Next, one calculates their endogenous or autodigestive oxygen uptake (oxygen demand). One can assume that the cells can be totally oxidized or that they are subject to only partial oxidation, i.e., a residual cell fraction may be assumed to exist. Often a value of 80 to 90 percent of total oxidation is assumed. In order to apply the value, one must know the total oxygen demand, the COD of the cells. This value is often estimated as a theoretical COD calculated using an empirical formula for activated sludge, e.g., $C_5H_7NO_2$.

Since many designs are based on BOD_5, one must now estimate how much of the total oxygen demand is exerted in 5 days. More often than not, it is assumed that 70 percent of the ultimate oxygen demand is exerted in 5 days. The basis for this assumption is the first-order decreasing-rate mode of expression of BOD with $K_1 = 0.23$ day^{-1}. Here we see assumption piled upon assumption in order to determine (estimate) the biochemical oxygen demand of the suspended solids, and without even knowing what percentage of these were organic in the first place. When one has made an estimate of the contribution of suspended solids to BOD, the remainder becomes the allowable soluble BOD that can be contained in the effluent. It is this value of S_e that one uses in the design calculation to determine the values of X_R, α, and $\bar{t}$, i.e., μ_n or Θ_c, needed to deliver the required S_e.

Depending on the numerical values used in the series of assumptions outlined above, the soluble BOD may amount to as little as 25 percent of the allowable effluent BOD. Thus, the bioreactor must, in the main, make up for the expectable lack of efficiency of the clarifier. This calls for extremely fine tuning of kinetic models and refined design and operational skills. Treatment plants are designed on the basis of "design assumptions"; these are certainly the most reasonable that existing knowledge allows, but plants function and are operated in the field in real situations. There is critical need for more information because the treatment plants do not function in the field in precise accord with the assumptions made in the design. The present lack of better information on the biological oxygen demand of both effluent suspended solids and soluble organic matter in the effluent can encourage the general application of tertiary treatment, which society may not be able to afford. Today if effluent standards lower than 20 mg/L BOD_5 and 20 mg/L of suspended solids are imposed, the tendency is to add on solids-removing unit processes or operations, e.g., chemical treatment, filtration, etc. These greatly increase costs to society. The paradox is that the price of ignorance may be higher than the cost of mounting the massive investigational effort needed to assess the biological effects of these dilute materials on the receiving body. Reliable information may well indicate that these materials can be safely left for nature to degrade or remove. The point is that we do not really know and it would be wise to find out before committing society to the enormous expense of treatment to achieve "zero" discharge.

The effluent suspended solids do exert a biochemical oxygen demand. In pilot plant studies in our laboratories, using both municipal sewage and synthetic media as feed, BOD_5 analyses on soluble and total clarifier effluent amounted to approximately one-fifth of the respective COD values. For example, for an effluent averaging 20 mg/L of soluble COD (i.e., COD_e) and 30 mg/L of total COD, the corresponding BOD_5 values were 5 and 8 mg/L on average. The amount of effluent suspended solids averaged less than 20 mg/L. Since the COD due to the effluent suspended solids was only 10 mg/L (i.e., $COD_{total} - COD_e$), a significant portion of these solids must have been inorganic or in any event highly oxidized, since they accounted for only 3 mg/L of the BOD_5. Analyses such as this indicate that it may be well to reevaluate the assumptions made in estimating the biochemical oxygen demand of the effluent; far more extensive sampling is needed, not only in laboratory studies, but particularly in the field at actual treatment plants in order to establish more realistic estimates for design purposes. Relationships between total and soluble BOD and COD and effluent suspended solids will, of course, vary for different wastes, degrees of treatment, and operational conditions.

It should be apparent that the organic materials causing biochemical oxygen demand in the influent are for the most part of an entirely different nature than those in the effluent. It seems fair, then, to ask whether the kinetics of oxygen uptake for treatment plant effluents will be different from those we have discussed previously in relation to untreated wastewaters. In a case where the oxygen uptake due to soluble substrate is lower than that due to organic suspended solids, which is usual under present effluent requirements, one cannot really tell which fraction of the suspended solids serves as seed organisms and which as substrate. That is, one has little idea of the S_0/X_0 ratio in such a system. However, some insight into the effect of the ratio of soluble to suspended organic matter on the kinetics of oxygen uptake is provided by the results described below. In experiments using a totally soluble substrate of known composition (glucose) in which oxygen uptake was measured by the open-jar technique (see Chap. 6), the S_0/X_0 ratio varied from 19 to 6, and it was found that the higher values tended to produce the

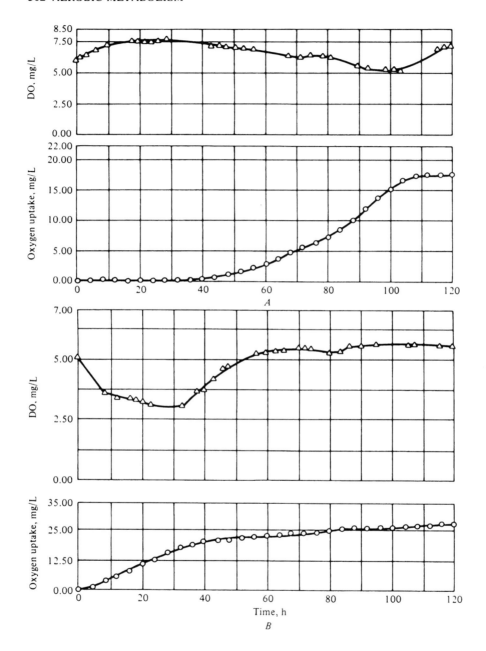

S-shaped oxygen uptake curve previously described. However, at the low values of S_0/X_0, the rate tended toward a linear and/or ever-decreasing rate. That is, as the system became more dominated by cells rather than soluble substrate, the oxygen uptake curve began to resemble an endogenous respiration or autodigestive curve. Thus, at low S_0/X_0 ratios, the data tended more toward an approximation of first-order, decreasing-rate kinetics. It is difficult to generalize in regard to the mode of oxygen uptake of an effluent because the nature of the substrate is not known, but some insights are obtained from Fig. 10-39A and B. Both portions of this figure represent dissolved oxygen sag curves and calculated oxygen uptake curves from open-jar BOD studies on effluents from a hydrolytically assisted extended aeration pilot plant. In part A the data were generated from a 1/1 dilution of effluent with tap water. The only seed organisms were those already in the effluent. The initial suspended solid concentration was 12 mg/L, and the initial filtrate COD was 49 mg/L. The effluent was highly nitrified and essentially no ammonia–nitrogen was present. There was a long lag period characteristic of a seed consisting of old cells at low concentration, and within the 5-day incubation period the oxygen uptake curve was S-shaped. The lag period may have been due in part to the need for acclimation to nitrate–nitrogen as nitrogen source for growth or for ac-

---

Figure 10-39 (A) DO profile and oxygen uptake curves for effluent from an extended aeration activated sludge pilot plant.

	Initial	Final
Substrate concentration filtrate COD	49 mg/L	—
Suspended solids concentration	12 mg/L	16 mg/L
Reaeration rate constant, $K_2 = 0.162\ h^{-1}$		
Dissolved oxygen saturation constant, $C_s = 7.55$ mg/L		
Oxygen uptake = 17 mg/L		
Dilution factor = 1/1		

(B) DO profile and oxygen uptake curves for effluent from an extended aeration activated sludge pilot plant. Mixed liquor suspended solids were purposely added. (*Data from A. F. Gaudy and M. P. Reddy, unpublished.*)

	Initial	Final
Substrate concentration filtrate COD	42 mg/L	40 mg/L
Suspended solids concentration	178 mg/L	160 mg/L
Reaeration constant, $K_2 = 0.185\ h^{-1}$		
Dissolved oxygen saturation constant, $C_s = 5.90$ mg/L		
Oxygen uptake = 27 mg/L		

climation to the organic matter in the effluent. In part B the initial soluble COD was 42 mg/L, and the initial suspended solids concentration was purposely made high by the addition of mixed liquor suspended solids from the aeration tank (178 mg/L). It is seen that the lag was reduced and the kinetic mode of oxygen uptake was changed from that shown in part A. Ammonia-nitrogen and nitrate-nitrogen were not checked during this period of pilot plant operation, and it is not known whether any of the 27 mg/L of oxygen uptake was due to nitrification. However, spot checks for nitrate-nitrogen at the beginning and end of the open-jar experiment indicated that no nitrates were present at either the beginning or end of this experiment. These results demonstrate two distinctly different types of kinetic expression for oxygen uptake, due presumably to the difference in S_0/X_0 ratios. However, they also indicate that more complete information is needed for analysis of the results. In addition to knowing oxygen uptake and S_0/X_0, one should know whether nitrogenous oxygen uptake occurred and possibly the nature of the limiting nutrient, since metabolism might be limited by the type and/or amount of nitrogen source rather than by the type and/or amount of COD. Clearly, the determination of BOD_5 on total (clarifier) effluent leaves serious gaps in one's knowledge of the effect of a treatment plant effluent on the receiving stream.

NITRIFICATION

In regard to both the treatment of wastes and the measurement of the effect of treated effluent on oxygen resources in the receiving stream, the oxygen uptake during nitrification of ammonia-level nitrogen is a very important consideration. A secondary plant effluent that contains 20 mg/L of ammonia-nitrogen (not an uncommon amount) can exert a BOD of approximately 90 mg/L. Equations (10-21) and (10-22) show the oxidation of ammonia-nitrogen to nitrite by *Nitrosomonas* and the subsequent oxidation of the nitrite to nitrate by *Nitrobacter*. Equation (10-23) shows the overall action.

$$NH_4^+ + 1.5O_2 \rightarrow 2H^+ + H_2O + NO_2^- \tag{10-21}$$

$$NO_2^- + 0.5O_2 \rightarrow NO_3^- \tag{10-22}$$

$$NH_4^+ + 2O_2 \rightarrow NO_3^- + 2H^+ + H_2O \tag{10-23}$$

The total oxidation of 20 mg/L of ammonia-nitrogen would, in accordance with Eq. (10-23), exert an oxygen demand of $20(64/14) = 91$ mg/L. Also note that nitrification tends to depress the pH value. In water containing carbonate alkalinity, the hydrogen ion reacts with biocarbonate ion to form H_2CO_3. Some of the NH_4^+ is incorporated into cells, which reduces the amount available for oxidation. Using average values for cell yields and oxygen consumption of these organisms and an empirical formula for cells $(C_5H_7NO_2)$, an overall reaction for the growth and oxidation of ammonia by both organisms can be written (U.S. EPA, 1975):

$$NH_4^+ + 1.83O_2 + 1.98HCO_3^- \rightarrow$$
$$0.021C_5H_7NO_2 + 1.041H_2O + 0.98NO_3^- + 1.88H_2CO_3 \qquad (10\text{-}24)$$

The oxygen demand of 20 mg/L of ammonia–nitrogen in accordance with this equation is $20(58.4/14) = 84$ mg/L, not including the nitrogen demand that might arise during the autodigestion of the nitrifying organisms that were grown. From the standpoint of wastewater treatment, nitrification of the ammonia originally in the waste or that produced from the decay of nitrogen-containing organic matter in the waste or of the biomass produced is desirable since this oxygen demand would otherwise be exerted in the receiving body.

The presence of ammonia–nitrogen and of nitrifying bacteria in the seed can cause variations in the results of the 5-day BOD test. In raw municipal waste or primary effluent, nitrification usually begins well after 5 days, and there is usually a clear separation between exertion of carbonaceous and nitrogenous oxygen uptake. Generally the nitrogenous BOD curve adds on to the carbonaceous curve as a distinct phase of oxygen uptake starting after approximately 8 to 10 days of incubation. Effluents from secondary municipal waste treatment plants generally contain more ammonia–nitrogen than is needed for growth on the carbon source they contain. Also, they may contain rather large numbers of organisms capable of nitrification; in fact, the effluent may be partially nitrified already and both carbonaceous and nitrogenous BOD can be exerted concurrently, leading to gross errors in assessing either carbonaceous or nitrogenous BOD in the effluent. If an effluent is already completely nitrified, the BOD_5 that is measured is an assessment of the carbonaceous oxygen demand remaining in the waste. Such an effluent can be expected to contain a high concentration of nitrifiers, but there is no energy source for them. They may, however, nitrify ammonia–nitrogen released during the autodigestive phase in the BOD bottle. Thus, they can still cause concern, depending largely on the amount of carbonaceous demand in the effluent. It is apparent that nitrification is a concern in the treatment plant, in the BOD test, and in the receiving stream. It has become accepted that not all nitrification should be allowed to occur in the receiving stream. This is as reasonable as not letting all of the carbonaceous demand be exerted in the receiving stream. It is not sound to take out organic matter and yet allow the dissolved oxygen level to drop below allowable limits in the stream because of nitrogenous oxygen demand. Thus, the production of nitrified effluent at public (or private) treatment plants is rapidly becoming an advanced treatment requirement.

Nitrification in the BOD Test

While we must be concerned with the total carbonaceous and nitrogenous oxygen demand of an effluent, it is also desirable to distinguish between the two, and this, as we have seen, cannot be accomplished simply by running BOD_5 tests on the effluent. A number of alternatives can be formulated. One can elongate the incubation time with the aim of providing time for both types of oxygen demand to be exerted completely. This has some merit, but it does not allow a distinction between carbonaceous and nitrogenous BOD. Also, much preliminary study is needed in order to decide on the best incubation period to use. Furthermore, the optimum conditions for growth of the heterotrophs are not necessarily the same as those for the slower growing nitrifiers. One must be able to distinguish between nitrification of nitrogenous material in the effluent and nitrification of nitrogenous material that may be present in the dilution water if the effluent sample requires dilution. Standard dilution water for the BOD_5 test contains ammonium ions in order to enhance balanced growth of the heterotrophs; this must be "calculated out" of the BOD results. If the dilution is made with water from the receiving stream, it is important to determine the kinds and amounts of nitrogen in this natural dilution water. If one measured the amounts of nitrite and nitrate at the beginning and end of the incubation period, one could calculate the oxygen that had to be used to increase the nitrite and nitrate concentrations and subtract the amount from the total oxygen uptake during incubation. This would distinguish between carbonaceous and nitrogenous BOD, but it requires extra analyses. These analyses and more will be required in any event, since it is folly to spend billions of dollars for treatment facilities and neglect operational and monitoring activities. A jug or jar technique, preferably using an open reactor with a K_2 value close to that expected in the near downstream reaches of the receiving stream, would be better than the standard 300-ml BOD bottle reactor, since it could be made large enough to permit periodic sampling and it simulates conditions in the receiving stream much more closely than does the BOD bottle. Electrometric measurement of oxygen and inorganic nitrogen rather than chemical analyses could facilitate generation of kinetic data.

Another alternative is to prevent one type of oxygen uptake by selectively killing or inhibiting the heterotrophs or the autotrophs in the seed. Nitrification is more easily prevented than is carbonaceous uptake, and various expedients have been suggested. Some of them include pasteurization and/or acidification and neutralization of the effluent, followed by reinoculation with seed devoid of nitrifiers. Addition of selected inhibitors of nitrification has also been tried, since nitrifiers are more easily inhibited than are the many different heterotrophs in the effluent. Nitrate rather than ammonium ion as nitrogen source in the dilution water has also been used as a means of curtailing nitrification. Nitrification is also inhibited by high concentrations of ammonia. Severe suppression of nitrification by an ammonia–nitrogen concentration of $0.1M$ (1400 mg/L) was reported by Meyerhof in 1917 (see Focht and Chang, 1975; Painter, 1970), and this concentration of ammonia–nitrogen was found to suppress nitrification under conditions of the standard 5-day BOD test (Siddiqi et al., 1967).

Incubation of samples for total BOD and for carbonaceous BOD exertion can be of some help in monitoring the separate oxygen demands of treatment plant effluents. In any event, one cannot negate consideration of nitrogenous BOD in effluents that contain mineralized nitrogen. In order to assess the effect of an effluent on the receiving stream, one must go further than determining the oxygen used in either the standard or an arbitrarily selected period of time. With nitrogenous oxygen demand as well as carbonaceous demand, some picture of the time course or rate curve for the process must be obtained and integrated with the reaeration characteristics (values of K_2) of the natural reactor into which the effluent is passed. Nitrification complicates the problem of producing a mathematical equation to describe the overall process adequately. Although exertion of the carbonaceous BOD of secondary treatment plant effluent tends to have a first-order, decreasing rate, exertion of nitrogenous BOD in a nonnitrified effluent tends to produce the usual S-shaped curve characteristic of growth. While it is desirable to seek mathematical expressions, the obtaining of a representative oxygen uptake curve and its numerical integration with the reaeration characteristics of the stream are practical methods that can be usefully employed now. The use of the open-jug technique and calculation of the expected DO profile downstream from the effluent outfall provide historical data during plant operations that can be compared with the actual DO profile observed in the stream. Once the similarity of the predicted and actual DO profiles has been established, any divergences can be used in analyzing the results of regulatory agency investigations, and they can be an aid in fine-tuning technological control of the aqueous environment. There is wisdom in being able to show, with data, the effect of one's effluent on the downstream dissolved oxygen resource, because in time of stress in the stream it is necessary to identify the offenders, and there may be other potential sources of pollution in the area. It is not enough to cast doubt in regard to these other sources; one will have to show that one's own effluent could not be among the potential offenders.

It must be emphasized that federal and state agencies regulate effluents and streams. This fact is sometimes forgotten during the present period when the federal agency, in particular, is functioning as a construction grant agency. This function is now necessary so that the regulating function, which is the main one, can be carried out in the future. Looking to the future, the public construction phase will give way to intensive monitoring and regulatory activities on the part of these government (social) pollution control agencies. Those responsible for producing purified effluent and operating and managing the receiving streams must, of necessity, maintain a continual and extensive monitoring program for aid in operation and for historical logging of both plant performance and the effect of the effluent on the natural reactor that the plant was installed to help control. This state of affairs will come about partly because these data will be required by the regulating agency and partly because it makes good sense for the plant owners to check on the results of the surveillance exercised by the regulatory agencies.

Nitrification in the Treatment Plant

We have dealt at some length with the growth of aerobic heterotrophs in treatment plant reactors, and much of the same discussion of the control of growth and its kinetic description pertains to the nitrifiers as well, although their carbon source is carbon dioxide and the initial energy source is ammonia–nitrogen. In general, the hyperbolic relationship between the concentration of growth-limiting nutrient and the specific growth rate μ can be applied to the nitrifiers. The carbon source, bicarbonate ion, is usually present in excess and the energy source is the limiting nutrient. The nitrifiers exhibit rather low values of μ_{max}. Values less than $1.0\,\text{day}^{-1}$ are not unusual for growth in sewage in the temperature range of 20 to 25°C, although μ_{max} values greater than $2.0\,\text{day}^{-1}$ have been reported for a pure culture of *Nitrosomonas*. Values for K_s generally fall between 1 and 5 mg/L of ammonia–nitrogen in this temperature range. Reported values for μ_{max} and K_s are higher in higher temperature ranges and lower than those cited above at lower temperatures. At temperatures in the range of 20 to 25°C, values for Y_t and k_d of 0.15 mg of cells per 1 mg of ammonia–nitrogen converted to nitrate–nitrogen and $0.07\,\text{day}^{-1}$, respectively, appear to be reasonable for nitrifying bacteria, although little actual information is available on k_d values for these organisms.

With the heterotrophs, most of the available evidence indicates that the rate of growth is not affected by the DO concentration unless it falls below a rather low critical level. However, with the nitrifiers, there is some evidence that nitrification is enhanced at increased levels of DO, in accordance with the hyperbolic function, i.e., a Monod-type relationship, and it would appear wise not to let the DO levels fall below 2 mg/L. This level of DO is recommended for the removal of carbonaceous substrate as well, in view of the fact that the cells exist as flocculated microbial communities. We do not really know where the nitrifiers live within the biological community, but their location may have much to do with the numerical values of the kinetic constants used to describe their growth. All available information indicates that nitrifiers grow best at pH values slightly on the alkaline side. A pH near 8 is optimum for growth of nitrifiers, but nitrification can proceed at pH values slightly below 6.

It is not necessary to grow nitrifiers at their maximum growth rate any more than it is to grow the heterotrophs at their fastest rate. In fact, as we shall see in Chap. 13 and have already seen in Chap. 6, it is essential to grow the organisms in any biological treatment system at well below the maximum specific growth rate. What is required is to grow the system at a rate commensurate with maximum purification capacity and stability, and this is definitely not the maximum growth rate. It is essential that one adjust the net specific growth rate to accommodate either or both types of organisms. If both carbonaceous and nitrogenous substrates are to be removed in one reactor, one must not operate at a μ_n so high (or Θ_c so low) as to wash out the nitrifying organisms; as μ_n increases, the μ_{max} of the nitrifiers is exceeded long before that of the heterotrophs. Alternatively, one can run separate reactors for removal of the carbonaceous substrate and ammonia–nitrogen. The latter alternative is sometimes preferred because one can optimize the operational conditions for each process separately. However, this requires an extra aerobic reactor and clarifier. The carbonaceous stage can be designed with fairly high μ_n because it is desirable to leave some carbonaceous substrate S_e in the inflow to the nitrification stage. The high μ_n or low mean cell residence time Θ_c tends to produce a large amount of excess sludge X_w from the first stage. On the other hand, removal of both carbonaceous and nitrogenous oxygen demand in one reactor tends to maintain a higher level of aeration solids X and lower excess sludge production X_w as a consequence of the low μ_n required to accommodate the nitrifiers. These aspects can be examined quantitatively using the growth model equations [Eqs. (6-75), (6-80), and (6-81)].

It is apparent that the low specific growth rate required for single-stage carbonaceous and nitrogenous BOD removal tends toward the rate characteristic of extended aeration plants. There is no easy guide to process selection, but is is essential to evaluate the overall costs for operation and construction of the plant and the reliability of accomplishing the goals of the plant—removal of both kinds of BOD and disposal of excess sludge—in making the final selection of the process scheme.

PROBLEMS

10-1 What types of macromolecular substrates are attacked by enzymes excreted by microorganisms, i.e., extracellular enzymes?

10-2 Consult a text on biochemistry and make a list of the specific bonds hydrolyzed by different proteinases. Compare enzymes of microbial and animal origin.

10-3 If waste A contained both carbohydrate and protein, and waste B contained protein but no carbohydrate, what would be the difference in the purpose for which protein was used in biological systems treating the two wastes? Which might foster the greatest diversity of species? Explain your answer.

10-4 What pathway is of major importance in the metabolism of fats?

10-5 In the text it was stated that 129 molecules of ATP can be synthesized when 1 molecule of palmitic acid is oxidized to carbon dioxide and water. Verify this calculation, assuming a P/O ratio of 3.0.

10-6 A certain industrial effluent exhibits the following characteristics:

COD = 12,000 mg/L, ΔCOD = 1000 mg/L
Nitrogen (as NH_3–N) = 600 mg/L
Phosphorus = 0
pH = 7.2 (titration curve shows it to be fairly well buffered)
$F = 2$ MGD
Effluent standard: 30 mg/L L_0, i.e., ultimate carbonaceous BOD; and 30 mg/L suspended solids

Prepare an outline for a proposal to the mill management in which you describe the information that is required in order to proceed with an approach to the functional design of an activated sludge process; this includes determination of the volume of the aeration tank V and estimation of the amount of excess sludge X_W. Describe how you would obtain the information, approximately how long it would take, and what your design procedures would be after you obtain the needed information. The treatment plant will probably cost approximately $3 million to construct. Try to estimate the cost for your "treatability and preliminary functional design study." Assume you are in the environmental engineering consulting business and have laboratory facilities and equipment. Defend the expense of the study versus proceeding with the functional design without it. How would you proceed without it?

10-7 A certain manufacturing plant produces a waste with a reasonably low BOD_5 of 280 mg/L. Nearly all of the BOD is soluble, and carbohydrates account for most of it. The waste is practically devoid of organic or inorganic sources of nitrogen. For several years the waste has been treated successfully in a well-mixed and well-aerated lagoon with a stilling basin between the aerated portion and the outfall. The mean hydraulic holding time $\bar{t}$ is 9 days. Phosphorus is added, *but* the operators have never added a source of nitrogen. The effluent has always been low in BOD and suspended solids. The plant managers have been told repeatedly by regulatory agency people and consulting engineers that they should be adding a nitrogen source. They call you in as a consultant to settle the question. How would you approach the situation? What questions would you ask? Do you feel that there may be a valid microbiological explanation for success without adding nitrogen?

10-8 The results of a long-term batch study are shown in the accompanying table. Glucose was fed to a small inoculum of acclimated cells (heterogeneous microbial population) in a growth medium. The experiment was conducted on the Warburg respirometer using 140-ml flasks. The oxygen uptake was measured manometrically. At each sampling time, a Warburg flask was removed and the contents were analyzed for soluble COD, mixed liquor COD, and biomass (suspended solids) concentration. (The data are unpublished results of A. F. Gaudy Jr., C. L. Goldstein, and P. A. Perkins.) The flask contents at the 22.5-h sample met with a not altogether uncommon misfortune. Alkali from the center well was spilled into the flask.

(a) Make materials and/or energy balances to determine the percentage recovery at each sampling time. Use as many calculation procedures as you can.

(b) If the recovery of mass or COD seems reasonable, estimate values for the missing data. State the basis for your estimates.

Time, h	Filtrate COD, mg/L	Mixed liquor COD, mg/L	Biomass, mg/L	O_2 uptake, mg/L
0	536	552	—	—
3.25	504	552	—	6
6.75	472	576	40	23
9.50	360	496	97	56
12.00	219	424	170	109
14.00	136	400	223	160
22.50	—	—	—	256
50.50	80	248	153	317
75.75	72	312	117	338
141.25	88	232	103	373
239.25	92	120	30	404
329.00	100	160	93	411
357.00	88	145	45	414

10-9 A wastewater treatment facility has been designed as an extended aeration (total oxidation) activated sludge process. The design conditions are as follows:

Mean hydraulic detention time $\bar{t} = V/F = 24$ h
Inflowing substrate, carbonaceous BOD_5, $S_i = 300$ mg/L
Inflowing nitrogen, ammonia–nitrogen, $N_i = 40$ mg/L

The applicable biokinetic constants are as follows:

Carbonaceous constants	Nitrification constants
$\mu_{max} = 5.0$ day^{-1}	$\mu_{max} = 0.5$ day^{-1}
$K_s = 75$ mg/L	$K_s = 2$ mg/L
$Y_t = 0.5$	$Y_t = 0.2$
$k_d = 0.10$ day^{-1}	$k_d = 0.05$ day^{-1}

Calculate (predict) the soluble carbonaceous BOD_5 in the effluent S_e and the ammonia–nitrogen concentration in the effluent N_e for the given design conditions.

10-10 Describe how the continuous oxidative assimilation process differs from the biosorption process.

10-11 A two-stage nitrification activated sludge plant has been designed so that the first stage decreases the carbonaceous BOD_5 to 50 mg/L. Find the volume of the second-stage aeration tank if the estimated ammonia–nitrogen concentration exiting the first stage is 30 mg/L and the effluent standard for ammonia–nitrogen is 1.0 mg/L.

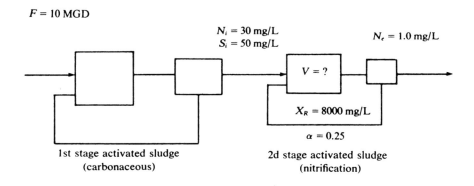

$F = 10$ MGD

$N_i = 30$ mg/L
$S_i = 50$ mg/L

$N_e = 1.0$ mg/L

$V = ?$

$X_R = 8000$ mg/L

$\alpha = 0.25$

1st stage activated sludge
(carbonaceous)

2d stage activated sludge
(nitrification)

The applicable biological constants are as follows:

Carbonaceous constants	Nitrification constants
$\mu_{max} = 5.0$ day^{-1}	$\mu_{max} = 0.6$ day^{-1}
$K_s = 100$ mg/L	$K_s = 2$ mg/L
$Y_t = 0.5$	$Y_t = 0.15$
$k_d = 0.08$ day^{-1}	$k_d = 0.07$ day^{-1}

10-12 Table 1 shows dissolved oxygen ("DO sag") data developed using the open-jar technique by placing a known dilution of wastewater in river water and subjecting it to gentle aeration. Table 2 shows reaeration curve data obtained after generating the DO sag data. The cells were killed with chlorine and sodium sulfite was added to remove the existing DO. The system was then monitored for DO. Table 3 shows the field data for the average cross section of the river downstream of the wastewater outfall. Plot the predicted DO profile in the river if the DO concentration at the point of entry of the waste is 7.0 mg/L.

Table 1

Time, h	DO, mg/L	Time, h	DO, mg/L	Time, h	DO, mg/L
0	6.55	38	0.70	76	5.30
4	5.40	40	1.30	80	5.35
8	4.25	44	2.10	84	5.50
12	3.25	48	2.80	88	5.55
16	2.35	52	3.35	92	5.70
20	1.40	56	3.75	100	5.80
24	0.90	60	4.25	104	5.80
26	0.35	64	4.65	108	5.85
28	0.48	68	4.75	112	5.55
32	0.60	72	5.10	120	6.10

Table 2

Time, h	DO, mg/L	Time, h	DO, mg/L	Time, h	DO, mg/L
0	0.2	2.50	3.05	6.50	5.50
0.17	0.38	3.00	3.45	7.00	5.75
0.42	0.68	3.42	3.70	7.50	6.00
0.70	1.08	4.42	4.40	8.50	6.30
0.92	1.39	5.00	4.70	9.17	6.40
1.42	1.90	5.50	5.00	11.42	7.00
2.00	2.60	6.00	5.25	14.67	7.40

Table 3 Velocities and depths for river cross section

Distance from left bank, ft	Depth, ft	Velocity, ft/s
0	0	0.4
2	1.1	0.5
4	0.9	0.66
6	0.9	1.06
7	0.7	1.00
8	0.2	0.70
10	0.2	0.66
12	0	0

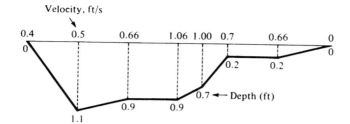

Velocity, ft/s

REFERENCES AND SUGGESTED READING

Anderson, R. L., and W. A. Wood. 1969. Carbohydrate metabolism in microorganisms. *Annu. Rev. Microbiol.* 23:539–578.

Angelbech, D. I., and E. J. Kirsch. 1969. Influence of pH and metal cations on aggregative growth of a non-slime-forming strain of *Zoogloea ramigera*. *Appl. Microbiol.* 17:435–440.

Atkinson, B., and I. S. Daoud. 1976. Microbial flocs and flocculation in fermentation process engineering. *Adv. Biochem. Eng.* 4:42–124.

Benedek, P., and I. Horvath. 1967. A practical approach to activated sludge kinetics. *Water Res.* 1:663–682.

Bhatla, M. N., and A. F. Gaudy, Jr. 1965a. Studies on the plateau in oxygen uptake during exertion of biochemical oxygen demand by pure cultures of bacteria. *Biotechnol. Bioeng.* 7:387–404.

Bhatla, M. N., and A. F. Gaudy, Jr. 1965b. Role of protozoa in the diphasic exertion of BOD. *J. Sanitary Eng. Div. ASCE* 91:SA3, 63–87.

Bhatla, M. N., and A. F. Gaudy, Jr. 1966. Studies on the causation of phasic oxygen uptake in high energy systems. *J. Water Pollut. Control. Fed.* 38:1441–1451.

Bonotan-Dura, F. M., and P. Y. Yang. 1976. The application of constant recycle solids concentration in activated sludge processes. *Biotechnol. Bioeng.* 18:145–165.

Brock, T. D. 1979. Biology of microorganisms, 3d ed. Prentice-Hall, Englewood Cliffs, N.J.

Busch, A. W. 1958. BOD progression in soluble substrates. *Sewage Ind. Wastes* 30:1336–1349.

Busch, A. W., and N. Myrick. 1960. Food population equilibria in bench scale biooxidation units. *J. Water Pollut. Control Fed.* 32:949–959.

Butterfield, C. T. 1935. Studies of sewage purification. II. A zoogloea-forming bacterium isolated from activated sludge. *Public Health Rep.* 50:671–684.

Butterfield, C. T., W. C. Purdy, and E. J. Theriault. 1931. Experimental studies of natural purification in polluted waters. IV. The influence of the plankton on the biochemical oxidation of organic matter. *Public Health Rep.* 46:393–426.

Costerton, J. W., G. G. Geesey, and K. J. Cheng. 1978. How bacteria stick. *Sci. Am.* 238(1):86–95.

Crabtree, K., W. Boyle, E. McCoy, and G. A. Rollick. 1966. A mechanism of floc formation by *Zoogloea ramigera*. *J. Water Pollut. Control Fed.* 38:1968–1980.

Dagley, S. 1971. Catabolism of aromatic compounds by microorganisms. *Adv. Microb. Physiol.* 6:1–46.

Dagley, S. 1975. Microbial degradation of organic compounds in the biosphere. *Am. Sci.* 63:681–689.

Dawes, E. A., and T. J. Senior. 1973. The role and regulation of energy reserve polymers. *Adv. Microb. Physiol.* 10:136–297.

Deinema, M. H., and L. P. T. M. Zevenhuizen. 1971. Formation of cellulose fibrils by Gram-negative bacteria and their role in bacterial flocculation. *Arch. Microbiol.* 78:42–57.

Doelle, H. W. 1975. Bacterial metabolism, 2d ed. Academic Press, New York.

Dugan, P. R. 1975. Bioflocculation and the accumulation of chemicals by floc-forming organisms.

EPA 600/2-75-032. Environmental Protection Technology Series, ORD. U.S. EPA, Cincinnati.

Dugan, P. R., and D. G. Lundgren. 1960. Isolation of the floc-forming organism *Zoogloea ramigera* and its culture in complex and synthetic media. *Appl. Microbiol.* **8**:357–361.

Eikum, A. S., D. A. Carlson, and A. Lundar. 1975. Phosphorus release during storage of aerobically digested sludge. *J. Water Pollut. Control Fed.* **47**:330–337.

Fisher, R. A., and R. A. Baker. 1976. Discussion—Application of kinetic models to the control of activated sludge processes. *J. Water Pollut. Control Fed.* **48**:1847–1853.

Focht, D. D., and A. C. Chang. 1975. Nitrification and denitrification processes related to wastewater treatment. *Adv. Appl. Microbiol.* **19**:153–182.

Frankel, D. G., and R. T. Vinopal. 1973. Carbohydrate metabolism in bacteria. *Annu. Rev. Microbiol.* **27**:69–100.

Friede, J., P. Guire, R. K. Gholson, E. Gaudy, and A. F. Gaudy, Jr. 1972. Assessment of biodegradation potential for controlling oil spills on the high seas. Report, Project DAT, U.S. Coast Guard.

Friedman, D. A., P. R. Dugan, R. M. Pfister, and C. C. Remsen. 1969. Structure of exocellular polymers and their relationship to bacterial flocculation. *J. Bacteriol.* **98**:1328–1334.

Gaden, E. L. Jr., Mary H. Mandels, E. T. Reese, and L. A. Spano (editors). 1976. Enzymatic conversion of cellulosic materials: Technology and applications. Biotechnology and Bioengineering Symposium No. 6. John Wiley, New York.

Gaudy, A. F., Jr. 1971. Comparative utility of research employing model synthetic systems and natural field systems. *In* Biological waste treatment, R. P. Canale (editor). Biotechnology and Bioengineering Symposium No. 2, 77–83, John Wiley, New York.

Gaudy, A. F., Jr. 1972. Biochemical oxygen demand. *In* Water pollution microbiology, R. Mitchell (editor). John Wiley, New York, pp. 305–332.

Gaudy, A. F., Jr. 1975. Prediction of assimilation capacity in receiving streams. *Water Sewage Works* **122**:62–65 (part I) and 78–79 (part II).

Gaudy, A. F., Jr. 1977. Chemically-assisted biological oxidation of wastes and excess sludge. *Water Sewage Works*, April 1977, **RN**:R48–R56.

Gaudy, A. F., Jr., M. N. Bhatla, and F. Abu-Niaaj. 1963. Studies on the occurrence of the plateau in BOD exertion. Proceedings 18th Industrial Waste Conference, Purdue University, Lafayette, Ind., pp. 183–193.

Gaudy, A. F., Jr., M. N. Bhatla, R. H. Follett, and F. Abu-Niaaj. 1965. Factors affecting the existence of the plateau during the exertion of BOD. *J. Water Pollut. Control Fed.* **37**:434–459.

Gaudy, A. F., Jr., M. N. Bhatla, and E. T. Gaudy. 1964. Use of chemical oxygen demand values of bacterial cells in wastewater purification. *Appl. Microbiol.* **12**:254–260.

Gaudy, A. F., Jr., and R. S. Engelbrecht. 1963. Basic biochemical considerations during metabolism in growing vs. respiring systems. *In* Advances in biological waste treatment, Vol. II. Pergamon Press, New York.

Gaudy, A. F., Jr., R. S. Engelbrecht, and B. G. Turner. 1961. Stripping kinetics of volatile components of petrochemical wastes. *J. Water Pollut. Control. Fed.* **33**:383–393.

Gaudy, A. F., Jr., and E. T. Gaudy, 1971. Biological concepts for the design and operation of the activated sludge process. U.S. Environmental Protection Agency, Report 17090FQJ 09/71.

Gaudy, A. F., Jr., and E. T. Gaudy. 1972. Mixed microbial populations. *Adv. Biochem. Eng.* **2**: Chap. 3. Springer-Verlag, Berlin.

Gaudy, A. F., Jr., K. C. Goel, and A. J. Freedman. 1969. Activated sludge process modification for nitrogen-deficient sludge. *Adv. Water Pollut. Res.* **613**–623.

Gaudy, A. F., Jr., K. C. Goel, and E. T. Gaudy. 1968. Continuous oxidative assimilation of acetic acid and endogenous protein synthesis applicable to treatment of nitrogen-deficient wastewaters. *Appl. Microbiol.* **16**:1358–1363.

Gaudy, A. F., Jr., K. C. Goel, and E. T. Gaudy. 1969. Application of continuous oxidative assimilation and endogenous protein synthesis to the treatment of carbonhydrate wastes deficient in nitrogen. *Biotechnol. Bioeng.* **11**:53–65.

Gaudy, A. F., Jr., S. R. Goswami, T. Manickam, and E. T. Gaudy. 1977. Formation of strippable metabolic products in biological waste treatment. *Biotechnol. Bioeng.* 19: 1239–1244.

Gaudy, A. F., Jr., T. S. Manickam, H. Saidi, and M. P. Reddy. 1976. Biological treatment of wastes with high ash content using hydrolytically-assisted extended aeration process. *Biotechnol. Bioeng.* 18: 701–721.

Gaudy, A. F., Jr., A. Obayashi, and E. T. Gaudy. 1971. Control of growth rate by initial substrate concentration at values below maximum rate. *Appl. Microbiol.* 22: 1041–1047.

Gaudy, A. F., Jr., M. Ramanathan, P. Y. Yang, and T. V. DeGeare. 1970. Studies on the operational stability of the extended aeration process. *J. Water Pollut. Control Fed.* 42: 165–179.

Gaudy, A. F., Jr., D. Scott, H. Saidi, and K. S. Murthy. 1976. Pilot plant studies on hydrolytically-assisted extended aeration processes at high organic loadings. Proceedings 31st Industrial Waste Conference, Purdue University, Lafayette, Ind., pp. 903–913.

Gaudy, A. F., Jr., and R. Srinivasaraghavan. 1974a. Effect of specific growth rate on biomass yield of heterogeneous populations growing in both continuous and batch systems. *Biotechnol. Bioeng.* 16: 423–427.

Gaudy, A. F., Jr., and R. Srinivasaraghavan. 1974b. Experimental studies on a kinetic model for design and operation of activated sludge processes. *Biotechnol. Bioeng.* 16: 723–738.

Gaudy, A. F., Jr., B. G. Turner, and S. Pusztaszeri. 1963. Biological treatment of volatile waste components. *J. Water Pollut. Control Fed.* 35: 75–93.

Gaudy, A. F., Jr., P. Y. Yang, R. Bustamante, and E. T. Gaudy. 1973. Exponential growth in systems limited by substrate concentration. *Biotechnol. Bioeng.* 15: 589–596.

Gaudy, A. F., Jr., P. Y. Yang, and A. W. Obayashi. 1971. Studies on the total oxidation of activated sludge with and without hydrolytic pretreatment. *J. Water Pollut. Control Fed.* 43: 40–54.

Glenn, A. R. 1976. Production of extracellular proteins by bacteria. *Ann. Rev. Microbiol.* 30: 41–62.

Goel, K. C., and A. F. Gaudy, Jr. 1968. regeneration of oxidative assimilation capacity by intracellular conversion of storage products to protein. *Appl. Microbiol.* 16: 1352–1357.

Goel, K. C., and A. F. Gaudy, Jr. 1969a. Studies on the relationship between specific growth rate and concentration of nitrogen source for heterogeneous microbial populations of sewage origin. *Biotechnol. Bioeng.* 11: 67–78.

Goel, K. C., and A. F. Gaudy, Jr. 1969b. Regulation of nitrogen levels for heterogeneous populations of sewage origin grown in completely mixed reactors. *Biotechnol. Bioeng.* 11: 79–98.

Gram, A. L. 1956. Reaction kinetics of aerobic biological processes. I.E.R. Series 90, Report #2. Sanitary Engineering Research Laboratory, University of California, Berkeley.

Gunsalus, I. C., and R. Y. Stanier. 1961. The bacteria. Vol. II: Metabolism. Academic Press, New York.

Gutnick, D. L., and E. Rosenberg. 1977. Oil tankers and pollution: A microbiological approach. *Annu. Rev. Microbiol.* 31: 379–396.

Harris, R. H., and R. Mitchell. 1973. The role of polymers in microbial aggregation. *Annu. Rev. Microbiol.* 27: 27–50.

Hawker, L. E., and A. H. Linton. 1971. Microorganisms. Function, form and environment. American Elsevier, New York.

Hawkes, H. A. 1963. The ecology of wastewater treatment. Macmillan, New York.

Heukelekian, H., H. E. Orford, and R. Manganelli. 1951. Factors affecting the quantity of sludge production in the activated sludge process. *Sewage Ind. Wastes* 23: 945–958.

Irgens, L. R., and H. O. Halvorson. 1965. Removal of plant nutrients by means of aerobic stabilization of sludge. Appl. Microbiol. 13: 373–386.

Isaacs, W. P., and A. F. Gaudy, Jr. 1967. Comparison of BOD exertion in a simulated stream and in standard BOD bottles. Proceedings 22nd Industrial Waste Conference, Purdue University, Lafayette, Ind., pp. 165–182.

Jenkins, D., and A. B. Menar. 1967. The fate of phosphorus in sewage treatment processes. I.

Primary sedimentation and activated sludge. SERL Report No. 67–6. University of California, Berkeley.

Jennelle, E. M., and A. F. Gaudy, Jr. 1970. Studies on the kinetics and mechanism of BOD exertion in dilute systems. *Biotechnol. Bioeng.* 12:519–539.

Keyes, T. W., and T. Asano. 1975. Application of kinetic models to the control of activated sludge process. *J. Water Pollut. Control Fed.* 47:2574–2585.

Klug, M. J., and A. J. Markovitz. 1971. Utilization of aliphatic hydrocarbons by microorganisms. *Adv. Microb. Physiol.* 5:1–43.

Komolrit, K., K. C. Goel, and A. F. Gaudy, Jr. 1967. Regulation of exogenous nitrogen supply and its possible applications to the activated sludge process. *J. Water Pollut. Control Fed.* 39:251–266.

Kountz, R. R., and C. Forney Jr. 1959. Metabolic energy balances in a total oxidation activated sludge system. *Sewage Ind. Wastes* 31:819–826.

Krishnan, P., and A. F. Gaudy Jr. 1968. Substrate utilization at high biological solids concentrations. *J. Water Pollut. Control Fed.* 40:1254–1266.

Lee, E. G., J. C. Mueller, and C. C. Walden. 1976. Ultimate disposal of biological sludges—a novel concept. *Pulp Paper Can.* 77:T104–T110.

Lehninger, A. L. 1975. Biochemistry, 2d ed. Worth, New York.

Levin, G. V., G. J. Topol, and A. G. Tarnay. 1975. Operation of full-scale biological phosphorus removal plant. *J. Water Pollut. Control Fed.* 47:577–590.

Long, D. A., and J. B. Nesbitt. 1975. Removal of soluble phosphorus in an activated sludge plant. *J. Water Pollut. Control Fed.* 47:170–184.

McCarty, P. L., and C. F. Brodersen. 1962. Theory of extended aeration activated sludge. *J. Water Pollut. Control Fed.* 34:1095–1103.

McKinney, R. E., and M. P. Horwood. 1952. Fundamental approach to the activated sludge process. I. Floc-producing bacteria. *Sewage Ind. Wastes* 24:117–123.

Metzler, D. E. 1977. Biochemistry. Academic Press, New York.

Middlebrooks, E. J., and C. F. Garland. 1968. Kinetics of model and field extended aeration wastewater treatment units. *J. Water Pollut. Control Fed.* 40:586–612.

Morgan, W. E., and E. G. Fruh. 1974. Phosphate incorporation in activated sludge. *J. Water Pollut. Control Fed.* 46:2486–2497.

Mortlock, R. P. 1976. Catabolism of unnatural carbohydrates by microorganisms. *Adv. Microb. Physiol.* 13:1–53.

Mulder, E. G., J. Antheuniser, and W. H. J. Cromback. 1971. Microbial aspects of pollution in the food and drug industries. *In* Microbial aspects of pollution, G. Sykes and F. D. Skinner (editors). Academic Press, New York, pp. 71–89.

Obayashi, A. W., and A. F. Gaudy, Jr. 1973. Aerobic digestion of extracellular microbial polysaccharides. *J. Water Pollut. Control Fed.* 45:1584–1594.

O'Herron, R. J. 1974. Literature survey of instrumental measurements of biochemical oxygen demand for control application. EPA-670/4-74-001. U.S. Environmental Protection Agency, Cincinnati.

Painter, H. A. 1970. A review of literature on inorganic nitrogen metabolism in microorganisms. *Water Res.* 4:393–450.

Payne, J. W. 1976. Peptides and microorganisms. *Adv. Microb. Physiol.* 13:55–113.

Peil, K. M., and A. F. Gaudy, Jr. 1975. A rational approach for predicting the DO profile in receiving waters. *Biotechnol Bioeng.* 17:69–84.

Phelps, E. B. 1944. Stream sanitation. John Wiley, New York.

Pike, E. B., and C. R. Curds. 1971. The microbial ecology of the activated sludge process. *In* Microbial aspects of pollution control, G. Sykes and F. D. Skinner (editors). Academic Press, New York, pp. 123–147.

Pipes, W. O. 1966. The ecological approach to the study of activated sludge. *Adv. Appl. Microbiol.* 8:77–103.

Pirt, S. J. 1965. The maintenance energy of bacteria in growing cultures. *Proc. R. Soc.* Series B. 163:224–231.

Porges, N., L. Jasewicz, and S. R. Hoover. 1953. Aerobic treatment of dairy wastes. *J. Appl. Microbiol.* 1:262–270.

Postgate, J. R., and J. R. Hunter. 1962. The survival of starved bacteria. *J. Gen. Microbiol.* 29:233–263.

Prather, B. V., and A. F. Gaudy, Jr. 1964. Combined chemical, physical and biological processes in refinery waste water purification. *Oil Gas J.* 62:96–99.

Saleh, M. M., and A. F. Gaudy, Jr. 1978. Shock load response of activated sludge operated with constant recycle sludge concentration. *J. Water Pollut. Control Fed.* 50:764–774.

Siddiqi, R. H., R. E. Speece, R. S. Engelbrecht, and J. W. Schmidt. 1967. Elimination of nitrification in the BOD determination with 0.10M ammonia nitrogen. *J. Water Pollut. Control Fed.* 39:579–589.

Sinclair, C. G., and D. N. Ryder. 1975. Models for the continuous culture of microorganisms under both oxygen- and carbon-limiting conditions. *Biotechnol. Bioeng.* 17:375–398.

Singh, T., and J. W. Patterson. 1974. Improvement of the aerobic sludge digestion process efficiency. *J. Water Pollut. Control Fed.* 46:102–112.

Sokatch, J. R. 1969. Bacterial physiology and metabolism. Academic Press, New York.

Srinivasaraghavan, R., and A. F. Gaudy, Jr. 1975. Operational performance of an activated sludge process with constant sludge feedback. *J. Water Pollut. Control Fed.* 47:1946–1960.

Stanier, R. Y., E. A. Adelberg, and J. Ingraham. 1976. The microbial world, 4th ed. Prentice-Hall, Englewood Cliffs, N.J.

Streeter, H. W., and E. B. Phelps. 1925. A study of the pollution and natural purification of the Ohio River. Bulletin #146. U.S. Public Health Service.

Surucu, G. A., R. S. Engelbrecht, and E. S. K. Chian. 1975. Thermophilic microbiological treatment of high strength waste waters with simultaneous recovery of single cell protein. *Biotechnol. Bioeng.* 17:1639–1662.

Swift, M. J. 1977. The ecology of wood decomposition. *Sci. Prog.* 64:175–199.

Sykes, G., and F. A. Skinner. 1971. Microbial aspects of pollution. Adademic Press, New York.

Symons, J. M., and R. E. McKinney. 1958. The biochemistry of nitrogen in the synthesis of activated sludge. *Sewage Ind. Wastes* 30:874–890.

Tago, Y., and K. Aida. 1977. Exocellular micropolysaccharide closely related to bacterial floc formation. *Appl. Environ. Microbiol.* 34:308–314.

Tenney, M. W., and W. Stumm. 1965. Chemical flocculation of microorganisms and biological waste treatment. *J. Water Pollut. Control Fed.* 37:1370–1391.

Thabaraj, G. J., and A. F. Gaudy, Jr. 1971. Effect of initial biological solids concentration and nitrogen supply on metabolic patterns during substrate removal and endogenous metabolism. *J. Water Pollut. Control Fed.* 43:318–334.

Thibodeaux, L. J. 1974. A test method for volatile component stripping of waste water. Environmental Protection Technology Series EPA-660/2-74-044, U.S. Environmental Protection Agency, Washington, D.C.

Thimann, K. V. 1963. The life of bacteria, 2d ed. Macmillan, New York.

Topiwala, H., and C. F. Sinclair. 1971. Temperature relationships in continuous cultures. *Biotechnol. Bioeng.* 13:795–813.

Unz, R. F., and J. A. Davis. 1975. Microbiology of combined chemical-biological treatment. *J. Water Pollut. Control Fed.* 47:185–194.

U.S. Environmental Protection Agency. 1975. Process design manual for nitrogen control. U.S. EPA Office of Technology Transfer, Washington, D.C.

Washington, D. R., L. J. Hetling, and S. S. Rao. 1965. Long-term adaptation of activated sludge organisms to accumulated sludge mass. Proceedings 19th Industrial Waste Conference, Purdue University, Lafayette, Ind., pp. 655–666.

Weddle, C. L., and D. Jenkins. 1971. The viability and activity of activated sludge. *Water Res.* 5:621–640.

Wells, W. N. 1969. Differences in phosphate uptake rates exhibited by activated sludges. *J. Water Pollut. Control Fed.* 41:765–771.

Wilson, I. S., and M. E. Harrison. 1960. The biochemical treatment of chemical wastes. *J. Inst. Sewage Purification* 3:261–275.

Yang, P. Y., and Y. K. Chen. 1977. Operational characteristics and biological kinetic constants of extended aeration process. *J. Water Pollut. Control Fed.* 4:678–688.

Yang, P. Y., and A. F. Gaudy, Jr. 1974a. Control of biological solids concentration in extended aeration. *J. Water Pollut. Control Fed.* 46:543–553.

Yang, P. Y., and A. F. Gaudy, Jr. 1974b. Nitrogen metabolism in extended aeration processes operated with and without hydrolytic pretreatment of portions of the sludge. *Biotechnol. Bioeng.* 16:1–20.

ELEVEN
ANAEROBIC METABOLISM

Anaerobic metabolism is metabolism in which oxygen plays no part, either as terminal electron acceptor or as a substrate for oxygenation reactions. In a variety of natural habitats, there is continually or intermittently a total absence of oxygen. Some of these, such as the depths of the oceans, lack nutrients and probably support little metabolic activity. Many anaerobic habitats, however, are rich in organic matter; in fact, it is often this abundance of organic matter that is responsible for the anaerobic conditions. In swamps, ponds, estuaries, and other bodies of water that contain decaying vegetation or in bodies of water that receive organic pollution from human or animal sources, large microbial populations develop. The organisms near the surface utilize oxygen as rapidly as it becomes available by diffusion from the air, and none reaches the lower depths. The anaerobic microorganisms, primarily bacteria, are thus a necessary link in the recycling of carbon from the anaerobic portions of the biosphere. Humans also utilize anaerobic metabolic processes in many important industrial applications and in some methods of waste treatment that we will discuss later in this chapter.

We have described several types of microorganisms that require, or are able to adapt to, anaerobic conditions. These may be divided into three broad groups based on their means of obtaining energy. As we saw in Chap. 7, there are three methods of generating ATP: oxidative phosphorylation, photophosphorylation, and substrate level phosphorylation. All three methods are used under anaerobic, as well as aerobic, conditions.

1. *Oxidative phosphorylation* can be utilized under anaerobic conditions by organisms that are able to use an external inorganic electron acceptor other

than oxygen. Bacteria belonging to several different genera are able to utilize nitrate when oxygen is unavailable. A few bacterial species, all strict anaerobes, use sulfate as an electron acceptor. A very important group of obligate anaerobes, the methane-forming bacteria, may be included in this group because they use carbon dioxide as electron acceptor, although the mechanism of formation of ATP in these organisms is not known.

2. *Photophosphorylation* can be used in anaerobic environments by any photosynthetic microorganism, since oxygen is not required for photosynthesis. Algal photosynthesis (including that of the cyanobacteria) produces oxygen and may aid in preventing anaerobiosis. However, such organisms utilize oxygen in the dark, and they flourish near the surface where oxygen is available. The photosynthetic bacteria are obligate anaerobes when growing photosynthetically, and they develop at depths where oxygen does not penetrate if other requirements for growth are available.

3. *Substrate level phosphorylation* supplies a minor portion of the ATP produced by organisms that utilize oxidative phosphorylation as their major source of ATP. However, in fermentative metabolism, substrate level phosphorylation is the sole means of producing ATP. In many anaerobic environments, fermentation probably accounts for most of the dissimilation of organic material. No external electron acceptor is required, since metabolic products formed by the organism from the energy source are used as electron acceptors. Microorganisms capable of fermentation include a great variety of bacterial species—both obligate and facultative anaerobes, the facultatively anaerobic fungi, and a few anaerobic protozoa. The greatest versatility in the fermentation of different types of organic compounds and in the formation of different fermentation products is found among the bacteria.

In this chapter we will examine the general principles that govern fermentation, the reactions involved in the formation of some of the more important fermentation products, and some of the natural and engineered processes that utilize anaerobic microorganisms.

SIMILARITIES IN ANAEROBIC AND AEROBIC METABOLISM

There is no a priori reason why most reactions that do not involve oxygen directly could not occur in both aerobic and anaerobic metabolism, and many of the reactions we have described previously do occur in both aerobes and anaerobes. The major difference between the two types of metabolism, as we discussed in Chap. 7, is in the final disposition of the hydrogen removed in oxidative reactions. This has a major effect only in organisms that do not utilize an external electron acceptor, i.e., in fermentative metabolism. Several categories of reactions are essentially identical in aerobes and anaerobes.

Anabolic Reactions

The pathways of biosynthesis are generally the same in both types of organism or in the same organism growing under aerobic or anaerobic conditions. Reactions involving the direct participation of oxygen are an exception, of course, We mentioned one example previously. Yeasts growing anaerobically require unsaturated fatty acids as organic growth factors, since the pathway used for their synthesis during aerobic growth involves oxygen. Obligate anaerobic organisms that are able to synthesize unsaturated fatty acids utilize a different pathway that does not require oxygen. Since most biosynthetic pathways do not involve the direct participation of oxygen, they can be utilized under either aerobic or anaerobic conditions. Prototrophy, the ability to grow without organic growth factors, synthesizing all cell material from a single carbon source, is more common among aerobic than among anaerobic microorganisms, but it is not unknown among anaerobes. Thus, there is no dividing line based on biosynthetic abilities or on biosynthetic pathways between aerobic and anaerobic metabolism.

Preparatory Reactions

The second area of similarity between aerobes and anaerobes is in the *preparatory reactions of catabolism*. These are the reactions by which various types of molecules, both polymers and small molecules, are prepared for entry into the central pathways. Again, only those reactions that require the direct participation of oxygen must necessarily be different under aerobic and anaerobic conditions. An example previously mentioned is the utilization of aliphatic hydrocarbons, which involves oxygenation by oxygen and is therefore restricted to aerobic microorganisms. The aerobic catabolism of aromatic compounds also involves the direct participation of oxygen, but recent work has shown that different methods of ring cleavage that do not require oxygen can be used by anaerobic bacteria (Evans, 1977). Organisms of the groups mentioned above are able to utilize aromatic compounds in this way, i.e., bacteria that grow photosynthetically (photoorganotrophs), those that utilize nitrate as terminal electron acceptor, and those that utilize fermentative pathways. Ring cleavage occurs by reduction followed by hydrolysis, rather than by oxygenation as in aerobes. It is probable, however, that a greater variety of aromatic compounds can be metabolized aerobically than anaerobically, since oxygenation reactions are used so commonly in the attack on natural and synthetic aromatics (Dagley, 1975).

Most commonly occurring polymers other than hydrocarbons are attacked hydrolytically by microorganisms. Hydrolytic enzymes such as amylases, proteinases, lipases, and nucleases are produced by both aerobic and anaerobic species. The strictly anaerobic genus *Clostridium* includes many species that are highly active in the production of extracellular hydrolytic enzymes. Thus, there is no difference in the methods used by aerobes and anaerobes to degrade the polymers synthesized by plant and animal cells. The monomers produced by hydrolysis can then be utilized, as in aerobes, for the synthesis of new polymers, or they can be metabolized to provide energy and starting material for the synthesis of other required compounds, Carbohydrates are the most commonly used substrates for fermentation, but there are some fermentative organisms that prefer compounds of other types, i.e., amino acids, organic acids, or nucleic acid bases. Many anaerobic organisms are quite limited in the variety of substrates they are capable of fermenting, sometimes being able to utilize only a few closely related compounds. However, organisms that can derive energy from only a limited number of compounds can generally utilize many different compounds as carbon source.

The preliminary steps required to convert various small molecules to intermediates of the central pathways generally do not require the participation of oxygen and can occur either aerobically or anaerobically. In some cases, such as the metabolism of tryptophan, tyrosine, phenylalanine, and hydroxyproline, ring cleavage requires oxygen in aerobic microorganisms such as *Pseudomonas*, and at least these steps must be carried out by different reactions if they occur in anaerobic bacteria. The cleavage of the ring in histidine is hydrolytic, however, in both *Pseudomonas* and *Clostridium*. Sugars are rather easily converted to intermediates of the central pathways by a few reactions such as phosphorylation and isomerization, with no involvement of oxygen. Anaerobes are therefore able to carry out most of the conversions of naturally occurring compounds to metabolic intermediates in the same way that they are carried out by aerobes.

Central Pathways

The third area in which there is similarity in aerobic and anaerobic metabolism is in the *central catabolic pathways*. In Chap. 9 we examined the pathways common to almost all organisms that utilize organic compounds as energy sources: the EMP pathway, the TCA cycle, the HMP pathway, and also the Entner-Douderoff pathway, which occurs in an important group of bacteria. The mode of utilization of these pathways in anaerobic metabolism is related to the means of ATP synthesis used by the organism. In bacteria that use nitrate as electron acceptor, the central catabolic pathways may be used in the same way and for the same purposes as in aerobic metabolism. The fact that the electrons removed in oxidative reactions are transferred to nitrate rather than to oxygen through an electron transport chain that may differ in its components from the chain that utilizes oxygen has no effect on the reactions by which the substrate is degraded. Since all such organisms are facultative anaerobes that utilize anaerobic respiration only in the absence of oxygen, the use of nitrate as terminal electron acceptor affects only their ability to metabolize substrates for which oxygenation reactions are required and probably, as pointed out in Chap. 7, the amount of ATP synthesized.

Substrates used for energy can be oxidized completely to carbon dioxide using either oxygen or nitrate as electron acceptor. None of the reactions of the central pathways requires the use of oxygen, since all oxidative reactions involve the removal of hydrogen. The reactions by which carbon dioxide is fixed to synthesize cellular material also do not involve oxygen and can occur either aerobically or anaerobically.

In fermentative bacteria, however, the requirement for use of the products of catabolism of the energy source as electron acceptors precludes the utilization of pathways that lead to complete oxidation of the energy source. The use of the EMP pathway is common in bacteria and fungi that ferment carbohydrates, but a complete TCA cycle cannot be used in fermentation. Cyclic use of the HMP pathway, resulting in complete oxidation to carbon dioxide, is also impossible in fermentative metabolism. The use of the TCA and HMP pathways in fermentative organisms is restricted to those reactions required for the synthesis of building blocks for biosynthesis. The HMP pathway may be used for the synthesis of ribose 5-phosphate and erythrose 4-phosphate, for the generation of reduced NADP, and for inter-conversions of sugars. The TCA cycle may be utilized for the synthesis of α-ketoglutarate and succinate. In organisms that use the TCA cycle for synthetic purposes only, i.e., fermentative organisms and autotrophs, cyclic operation is prevented by the lack of an enzyme, usually α-ketoglutarate dehydrogenase. Succinate, which is required for the synthesis of the heme group, is synthesized by reversal of the remaining reactions of the cycle from oxalacetate, which is formed by the addition of carbon dioxide to pyruvate (see Chap. 9). The Entner-Douderoff pathway can be used for the fermentation of sugars but is used in this way by only one genus of bacteria, *Zymomonas*, since all other known organisms that utilize this pathway are strict aerobes. Thus, of the central catabolic pathways described in Chap. 9, only the EMP pathway is of central importance in the energy metabolism of fermentative microorganisms, whereas other central pathway reactions are used for biosynthesis or for the conversion of substrates to intermediates of the EMP pathway.

The restrictions placed on the fermentative degradation of organic compounds by the fact that an external electron acceptor is not used determine the pathways that can be used and the products that can be formed. An understanding of the principles governing fermentation is essential for understanding the anaerobic processes by which organic compounds are degraded, and we shall now consider these general principles.

FERMENTATION BALANCES

We discussed in Chap. 7 the importance to the cell of reoxidizing reduced electron carriers such as NAD. The electron carriers that serve as primary electron acceptors, i.e., as cofactors, for oxidative reactions are present at low concentrations in the cell, and the energy metabolism of the cell would come to a halt if the reduced electron carriers, primarily NAD, were not continually reoxidized by passing electrons to another carrier or to a terminal electron acceptor. In organisms that utilize an external inorganic electron acceptor, the reoxidation of reduced NAD presents no problem. However, the synthesis of electron acceptors is a major problem in fermentative metabolism, and the mix of products resulting from the use of different electron acceptors is one of the distinguishing characteristics of the different genera and species of fermentative microorganisms.

The fact that all hydrogen removed from the energy source in the oxidative reactions that release energy must be used to reduce some degradative product of the energy source (unless it can be released as hydrogen gas) means that *no net oxidation can be achieved in fermentation*. The oxidation level of all the products taken together must be the same as that of the starting material, the energy source. Fermentation can, and usually does, occur in the total absence of oxygen. Any oxygen introduced into the products of fermentation is derived from water. The addition or removal of water does not alter the oxidation level of the substrate. (Addition of water is, in effect, simultaneous and equivalent oxidation and reduction.) The oxidation-reduction state of an organic compound is calculated by comparing it with that of water. Hydrogen is arbitrarily assigned a value of -0.5 and oxygen a value of $+1$. The oxidation state of water, or of any compound containing hydrogen and oxygen in a ratio of $2:1$, e.g., carbohydrates, is thus 0. Nitrogen is assigned a value of $+1.5$ since, in the catabolism of nitrogen-containing substrates, nitrogen is released as ammonia, which accounts for the removal of three atoms of hydrogen (-1.5). The oxidation state of ammonia, like that of water, is thus 0.

In studies of fermentation, it is necessary to recover, identify, and quantitate all products formed from the energy source. A fermentation balance is calculated as a check on the recovery of products. In anaerobic metabolism, such a balance is analogous to the aerobic balances discussed in Chap. 10, except that in obtaining data for a fermentation balance, the experiment is conducted under conditions that preclude the use of the energy source for the synthesis of cell material. This is accomplished by using washed nonproliferating cells in a medium that does not allow growth, e.g., one containing no nitrogen source, or more commonly by adding a mixture of organic growth factors in excess, e.g., tryptone and yeast extract, to serve as carbon source. Under these conditions, it has been shown by using radioactive glucose as energy source that none of the glucose is used for the synthesis of cell material since no radioactivity is found in the cells. The amount of energy source used is determined analytically. The recovery of energy source as products is calculated on the basis of carbon recovered. The oxidation state of each product is also calculated, and the overall oxidation/reduction (O/R) balance is determined. If the carbon recovery is low, the calculated O/R balance provides a clue to the nature of the missing product by indicating whether its oxidation state is higher or lower than that of the substrate.

The principle of the fermentation balance can be illustrated with several examples of fermentations, for which we can write simplified equations using only the principal products.

Example 1 The lactic acid fermentation of *Streptococcus*:

$$C_6H_{12}O_6 \rightarrow 2C_3H_6O_3 \tag{11-1}$$

glucose	lactic acid	
6	6	carbon atoms
0	0	oxidation state

Example 2 The alcoholic fermentation of yeast:

$$C_{12}H_{22}O_{11} + H_2O \rightarrow 4CO_2 + 4C_2H_5OH \tag{11-2}$$

sucrose	water	carbon dioxide	ethanol	
12	0	4	8	carbon atoms
0	0	4(+2)	4(−2)	oxidation state

Example 3 The mixed acid fermentation of *Escherichia coli*:

$$2C_6H_{12}O_6 + H_2O \rightarrow 2C_3H_6O_3 + 2CO_2 + 2H_2$$

glucose	water	lactic acid	carbon dioxide	hydrogen	
12	0	6	2	0	carbon atoms
0	0	0	2(+2)	2(−1)	oxidation state

$$+ C_2H_4O_2 + C_2H_5OH \tag{11-3}$$

acetic acid	ethanol	
2	2	carbon atoms
0	−2	oxidation state

In each of the examples, the equation is balanced, which means, of course, that all the carbon of the energy source is accounted for in the products and that the O/R level of the products is the same as that of the substrate, because there are equal amounts of oxygen and hydrogen on both sides of the equation.

Actual fermentations can seldom be described by such simple equations as those we have used for illustrative purposes. Small amounts of minor products are usually formed and molar ratios are not integral. Table 11-1 shows the products formed in eight actual fermentations of glucose by various microorganisms. The first three columns show data for the fermentations used in the examples above [Eqs. (11-1), (11-2), and (11-3)]. In each case, the actual fermentation is more complex than the simplified equation indicates because of the formation of relatively small amounts of a variety of products other than the major ones that are typical of each fermentation type. Formation of these minor products varies with the individual strain and often with the pH and temperature.

Table 11-1 Products formed in fermentation of glucose by microorganisms

Millimoles of product per 100 mmol of glucose fermented by organism

Products	Yeast[1]	Streptococcus faecalis[2]	Escherichia coli[3]	Aerobacter indologenes[4]	Propionibacterium arabinosum[5]	Butyribacterium rettgeri[6]	Clostridium butylicum[7]	Clostridium acetobutylicum[8]
Butyric acid	0.21	—	—	—	—	29	17.2	4.3
Lactic acid	1.37	146	79.5	2.9	—	107	—	—
Acetic acid	15.1	18.8	36.5	0.47	10.0	88	17.2	14.2
Formic acid	0.49	33.6	2.43	17.0	—	—	—	—
Succinic acid	0.68	—	10.7	—	7.8	—	—	—
Butanol	—	—	—	—	—	—	58.6	56
Ethanol	129.9	14.6	49.8	69.5	—	—	—	7.2
Acetone	—	—	—	—	—	—	—	22.4
Isopropanol	—	—	—	—	—	—	12.1	—
Carbon dioxide	148.5	—	88.0	172.0	63.6	48	203.5	221
Hydrogen	—	—	75.0	35.4	—	74	77.6	135
2,3-Butanediol	0.68	—	0.3	66.4	—	—	—	—
Glycerol	32.3	—	1.42	—	—	—	—	—
Acetoin	0.19	—	0.059	—	—	—	—	6.4
Propionic acid	—	—	—	—	148.8	—	—	—

[1] A. C. Neish and A. C. Blackwood. 1951. *Can. J. Technol.* 29:123.
[2] T. B. Platt and E. M. Foster. 1958. *J. Bacteriol.* 75:453.
[3] J. L. Stokes. 1949. *J. Bacteriol.* 57:147.
[4] H. Reynolds and C. H. Werkman. 1937. *J. Bacteriol.* 33:603.
[5] H. G. Wood and C. H. Werkman. 1936. *Biochem. J.* 30:618.
[6] L. Pine, V. Haas, and H. A. Barker. 1954. *J. Bacteriol.* 68:227.
[7] O. L. Osburn, R. W. Brown, and C. H. Werkman. 1937. *J. Biol. Chem.* 121:685.
[8] J. B. van der Lek. 1930. Ph.D. Thesis. Technische Hoogeschool, Delft, Holland.

We will use the fermentation by *Clostridium acetobutylicum* to illustrate the calculation of a fermentation balance (see Table 11-2).

1. The numbers in col. (1) of the table were taken from Table 11-1 and were obtained by dividing the concentration of each product, as determined by the appropriate analytical procedure, by the difference in the initial and final concentrations of glucose, also determined by analysis, and multiplying by 100. These numbers represent the number of millimoles of each product formed per 100 mmol of glucose fermented.
2. The carbon recovery is calculated by converting millimoles of product to millimoles of carbon by multiplying each number in col. (1) by the number

Table 11-2 Calculation of a fermentation balance*

Products	(1)	(2)	(3)	(4)	(5)
Butyric acid	4.3	17.2	-2		-8.6
Acetic acid	14.2	28.4	0		
Carbon dioxide	221	221	$+2$	$+442$	
Hydrogen	135	—	-1		-135
Ethanol	7.2	14.4	-2		-14.4
Butanol	56	224	-4		-224
Acetone	22.4	67.2	-2		-44.8
Acetoin	6.4	25.6	-3		-19.2
		597.8		$+442$	-446

Carbon recovery $= 597.8/600 \times 100 = 99.6\%$

O/R balance $= 442/446 = 0.99$

*Columns:

(1) mmol of product formed/100 mmol of glucose fermented

(2) mmol of carbon = mmol of product × no. of carbon atoms in molecule

(3) O/R state of product = (no. of oxygen atoms in molecule) -0.5(no. of hydrogen atoms in molecule)

(4) oxidized products; col. (1) × col. (3), for products with positive values

(5) reduced products; col. (1) × col. (3), for products with negative values

of carbon atoms in the molecule. For example, butyric acid is a four-carbon acid, $CH_3CH_2CH_2COOH$, so the number of millimoles of carbon present as butyric acid can be calculated to be $4 \times 4.3 = 17.2$. The sum of col. (2) represents the total amount of carbon recovered as products. Since 100 mmol of glucose contains 600 mmol of carbon, the percentage recovery of carbon is calculated by dividing the total number of millimoles of carbon found in the products by 600.

3. The oxidation state of each product is calculated, as described previously, by assigning a value of $+1$ to each oxygen atom in the molecule and -0.5 to each hydrogen atom. These values are listed in col. (3).

4. The overall oxidation-reduction state of the products is calculated by multiplying the number of millimoles of each product by its O/R value, i.e., col. (1) × col. (3). The values are entered in separate columns—those for oxidized products (positive values) in col. (4) and those for reduced products (negative values) in col. (5), and the columns are totaled. Since the starting material, glucose, had an oxidation state of 0, the totals for oxidized and reduced products should agree; i.e., the ratio oxidized/reduced should be 1.0. This ratio is calculated as the O/R value for the overall product mix.

The fermentation used as an example for calculation meets both the requirements for a satisfactory balance: (1) essentially all of the carbon of the energy source that disappeared from the medium was recovered as identified products, and (2) no net oxidation occurred, as shown by the near equality of oxidized and reduced products.

Additional calculations are sometimes made. Fermentation balances are valuable in providing preliminary evidence concerning the metabolic pathway used by the organism. Since the predominant pathway for fermentation of sugars is the EMP pathway, which involves the splitting of a hexose into two C_3 fragments, calculations are sometimes made on the basis of a three-carbon unit. Amounts of products are expressed as the number of millimoles per 100 mmol of C_3. This allows prediction of the millimoles of C_1 product expected, since a C_2 product should be formed by a split of $C_3 \rightarrow C_1 + C_2$. Similarly, a C_4 product would usually be formed from two C_3 units, with the accompanying formation of two C_1 products. If the amounts of C_1 products found do not agree with the calculated expected amounts in a balance that is otherwise satisfactory, it can be concluded that the pathway involved in formation of the products is not the usual one, and further investigation is required.

FORMATION OF COMMON PRODUCTS IN FERMENTATION OF SUGARS

Sugars are the energy sources most commonly used by fermentative microorganisms, and the products listed in Table 11-1 are those most frequently formed in these fermentations. Before we examine the overall fermentation patterns of the major groups of fermentative microorganisms, it is important to describe the reactions by which each of these fermentation products is formed. Some major differences are known to exist in the pathways used by different organisms to form the same product. Most products, however, are probably formed by identical reactions in a variety of species. We will examine some of the known differences because they show the evolutionary development of different mechanisms for accomplishing the same result. Some of these differences are thought to have ecological significance in that the fermentation pattern of the microorganism may represent an adaptation that is advantageous for survival in its normal habitat.

Formation of Pyruvate from Glyceraldehyde 3-Phosphate

Most of the products formed in the fermentation of sugars originate from pyruvate, and it is desirable to set the stage for these further reactions by reviewing the essential features of the lower portion of the EMP pathway. In Chap. 9 we described three central catabolic pathways, the EMP, HMP, and Entner-Douderoff (E-D) pathways, all of which lead to the formation of glyceraldehyde 3-phosphate. All three pathways use a common reaction sequence, the lower EMP, to convert glyceraldehyde 3-phosphate to pyruvate, although in the E-D pathway, one molecule of pyruvate is formed directly from the hexose 2-keto-3-deoxy-6-phosphogluconate. We shall describe below another pathway by which glyceraldehyde 3-phosphate is formed—the heterolactic fermentation.

The two considerations of particular relevance in fermentation are the reactions involving ATP and NAD. Regardless of the reactions by which glyceraldehyde 3-phosphate is formed, the phosphate group represents the investment of one molecule of ATP. In the conversion of glyceraldehyde 3-phosphate to pyruvate, that ATP is regained and a second ATP is synthesized, yielding a net gain of one ATP molecule per molecule of glyceraldehyde 3-phosphate. The pyruvate formed from hexose in the E-D pathway represents no gain or loss of ATP. It does, however, result from an oxidation, as does the pyruvate formed from glyceraldehyde 3-phosphate. The formation of pyruvate from a sugar necessarily represents an oxidation, i.e., a reduction of NAD, because the oxidation state of pyruvate is $+1$, whereas that of all sugars is 0. Thus, beginning from a hexose such as glucose, for each molecule of pyruvate produced, one molecule of NAD has been reduced. If the lower EMP pathway was used to form the pyruvate, one molecule of ATP has been gained for each molecule of pyruvate produced. This is the situation in almost all organisms that ferment sugars, since most use the EMP pathway in whole or in part.

As we have stated, pyruvate is the pivotal compound in the catabolism of sugars. All the products listed in Table 11-1 except glycerol can originate from pyruvate. The reactions leading from pyruvate to the final fermentation product accomplish two results: (1) they oxidize the $NADH_2$ produced in the reactions that lead to pyruvate, and (2) they *may* generate additional ATP by substrate level phosphorylation. The first result is essential; the second is desirable since anaerobic growth is limited by ATP. It will be recalled that growth in a fermentative organism is strictly proportional to the amount of ATP produced (see Chap. 7). Y_{ATP} is 10 mg of cells (dry weight) per millimole of ATP produced. Not all fermentations produce ATP beyond that generated in forming pyruvate, and those that utilize pathways other than the EMP generally form less ATP and therefore allow less growth than do fermentations that form pyruvate through the EMP reactions. In any fermentation it is only necessary that there be at least one reaction that generates net ATP by substrate level phosphorylation and at least one set of coupled reactions by which NAD is reduced and reoxidized. The reduction of NAD in most fermentations occurs when glyceraldehyde 3-phosphate is oxidized to 3-phosphoglycerate. We shall now examine the reactions used by various microorganisms to reoxidize the reduced NAD.

Formation of Lactic Acid

Lactic acid is possibly the most common product of the fermentation of sugars. It is produced from a variety of sugars by many different types of microorganisms. It is the major product in most fermentations utilized by the dairy industry in the manufacture of such products as buttermilk, sour cream, yogurt, and some types of cheese.

The production of lactic acid is the simplest solution to the problem of finding an acceptor for the hydrogen removed in the pathway that leads to pyruvate, since pyruvate itself is the acceptor.

$$
\begin{array}{ccc}
\overset{\displaystyle O}{\overset{\displaystyle \parallel}{C}}\!\!-\!\!OH & & \overset{\displaystyle O}{\overset{\displaystyle \parallel}{C}}\!\!-\!\!OH \\
| & \text{NADH}_2 \quad \text{NAD} & | \\
C\!\!=\!\!O & \longrightarrow & HCOH \\
| & & | \\
CH_3 & & CH_3 \\
\text{pyruvic acid} & & \text{lactic acid}
\end{array}
\qquad (11\text{-}4)
$$

Lactic acid is a major or minor product of many fermentations, but it is especially important in the group of Gram-positive cocci commonly called the *lactic acid bacteria* and in several closely related genera, one of which, the rod-shaped *Lactobacillus*, is also included among the lactic acid bacteria. These organisms are divided into two subgroups: the *homolactics* or homofermenters and the *heterolactics* or heterofermenters. The homolactics include all species of *Streptococcus* and some species of *Lactobacillus*, and it is these organisms that are used in most dairy industry fermentations. They metabolize sugars through the EMP pathway and convert pyruvate totally or almost totally to lactic acid. Under carefully controlled conditions, some strains of *Streptococcus* convert 98 percent of the available sugar to lactic acid. Most species form variable amounts of other products such as acetic acid, formic acid, and ethanol (see *S. faecalis* in Table 11-1).

The heterolactics include all species of *Leuconostoc* and some species of *Lactobacillus*. These organisms possess a unique pathway found only in the lactic acid bacteria. This pathway is sometimes called the *anaerobic pentose pathway* or the *phosphoketolase pathway*, the latter name derived from the unique enzyme of the pathway. The pathway is shown in Fig. 11-1. The metabolism of hexoses through this pathway results in the formation of only one molecule of ATP per molecule of glucose, and it is thus only half as efficient in the production of ATP as is the homofermentative path through the EMP route. Acetylphosphate represents a potential ATP, but the phosphate bond energy is not conserved since acetylphosphate must be used as an acceptor for the hydrogen removed in the oxidation of glucose 6-phosphate to ribulose 5-phosphate. When pentoses are fermented via the phosphoketolase pathway, acetyl phosphate is not required as hydrogen acceptor and two molecules of ATP are gained per molecule of pentose. At least some of the homolactics are able to use the phosphoketolase pathway for the fermentation of pentoses, but they do not use it for hexoses. The products of the heterolactic fermentation of hexose sugars are lactic acid, carbon dioxide, and ethanol, 1/1/1, while pentoses yield a 1/1 ratio of lactic acid and acetic acid.

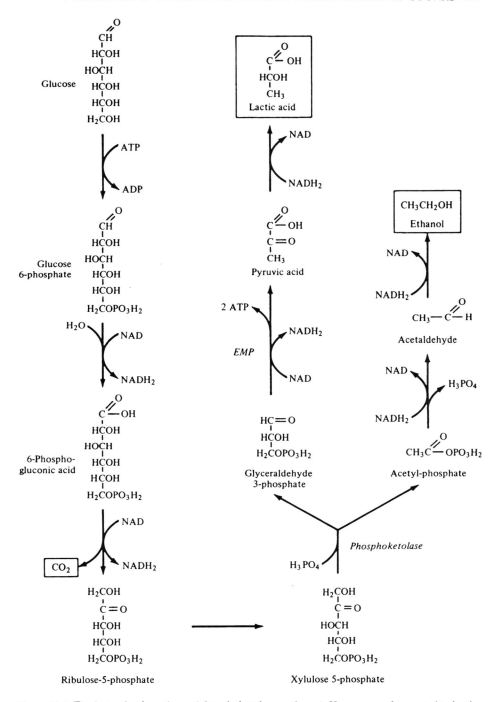

Figure 11-1 The heterolactic pathway (phosphoketolase pathway). Hexoses are fermented to lactic acid, ethanol, and carbon dioxide.

The formation of ethanol, acetic acid, and carbon dioxide in the phosphoketolase pathway involves an unusual mechanism of formation of the C_2 products. In most fermentations, these products are formed as a result of the decarboxylation of pyruvate by one of several mechanisms.

Decarboxylation of Pyruvate

In *aerobic* metabolism, pyruvate is decarboxylated by an oxidative reaction in which NAD is reduced and carbon dioxide and acetyl-CoA are formed.

$$
\begin{array}{c}
\overset{O}{\overset{\|}{C}-OH} \\
| \\
C=O \\
| \\
CH_3
\end{array}
\xrightarrow[\text{CoASH} \quad \text{CO}_2]{\text{NAD} \quad \text{NADH}_2}
CH_3-\overset{O}{\overset{\|}{C}}-SCoA \qquad (9\text{-}15)
$$

pyruvic acid   acetyl-CoA

In *anaerobic* metabolism, the major objective of reactions that convert pyruvate to various fermentation products is the reoxidation of the NAD that has already been reduced. Reaction (9-15) therefore seldom occurs in fermentative organisms, and the C_1–C_2 split of pyruvate is accomplished by other means.

In yeast (and in some bacteria), pyruvate is decarboxylated without oxidation to form acetaldehyde.

$$
\begin{array}{c}
\overset{O}{\overset{\|}{C}-OH} \\
| \\
C=O \\
| \\
CH_3
\end{array}
\xrightarrow{\text{CO}_2}
CH_3-\overset{O}{\overset{\|}{C}}H \qquad (11\text{-}5)
$$

pyruvic acid   acetaldehyde

In reaction (11-5), the hydrogen remains with the C_2 product.

At least two mechanisms are utilized by some bacteria to remove two hydrogen atoms along with the C_1 fragment without reducing an electron carrier. In the enterics, the products of the C_1–C_2 split are formic acid and an acetyl fragment [Eq. (11-6)]. The C_2 product has been oxidized from the aldehyde to the acid level by the removal of two hydrogen atoms, and acetyl-CoA can be formed as it is in the aerobic reaction but without the reduction of NAD.

$$
\begin{array}{c}
\overset{O}{\overset{\|}{C}-OH} \\
| \\
C=O \\
| \\
CH_3
\end{array}
\xrightarrow{\text{CoASH}}
H\overset{O}{\overset{\|}{C}}-OH \; + \; CH_3-\overset{O}{\overset{\|}{C}}-SCoA \qquad (11\text{-}6)
$$

pyruvic acid   formic acid   acetyl-CoA

In *Clostridium*, the C_2 fragment is also oxidized with the resultant formation of an acetyl fragment. Hydrogen is removed in this case as H_2 (gas), and acetyl-CoA or acetylphosphate is produced.

$$\underset{\text{pyruvic acid}}{\overset{\displaystyle \overset{\displaystyle O}{\underset{\displaystyle |}{\overset{\|}{C}-OH}}}{\underset{\displaystyle CH_3}{\underset{|}{C=O}}}} \quad \xrightarrow{\text{CoASH}} \quad CO_2 \;+\; H_2 \;+\; \underset{\text{acetyl-CoA}}{CH_3\overset{\displaystyle O}{\overset{\|}{C}}-SCoA} \qquad (11\text{-}7)$$

The ability to remove hydrogen, i.e., to oxidize, without the reduction of NAD provides the organism with an acceptor for an additional two hydrogen atoms, which can be used to advantage by often allowing the formation of additional ATP [see Eq. (11-10)].

Formation of ethanol In yeast, the acetaldehyde formed by decarboxylation of pyruvate is used as the hydrogen acceptor for reoxidation of the $NADH_2$ formed in the pathway to pyruvate.

$$\underset{\text{acetaldehyde}}{CH_3-\overset{\displaystyle O}{\overset{\|}{C}}-H} \quad \xrightarrow{\quad NADH_2 \quad NAD \quad} \quad \underset{\text{ethanol}}{CH_3CH_2OH} \qquad (11\text{-}8)$$

The two major fermentation products are thus accounted for by reactions (11-5) and (11-8); the fermentation is balanced when glucose or another hexose is converted to two molecules each of carbon dioxide and ethanol.

In bacteria that are able to remove two hydrogen atoms from pyruvate as either H_2 or formic acid, one molecule of acetyl-CoA or acetylphosphate can accept four hydrogen atoms; that is, one molecule can accept the hydrogen from two molecules of $NADH_2$. As we noted previously in the heterolactic fermentation, the energy of the thiolester or phosphate bond cannot be conserved when the molecule is reduced. The reactions forming ethanol are the same as those in the heterolactic pathway (see Fig. 11-1).

$$\underset{\text{acetyl-CoA}}{CH_3-\overset{\displaystyle O}{\overset{\|}{C}}-SCoA} \xrightarrow{+2H} \underset{\text{acetaldehyde}}{CH_3-\overset{\displaystyle O}{\overset{\|}{C}}H} \xrightarrow{+2H} \underset{\text{ethanol}}{CH_3CH_2OH} \qquad (11\text{-}9)$$

Formation of acetic acid When glucose is converted to two molecules of pyruvate, with the reduction of two molecules of NAD, both molecules of pyruvate are then converted to acetyl-CoA or acetylphosphate, and one molecule of the latter is reduced twice to form ethanol, the other molecule is not needed as a hydrogen acceptor. This allows the organism to conserve the bond energy of acetylphosphate by synthesizing ATP. The CoA and phosphate derivatives are equivalent and interconvertible, as we have seen previously. The acetyl-CoA formed by the enterics is converted to acetyl-phosphate, which is then used as phosphate donor for the synthesis of ATP. The acetyl-CoA or acetylphosphate formed by *Clostridium* can be used for the same purpose.

$$CH_3-C\overset{O}{\diagup}-SCoA \quad \xrightarrow[\quad CoASH \quad]{H_3PO_4} \quad CH_3C\overset{O}{\diagup}-OPO_3H_2$$

acetyl-CoA   acetylphosphate

$$\xrightarrow[\quad ADP \quad ATP \quad]{} \quad CH_3C\overset{O}{\diagup}-OH \qquad (11\text{-}10)$$

acetic acid

Each mole of acetic acid that appears as a product of fermentation may thus be assumed to represent an additional net gain of 1 mol of ATP, beyond the 2 mol of ATP per mole of glucose formed in the pathway to pyruvate.

Formic acid, carbon dioxide, and hydrogen In Eq. (11-3), the basic fermentation pattern of *E. coli* (excluding minor products) shows the production, from 2 mol of glucose, of 2 mol each of lactic acid, carbon dioxide, and hydrogen, and 1 mol each of ethanol and acetic acid. These products might be expected if the fermentation were carried out at a pH slightly below 7. At a pH above neutral, the production of carbon dioxide and hydrogen would be drastically decreased and that of formic acid would be proportionately increased. The reciprocal relationship between the amounts of carbon dioxide plus hydrogen and formic acid results from the fact that carbon dioxide and hydrogen are produced exclusively from formic acid in the mixed acid fermentation that is carried out by *E. coli* and some other enteric bacteria. The enzyme *formic hydrogenlyase* acts upon formic acid to produce carbon dioxide and hydrogen.

$$HC\overset{O}{\diagup}-OH \rightarrow H_2 + CO_2 \qquad (11\text{-}11)$$

formic acid

Depending on the strain and the pH, variable amounts of formic acid may be degraded, with the remainder appearing as a fermentation product. Some of the enterics lack the ability to synthesize formic hydrogenlyase and therefore accumulate formic acid. The formation of hydrogen is detected in fermentation tests by trapping the gas in a tube inverted in the fermentation broth. "Gas production" in a sugar fermentation test means hydrogen production, since carbon dioxide is too soluble in ordinary media to be detected as a gas. The production of gas is generally used as one criterion for distinguishing between the nonpathogenic gas-forming enterics, such as *E. coli* and *Enterobacter*, and the pathogenic lactose-fermenting *Shigella* species and *Salmonella typhi*, which do not produce gas. This is not a totally dependable criterion, since strains of *E. coli* that do not produce hydrogen have been isolated from natural sources and can be isolated readily in the laboratory after treatment with mutagenic agents. Since the production of hydrogen is dependent on the ability to synthesize the single enzyme, formic hydrogenlyase, mutation causing loss of the enzyme can occur. This will have no effect on the survival of the organism in its natural habitat, the well-buffered intestine (loss of the enzyme does increase the amount of acid produced during fermentation).

Formation of Acetoin and Butanediol

The products whose formation has been described thus far are typical of the mixed acid fermentation of *E. coli* and certain other enterics. In Chap. 8 we discussed the importance of the fermentation type in identifying the enterics, particularly in distinguishing between *E. coli* and *Enterobacter*. The two types of fermentation are readily distinguished by the MRVP tests. Organisms with a mixed acid fermentation pattern produce sufficient acid to reach the pH at which methyl red changes color, approximately 4.5, and are thus MR^+. These organisms also produce a green sheen on colonies grown on eosin–methylene blue–lactose (EMB-lactose) agar due to precipitation of the dyes at low pH. The second type of fermentation found among the enterics is the butanediol, or butylene glycol, fermentation, in which much of the energy source is converted to 2,3-butanediol rather than to acids (see *Aerobacter indologenes* in Table 11-1; this species is no longer recognized but the fermentation pattern is typical of the species formerly classified as *Aerobacter* and now placed in the genera *Enterobacter* and *Klebsiella*). Because much of the carbon is diverted to butanediol formation, reducing the amount of acids formed, the pH does not drop sufficiently to be detected by a methyl red indicator or to produce a green sheen on EMB-lactose agar. Butanediol producers are therefore typically MR^-. The Voges-Proskauer test detects butanediol or acetoin, and organisms with a butanediol fermentation are therefore VP^+. Organisms that have the mixed acid fermentation do not produce butanediol and are VP^-. Thus, *E. coli* is MR^+VP^-, while *Enterobacter* is MR^-VP^+.

The pathway of formation of butanediol is shown in Fig. 11-2. Two molecules of pyruvate are condensed with the elimination of carbon dioxide. A second decarboxylation results in the formation of acetoin (acetylmethyl carbinol), which serves as the hydrogen acceptor in the reoxidation of one molecule of $NADH_2$. Acetoin may appear in small amounts as a final fermentation product if $NADH_2$ is reoxidized in other reactions.

The formation of butanediol is not restricted to the enterics. Several of the facultative anaerobic species of *Bacillus* produce butanediol as a major product of sugar fermentations. Small amounts are also produced by other microorganisms. Yeasts probably use a different pathway for its formation (Doelle, 1975).

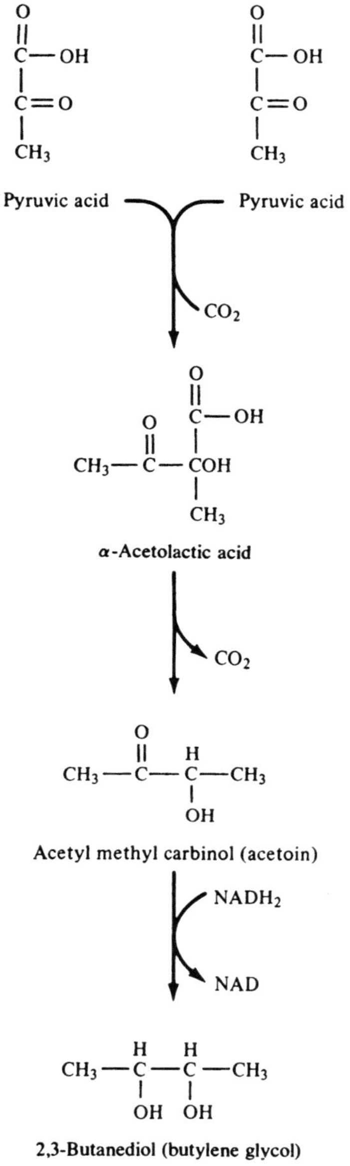

$$
\begin{array}{c}
O \\
\parallel \\
C-OH \\
\mid \\
C=O \\
\mid \\
CH_3
\end{array}
\qquad
\begin{array}{c}
O \\
\parallel \\
C-OH \\
\mid \\
C=O \\
\mid \\
CH_3
\end{array}
$$

Pyruvic acid   Pyruvic acid

CO_2

$$
\begin{array}{c}
\qquad\quad O \\
\qquad\quad \parallel \\
\;\; O \quad C-OH \\
\;\; \parallel \quad \mid \\
CH_3-C-COH \\
\qquad\quad \mid \\
\qquad\quad CH_3
\end{array}
$$

α-Acetolactic acid

CO_2

$$
\begin{array}{c}
O \qquad H \\
\parallel \qquad \mid \\
CH_3-C-C-CH_3 \\
\qquad\quad \mid \\
\qquad\quad OH
\end{array}
$$

Acetyl methyl carbinol (acetoin)

$NADH_2$

NAD

$$
\begin{array}{c}
H \quad H \\
\mid \quad \mid \\
CH_3-C-C-CH_3 \\
\mid \quad \mid \\
OH \quad OH
\end{array}
$$

2,3-Butanediol (butylene glycol)

Figure 11-2 The formation of 2,3-butanediol from pyruvate.

Formation of Acetone and Isopropanol

A number of saccharolytic species of *Clostridium* and at least one species of *Bacillus* form acetone in the fermentation of sugars. One species, *C. butylicum*, is able to use all or part of the acetone as hydrogen acceptor, converting it to isopropanol.

The reaction sequence begins with the condensation of two molecules of acetyl-CoA, which are formed from pyruvate by the reaction shown in Eq. (11-7). The entire pathway is shown in Fig. 11-3. This pathway results in the conversion of two molecules of pyruvate to three molecules of carbon dioxide, two of hydrogen, and one of acetone. It wastes energy because

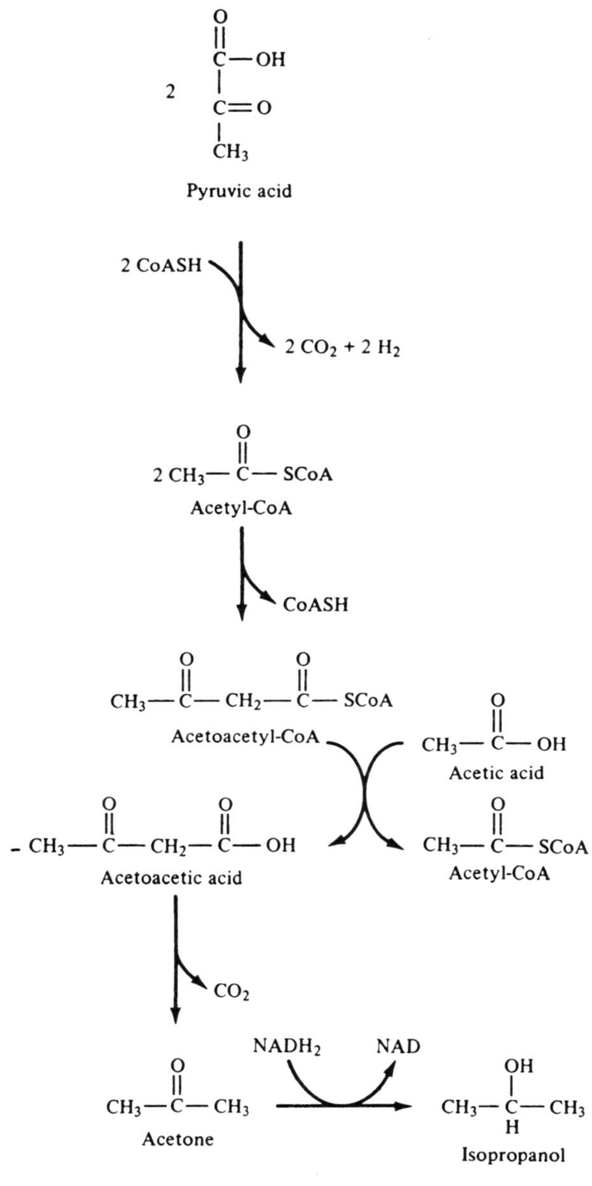

Figure 11-3 The formation of acetone from pyruvic acid. In *C. butylicum*, acetone may serve as electron acceptor and be reduced to isopropanol.

acetyl-CoA represents a potential ATP. It also produces no electron acceptor except in *C. butylicum*, which can reduce acetone to isopropanol. The only apparent advantage to the organism of forming acetone rather than acetic acid is that acetone is a neutral product. Both acetone and butanol, the other product of the "solvent" fermentation of clostridia, are formed during the latter part of the fermentation in a batch system when the pH has dropped to about 4.0. The acids already produced can be converted to neutral products, butanol, acetone, and sometimes ethanol, thus raising the pH.

The acetone-butanol fermentation of clostridia was an important source of these solvents for industrial use during World War I and is still used to some extent in their production. As the present source of solvents, petroleum, becomes more scarce and expensive, the microbiological production of these and other fermentation products may be expected to increase in importance.

Formation of Butyric Acid and Butanol

A pathway that utilizes the same initial steps as those used in the formation of acetone is that leading to the formation of butyric acid. The reaction sequence shown in Fig. 11-4 is similar to that used for the synthesis of fatty acids in mammalian cells. In discussing the reactions by which succinate is converted to oxalacetate in the TCA cycle, we pointed out the utility of a series of reactions in which double bonds are created by the removal of two hydrogen atoms or of water and the possibility of creating hydrogen acceptors by this means or of introducing oxygen into a molecule by adding water to a double bond. In fermentation, we can imagine that the organism looks at any double bond, whether a C=C bond or a C=O bond, as a site for the disposition of two hydrogen atoms. In the pathway to butyrate, two such opportunities occur so that both molecules of $NADH_2$ formed in the pathway to the two pyruvates used to synthesize butyrate are reoxidized.

When the pH reaches about 4.0, due to the formation of acetic and butyric acids, the enzymes of the pathway leading to acetone become active. The use of acetoacetyl-CoA for the formation of acetone competes with the use of the same compound for the synthesis of butyric acid. Since no $NADH_2$ is reoxidized in the synthesis of acetone, another hydrogen acceptor must be found; butyric acid is used for this purpose and is reduced to butanol. This reduction and the use of some of the acetic acid to form acetone or ethanol causes an increase in pH as the acids are removed from the medium.

The thiolester bond energy of acetyl-CoA and butyryl-CoA is at least partially conserved by transferring the CoA from one molecule to another. It is probable that butyric acid is converted to butyryl-CoA before being reduced to butyraldehyde. Conversion of butyrate to butyryl-CoA would require the use of an ATP molecule to phosphorylate butyric acid to butyryl-phosphate, which could then exchange phosphate for CoA. However, a transfer of CoA has been shown to occur in which acetoacetyl-CoA donates its CoA to butyric acid.

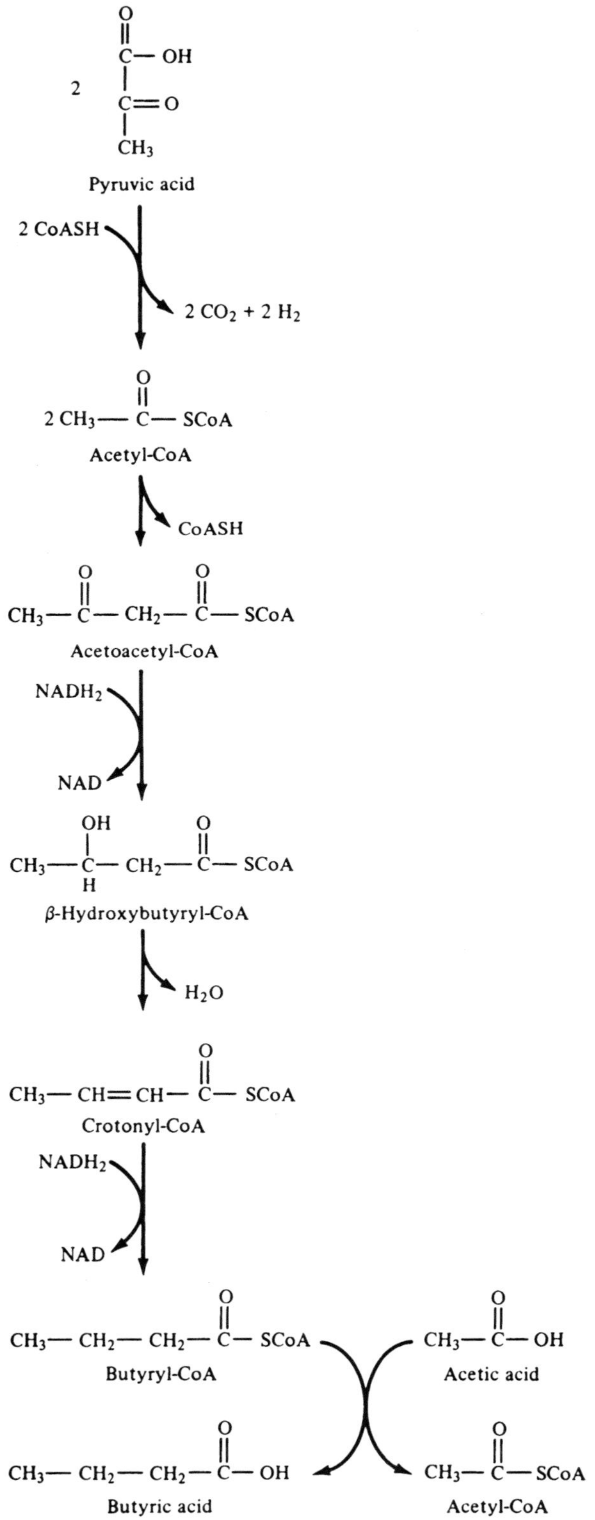

Figure 11-4 The formation of butyric acid from pyruvic acid.

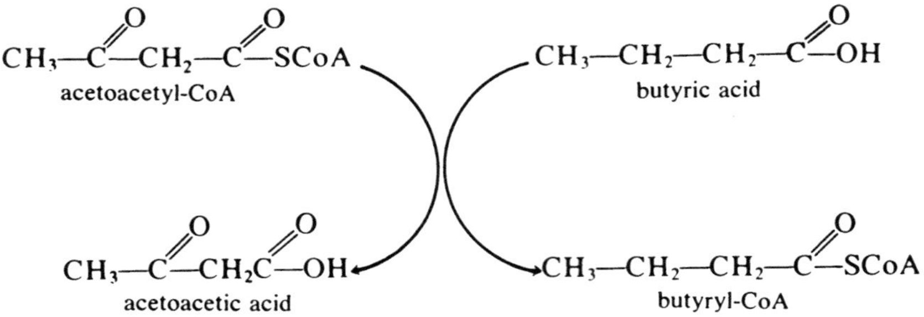

$$CH_3-\overset{\displaystyle O}{\overset{\|}{C}}-CH_2-\overset{\displaystyle O}{\overset{\|}{C}}-SCoA \qquad CH_3-CH_2-CH_2-\overset{\displaystyle O}{\overset{\|}{C}}-OH$$

acetoacetyl-CoA   butyric acid

(11-12)

$$CH_3-\overset{\displaystyle O}{\overset{\|}{C}}-CH_2\overset{\displaystyle O}{\overset{\|}{C}}-OH \qquad CH_3-CH_2-CH_2-\overset{\displaystyle O}{\overset{\|}{C}}-SCoA$$

acetoacetic acid   butyryl-CoA

Since the CoA must be removed from acetoacetyl-CoA before it is decarboxylated to acetone, this transfer reaction is a mechanism for conservation of energy that would otherwise be wasted. A similar exchange occurs in the formation of butyric acid from butyryl-CoA. It is theoretically possible that butyryl-CoA could be converted to butyrylphosphate, which could serve as the phosphate donor for ATP formation, and this may occur in some species. Another method of conserving the bond energy of butyryl-CoA is transfer of the CoA to acetic acid to form acetyl-CoA [Eq. (11-13)], which

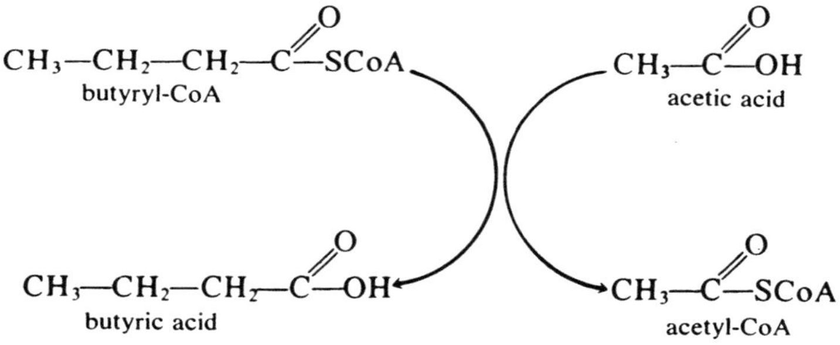

$$CH_3-CH_2-CH_2-\overset{\displaystyle O}{\overset{\|}{C}}-SCoA \qquad CH_3-\overset{\displaystyle O}{\overset{\|}{C}}-OH$$

butyryl-CoA   acetic acid

(11-13)

$$CH_3-CH_2-CH_2-\overset{\displaystyle O}{\overset{\|}{C}}-OH \qquad CH_3-\overset{\displaystyle O}{\overset{\|}{C}}-SCoA$$

butyric acid   acetyl-CoA

can be used to synthesize ATP through acetylphosphate [Eq. (11-10)] or can recycle through acetoacetyl-CoA.

Formation of Propionic Acid

The propionic acid fermentation, carried out by bacteria of the genus *Propionibacterium*, is important industrially because of its use in the manufacture of Swiss cheese. The major products are propionic acid and carbon dioxide, with smaller amounts of acetic acid and succinic acid. The acids, primarily propionic, are responsible for the distinctive flavor and odor of Swiss cheese, and the holes are formed by the production of large quantities of carbon dioxide. Sugars are readily fermented by these organisms, but in the making of Swiss cheese it is the lactic acid produced by the lactic acid bacteria that is

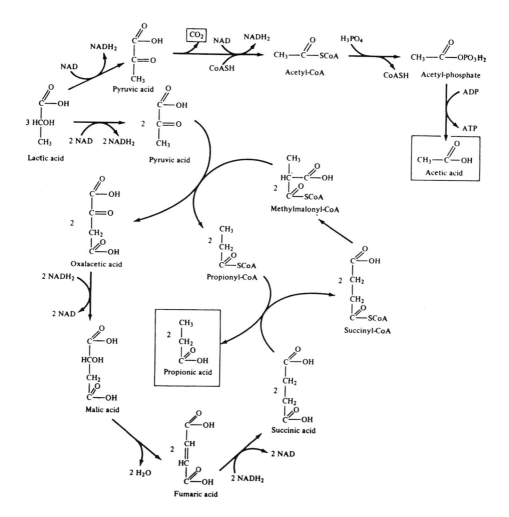

Figure 11-5 The propionic acid fermentation of *Propionibacterium*.

fermented by the propionibacteria. The pathway, shown in Fig. 11-5, is a very interesting and unusual one, which was elucidated only after many years of research. The pathway to propionic acid furnishes excess capacity for the reoxidation of $NADH_2$, allowing acetyl-CoA to be formed from pyruvate by oxidative decarboxylation. The acetyl-CoA is then converted to acetyl-phosphate and used to synthesize ATP.

Propionic acid is also a fermentation product of one species of *Clostridium*, using a different pathway that has not been studied completely. The Gram-negative coccus *Veillonella* also produces propionic acid by a pathway similar to that used by *Propionibacterium*.

Propionic acid is a significant component of the volatile acids in an anaerobic digester. It may be formed from carbohydrates by organisms using the pathway shown in Fig. 11-5 or by the clostridial pathway. Propionic acid also may be formed in the digester as an end product of the metabolism of fatty acids that contain odd numbers of carbon atoms. In Chap. 10 we saw that after the metabolism of odd-numbered fatty acids by β-oxidation, the final three carbons remaining as propionyl-CoA are converted to succinyl-CoA and are oxidized to carbon dioxide through the TCA cycle. Since anaerobes cannot use the TCA cycle in its entirety, anaerobic metabolism of odd-numbered fatty acids would produce propionic acid as an end product.

Formation of succinic acid Succinic acid is formed as an intermediate in the cyclic pathway that leads to propionic acid, and small amounts of it usually appear as a product of the fermentation. Succinic acid is also produced in relatively small amounts in other fermentations, e.g., the mixed acid fermentation of the enterics. Succinic acid is formed by the reversal of the TCA cycle reactions between succinate and oxalacetate. When the cyclic pathway producing propionate is in operation, oxalacetate is formed from pyruvate by transcarboxylation, i.e., the transfer of a C_1 fragment from methylmalonyl-CoA to pyruvate. However, if succinate is removed from the cycle to appear as a fermentation product, pyruvate must be converted to oxalacetate by a different reaction to replace the succinate removed. Also, when the pathway is initiated, there must be a means of synthesizing oxalacetate from pyruvate. Other organisms that produce succinate as a fermentation product must synthesize oxalacetate by some means other than that shown in Fig. 11-5. The five known reactions by which carbon dioxide can be fixed to produce either oxalacetate or malate were shown in Fig. 9-7.

Formation of Glycerol

Glycerol, unlike most products of sugar fermentations, is not formed from pyruvate. Its synthesis from dihydroxyacetone phosphate, an EMP intermediate, is shown in Fig. 11-6. Glycerol is a major product of the fermentation of sugars by some species of *Bacillus* and is formed in variable

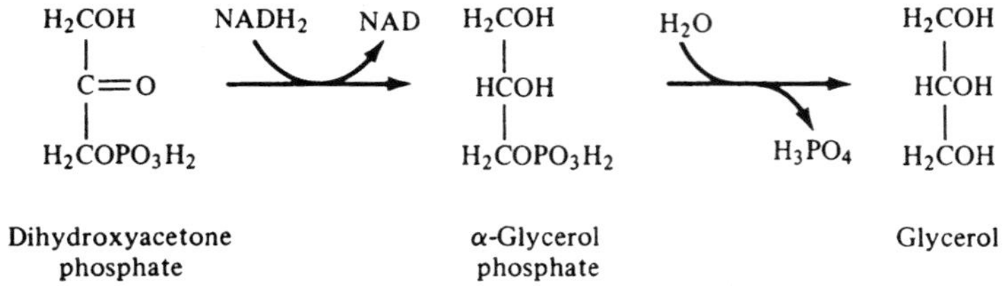

Figure 11-6 Formation of glycerol in fermentation.

amounts by many fermentative microorganisms. The enzymes required for its formation are probably present in most microorganisms. The immediate precursor of glycerol, α-glycerol phosphate, is used in the synthesis of phospholipids, which are an essential component of all cellular membranes. A phosphatase capable of removing the phosphate group of α-glycerol phosphate and of a number of other phosphorylated compounds, i.e., an enzyme of relatively low specificity, is found in many cells.

The formation of glycerol wastes energy, since the phosphate group removed by hydrolysis was derived from ATP, and the conversion of dihydroxyacetone phosphate to pyruvate would result in the synthesis of two molecules of ATP. Thus, the real cost to the cell is three ATP molecules per molecule of glycerol, since one ATP is consumed and two potential ATPs are not formed. In yeast and probably also in other organisms, the formation of glycerol as an end product of fermentation is influenced by pH. Under alkaline conditions, less ethanol and more glycerol appear as products. Glycerol formation can be forced by adding sodium sulfite or other compounds that react with acetaldehyde, preventing its use as an electron acceptor in the yeast fermentation. This process has been used industrially in the manufacture of glycerol.

MAJOR SUGAR FERMENTATIONS

The pathways to the individual end products of the fermentation of sugars we have described are used in fairly specific combinations by different groups of

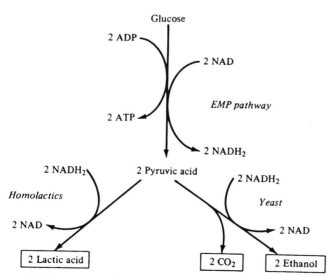

Figure 11-7 The fermentation patterns of the homolactics (lactic acid fermentation) and of yeasts (alcohol fermentation). Only the major products are shown.

microorganisms. These typical fermentation patterns have been given names that are usually taken from the major product or a product unique to that fermentation. Thus, we speak of the alcoholic fermentation of yeast, the mixed acid fermentation of *E. coli*, the butanediol fermentation, or the acetone-butanol fermentation. Figures 11-7 to 11-11 show the product formations typical of some well-studied fermentations. Others were shown in Figs. 11-1 and 11-5. Many other possible fermentation patterns exist; these are presented only as examples of the ways in which the reactions described previously can be combined.

It is apparent from the list of fermentation products in Table 11-1 and from much published data of the same type that fermentations are generally rather complex, producing a few major products and often a variety of minor

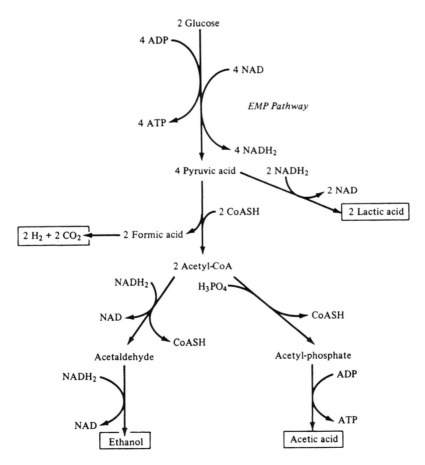

Figure 11-8 The mixed acid fermentation pattern of *E. coli* and other enteric bacteria. Only the major products are shown.

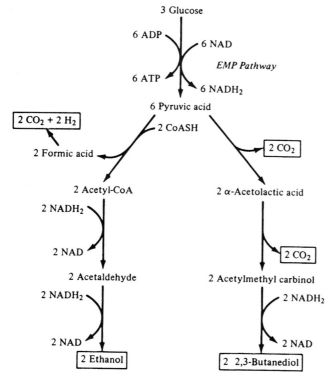

Figure 11-9 The butylene glycol (butanediol) fermentation pattern of *Enterobacter* and other enteric bacteria. Only the major products are shown.

products. Each microbial species possesses the genetic information for making a specific set of fermentation products and is incapable of making others. The products that it actually makes at any time depend on the prevalent conditions, such as the specific energy source used, the initial pH, the changes in pH that occur as a result of fermentation, and sometimes the presence of specific inorganic ions or organic growth factors. However, the typical major products are less subject to variation than are the minor products, so we can use the fermentation pattern, as determined by analysis of the major or unique products, as a tool in the identification of bacteria.

Whatever the product mixture formed by an organism in a particular situation, if no external oxidant is used, the requirement that all $NADH_2$ formed be reoxidized dictates certain combinations of products. This is apparent when one examines the oxidations and reductions involved in the formation of each individual product. These are summarized in Table 11-3 for the common C_2, C_3, and C_4 products. It is assumed that each is formed via the EMP pathway (at least the lower portion), as is usually the case except for glycerol. Thus, the number of molecules of pyruvate used to synthesize the

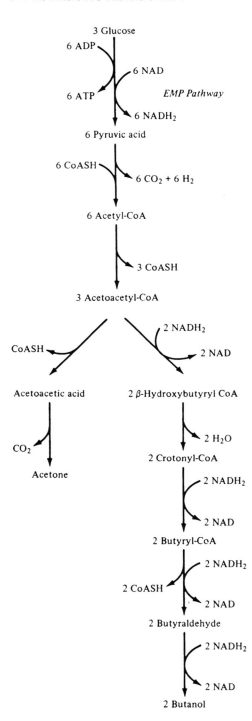

Figure 11-10 The acetone-butanol (solvent) fermentation of the clostridia. Only the major products are shown, but note that more $NADH_2$ is oxidized than is formed. This will allow some additional ATP to be made from acetyl-CoA.

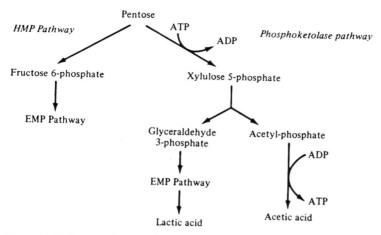

Figure 11-11 Two mechanisms for fermentation of pentoses. In most microorganisms, pentoses are converted to fructose 6-phosphate by the transaldolase-transketolase reactions of the HMP pathway (see Chap. 9). In the heterolactics, pentoses are metabolized through the phosphoketolase pathway, forming lactic acid and acetic acid.

Table 11-3 Role of pathways to individual fermentation products in accepting hydrogen or generating ATP

Products	(1) Pyruvates used	(2) NADH₂ formed	(3) NADH₂ reoxidized	(4) ATP per mole
Lactic acid	1	1	1	0
Ethanol	1	1	2	0
Acetic acid	1	1	0	1
Butanediol	2	2	1	0
Butyric acid	2	2	2	(1)*
Butanol	2	2	4	0
Acetone	2	2	0	0(1)*
Isopropanol	2	2	1	0(1)*
Glycerol	—	—	1	0
Succinic acid	1	1	2	0
Propionic acid	1	1	2	0

Note: Data are based on the assumption that the products are derived from a hexose fermented through the EMP pathway.

*ATP may be formed in the pathway to butyric acid in some organisms. ATP may also be formed indirectly by the transfer of CoA to acetate in the pathway to butyric acid, acetone, or isopropanol (see text).

product corresponds to the number of molecules of NADH₂ formed in its synthesis. Comparison of cols. (2) and (3) indicates which pathways are deficient in accepting hydrogen atoms. Only lactic and butyric acid are formed by a series of reactions in which oxidation and reduction are balanced. Those products for which the number in col. (2) is greater than that in col. (3) can be formed only if a product that accepts excess hydrogen is also formed, i.e., one for which the number in col. (2) is less than that in col. (3). The combinations of products formed must be such that the molar amounts of hydrogen removed and hydrogen accepted are equal. Within this restriction, ratios of products can vary with the strain and with the prevalent conditions. The ability to form hydrogen gas allows the formation of more oxidized products. Thus, in organisms that have the genetic information for synthesizing a variety of products, no typical balanced equation can be written to describe the fermentation, and a fermentation balance must be performed to determine the specific course followed by a particular strain under specific conditions. If, as in industrial applications, it is desired to maximize the production of one or more products, use of different strains and different conditions in fermentation balance studies provides the data for choosing the best microbial strain and the optimum conditions for the fermentation process.

FERMENTATION OF NONCARBOHYDRATE SUBSTRATES

Many microorganisms that are able to utilize compounds other than carbohydrates as energy sources in the presence of oxygen or nitrate are unable to derive energy from them through fermentation. The enterics, for example, grow well in media containing proteins, peptides, or amino acids as sole sources of carbon and energy if oxygen or nitrate is available as electron acceptor. In the absence of an inorganic electron acceptor, growth in the same media is possible only if a fermentable carbohydrate is added. The amino acids are not fermented but are taken up and used in the synthesis of cellular protein.

Several genera of bacteria that are pathogenic to humans and animals or are normal inhabitants of the intestinal or respiratory tract ferment amino acids, and some of these also may be found in anaerobic digesters. However, the anaerobic decomposition of proteins in nature and in anaerobic biological processes of waste treatment may be primarily the work of the proteolytic species of *Clostridium*. We have stated that these organisms generally are active producers of proteolytic enzymes. They ferment the resulting amino acids in two ways. Some amino acids are fermented individually by pathways specific for each compound. Not all amino acids can be fermented singly, or at least no organism capable of fermenting certain amino acids has been isolated. The products of fermentation of amino acids are generally ammonia, carbon dioxide, hydrogen, acetic acid, and butyric acid. Propionic and other low molecular weight acids and ethanol may be formed also, depending on the amino acid fermented. Few of these pathways have been studied in detail.

The second mechanism used by many species of *Clostridium* for the fermentation of amino acids is the Stickland reaction. Pairs of amino acids are fermented; one is oxidized and the other is reduced (see Fig. 11-12). This

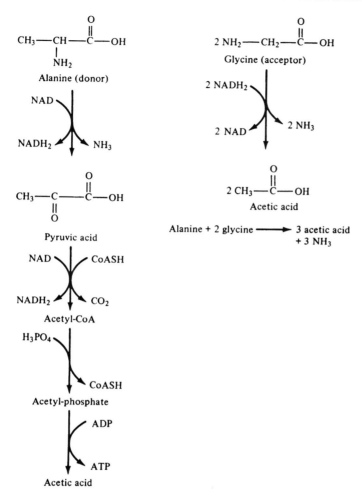

Figure 11-12 The Stickland reaction for fermentation of pairs of amino acids by clostridia. One amino acid acts as electron donor, i.e., is oxidized. The hydrogen atom removed from the electron donor is used to reduce the amino acid which acts as electron acceptor to regenerate NAD.

method of fermentation allows the amino acids that cannot be fermented individually to be used as energy source. Table 11-4 lists the amino acids fermented by the two mechanisms in one or more species of *Clostridium*. Only one amino acid that cannot be fermented singly by at least one species of *Clostridium* is fermented by another bacterium. This is glycine, which is fermented to carbon dioxide, ammonia, and acetic acid by *Peptococcus anaerobius*. No other amino acid is fermented by this organism, which was previously given several different names (see *Bergey's Manual*). Thus, members of the genus *Clostridium* are very active in the anaerobic decomposition of proteins.

Table 11-4 Fermentation of amino acids by one or more species of *Clostridium*

| Amino acid | Fermented singly | Stickland reaction | |
		Donor	Acceptor
Alanine	+	+	
Arginine	+		+
Aspartic acid	+		+
Cysteine	+		+
Glutamic acid	+		
Glycine			+
Histidine	+	+	
Hydroxyproline			+
Isoleucine		+	
Leucine	+	+	
Lysine	+		
Methionine	+		+
Phenylalanine	+	+	
Proline			+
Serine	+	+	
Threonine	+		
Tryptophan	+	+	+
Tyrosine	+	+	+
Valine		+	

Data from the chapter by Barker in Gunsalus and Stanier, 1961.

Two species of *Clostridium* also are able to ferment purines: *C. acidiurici* and *C. cylindrosporum*. Fermentation of purines by pure cultures is rare, being restricted to *Clostridium* and a few anaerobic cocci that are probably species of *Peptococcus*. Fermentation of pyrimidines is even more rare. Only one species of *Clostridium* (formerly classified as *Zymobacterium*) has been shown to obtain energy from pyrimidine fermentation. As in aerobic organisms, both purines and pyrimidines are undoubtedly used preferentially as growth factors and are taken up for use in the synthesis of nucleic acids.

The fermentation of lipids is thermodynamically unfavorable. Although the β-oxidation pathway for the conversion of long-chain fatty acids to acetyl-CoA does not involve the direct participation of oxygen, as does the oxidation of hydrocarbons, these molecules are essentially hydrocarbon in nature. An equation can be written for the conversion of a fatty acid such as palmitic or stearic to acetic acid and butanol, for example, but only if much of the hydrogen removed (four hydrogen atoms for each molecule of acetyl-CoA formed) can be converted to hydrogen gas. Many anaerobic bacteria form the enzyme hydrogenase, which catalyzes the reversible formation of hydrogen from a reduced high-energy electron carrier such as ferredoxin (Benemann and Valentine, 1971).

$$2H^+ + 2e^- \rightleftharpoons H_2 \qquad (11\text{-}14)$$

As Barker has pointed out, however, the problem in the fermentation of fatty acids lies in the electrode potential of the first dehydrogenation reaction (see chapters by Barker and Dolin in Gunsalus and Stanier, 1961). The electron acceptor for this dehydrogenation in aerobic organisms is flavoprotein (see Fig. 10-3); the reaction has a less negative electrode potential than does the reaction in Eq. (11-14) at pH 7.0, and electron flow to hydrogen is therefore unfavorable. However, lipids are decomposed in anaerobic digesters under strictly anaerobic conditions. The explanation for the occurrence of this and other reactions that are not observed in pure cultures undoubtedly lies in the ability of mixtures of organisms to accomplish metabolic transformations of which none acting alone is capable. Recent studies of methanogenesis are beginning to shed some light on the interrelationships of organisms in the anaerobic environment.

METHANOGENESIS AND THE ANAEROBIC DIGESTION OF WASTES

We have said that the anaerobic bacteria perform an essential function in the recycling of carbon from the anaerobic habitats of the biosphere to the atmosphere. However, an examination of the pathways of carbohydrate fermentation (see Figs. 11-1, 11-5, and 11-7 to 11-11) demonstrates the inefficiency of fermentation as a means of recycling organic carbon. The major function of the fermentative bacteria is the breakdown of biologically synthesized polymers to monomeric units and the conversion of these to even simpler compounds.

Some anaerobic environments contain electron acceptors that can be used instead of oxygen. If nitrate is available, facultative anaerobes that are able to use it as terminal electron acceptor can carry out the total oxidation of organic carbon to carbon dioxide in the absence of air. However, in most truly anaerobic environments, little nitrate would be expected to be formed because nitrification is a strictly aerobic process. Nitrate may, of course, be supplied by agricultural runoff in some ponds, lakes, and swamps. Organisms that can use sulfate as terminal electron acceptor also convert some organic carbon to carbon dioxide but often do not carry out complete oxidations. Sulfate, like nitrate, is not formed under anaerobic conditions, but it may be present in many waters, particularly in brackish waters.

Some carbon dioxide is formed in fermentation and much of this is recycled to the atmosphere, but most of the organic carbon metabolized by fermentation remains in the environment in altered form. As we shall see later, fermentation removes no COD. The energy source is simply converted to other compounds. Some COD is "removed" by incorporation into cells, i.e., by use as carbon source, but this removal depends on the separation of the cells. When the feed to the digester consists of excess sludge (cells) from an aerobic treatment unit, conversion of the original cell material to new cells represents no real progress toward recycling of carbon. The low growth yield of fermentative cells means that the percentage of total carbon removed by cell synthesis is low. Most fermentations of sugars produce 2.0 to 2.5 mmol of ATP per 1 mmol of sugar. The Y_{ATP} value of 10 mg dry weight can be used to calculate the cell yield as 20 to 25 mg. Thus, for each 180 mg of glucose fermented, 20 to 25 mg of other organic material can be converted to cell substance. One cannot calculate a percentage yield, as is done with a minimal medium in an aerobic system where a single compound is used as both carbon and energy source, but, assuming conversion of compounds used as carbon source to cell material without a change in weight, a percentage yield on a weight basis could be calculated as $(2 \times 10)/(180 + 20) \times 100 = 10$ percent for an organism with a fermentation yielding two molecules of ATP for each molecule of glucose, e.g., the homolactics.

Fermentation, then, even allowing for conversion of some of the energy source to carbon dioxide and for conversion of perhaps 10 to 15 percent of the total organic matter to cells, is not an adequate process for either returning organic carbon to the atmosphere or removing it from a waste stream. The activities of the fermentative bacteria are merely the first step in the recycling process, and they must be supplemented by the activities of other microorganisms that complete the process. These further metabolic activities may be aerobic or anaerobic depending on the locale in which the fermentation occurs. The organic matter in many natural anaerobic environments is often largely particulate, as it is in the anaerobic digester, or it may be in larger solid masses, e.g., plants or leaves, in ponds and swamps. When fermentative microorganisms convert this solid or particulate material to small, soluble molecules in a natural body of water, these may diffuse upward into the aerobic layer where they can be converted to carbon dioxide by aerobic microorganisms. The soluble organic material in the digester supernatant is generally treated aerobically by recycling it to the head end of the treatment plant. This is a small volume of liquid with a very high BOD. Its BOD would be much higher, however, if the reactions we have described thus far were the only ones that had occurred in the digester. In the digester and in anaerobic portions of ponds, lakes, and other water bodies, many of the acids and alcohols produced by primary fermentations are converted to an insoluble gas, methane, which escapes to the atmosphere, thus providing the major mechanism for the recycling of carbon under anaerobic conditions. It should be noted, however, that the formation of methane and its transfer to the atmosphere do not complete the carbon cycle. Methane in the atmosphere can be regarded only as a pollutant, and its formation and escape to the atmosphere only exchange air pollution for water pollution. The methylotrophs and the heterotrophs, which are able to oxidize methane to carbon dioxide, require oxygen, and if methane is evolved slowly enough in a body of water, much of it may be oxidized to carbon dioxide by aerobes near the surface.

Unless methane is oxidized to carbon dioxide, either biologically or by incineration, it is not available for recycling through photosynthesis.

Formation of Methane

The past decade has seen rapid progress toward understanding the process of methane formation. Much of this progress was made possible by the demonstration that a culture that had been used in research on methane formation for many years was actually an association of two organisms (Bryant et al., 1967). One, an unidentified Gram-negative, anaerobic rod, "S" organism, ferments ethanol to acetic acid and hydrogen and does not form methane. The other, *Methanobacterium* MOH, cannot metabolize ethanol but is able to utilize hydrogen as its energy source and electron donor and carbon dioxide as its carbon source and electron acceptor, forming methane. In the ethanol-carbonate medium used for many years for growth of this culture, each organism is totally dependent on the other. The organism that ferments ethanol cannot do so unless the hydrogen is removed by the methanogenic organism, and the methanogen requires the hydrogen as energy source and electron donor.

A number of pure cultures of methanogenic bacteria have now been isolated. Studies of methane formation have shown that the range of substrates used by these methanogenic bacteria is very limited. All methanogens isolated thus far are able to form methane from hydrogen and carbon dioxide, and some species are able to use only hydrogen and carbon dioxide for growth and methane formation. Others are able to use formic acid, HCOOH, which is probably first converted to carbon dioxide and hydrogen. At least two species of *Methanosarcina* are able to form methane from methanol or acetic acid (Mah, Smith, and Baresi, 1978). The use of acetic acid is of particular importance because approximately 70 percent of the methane produced in sludge digestion is formed from acetic acid and the remaining 30 percent from carbon dioxide and hydrogen. In the absence of hydrogen, the methyl group of acetic acid is reduced to methane and the carboxyl group is oxidized to carbon dioxide.

$$*CH_3\overset{O}{\overset{\|}{C}}OH \rightarrow *CH_4 + CO_2 \qquad (11\text{-}15)$$

If hydrogen is available, carbon dioxide is reduced to methane:

$$CO_2 + 4H_2 \rightarrow CH_4 + 2H_2O \qquad (11\text{-}16)$$

In the early literature on methane formation, many alcohols and organic acids were listed as substrates for methanogenesis (Barker, 1956). However, it is now apparent that the cultures used in earlier studies were not pure and that the formation of methane from complex organic molecules involves the conversion of these molecules by other bacteria to a few simple substrates that can be used by the methanogenic bacteria. Sludge digestion has long been considered to be a two-step process, usually described as follows. In the first step, a variety of fermentative bacteria hydrolyzed polymers and fermented the hydrolysis products and other small molecules, using the pathways described previously, to alcohols and fatty acids. In the second step, the methane bacteria fermented these products of the primary fermentations to methane. Since the substrates that can be used for methane formation now appear to be so limited, it is apparent that either (1) another step, intermediate between the two previously recognized, must be required to convert fermentation products other than acetate, formate, methanol, carbon dioxide, and hydrogen to these compounds, or (2) the known fermentations must be altered in the presence of the methanogens. Both of these alterations of the traditional scheme probably occur.

The overall process of anaerobic digestion requires an interaction between the fermentative bacteria and the methanogenic bacteria that is more complex than the simple sequential metabolism originally envisioned. This relationship is based on *interspecies hydrogen transfer*, a process first demonstrated with the symbiotic association of the two organisms maintained as *Methanobacillus omelianskii*. The anaerobic digestion of sludge represents a delicately balanced ecological system in which each type of organism plays an essential role. The methanogenic bacteria perform two essential functions. They form the insoluble gaseous product methane, which accomplishes the removal of organic carbon from the anaerobic environment, and they utilize hydrogen, allowing the fermentative bacteria to ferment compounds not otherwise fermentable and to form acetic acid, which can be converted to methane.

We have noted previously that certain types of compounds are not fermentable by any known organism in pure culture. The fatty acids derived from the hydrolysis of lipids are a prominent example of these nonfermentable substrates. These compounds constitute a very significant proportion of the feed to the digester at a municipal waste treatment plant, sometimes as much as 50 percent of the biodegradable COD of raw sewage sludge. Although fats are a problem in the digester because of their tendency to collect in large balls and to form a heavy scum layer on the surface of the digesting mass, they are metabolized in the digester. Several studies have indicated that fatty acids are probably metabolized by a process similar to the aerobic β-oxidation pathway (see Fig. 10-3). This pathway converts even-numbered, saturated fatty acids to acetyl-CoA, with the removal of four hydrogen atoms for each acetyl-CoA molecule formed. Two difficulties arise in the use of this pathway under anaerobic conditions. First, as we have pointed out, the initial oxidative step is thermodynamically unfavorable except in the presence of oxygen (Barker, 1961). Second, according to the rule that governs fermentation, i.e., no net oxidation, the hydrogen atoms removed in the formation of acetyl-CoA would have to be used to reduce the acetyl-CoA to ethanol [Eq. (11-9)]. This would mean that no energy would be available to the organism. If the hydrogen atoms could be used to form hydrogen gas, the acetyl-CoA could be used for the synthesis of ATP [Eq. (11-10)]. This is believed to be made possible by interspecies hydrogen transfer in the mixed populations of the digester and the anaerobic sediments in which methane formation occurs. The removal of hydrogen by the methane bacteria pulls the reaction forming hydrogen and makes thermodynamically unfavorable reactions possible. Thus, amino acids, fatty acids, and other compounds that are metabolized to acetyl-CoA in aerobic systems but not in anaerobic pure cultures can be fermented to acetate and hydrogen, yielding energy for the fermentative organism and hydrogen for use in methane formation.

Evidence that the pattern of fermentation is altered in the presence of methanogenic bacteria has been obtained in several experiments in which fermentative bacteria have been grown in pure culture and in mixed culture with a methanogenic species (e.g., see Iannotti et al., 1973). These organisms produce the fermentation products we have discussed previously when grown in pure culture, i.e., formic, acetic, lactic, butyric, and succinic acids, ethanol, carbon dioxide, and hydrogen. When grown together with a methane-producing organism, the amounts of the reduced products are decreased and the amount of acetate is increased. Thus, the utilization of hydrogen for methane production provides fermentative organisms with an alternative means of reoxidizing at least some of the $NADH_2$ that would otherwise have to be used to form reduced products. In the absence of the hydrogen-utilizing methanogen, the hydrogen would accumulate in sufficient concentrations to be toxic or to prevent its further formation by mass action due to product accumulation.

It seems certain, based on experiments with mixed cultures, that the products of a fermentation can be altered in the presence of methanogenic bacteria. It is not known yet whether a third group of organisms, carrying out a step intermediate between the primary fermentations and the formation of methane, is of real importance in the overall conversion of organic matter to methane. The "S" organism in the symbiosis, which was long called *Methanobacillus omelianskii*, ferments ethanol, a product of many primary carbohydrate fermenters. Ethanol is not fermentable in pure culture. Organisms such as this, which can ferment such products as alcohols with the aid of the methane formers, may be an important link in the ecological chain responsible for anaerobic digestion. If the range of substrates utilizable by methanogens is truly as narrow as it now appears to be, and if 70 percent of the organic matter in a digester must be converted to acetate for use by the methanogens, the primary fermentations may have to be altered toward greater amounts of more oxidized products and another group of bacteria must also act to convert any fermentation products not utilizable for methane formation to acetate, formate, methanol, carbon dioxide, and hydrogen. On the other hand, it is possible that methanogens may be found that are capable of using other compounds. The methods for isolation and study of these extremely oxygen-sensitive organisms have been developed only recently.

Application of techniques for culture of strict anaerobes to studies of digester populations has now shown that strict anaerobes far outnumber the facultative anaerobes, which were once considered to constitute the majority of the fermentative bacteria (Hobson et al., 1974). Much work remains to be done before we achieve a reasonably good understanding of the ecological system that results in methane production. It is important to achieve this understanding, not only from the viewpoint of the microbiologist who desires to increase our knowledge of microbial physiology and ecology, but also from the practical viewpoint of the engineer who needs to be able to make more informed decisions in designing and operating the process of anaerobic digestion. In 1962, McKinney wrote "Anaerobic digestion is the uncharted wilderness in sanitary engineering." It was an apt description. The wilderness is only now beginning, gradually, to yield to exploration.

ANAEROBIC TREATMENT PROCESSES

Two major processes for the treatment of wastes utilize anaerobic metabolism. The first is anaerobic digestion of sludges removed in primary and secondary sedimentation operations at municipal treatment plants. Anaerobic treatment is also used, less frequently, for treating certain high-strength industrial wastes. These processes are applications of the biochemical reactions described in this chapter. The second process is denitrification for the removal of nitrate produced from excess reduced nitrogen compounds at aerobic biological treatment facilities. Denitrification may also be used for the treatment of industrial wastes that contain high concentrations of nitrate, e.g., munitions plant wastes.

Anaerobic digestion is a more complex process than is denitrification; it involves hydrolysis, fermentation, and the production of highly volatile materials such as methane and hydrogen, whereas denitrification involves anaerobic respiration, i.e., the replacement of oxygen by nitrate as terminal electron acceptor. The biochemistry and physiology of both processes are already familiar, and the material that follows describes the treatment processes, possible modes of operation, and effectiveness of the treatments.

Anaerobic Digestion

Anaerobic digestion has been the traditional method for preparing sludge from municipal waste treatment plants for ultimate disposal. Approximately 50 percent reduction in organic content can be attained, and that which remains is much less putrescible than was the original organic matter. The reduction in organic content of the sludge arises because the more easily degraded material is converted to volatile materials such as carbon dioxide and methane, which are stripped from the reaction mixture. A portion of the sludge is converted to soluble organic matter, which remains in the supernatant liquid upon decantation from the digested sludge.

Anaerobic reactors As with aerobic treatment of wastes, there are various modifications of anaerobic digestion processes. All of them involve either quiescent or mixed fluidized conditions. Fixed-bed reactors, i.e., anaerobic trickling filters, are not generally used for the anaerobic digestion of wastes that contain high concentrations of suspended organic matter.

The most simple reactor is a covered holding tank into which excess sludge is intermittently pumped, generally two to four times daily. Distinct zones develop, i.e., gas, supernatant, and sludge zones, as shown in Fig. 11-13. The influent organic solids settle in the sludge layer. Active decomposition of the organic solids occurs in the upper portion of this layer. Significant portions of the solids are liquefied and further metabolized to smaller molecules, some of which are volatile and escape to the gas zone where they can be drawn off. The less putrescible (stabilized) sludge solids migrate to the lower portion of the sludge zone and are drawn off periodically for further treatment (dewatering or draining) and final disposal, e.g., incineration or land disposal (see Table 2-3). The gas consists largely of methane and carbon dioxide with small amounts of hydrogen. The oxygen demand of the methane and hydrogen can be dissipated to carbon dioxide and water by burning the gas with oxygen. The heat value of the off-gas is more than sufficient to supply fuel for heating the reactor to the slightly elevated temperature (35°C) found to give reasonably good methane production and to provide some of the other energy needs of the treatment plant. When a plant is dependent on the production of methane for some of its energy needs, there is a necessity for very close monitoring of performance and for the development of a system of operational guidelines to ensure reliable delivery of the required gas. As with most of the aerobic microbiological processes, the reliability and stability of this anaerobic process usually translate in one way or another to a provision for slow growth rate. Since the fermentative and methanogenic microorganisms are very slow-growing organisms to begin with, and since the batch reactor shown in Fig. 11-13 could be operated as a once-through continuous flow reactor, the only way to engineer slow growth is to provide a

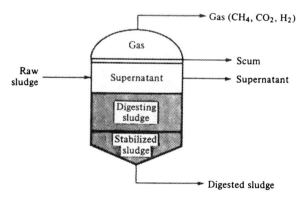

Figure 11-13 Single-stage unmixed anaerobic sludge digester, intermittent feeding and withdrawal, heated or unheated. Note gas, supernatant, and sludge zones. The zone of actively digesting sludge occupies a relatively small portion of the reactor.

rather long hydraulic retention time $\bar{t}$ of 30 to 60 days, i.e., a large volume since $\bar{t} = V/F$. Thus, the initial cost of construction is rather high.

The supernatant layer is a rather turbid liquid that contains high concentrations of colloidal and soluble organic matter that would exert a significant biochemical oxygen demand if discharged along with the plant effluent. Digestion of primary sludge alone leads to digester supernatant BOD_5 values ranging from 500 to 3000 mg/L, whereas for the digestion of combined primary and excess activated sludge, the supernatant BOD_5 may range from 1000 to 10,000 mg/L. The supernatant is usually recycled to the plant inflow where it undergoes aerobic treatment. At many municipal treatment plants, anaerobic digestion is used for only the primary sludge and aerobic digestion is used for the excess activated sludge, since supernatant BOD_5 concentrations are much smaller for aerobic than for anaerobic digestion. Most of

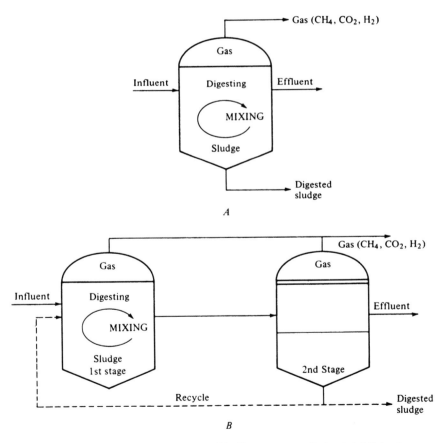

Figure 11-14 (*A*) Single-stage mixed anaerobic digester, generally heated. Mixing increases rate of digestion. The zone of actively digesting sludge occupies a larger portion of the tank volume than in an unmixed digester, but a second stage (*B*) is needed to separate the anaerobic mixed liquor.

the supernatant BOD from aerobic digestion consists of suspended organic matter rather than soluble materials.

The rate of reaction can be increased if the supernatant and solid zones are mixed, as shown in Fig. 11-14A. This increases the volume of the active zone for decomposition reactions and provides a more homogeneous reaction mixture. If the reactor is completely mixed and if the inflow and outflow rates are equal and the gas pressure is held constant, the reactor can be operated as a once-through anaerobic chemostat. A separate decantation and/or thickening operation must be used if the liquid and solid zones are mixed. Sometimes the postdigestion operations are accomplished in a second unmixed digestion tank placed in sequence with the mixed tank, as shown in Fig. 11-14B.

An even more flexible process includes the recycle of sludge underflow from the unstirred tank to the mixed reactor (see dashed lines in Fig. 11-14B). This reactor arrangement has been termed *anaerobic contact digestion*. It is in many respects an anaerobic activated sludge system. The separation tank need not be an unstirred digestion tank; uncovered settling tanks have been used. The separation of the mixed liquor from the reaction tank is usually enhanced if the liquor is degasified prior to entry into the separation tank (see Fig. 11-15). The system may be designed and operated in accord with models for activated sludge processes [e.g., Eqs. (6-76) and (6-77), Table 6-4], provided one determines the values of the four biokinetic constants and can satisfy the requirements and assumptions made in the derivation of the steady state model equations, e.g., complete mixing of the substrate and cells in the stirred reactor as well as little or no biological activity in the cell separator. In any event, care should be taken to determine whether biological activity in the separator changes the values of the biokinetic constants for the sludge significantly, due to its residence time in the quiescent tank. The system has the advantage that one can control, i.e., slow down, the specific growth rate or increase the sludge age using the engineering control parameters α and X_R as well as D, or $\bar{t}$, which is the only means available in a once-through reactor.

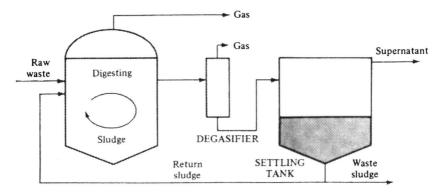

Figure 11-15 The anaerobic contact process, an anaerobic activated sludge process.

Energy considerations Fewer data are available on the values of the kinetic constants for anaerobic treatment than for aerobic treatment. Although the anaerobic digestion process represents rather complicated interrelationships of perhaps three main groupings of organisms—the fermenters, the hydrogen producers, and the methane formers—the hydrogen and methane formers are slower growers than the fermentative bacteria, and therefore their kinetic characteristics ought to govern the values for the biokinetic constants at the mesophilic temperature 35°C. Based on available literature, values of cell yield Y_t of 0.04 to 0.05 mg of biomass per milligram of COD converted to methane and k_d values of 0.010 to 0.040 day^{-1} seem reasonable (Lawrence and McCarty, 1969). A value of maximum substrate utilization rate U [$U = (dS/dt)(1/X)$] of 6.7 day^{-1}, which translates to μ_{max} values of 0.268 to 0.335 day^{-1} ($\mu = UY_t$) and K_s values of approximately 2000 mg/L seem reasonable based on considerations presented by Lawrence (1971). However, much more experimentation is needed before values of the kinetic constants and the factors that determine these values are well defined.

The cell recycle scheme shown in Fig. 11-15 would appear to be an ideal way to run an anaerobic digester for the treatment of high-strength industrial wastes, e.g., feedlot wastes or piggery wastes. However, as with the activated sludge process, merely providing a cell recycle line does not guarantee the advantages of process control that recycle can bestow. That is, α and X_R can be used as design and operational control parameters only if provisions are made in the design and construction of the system that permit these factors to be used by the operator of the process. For example, positive control of X_R and possible preconditioning of the sludge prior to recycle could be provided by the use of a sludge consistency or dosing tank similar to that recommended in Chap. 6 for aerobic activated sludge processes. In view of the urgent need to conserve energy as well as the need for the reliable delivery of high-quality effluent, more research on the anaerobic activated sludge mode of digester operation is warranted, since it appears capable of delivering the highest possible methane yields for suspended or soluble substrates. The recent strides being made in understanding the biological mechanisms of methane production, coupled with continued kinetic studies to gain more definitive information on the kinetic constants, augur well for anaerobic bioconversion processes.

Fixed-bed reactors, i.e., anaerobic filters, may also be used to treat wastes. They are best adapted to wastes of much lower organic content than municipal treatment plant sludge or feedlot wastes but nonetheless can handle rather strong wastes with a BOD$_5$ of 2000 mg/L or more. Rather than trickling downward, as is usually the case for aerobic fixed-bed reactors like the type shown in Fig. 2-4, the waste flows upward through the tower, as shown in Fig. 11-16. This type of filter finds its most immediately useful application in denitrification processes. Note that the medium in this type of fixed-bed reactor is submerged. This means that the liquid retention time can be controlled by the liquid flow rate, unlike the downward flowing (trickling)

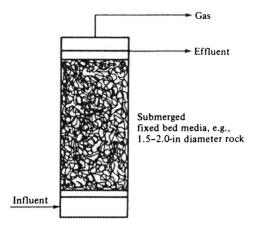

Figure 11-16 Anaerobic contact bed.

filter used in aerobic treatment for which the liquid retention time is, of necessity, very short.

One can utilize the energy and fermentation balance concepts in this chapter and in Chaps. 3 and 10 to examine the energy-conserving characteristics of anaerobic processes. For example, consider the anaerobic digestion of a polysaccharide. Upon hydrolysis, the hexose units could be fermented to ethanol and carbon dioxide as we have seen previously (see Fig. 11-7).

$$C_6H_{12}O_6 \rightarrow 2CH_3CH_2OH + 2CO_2$$
$$\text{hexose} \qquad \text{ethanol}$$

No COD is removed from the system in this reaction. The COD of the carbon dioxide is 0; i.e., the carbon in carbon dioxide is already in its most oxidized state and therefore it exerts no oxygen demand. It is important to note that the amount of organic matter has been reduced, but the original COD is now contained in the ethanol that has been produced. Assuming an initial glucose concentration of 1000 mg/L:

$$\text{COD of glucose} = 1000 \times \frac{192}{180} = 1067 \text{ mg/L}$$

$$\text{Ethanol produced} = 511 \text{ mg/L}$$

$$\text{COD of ethanol} = 511 \times \frac{192}{92} = 1067 \text{ mg/L}$$

Now suppose that the ethanol produced from the glucose is acted upon by other microorganisms to produce methane as follows:

$$2CH_3CH_2OH \rightarrow 3CH_4 + CO_2$$

$$\text{Methane produced} = 511 \times \frac{48}{92} = 267 \text{ mg/L}$$

$$\text{COD of methane} = 267 \times \frac{192}{48} = 1067 \text{ mg/L}$$

Thus, the only way COD is removed from the mixed liquor is by the stripping of the volatile product methane.

Although no oxygen demand is exerted in anaerobic processes, the capture of a burnable gaseous product such as methane and the exertion of its oxygen demand accompanied by utilization of the heat energy released can be a valuable asset. Unfortunately, one generally finds that most of the gas produced by the digester is burned without capturing and utilizing the released energy. Some of the reasons for this are the need to process the off-gas to remove impurities and the fact that the gas produced is usually not enough to supply all the energy needs of the plant. It is envisioned as a supplementary energy source only. Also, a digester is not very reliable in producing a steady supply of methane in view of the general paucity of fundamental and engineering knowledge regarding biochemical mechanisms and kinetics as well as their control in the rather complex, finely balanced ecosystem in anaerobic digesters. In specific cases in which large amounts of concentrated organic matter are available, for example, stockpiles of animal manure from feedlot operations, and where the anaerobic digestion process is operated at sufficiently low growth rates and carefully controlled loading conditions, excess marketable gas can be produced and mixed with drilled natural gas. The anaerobic digestion that has been used for so many years at municipal treatment facilities seldom reliably produces enough excess energy needs at the plant to warrant the special precautions and backup systems needed for its use for anything but heating the digester. The burning off of excess gas at treatment facilities is as common as it is in many of the oil fields in the drilling areas of the world. However, in the future, as the dwindling energy supply from traditional sources becomes an even more acute problem than it already is, more earnest efforts to finely tune anaerobic digestion and use the off-gases will be necessary. It is not entirely improbable to expect that power from the burning of digester gas will provide significant portions of the compressed air needed for aerobic fluidized secondary biological treatment processes at the head end of the plant.

One might argue that since the energy of the digestible material is retained in the off-gas, one could burn (incinerate) the original material without digestion. In some cases it proves economical to incinerate the sludge solids directly, but the water content first must be drastically reduced. Estimation of the costs of the various alternatives is a design aspect that we shall pass over here because it is outside the scope of a text on microbiology. It is important, however, to emphasize that there is no one best way to treat and dispose of the sludge produced in wastewater treatment operations or as waste products of manufacturing processes. Some approaches that seem to provide ideally simplified solutions often do not save either energy or taxpayers' dollars when put to the test of cost analysis. In regard to the methane produced at municipal wastewater treatment plants, it seems that its best use is for its heat value rather than, for example, as a raw material for bioconversion to cells (for example, for protein feedstock). As in the example just given, the lack of change in COD when 1000 mg/L of glucose is converted to 267 mg/L of methane attests to the conservation of energy; the heat of combustion ΔH of 1 mol of glucose (solid) equals 669,000 cal and that for 3 mol of methane (gas) equals 638,400 cal. More than 95 percent of the heat value of the original substrate is conserved in the product. There has been a reduction in heat value of only 5 percent if all the methane is captured. Hence, if one utilizes the calories directly for this heat value, the return is very high—assuming the gas-cleaning operation can be performed economically.

One could use the methane in bioconversion projects; for example, one could grow microorganisms on the hydrocarbon to harvest the cells as a protein supplement for cattle diets. However, considering the example cited above, the yield of bacterial organic matter on the methane would be lower than on the original hexose. This situation comes about because the original amount of carbon source, 1000 mg/L of hexose, has been lessened to 267 mg/L of methane. The COD values and caloric contents are essentially the same and, although the cell yield on methane is higher per gram of carbon than on hexose [1.4 versus 1.26 g of cells per gram of carbon, according to Linton and Stephenson (1978)], growth on methane is limited by the available carbon. The yield on 1000 mg of hexose is:

$$\frac{72}{180} \times 1000 \times 1.26 = 502 \text{ mg of cells}$$

whereas on the methane produced from the hexose, it is:

$$\frac{12}{16} \times 267 \times 1.4 = 280 \text{ mg of cells}$$

Thus, from the standpoint of energy conservation by bioconversion to cellular materials, one would appear to be better off going not to the methane stage but only to the hexose stage in the breakdown of waste polysaccharides. One might use chemical hydrolysis of excess sludge (as described for the hydrolytically assisted extended aeration process) to break down complex macromolecules and sterilize the sludge. The resulting rich medium could be used for growing cells aerobically, thus achieving higher cell yields. Again, there is great need for caution against oversimplification because municipal sludge can be expected to contain some complement of heavy metals and other materials potentially toxic or hazardous to animals and humans. Thus, gasification and the partial recovery of the carbon as methane could provide an essential purification process prior to bioconversion.

There are no easy choices of alternatives for biological-chemical-physical processing of the massive amounts of sludge generated from municipal wastes. Ultimate oxidation to carbon dioxide and water and recycle to the carbon-oxygen-water cycle via biological digestion (aerobic or anaerobic), using processes as economical and safe as is possible, seem the best approach for now. However, if the rate of waste generation continues to increase, recovery and/or more immediate recycle schemes such as those outlined above will become more attractive.

Denitrification

Nitrates in wastewaters can be removed by denitrifying organisms in the absence of dissolved oxygen provided a source of carbon is available. While endogenous carbon sources may be sufficient for the removal of low nitrate concentrations by rather high concentrations of autodigesting cells, the addition of an external carbon source is desirable. The genus *Pseudomonas* makes up a large fraction of the denitrifying microbial population. Many of the *Pseudomonas* species known to accumulate nonnitrogenous storage products do not denitrify, whereas many of the species capable of denitrification do not make the major nonnitrogenous storage products, i.e., poly-β-hydroxybutyrate or polysaccharides; thus, they oxidize protein and release ammonia in endogenous or autodigestive stages, thereby defeating the purpose of biological denitrification processes.

Commercially available carbon sources and/or waste carbon sources may be used in denitrification processes. Brewery wastes have been successfully used. Sugar refining waste should also make a suitable carbon source, since it is high in carbohydrate and essentially devoid of ammonia or organic nitrogen. Chemically or biologically hydrolyzed waste cellulose, e.g., wastepaper or possibly sawdust, would also make a suitable carbon source. Molasses, volatile acids, and methanol are among the commercially available carbon sources that have been used. Highly reduced compounds are better from the standpoint of minimizing excess sludge production, since the cell or sludge yield is generally lower (measured in milligrams of cells per milligram of COD removed) for compounds such as methanol or ethanol than for carbohydrates. Most of the experimental work has been accomplished using methanol; it has generally been the substrate of choice, but it is rather costly. Some of the denitrifiers that use methanol may also use methane; thus in plants using anaerobic sludge digestion, some of the methane that often is burned off might provide the electron donor for biological denitrification. Selection of the carbon source is an important aspect, since it can exert a selective pressure on the species in the denitrifying biomass. That is, the population is already narrowed to the denitrifying species, and one carbon source may cause a much narrower band of species variation than another one. Adequate ecological studies have not yet been undertaken on denitrifying heterogeneous populations, but the expected further tightening of advanced treatment requirements would seem to increase the usefulness of such studies.

When methanol is used as the carbon source and electron donor, there is evidence that the K_s with respect to this substrate is very low; thus, the Monod relationship between μ and the carbon source concentration usually does not govern the rate of denitrification because methanol concentration is limiting at only very low concentrations. The growth rate and hence the denitrification rate can, however, be related in an equation of the Monod type (rectangular hyperbola) to the concentration of the electron acceptor nitrate. The kinetic model presented in Chap. 6 can be used to predict the performance of denitrification processes.

With methanol as the substrate, values of μ_{max} ranging between 0.2 and 2.0 day^{-1} and K_s values between 0.06 and 0.16 mg/L of nitrate–nitrogen have been observed. Cell yield values Y_t between 0.6 and 1.2 mg/mg of nitrate–nitrogen and k_d values of approximately 0.04 day^{-1} have been reported (U.S. EPA, 1975). As with any biological system, the values of the kinetic "constants" are affected by environmental conditions such as pH and temperature. The range of optimum pH values is somewhat more narrow than those for aerobic growth, presumably because the spectrum of microbial species in the population is more narrow due to the selective pressure of the electron donor and electron acceptor used. The pH values of 7 to 7.5 seem optimum. Fairly good denitrification is observed at 20°C, and a significant increase can be expected for a modest increase in temperature to 25 to 30°C. While there appears to be little to be gained by adding energy to heat the reactor above 20 to 25°C, there is some evidence (see Focht and Chang, 1975) that the rate of denitrification is affected proportionately more than would be predicted by the Arrhenius equation when temperatures are below the 10 to 15°C range. Thus, there may be some advantage in protecting the process against extreme cold during winter months and providing higher concentrations of denitrifying biomass to take up the slack in the removal rates due to lower temperatures.

In a fluidized (suspended growth) reactor system, i.e., an activated sludge system with nitrate rather than DO as the electron acceptor, a higher biomass concentration (hence, a slower growth rate) can be provided by controlling the recycle sludge concentration X_R and the flow rate F_R, as in the mode of operation recommended in Chap. 6. Since the mixed liquor contains escaping nitrogen gases (one of the causes of rising sludge), an aerated stripping chamber with approximately 5 min of holding time is placed between the reactor and the sedimentation tank. In some instances, an aerated stabilization tank with approximately 1 h of holding time is inserted between the reactor and the aerated stripping chamber. An advantage claimed for this procedure is that the endogenous respiration rate (cell autodigestion) is much higher under aerobic than anoxic conditions. Thus, the net or observed cell yield Y_o and hence the amount of excess sludge are lower. This seems worthy of more investigation. It would seem that to autodigest a significant amount of cells in a relatively short aerobic stabilization time, most of the endogenous carbon source would be derived from short-term oxidative assimilation stores. The ability to store carbon is not a common characteristic of denitrifiers. It is conceivable that the quick return of the cells to an aerobic condition may cause some of them to undergo lysis, thus enhancing sludge reduction. The lysis products would have to be oxidized rapidly in order to minimize leakage of organic compounds into the effluent. In general, our own experience on changing cultural conditions from aerobic to anaerobic to aerobic for an activated sludge containing an ample complement of denitrifying organisms is that the shock of the switch from one electron acceptor to another does not appear to hamper the metabolic efficiency of an activated sludge when returned to an aerobic condition. Presumably such is also the case when the cycle is anaerobic to aerobic to anaerobic. However, in experiments with some denitrification schemes in which a nitrified activated sludge mixed liquor is subjected to denitrification under endogenous anaerobic conditions and then returned to aerobic conditions, sludge bulking problems have developed (U.S. EPA, 1975). In such cases, chemically enhanced flocculation may be required.

Denitrification can be effected also in attached growth or fixed-bed reactors, i.e., submerged filters or enclosed filters. Various types of coarse and fine solid media are available.

In Chap. 7 we discussed the interconversions of the various forms of inorganic nitrogen by microorganisms and some of the ecological effects of the presence of inorganic nitrogen compounds in the environment. Nitrification of a treatment plant effluent is necessary to prevent the exertion of a nitrogenous BOD in the receiving water. Discharge of a nitrified effluent, however, can lead to eutrophication because nitrate is an excellent nitrogen source for the growth of algae and other types of microorganisms. Allowable levels of nitrate in public water supplies are low because of two potential health hazards: (1) methemoglobinemia and (2) formation of carcinogenic nitrosamines when nitrate is ingested (see Chap. 7). Thus, it may be expected that denitrification processes, which have not been traditionally included in waste treatment facilities, may assume greater importance in the future. Much more research is needed to develop economical and reliable design and operational criteria for denitrification.

PROBLEMS

11-1 For an organism that uses the Embden-Meyerhof-Parnas pathway from glucose to pyruvate, the following products are formed from each 100 mmol of glucose fermented: 60 mmol of lactic acid, 20 mmol of 2,3-butanediol, 140 mmol of carbon dioxide, 100 mmol of hydrogen, 60 mmol of ethanol, and 40 mmol of acetic acid.

    (*a*) Write the pathway by which each product is formed from pyruvate.

    (*b*) Calculate the following for this fermentation:

    1. Millimoles of NAD reduced in converting glucose to pyruvate

    2. Millimoles of $NADH_2$ oxidized in converting pyruvate to 2,3-butanediol

    3. Millimoles of $NADH_2$ oxidized in converting pyruvate to ethanol

    4. Millimoles of $NADH_2$ oxidized in converting pyruvate to acetic acid

    5. Net production of ATP in millimoles

    6. Milligrams of cells produced

    (*c*) Calculate the carbon recovery and the O/R balance (see Table 11-2 for the form to be used).

11-2 If an organism that uses the EMP pathway for the fermentation of glucose produces acetone, carbon dioxide, and hydrogen, could lactic acid be the only other product? Explain your answer.

11-3 *E. coli* can form acetyl-CoA from pyruvate under either aerobic or anaerobic conditions, but the reactions are not the same.

(*a*) Write the equation for conversion of pyruvate to acetyl-CoA by *E. coli* growing aerobically.

(*b*) Write the equation for conversion of pyruvate to acetyl-CoA by *E. coli* fermenting a sugar.

(*c*) Explain why the reactions in parts (*a*) and (*b*) are different.

11-4 Name an organism that might produce each of the product mixtures listed below if it were fermenting a sugar.

(*a*) Carbon dioxide, ethanol

(*b*) Carbon dioxide, ethanol, lactic acid

(*c*) Carbon dioxide, hydrogen, ethanol, lactic acid, acetic acid

(*d*) Carbon dioxide, hydrogen, ethanol, lactic acid, acetic acid, butanediol

(*e*) Carbon dioxide, hydrogen, ethanol, lactic acid, acetic acid, butanol, butyric acid, acetone

11-5 A heterolactic fermenting a hexose can form only one molecule of ATP per molecule of glucose (net). Assuming that the pathway shown in Fig. 11-1 is used for the fermentation of pentose and that ribose could be phosphorylated by ATP and converted to ribulose 5-phosphate, what would be the products of fermentation and how many molecules of ATP per molecule of glucose could be formed? Write the pathway as it would be used for the fermentation of ribose.

11-6 For what two purposes do the methanogens use carbon dioxide?

11-7 In what ways are the methanogens different from all other bacteria?

11-8 What is the unique function of the methanogens in the carbon cycle?

11-9 State why and under what conditions you would justify considering denitrification as a necessary treatment process prior to discharge into a receiving stream.

11-10 Why is the supernatant from anaerobic digestion of secondary sludge usually so much higher in organic content (BOD) than supernatant from anaerobic digestion of primary sludge?

11-11 Why is the supernatant from aerobic digestion of secondary sludge usually so much lower in organic content than supernatant from anaerobic digesters?

11-12 Discuss the possible value of fermentation as a treatment before putting a waste through aerobic treatment.

11-13 Discuss the possible use and the significance of hydrogen analysis as an operational tool for anaerobic digesters.

11-14 Waste cellulose, e.g., used paper, might be used as a starting material for petrochemical manufacture. Discuss the biological and economic aspects of this use as compared to use of the material for its energy content, e.g., as fuel.

GLOSSARY

Active fraction The fraction of a biomass that is capable of metabolizing the substrate. The cells may be viable or nonviable. The ratio of the mass of the metabolically active mass to the total mass of volatile suspended solids is the active fraction.

Adsorption A surface phenomenon in which colloidal material and soluble compounds carrying an electric charge are held to the surface of an adsorbent material such as activated carbon.

Advanced waste treatment The removal of noncarbonaceous materials such as excess phosphorus and nitrogen. The term implies treatment beyond secondary treatment, and advanced treatment is most effective after the organic matter has been removed.

Assimilation capacity The ability of a body of water that receives waste components to permit the aerobic metabolism of the waste to proceed without limitation due to oxygen supply.

Autodigestion A decrease in the amount of biomass due to metabolism of cellular components or storage products. In a heterogeneous microbial population, autodigestion may include both endogenous metabolism and predatory activity, i.e., utilization of some cells as food material by other cells.

Autotroph An organism that uses carbon dioxide as its source of carbon for the synthesis of cell material.

Auxotroph An organism that requires one or more organic growth factors, e.g., amino acids, that it cannot synthesize. An auxotrophic mutant is one that has lost the ability to synthesize a metabolic product that can be synthesized by the wild-type (prototrophic) parent.

Batch system A system that is not fed continuously. A batch system may be one in which a volume of medium is inoculated with cells that are allowed to grow without further addition of nutrients. A "fill-and-draw" system, in which a fraction of the culture is removed at regular intervals, e.g., once daily, and replaced with fresh nutrient medium, is also considered to be a batch system. In both cases, growth occurs in an environment in which the concentrations of nutrients are continually decreasing.

Biological constant (biokinetic constant) A velocity constant or a proportionality factor used to describe the rate or total amount of growth in a biological system. Such constants are affected by environmental conditions such as pH, temperature, and chemical composition of the growth environment, and they are characteristic of the microbial species.

Biomass A heterogeneous microbial population, such as an activated sludge or a trickling filter slime.

Biosphere Those regions of the earth, i.e., the land (lithosphere), water (hydrosphere), and air (atmosphere), that constitute the life-support system of plants and animals.

Carbon source An organic or inorganic compound from which organic constituents of the cell are synthesized.

Catabolism Metabolic reactions by which a substrate is degraded to simpler compounds, yielding energy and usually also building blocks for synthetic reactions.

Catabolite repression The process by which synthesis of the inducible enzymes of a catabolic pathway is prevented when both the substrate (inducer) for the pathway and another preferred substrate, often glucose, are available to the cell.

Cell recycle The return of a concentrated biomass to a continuous flow reactor.

Centrifugation A process wherein gravitational force in excess of natural gravity is maintained due to rotary motion.

Chemotroph An organism that uses a chemical energy source.

Chlorine demand An amount of chlorine that must be administered to a sample containing microorganisms and organic and inorganic chemicals before free chlorine can be detected in the aqueous sample.

Clarifier A settling tank.

Completely mixed system A hydraulic regime in which the components flowing into a continuous flow reactor are instantaneously mixed with the contents of the reactor.

Contact stabilization A modification of the activated sludge process in which the wastewater is held in contact with a very high concentration of activated sludge for a short period of time. This is followed by quiescent settling and reaeration of a portion of the sludge prior to its recycling.

Continuous flow system A biological system in which the feed substrates are fed to the reactor continuously. Usually the system is operated at a constant volume so that the rate of outflow is equal to the rate of inflow.

Control parameters Factors used in the design and operation of a process to control the rate of the process. For example, mean hydraulic retention time, hydraulic recycle ratio, and concentration of sludge in the recycle to an activated sludge process are control parameters for the system.

Denitrification A biological process in which nitrate and nitrite are reduced to nitrogen gas. Denitrification plays a vital role in the nitrogen cycle, and it is an important engineering process for the removal of nitrogen from wastewaters.

Disinfection The application of microbicidal chemicals to materials (surfaces as well as water), which come into contact with or are ingested by humans and animals, for the purpose of killing pathogenic microorganisms.

Electron acceptor A chemical entity that accepts electrons transferred to it from another compound. It is an oxidizing agent that, by virtue of its accepting electrons, is itself reduced in the process. Oxygen is the terminal electron acceptor in aerobic metabolism.

Electron carrier A molecule that accepts electrons (is reduced) and passes them to another carrier or to a final electron acceptor. NAD, nicotinamide adenine diphosphate, is the most common electron carrier in biological oxidations.

Electron donor A compound or element that furnishes electrons for reductive reactions. Electron donors are especially important for autotrophs that must reduce carbon dioxide to synthesize cellular components.

Endogenous Internal; e.g., an endogenous substrate is one present within the cell, often a stored product of cellular metabolism.

Endogenous respiration The oxidation of internal carbon resources in a cell. The term is also used to describe the process of autodigestion in a microbial population.

Energy source Light or an oxidizable compound or element utilized by the cell in the synthesis of ATP.

Enteric microorganisms Bacteria, viruses, and protozoa that multiply in the intestinal tract and are excreted in the feces.

Eucaryotic cell A cell with a true nucleus and other internal organelles bounded by unit membranes.

Eutrophication The enrichment of a body of water with various nutrients, particularly nitrogen and phosphorus, which can lead to excessive growth of algae.

Exogenous External to the cell; e.g., an exogenous substrate is one that is available in the environment of the cell.

Extended aeration process A modification of the activated sludge process in which all of the sludge from the secondary clarifier is returned to the aeration chamber. The process is intended to achieve total oxidation of both the inflowing waste and the biomass produced in the process.

Facultative Used to describe organisms that are able to grow in either the absence or presence of a specific environmental factor; e.g., a facultative anaerobe can grow in the presence or absence of oxygen.

Feedback inhibition Control of a biosynthetic pathway through inhibition of the activity of the initial enzyme of the pathway by the product of the pathway.

Fermentation Energy-yielding metabolism in which there is no electron transport system or inorganic external electron acceptor. The substrate is oxidized and the electrons removed are accepted by organic metabolic intermediates. No net oxidation occurs. Substrate level phosphorylation is the sole mechanism of ATP synthesis.

Filtration Passage of an aqueous or gaseous carrier, water or air, through a porous medium (sand, charcoal, etc.) for the purpose of trapping undesirable materials, usually in suspension, in the water or air.

Flocculation A process by which colloidal materials, living or inanimate, are brought together and held in a clumped or aggregate form, which is larger and heavier than the individual particles. Flocculation may come about in the natural course of a process (autoflocculation), or it may be enhanced by the addition of chemicals or flocculent aids (chemical flocculation). Autoflocculation of microorganisms is a vital biological unit process and is not yet fully understood.

Flotation The process in which immiscible liquid particles, such as grease and oil, or particulate solids, such as clays or bacteria, are separated from aqueous suspension due to a density difference. The materials may be naturally flotable, e.g., oils, or they may be entrapped in or transferred to rising minute gas bubbles, which are added to the fluid for this purpose (pressure filtration) or are formed in the liquid due to the creation of a vacuum above the liquid (vacuum filtration).

Heterogeneous microbial population A microbial population, for which the initial seed was obtained from some natural source expected to contain a diversity of microorganisms, developed in a system in which no attempt is made to control the species present other than the selection imparted by the environmental or operational conditions under which the cells are being grown. (Note: Ecologists define a *population* as a collection of individuals belonging to a single species; however, the term as defined above is widely used in the pollution control literature.)

Heterotroph An organism that requires organic material as its source of carbon for making cell materials.

Induction The process by which synthesis of the enzymes required to metabolize a substrate is initiated by the inactivation of a repressor. The substrate acts as inducer and binds to the repressor.

Lagoon An earthen holding pond for wastewater, usually used for biological treatment of the waste. Lagoons may be mechanically aerated (aerated lagoons), they may be designed to operate by biological aeration (oxidation ponds), or they may be designed to enhance anaerobic metabolism (anaerobic lagoons).

Lithotroph An organism that uses an inorganic electron donor.

Logarithmic growth The phase of growth in which the time required for doubling of the cell mass is constant.

Maintenance The use of a portion of the exogenous substrate for purposes other than growth, i.e., the use of substrate simply to maintain the status quo or to maintain viability. The substrate used for maintenance may be considered also to be taken from internal or endogenous stores in the presence of an exogenous substrate.

Mineralization The release of inorganic chemicals from organic matter during the process of aerobic or anaerobic decay.

Monomolecular reaction A reaction in which one constituent undergoes a change, e.g., the diffusion of oxygen gas into a liquid or radioactive decay. Such reactions follow first-order reaction kinetics.

Nitrification The oxidation of ammonia to nitrite and nitrate by chemolithotrophic bacteria such as *Nitrosomonas* and *Nitrobacter*.

Nitrogen fixation The reduction of nitrogen gas, N_2, to the level of ammonium ion. The term usually refers to the biological process performed by nitrogen-fixing microorganisms; however, a large amount of nitrogen gas is reduced to ammonia by the chemical industry.

Once-through system A continuous flow reactor system in which the flow of material passes through the system only once; there is no recycle.

Organotroph An organism that uses an organic compound as the electron donor.

Oxidation pond An aerobic lagoon (see **Lagoons**) that is maintained in the aerobic state by the biological (photosynthetic) production of oxygen.

Oxidative phosphorylation The synthesis of ATP coupled to electron transport in an organism using an organic or inorganic compound as energy source.

Pathogen A microorganism capable of causing disease or damage when it infects a host.

pH A means of expressing the hydrogen ion concentration in water. It is the negative logarithm of the molar concentration of hydrogen ion. pH values below 7 are considered to be acid, whereas pH values from 7 to 14 are considered to be alkaline.

Photophosphorylation The synthesis of ATP coupled to electron transport in an organism using light as energy source.

Photosynthesis The conversion of light energy to chemical energy and use of the chemical energy to form organic matter from carbon dioxide.

Phototroph An organism that uses light as energy source.

Plug flow system A hydraulic regime in which the components flowing into a reactor experience little or no mixing as they proceed through the reactor from inlet to outlet.

Primary treatment The removal of separable materials from wastewaters by sedimentation.

Procaryote A cell that does not contain a true nucleus. Procaryotic micro-organisms include bacteria and blue-green algae (cyanobacteria).

Prototroph An organism capable of synthesizing all of its cellular components from a single organic or inorganic source of carbon.

Replication Synthesis of a copy. Cells replicate by increasing in size and dividing to produce two daughter cells identical with the original cell. Replication of DNA occurs when each of the two strands is used as the template for synthesis of a complementary strand, resulting in two double-stranded molecules identical with the original molecule.

Repression The process by which the synthesis of unneeded proteins is prevented by preventing transcription of the genetic information that specifies their structure.

Respiration The biological process by which organisms pass electrons to a terminal electron acceptor when they oxidize organic or inorganic substrates. During this process, energy is trapped in the form of ATP. Aerobic respiration refers to the use of molecular oxygen as terminal electron acceptor. Anaerobic respiration refers to the use of a compound other than oxygen as terminal electron acceptor, e.g., nitrate in the process of denitrification.

Secondary treatment The removal of colloidal and soluble organic material from wastewaters. Settleable material is usually removed prior to secondary treatment. Secondary treatment processes are usually biological in nature, e.g., activated sludge or trickling filtration, but may be chemical and physical in nature. The term *secondary treatment* is sometimes used to indicate a certain level of removal of biochemical oxygen-demanding materials.

Sedimentation The subsidence of particulate materials in aqueous suspension; the process usually takes place under quiescent conditions.

Settling tank (clarifier) A tank or vat in which sedimentation or subsidence takes place. It may contain a scraper for removing flotables, a scum skimmer, and a mechanism in the tank bottom for scraping and removing the settleable material.

Shock loading Any of several types of abrupt change in the environment in a biological treatment process. The shock may involve a change in the nature of the waste (qualitative), in its concentration (quantitative), in its rate of flow (hydraulic), or in other characteristics such as pH, temperature, or concentration of toxic components.

Sludge disposal Various processes for the terminal disposition of sludges created during the treatment of wastewaters.

Sludge treatment Various processes for reducing the volume and putrescibility of sludges developed from the treatment of waters and wastewaters prior to ultimate disposal of the sludge.

LIST OF SYMBOLS

BOD   The biochemical oxygen demand, measured as the amount of oxygen used for respiration during the aerobic metabolism of an energy source by acclimated microorganisms. Carbonaceous BOD is a measure of the amount of oxygen used during the metabolism of an organic substrate and represents the amount of ΔCOD that has been oxidized biologically at any time. Nitrogenous BOD is a measure of the amount of oxygen required for the biological oxidation of inorganic nitrogen compounds (nitrification).

$$\text{weight} \times \text{volume}^{-1}$$

c The cell concentration factor; the ratio of the concentration of cells in the recycle X_R to the concentration of cells in the reactor X; $c = X_R/X$. It is used as a control parameter in Herbert's treatment of continuous culture theory for completely mixed cell culture systems.

C   Coulomb, coulombs.

C_s The saturation value for dissolved oxygen under the prevailing operational conditions during a reaeration experiment.

$$\text{weight} \times \text{volume}^{-1}$$

C_t The concentration of dissolved oxygen at any time t during a reaeration experiment.

$$\text{weight} \times \text{volume}^{-1}$$

COD   Chemical oxygen demand; the amount of oxygen required to oxidize the organic matter in a given sample under the best possible analytical conditions for maximum oxidation of the organic matter to carbon dioxide and water. The theoretical COD

(COD_{th}) is the COD that can be calculated from a balanced equation for total oxidation of the organic matter to carbon dioxide and water; for this, the empirical formula for the organic matter must be known.

$$\text{weight} \times \text{volume}^{-1}$$

COD_e The soluble COD in the effluent from a continuous flow reactor. The same symbol is used to designate the concentration of COD remaining at the end of a batch experiment.

$$\text{weight} \times \text{volume}^{-1}$$

COD_i The concentration of soluble COD in the influent to a continuous flow reactor.

$$\text{weight} \times \text{volume}^{-1}$$

COD_0 The initial concentration of COD in a batch experiment.

$$\text{weight} \times \text{volume}^{-1}$$

ΔCOD The maximum difference obtainable between the initial and final COD in a batch experiment in which the organic matter has been contacted with an acclimated biomass; $\Delta COD = COD_0 - COD_e$. ΔCOD is a measure of the metabolizable carbon source expressed in terms of oxygen demand.

$$\text{weight} \times \text{volume}^{-1}$$

ΔCOD_t The difference between the initial COD and the COD remaining at any time t in a batch reactor.

$$\text{weight} \times \text{volume}^{-1}$$

CV   Coulomb-volts.

D Dilution rate; ratio of the rate of inflow F to the volume of reaction liquor V in the bioreactor. It is equal to the reciprocal of the mean hydraulic residence time $\bar{t}$ in a once-through completely mixed reactor; $D = F/V = 1/\bar{t}$.

$$\text{time}^{-1}$$

D_c The critical dilution rate; the dilution rate at which cells will be washed out of a continuous flow reactor.

$$\text{time}^{-1}$$

D_t The dissolved oxygen deficit at any time t; $D_t = C_s - C_t$

$$\text{weight} \times \text{volume}^{-1}$$

D_1 Dilution rate referred to the reactor in a cell recycle system. It is always greater than D by a factor of $(1 + \alpha)$; $D_1 = D(1 + \alpha) = 1/\bar{t}_1$.

$$\text{time}^{-1}$$

DO   Dissolved oxygen; the concentration of molecular oxygen dissolved in a liquid sample.

$$\text{weight} \times \text{volume}^{-1}$$

DO.   The initial concentration of dissolved oxygen in a batch experiment. It is also used to designate the dissolved oxygen concentration in a receiving stream at the point of entry of a source of pollution.

weight × volume^{-1}

DO_t The dissolved oxygen concentration at a given time t.

weight × volume^{-1}

F Hydraulic inflow rate for the feed stream containing the substrate.

volume × time^{-1}

F_R Hydraulic flow rate of biomass recycled to a bioreactor.

volume × time^{-1}

F_w The flow rate of excess sludge (underflow) from a cell separator (clarifier); the underflow taken away from the reactor–clarifier system, i.e., does not include the cell recycle flow F_R.

volume × time^{-1}

g The generation time; the time required for the number of viable cells to double.

time^{-1}

ΔG Change in free energy for a given reaction.

kcal × mol^{-1}

$\Delta G°$ Change in free energy for a given reaction referred to specified standard conditions, i.e., temperature of 298 K, 1 atm pressure, and reactants at initial concentrations of 1.0 molar.

kcal × mol^{-1}

ΔH Change in heat energy for a given reaction.

kcal × mol^{-1}

$\Delta H°$ Change in heat energy for a given reaction referred to standard conditions of temperature, pressure, and concentration.

kcal × mol^{-1}

k' First-order, decreasing specific rate constant for cells undergoing autodigestion in a batch system.

time^{-1}

K_a Equilibrium constant for an ionization reaction.

k_d The maintenance or endogenous coefficient; the unit rate at which an exogenous substrate is used for purposes other than growth or the rate of autodigestion during growth.

time^{-1}

K_i The inhibitor constant; used in the Haldane expression for the effect of concentration of an inhibitory substrate on the specific growth rate.

weight × volume^{-1}

K_m The Michaelis-Menten constant; the substrate concentration at which the velocity of an enzymic reaction is half the maximum velocity.

$$\text{weight} \times \text{volume}^{-1}$$

K_s The saturation constant; the shape factor in the rectangular hyperbola form of the Monod relationship. It is defined as the concentration of the limiting substrate at which $\mu = 0.5\mu_{max}$.

$$\text{weight} \times \text{volume}^{-1}$$

K_2 The reaeration rate constant for a reactor or a "reach" of river. The numerical value is calculated by using natural logarithms (base e).

$$\text{time}^{-1}$$

k_2 Reaeration rate constant calculated using common logarithms (base 10); $k_2 = K_2/2.3$.

$$\text{time}^{-1}$$

L_0 The total or ultimate amount of carbonaceous biochemical oxygen uptake (BOD) manifested for a given sample.

$$\text{weight} \times \text{volume}^{-1}$$

L_t The biochemical oxygen demand which remains to be expressed at any time t. It is equal to the total amount of oxygen which will be used, L_0, minus the accumulated oxygen demand already expressed; $L_t = L_0 - y_t$.

$$\text{weight} \times \text{volume}^{-1}$$

N The concentration of viable microorganisms.

$$\text{number} \times \text{volume}^{-1}$$

OD   Optical density (absorbance); a measure of the absorbance of light from a monochromatic source. It is related to the fraction of light transmitted (transmittance T) as follows: $OD = \log(1/T)$. It is a useful measurement for the growth of dispersed cells when correlated to concentration X.

ΔO_2 The amount of oxygen utilized in a given time interval; $\Delta O_2 = DO_i - DO_t$.

$$\text{weight} \times \text{volume}^{-1}$$

S A general symbol for microbial substrate; usually designates an organic carbon source. Often, as in the Monod equation, it designates the growth-limiting nutrient, e.g., carbon or nitrogen source.

$$\text{weight} \times \text{volume}^{-1}$$

S_e The concentration of the substrate in solution in a completely mixed reactor, or in the effluent from a completely mixed reactor or from the cell separator when a system is operating in steady state.

$$\text{weight} \times \text{volume}^{-1}$$

S_i Inflowing concentration of substrate in a continuous flow reactor; usually refers to the carbon source.

weight × volume^{-1}

S_r Substrate concentration that has been removed; also designated by the symbol ΔS.

weight × volume^{-1}

S_R Substrate concentration in the recycle to an activated sludge system.

weight × volume^{-1}

S_0 The initial concentration of substrate in a batch experiment.

weight × volume^{-1}

ΔS Change in entropy, which is the capacity of a system for containing energy unavailable for doing work.

kcal × mol^{-1}

T Temperature measured on the absolute scale, i.e., in degrees centigrade plus 273.

degrees

t Time.

$\bar{t}$ The mean hydraulic retention time in a biological reactor. It is the reciprocal of the dilution rate D

time

t_d The time required for the biomass concentration X to double. The doubling time is constant in an exponential growth phase.

time

$\bar{t}_1$ The mean hydraulic retention time referred to the reactor in a cell recycle system. It is always less than $\bar{t}$ by a factor of $1/(1 + \alpha)$; $\bar{t}_1 = 1/[D(1 + \alpha)]$.

time

TOC   Total organic carbon; the amount of carbon in the organic matter in a given sample; the carbon released as carbon dioxide when a sample is oxidized under conditions assuring maximum oxidation of the sample. The theoretical TOC can be calculated if the empirical formula for the organic matter in the sample is known.

weight × volume^{-1}

TOD   Total oxygen demand; the amount of oxygen required to oxidize the organic and inorganic materials in a sample under specified conditions. The conditions vary, depending on the proprietary device used. In some, the oxidizing conditions are such that inorganic nitrogen, for example, is partially oxidized.

weight × volume^{-1}

ΔTOC These terms are similar to ΔCOD except that TOC or TOD is used
ΔTOD for initial and final measurements.

weight × volume^{-1}

U The specific rate of substrate utilization, i.e., the rate of substrate utilization per unit concentration of cells; $U = (dS/dt)(1/X)$, or $(\Delta S/\Delta t)(1/X)$.

time^{-1}

V Volume of mixed liquor undergoing reaction; the effective volume of a fluidized bioreactor; does not include freeboard.

volume

v Velocity or rate of reaction, e.g., rate of increase in cell mass, dX/dt.

time^{-1}

X Biomass concentration; can be expressed as the concentration of total suspended solids or volatile suspended solids.

weight $\times$ volume^{-1}

X_e Biomass concentration in the effluent of a clarifier or cell separator. The same symbol is also used for biomass concentration at the end of a batch experiment.

weight $\times$ volume^{-1}

X_R The concentration of cells in the recycle flow to a reactor system for which cell recycle or cell feedback is practiced. It is used as a control parameter in Gaudy's treatment of continuous culture theory for completely mixed cell culture systems.

weight $\times$ volume^{-1}

X_t The concentration of biomass (cells) present at any time t in a batch experiment.

weight $\times$ volume^{-1}

X_w In calculating the amount of excess sludge X_w, the mass rate of cells in the clarifier overflow (effluent), $X_e(F - F_w)$, should also be included. This is usually done in making accurate mass balances for the system; however, it is not included for purposes of estimating the amount of excess sludge produced at the reactor site [see Eqs. (6-83) and (6-84)].

weight $\times$ time^{-1}

X'_w The amount of excess sludge from a continuous flow growth system; the weight of the cells in the underflow from the clarifier that is in excess of that needed to supply the cells for recycle. It is usually given as the mass rate of cell output, i.e., the product of the concentration of cells in the clarifier underflow and the rate of flow of excess sludge F_w.

weight $\times$ time^{-1}

X_0 The initial biomass concentration in a batch experiment.

weight $\times$ volume^{-1}

ΔX The difference between the amounts of initial biological solids and biological solids at any time due to growth in a batch experiment.

weight $\times$ volume^{-1}

INDEX

LaVergne, TN USA
30 August 2010
195090LV00001B/12/A

9 780910 554671